T0393264

Applications of Functional Foods and Nutraceuticals for Chronic Diseases

While disease is inevitable in humankind, the current century has been burdened with many chronic diseases, most of which are lifestyle mediated and which, in part, can be controlled by consuming foods with specific functions. Functional foods are a special category of natural food or nutrient-derived pharmaceutical products containing beneficial biochemicals and phytochemicals beyond their basic nutritional functions.

The first of two volumes, ***Applications of Functional Foods and Nutraceuticals for Chronic Diseases***, collects information on the association between functional foods and chronic diseases. The burden of escalating chronic diseases is discussed in the first chapter, and the remaining fourteen chapters summarize the effect of functional foods on a range of chronic diseases.

Key Features:

- Discusses the clinical application of functional foods for the management of a wide range of chronic diseases
- Covers chronic diseases, including obesity, arthritis, cardiovascular diseases, endocrinal and hormonal diseases, among others
- Explores beneficial effects of nutraceuticals on chronic diseases

Contributors hail from different geographical locations around the world and possess many years of research and scholarly experience in functional foods, nutraceuticals, and biology. The world's leading wellness centers for chronic diseases are using functional foods and nutraceuticals in their practice and discovering their useful applications, and this book is a great reference for practitioners, scientists, and clinicians in the management of chronic diseases.

Nutraceuticals: Basic Research and Clinical Applications
Series Editor: Yashwant Pathak, PhD

Marine Nutraceuticals: Prospects and Perspectives
Se-Kwon Kim

Nutrigenomics and Nutraceuticals: Clinical Relevance and Disease Prevention
edited by Yashwant Pathak and Ali M. Ardekani

Food By-Product Based Functional Food Powders
edited by Özlem Tokuşoğlu

Flavors for Nutraceuticals and Functional Foods
M. Selvamuthukumaran and Yashwant Pathak

Antioxidant Nutraceuticals: Preventive and Healthcare Applications
Chuanhai Cao, Sarvadaman Pathak, and Kiran Patil

Advances in Nutraceutical Applications in Cancer: Recent Research Trends and Clinical Applications
edited by Sheeba Varghese Gupta, Yashwant Pathak

Flavor Development for Functional Foods and Nutraceuticals
M. Selvamuthukumaran and Yashwant Pathak

Nutraceuticals for Prenatal, Maternal and Offspring's Nutritional Health,
Priyanka Bhatt, Maryam Sadat Miraghajani, Sarvadaman Pathak, Yashwant Pathak

Bioactive Peptides: Production, Bioavailability, Health Potential and Regulatory Issues,
edited by John Oloche Onuh, M. Selvamuthukumaran, Yashwant Pathak

Nutraceuticals for Aging and Anti-Aging: Basic Understanding and Clinical Evidence,
edited by Jayant Lokhande, Yashwant Pathak

Marine-Based Bioactive Compounds: Applications in Nutraceuticals,
edited by Stephen T. Grabacki, Yashwant Pathak, Nilesh H. Joshi

Applications of Functional Foods and Nutraceuticals for Chronic Diseases, Volume I,
edited by Syam Mohan, Shima Abdollahi, and Yashwant Pathak

Flavonoids and Anti-Aging: The Role of Transcription Factor Nuclear Erythroid 2-Related Factor2,
edited by Karam F.A. Soliman and Yashwant Pathak

For more information about this series, please visit: https://www.crcpress.com/Nutraceuticals/book-series/CRCNUTBASRES

Applications of Functional Foods and Nutraceuticals for Chronic Diseases

Volume I

Edited by
Syam Mohan, Shima Abdollahi,
and Yashwant V. Pathak

CRC Press is an imprint of the
Taylor & Francis Group, an **informa** business

First edition published 2023
by CRC Press
6000 Broken Sound Parkway NW, Suite 300, Boca Raton, FL 33487-2742
and by CRC Press

4 Park Square, Milton Park, Abingdon, Oxon, OX14 4RN

CRC Press is an imprint of Taylor & Francis Group, LLC

Library of Congress Cataloging-in-Publication Data

Names: Mohan, Syam, editor. | Abdollahi, Shima, editor. | Pathak, Yashwant, editor.
Title: Applications of functional foods and nutraceuticals for chronic diseases / edited by Syam Mohan, Shima Abdollahi, and Yashwant V. Pathak.
Description: First edition. | Boca Raton : CRC Press, 2023. | Series: Nutraceuticals | Includes bibliographical references and index.
Identifiers: LCCN 2022032399 (print) | LCCN 2022032400 (ebook) | ISBN 9781032072951 (volume 1 ; hardback) | ISBN 9781032114774 (volume 1 ; paperback) | ISBN 9781003220053 (volume 1 ; ebook)
Subjects: LCSH: Functional foods--Therapeutic use. | Chronic diseases--Diet therapy.
Classification: LCC RM217 A745 2023 (print) | LCC RM217 (ebook) | DDC 613.2--dc23/eng/20221021
LC record available at https://lccn.loc.gov/2022032399
LC ebook record available at https://lccn.loc.gov/2022032400

ISBN: 9781032072951 (hbk)
ISBN: 9781032114774 (pbk)
ISBN: 9781003220053 (ebk)

DOI: 10.1201/9781003220053

Typeset in Garamond
by KnowledgeWorks Global Ltd.

This book is dedicated to the students, professors, researchers, and clinicians who use this text. We also appreciate the authors for sharing their wisdom, knowledge, experience, and insights. We also thank CRC Press, Dr. Steve Zollo, other T&F staff, our families, and our institutions.

—The Editors

Contents

Preface xi

About the Editors xiii

List of Contributors xv

SECTION I Burden of Chronic Diseases

1 Global Trends in Growth of Chronic Diseases 3
Janefar Eva and Yashwant Pathak

2 Importance of Integrative Health Sciences in the Therapy of Chronic Diseases 17
Ghazaleh Eslamian, Arman Ghorbani, and Morvarid Noormohammadi

SECTION II Effect of Functional Foods in Treating Chronic Diseases

3 Safety and Efficacy Determination in Functional Foods and Nutraceuticals 37
Arun Soni, Tulsi Patil, Sanjeev Acharya, and Niyati Acharya

4 Global Burden of Disease, the Heavy Cost of Sun Deprivation: Implications for Mass Food Fortification with Vitamin D. 67

Zahra Yari, Bahareh Nikooyeh, Samira Ebrahimof, and Tirang R. Neyestani

5 Functional Vegetables and Medicinal Uses: Cure and Curse 117

Sivakumar S. Moni

6 Dietary Supplements and Functional Foods in the Management of Endocrine Disorders 137

Elham Razmpoosh, Shima Abdollahi, and Sepideh Soltani

7 Changes in the Regulation of Energy Metabolism in Chronic Diseases Using Functional Foods and Nutraceuticals 167

Aparoop Das, Manash Pratim Pathak, Kalyani Pathak, Urvashee Gogoi, and Riya Saikia

8 Functional Foods for the Prevention and Treatment of Cardiovascular Diseases Including Hypertension 197

Nahid Ramezani-Jolfaie and Mohammad Mohammadi

9 Functional Foods and Natural Products for Obesity Management 231

Idris Adewale Ahmed, Najihah Mohd Hashim, and Rozana Othman

10 Noncommercial Plant-Based Edible Oil for Prevention and Treatment of Chronic Diseases 249

Der Jiun Ooi, Yun Ping Neo, Yin Sim Tor, and Jhi Biau Foo

11 Oral Health Challenges during Chronic Diseases: Prevention and Treatment Using Functional Foods and Nutraceuticals 293

Sangeeta Jayant Palaskar, Rasika Balkrishna Pawar, and Darshana Rajesh Shah

12 Urogenital System Disorders and Functional Foods and Nutraceuticals 317

Manish P. Patel, Arya S. Vyas, Praful D. Bharadia, Jayvadan K. Patel, and Dipti H. Patel

13 Functional Foods and Nutraceuticals and Respiratory Diseases 341

Azadeh Dehghani and Mehran Rahimlou

14 Strategies Using Functional Food and Nutraceuticals to Prevent and Treat Arthritis. 361

Shikha Sharma, Ramesh Bhonde, and Kalpana Joshi

15 Functional Foods of Polyphenolics for Alzheimer's Disease. 375

Hanish Singh Jayasingh Chellammal and Dhani Ramachandran

Index 409

Preface

The prevalence of chronic diseases is increasing worldwide, partly related to the increase in life expectancy and an aging population. Many studies are being done to help people age healthily. It has long been recommended to follow a healthy, balanced diet, ensuring adequate intake of essential macro- or micronutrients. However, recently, the focus has been shifted to identify food agents that improve immunity, well-being, quality of life, and increase life expectancy. One of the areas that has recently been of interest to researchers is related to "functional foods/products." Functional foods are a special category of natural food or nutrient-derived pharmaceutical products containing beneficial biochemical and phytochemicals beyond their basic nutritional functions. Functional foods and nutraceuticals play a promising role in maintaining and developing health and helping the body fight inflammation and chronic diseases or delaying their onset. Recently, more attention has been paid to these products, and the term "functional foods" has been added to the product labels in the food industry. Moreover, growing evidence on the benefits of functional foods and nutraceuticals has led to the development of pharmaceutical products and increased sales of these supplements.

Applications of Functional Foods and Nutraceuticals for Chronic Diseases addresses the effects of functional foods/nutraceuticals in relation to chronic diseases such as obesity, cardiovascular diseases, diabetes, cancer, and so on. This book is divided into two parts: the first part focuses on the clinical evidence of functional foods and nutraceuticals and chronic diseases; the second part discusses the molecular mechanisms and the research on the roles of functional foods/nutraceuticals in preventing and treating chronic diseases through epigenetic modulation. Chapter 1 of the book looks at the burden of escalating chronic diseases, describes the related risk factors, and emphasizes the importance of integrative medicine in promoting health-related behavior changes. The remaining chapters summarize the current knowledge about the effect of functional foods on a range of chronic diseases, including obesity, metabolic syndrome, diabetes, endocrine disorders, urogenital system disorders, neurodegeneration diseases, cardiovascular diseases, and inflammatory diseases.

We sincerely hope this book can stimulate new thought processes and paradigm shifts in research and development to explore the beneficial effects of nutraceuticals in chronic diseases and improving quality of life in patients with chronic diseases.

About the Editors

Dr. Syam Mohan is currently working as an associate professor at Jazan University, Saudi Arabia. Before that, he was at the University of Malaya, the oldest and premier public university in Malaysia. He graduated from TN DR MGR Medical University, India, in 2003. In 2007 he joined MS Pharmacology. His research outcomes were so outstanding that his Master's degree was upgraded to Ph.D. with a fellowship, awarded by University Putra Malaysia. He undertook a year of senior research fellowship (Postdoctoral) at the University of Malaya, later joined the Department of Pharmacy as associate professor, and established the pharmaceutical biology research group in the Pharmacology Unit. His specific expertise is focused on the pharmacology of cancer and inflammation. He acquired an overall H-index of 40 with his high number of research publications (120) in ISI-cited journals. Dr. Syam, the author of one book, also holds two patents. He was also one of the two selected participants from Asia who received a fellowship to attend the FEBS Advanced Lecture Course on Translational Cancer Research at Portugal organized by the Federation of European Biochemical Society, UK, and was a member of the European Association of Cancer Research (EACR) and is an active member of Malaysian Biochemistry and Molecular Biology Society (MBMBS). He has organized several workshops and seminars and has attracted over US$ 250 000 as research funding so far, which includes the prestigious High Impact Research grant from the Ministry of Higher Education, Malaysia, and the Ministry of Education funding, Saudi Arabia. Recently he has been positioned among the Top 2% of scientists of the world for the consecutive two years (2020 & 2021) by a research study conducted by Stanford University. He also plays his role as a member of several international scientific organizations, editorial board, and reviewer for many international peer-reviewed journals. Recently, he has been appointed as an International Grant reviewer for the South African Medical Research Council. As a conscientious and potentially valuable scientist, he has been acknowledged and recognized by the Governments of various countries, including Iran and Malaysia.

Shima Abdollahi, PhD, is currently an assistant professor in North Khorasan University of Medical Sciences, Bojnurd, Iran. She earned her BS, MS, and PhD degrees in nutrition sciences from Tabriz, Tehran, and Shahid Sadoughi universities of medical sciences, respectively. She received a research grant from Iran national science foundation (INSF) for her doctoral thesis on the effect of resveratrol supplementation in patients with diabetes. Dr. Abdollahi has experience in gene expression studies, as well as systematic review and meta-analysis studies.

Yashwant V. Pathak, PhD, is currently the associate dean for faculty affairs at the College of Pharmacy, University of South Florida, Tampa, Florida. Dr. Pathak earned his MS and PhD degrees in pharmaceutical technology from Nagpur University, Nagpur, India, and EMBA and MS degrees in conflict management from Sullivan University in Louisville, Kentucky. With extensive experience in academia and industry, Dr. Pathak has more than 150 publications, research papers, abstracts, book chapters, and reviews to his credit. He has presented over 180 presentations, posters, and lectures worldwide in the field of pharmaceuticals, drug delivery systems, and other related topics. He has received several national and international awards including Fulbright Senior Scholar fellowship for Indonesia, Endeavour Executive fellowship from the Australian government, CNPQ research award from the Brazilian government and also was recognized by USF as outstanding faculty award and global engagement achievement award.

Contributors

Niyati Acharya
Nirma University
Ahmedabad, India

Sanjeev Acharya
SSR College of Pharmacy
Silvassa, India

Idris Adewale Ahmed
University Malaya
Kuala Lumpur, Malaysia

Praful Bharadia
L.M. College of Pharmacy
Ahmedabad, India

Ramesh Bhonde
D.Y. Patil Vidyapeeth University,
Pune, Maharashtra, India

Hanish Singh Jayasingh Chellammal
UiTM
Puncak Alam, Selangor, Malaysia

Aparoop Das
Dibrugarh University
Dibrugarh, Assam, India

Azadeh Dehghani
Tabriz University of Medical Sciences
Tabriz, Iran

Samira Ebrahimof
Shahid Beheshti University of Medical Sciences
Tehran, Iran

Ghazaleh Eslamian
Shahid Beheshti University of Medical Sciences
Tehran, Iran

Janefar Eva
University of South Florida
Tampa, Florida

Jhi Biau Foo
Taylor's University
Kuala Lumpur, Malaysia

Arman Ghorbani
Shahid Beheshti University of Medical Sciences
Tehran, Iran

Urvashee Gogoi
Dibrugarh University
Dibrugarh, Assam, India

Najihah Mohd Hashim
University Malaya
Kuala Lumpur, Malaysia

Ooi Der Jiun
MAHSA University
Jenjarom, Selangor, Malaysia

Nahid Ramezani-Jolfaie
Hormozgan University of Medical Sciences
Bandar Abbas, Iran

Kalpana Joshi
Savitribai Phule Pune University
Pune, India

Mohammad Mohammadi
Hormozgan University of Medical Sciences
Bandar Abbas, Iran

Tirang R. Neyestani
Shahid Beheshti University of Medical Sciences
Tehran, Iran

Bahareh Nikooyeh
Shahid Beheshti University of Medical Sciences
Tehran, Iran

Morvarid Noormohammadi
School of Public Health Iran University of Medical Sciences
Tehran, Iran

Rozana Othman
University Malaya
Kuala Lumpur, Malaysia

Sangeeta Jayant Palaskar
Sinhgad Dental College and Hospital
Pune, India

Dipti H. Patel
Parul University
Vadodara, Gujarat, India

Jayvadan K. Patel
Sankalchand Patel University
Visnagar, Gujarat, India

Manish P. Patel
L.M. College of Pharmacy
Ahmedabad, Gujarat, India

Kalyani Pathak
Dibrugarh University
Dibrugarh, Assam, India

Manash Pratim Pathak
Assam Downtown University
Guwahati, Assam India

Tulsi Patil
SSR College of Pharmacy
Silvassa, India

Rasika Balkrishna Pawar
Sinhgad Dental College and Hospital
Pune, India

Neo Yun Ping
Taylor's University
Subang Jaya, Selangor, Malaysia

Dhani Ramachandran
International Medical School, Management and Science University
Shah Alam, Malaysia

Mehran Rahimlou
Zanjan University of Medical Sciences
Zanjan, Iran

Elham Razmpoosh
Shahid Beheshti University
Tehran, Iran

Riya Saikia
Dibrugarh University
Dibrugarh, Assam, India

Darshana Rajesh Shah
Sinhgad Dental College and Hospital
Pune, India

Shikha Sharma
Institute of Stem Cell Science & Regenerative Medicine
Bangalore, Karnataka, India

Tor Yin Sim
Taylor's University
Selangor, Malaysia

Sivakumar Sivagurunathan Moni
Jazan University
Jazan, KSA

Arun Soni
SSR College of Pharmacy
Silvassa, India

Arya S. Vyas
L.M. College of Pharmacy
Ahmedabad, Gujarat, India

Zahra Yari
Shahid Beheshti University of Medical Sciences
Tehran, Iran

I

Burden of Chronic Diseases

Global Trends in Growth of Chronic Diseases

Janefar Eva and Yashwant Pathak

Contents

Introduction 3
Global and regional burden of disease 4
Population trends 5
Relevance of chronic diseases 6
Chronic diseases and the humanistic perspective 8
Implications of globalization 9
Result of pandemics 10
Age of medicalization and the sick role 12
Conclusion 13
References 15

Introduction

To address the problems of global health infrastructure, an issue of such caliber, one lifetime is too short of a timeline. What is more probable is to riddle or interpret the problems that exist. Global health is a matter of equity, and not merely equality, that will begin paradigm shifts. Doing so will systematically create sustainable practices that extend beyond the lifetime of those implementing. Not enough people have taken on the project of widening healthcare accessibility for the world, as it is a rather difficult area to make profitable. Moreover, as a public health endeavor, it is hard to make these ideas profitable on a large enough scale. Initially, global health does not lend itself to being marketable and inviting. Stakeholders appreciate their autonomy, and investing in non-profit organizations that target health reform is not appealing on market scales. This prefaces the value in innovation that optimizes the current environment, so stakeholders recognize their vested interest. It's about taking what's wrong with healthcare and making it right or, in other words, *seeking opportunity in adversity*. Global trends in the growth of chronic diseases will be identified through the burden of disease, population trends, humanistic perspectives, impacts of globalization, result of pandemics, effects of medicalization, and discussion of the sick role.

DOI: 10.1201/9781003220053-2

The depth and profound nature of chronic diseases is vast; they are notably the largest cause of death in the world. They are not only identified in underdeveloped countries but also in established countries, topping gross domestic product (GDP) charts. Mortality is a shared attribute among every individual, without barriers of culture, nationality, socioeconomic status (SES), or genetic predispositions. It is the quality that makes us innately human and intrinsically living. Globally recognized, chronic diseases are identified as persistent conditions that cause disability and lessen the quality of life. Public health endeavors, private sector institutes, 501c organizations, governing entities, grassroots organizations, and, now more pertinent, global pandemics have retarded attempts to address the factors that cause chronic diseases. There are in fact social problems on the institutional and social levels that need intervention. Society is severely misguided in its belief that there are enough opportunities and substantial infrastructure in the United States to remove oneself from a poor socioeconomic class in a lifetime. For this very reason, we should not blame the individuals affected by poverty, but rather assess the conditions that caused their impoverished status in society. To provide a sustainable solution, changing the status quo will hold us accountable to advocate for widespread reform.

Global and regional burden of disease

In reference to global metrics, the burden of chronic diseases is far too severe for inaction. Discussing comorbidities associated with chronic diseases, many of which are due to side effects of prescription medications, a third of adults suffer from multiple chronic conditions. Certain comorbidities occur in clusters; this is an explanation for why two or more conditions occur frequently together. Examples of these include depression and stroke, anxiety and heart attacks, and tuberculosis with HIV/AIDS. These co-occurring disorders place burdens on individuals, families, and societies with each additional diagnosis. Global trends suggest it is not only SES that influences the prevalence of chronic diseases, as high-income countries have their share of common ailments, while lower-income countries also have their own common illnesses. What can be observed across metrics, however, is the presence of lower back pain, arthritis, and mental health conditions, for example. As societies become advanced and industrious, the world witnesses aging populations that are outliving the generations prior. This shift in demographic gives rise to chronic diseases. In other words, when people live longer and overcome acute illnesses, the probability they will develop other chronic diseases is more likely (1).

Referencing the research paper *Estimating the Economic Cost of Childhood Poverty in the United States*, the authors Michael McLaughlin and Mark Rank quantified the future implications of this topic. The three categories assessing childhood poverty included effects on future earnings, crime engagement, and the quality of health with a total cost of $500 billion each year (2). This estimation is an apparent call to action in terms of the severe economic losses incurred without any reform. It is a necessity to analyze the factors that lead to poor health outcomes for children. Auxiliary factors such as poor school systems contribute to healthcare behavior later in life. Providing much needed support to school systems and programs in lower-SES neighborhoods and schools would lessen future financial burdens. Increasing opportunities for school children could allow for higher education to become a possibility for these students. Moreover, school retention rates would directly be increasing the chances for students to qualify for higher paying jobs and thus better-quality healthcare. These changes may, then, allow families and individuals to be removed from the cycle of generational poverty. The current ineffective social policies and production of low-paying jobs are only negatively impacting students and children, with the backdrop of growing inflation. With more opportunity given to these children, it becomes more probable that they will receive better quality healthcare and thus more access to chronic disease preventative measures.

Taking a broader look, although these numbers continue to increase, an estimated 60% of deaths globally are attributed to noncommunicable diseases (NCDs) and 80% of this burden of mortality is from low- and middle-income countries (3). These NCDs are projected to cost US$47 trillion by 2030 yet were excluded from the Center of Global Development's 2010 program (4). The burden of disease is a threat to global development, as a study in 2010 found that less than 3% of aid meant for global health endeavors was spent on NCDs. This prefaces discussion of the increased price of healthcare altogether. Chronic diseases, the focus of this chapter, have become an epidemic, with 50% of the US population affected as of 2020 (5). The chairman of the Medicare Payment Advisory Commission stated "U.S. health care is too expensive and its quality too inconsistent. To ensure that health care will be affordable for future generations and appropriate for our burgeoning geriatric population, its delivery and organization must change. Physicians should be in the vanguard of this change and transforming medical education will be instrumental in preparing tomorrow's physicians to lead the way" (6). In terms of the global cost of healthcare, the World Health Organization's 2020 analysis showed that global healthcare costs reached 10% of the world's GDP, with out-of-pocket expenses from low- and middle-income countries of more than 40% in 2018 (7).

Population trends

The myriad of ailments increasing with age is an indication of mortality and prompts investigation into aging healthily. Historically, according to trends in census data, population growth was linear until the 1900s (8). After the early 1900s, populations grew exponentially, with contributions from medical, technological, and public health innovations. Refrigeration, hygiene, antiseptics, vaccinations, environmental regulation, and education were notable developments in the 20th century. "In the developed world in the 21st century, communicable diseases now account for fewer than 10% of years of life lost compared with 20–80% in less developed countries" (9). With an improved quality of life, women give birth to children that survive longer. Signifying a more current shift, within our current lifetimes, aging pyramids for both less-developed and more-developed countries show older populations growing faster than other age groups (1). This suggests women are having fewer children and therefore the working age population, people 15 to 65 in age, will continue to shrink. Now, with slowed population trends, lower fertility is contributing to aging populations that consume more resources. These statistics preface the concept of disability-adjusted life years (DALYs). This concept not only looks at premature death from a disease but also considers the years people spend living with the disease (1). A global burden of disease study found that from 1990 to 2010, 52.8 million people died in 2010 compared to 46.5 million deaths in 1990 (10). Although trends in mortality have decreased, this statistic refutes the false narrative suggesting people are living longer and healthier lives. Realistically, DALYs stats indicate illnesses such as chronic obstructive pulmonary disease (COPD), interstitial lung disease (ILD), cardiovascular disease, stroke, and overnutrition are risk factors with severe growth. More specifically, looking at ILD data, there has been a 77% increase in the number of deaths from 1990 to 2010 (10). When researchers attach a numerical value to global health trends, the depth of disease is personified. Consolation to note, however, is the understanding that as modern imaging technology and the quantity of machines become available in less-developed countries, ILD detection will increase and contribute to fewer deaths. Although organizations such as the American Association of Physicists in Medicine are promoting these machines in underserved countries, the progress is slow. Therefore, with the delay in global health efforts, healthcare management on an individual scale becomes relevant. Population trends and DALY statistics suggest that more people are dying of morbidities attributed to NCDs compared to the communicable diseases in the past.

Relevance of chronic diseases

The increase in NCDs is driving current healthcare infrastructure to capacity. As globalization continues and reaches underserved communities, the prevalence of chronic conditions will equally increase. As previously stated, the burden of daily life is physiological consequences such as unhealthy diets and physical inactivity. Lower-income countries face twice the burden, as they not only face problems associated with hyperactive lifestyles but also face the burden of infection and NCDs. Fundamental determinants of health such as genetics, external and internal influences, behavior, healthcare, and social factors have a strong association with absolute poverty (11). Countries that remain facing problems of absolute poverty are burdened far beyond their means in the modern world.

Epidemiological transitions demonstrate not only aging populations but also ground observations of morbidity shifting from communicable to noncommunicable disease. To identify these shifts, comorbidities and multimorbidity must be analyzed. Primarily looking at high-income countries, multimorbidity is more common for older people, females, and those with lower SES (12). Multimorbidity for older populations and lower SES seems explanatory. However, females may show this increased recording due to a greater health-seeking behavior than men (13). Women are more likely to report their illnesses and symptoms than men, who tend to avoid social stigma of appearing weaker. Taking a closer look at SES in countries around the world, comorbidity and chronic diseases show a weak correlation with SES, which is not readily explanatory. Bangladesh is a low-SES country in northeastern South Asia. A cross-sectional study showed that the wealthy, or those of higher SES, in Bangladesh have increased chronic diseases and multimorbidity than the lower quintiles of the population (14). With associations to the World Health Organization and the World Bank, a study found that, "the relative risks were twofold to threefold greater for those living on < US$1.00 per day compared with those living on > US$2.00 per day" (15). This further justifies the point that the unequal distribution of wealth in the world is promoting poor health outcomes.

Quite opposingly, Scotland, known to be a high-income country, had multimorbidity correlate more closely with lower SES (16). In terms of Bangladesh as a lower-income country, richer foods and fatty ingredients are costlier and therefore reserved for those who can afford them, while the poor eat less-dense foods. This can stand as an explanation for the presence of diabetes and blocked arteries leading to heart conditions due to food intake; thus, resulting in higher levels of chronic diseases in those of higher SES in Bangladesh. In reference to Scotland, increased education and access to healthier foods for the wealthy stand as an explanation for their data. Hence, their wealthy have fewer chronic diseases than their poorer counterparts. Regardless of country GDP, people are becoming sicker with each additional diagnosis, regardless of SES.

The nature of the increase in chronic diseases is attributable to various factors. The UN World Summit in 2011 addressed with extreme priority that intervention must be taken on NCDs, those being tobacco product usage, unhealthy diets, use of alcohol, and physical inactivity (17). There are well-established determinants of chronic diseases apart from family history and genetic predisposition, notably, age, gender, SEC, weight, and preexisting conditions. These risk factors can influence not only the way the body fights new infection but also the way the body is able to maintain itself. For example, the presence of mental health illnesses can lower the body's immune response, alter hormonal responses, and negatively impact the body's sleep cycle.

Chronic diseases and the humanistic perspective

Analysis of the human experience reveals trends and new patterns. Sleep is often undervalued, although it is a crucial mechanism of repair functions and growth. On the surface level, chronic diseases can impact sleep cycles, and on a deeper level make the body more susceptible to developing other chronic conditions. On the flip side, insufficient sleep has been linked to the development of many chronic conditions, including type 2 diabetes, obesity, depression, and cardiovascular diseases (18). Essentially, sleep is needed to prevent and manage chronic diseases. From a sociological perspective, sleep deprivation has increasingly become more prominent over time, as people are working longer hours. Moreover, with the rise of women in the workforce, there are more people living goal-driven lives. The emergence of hyper-productivity lifestyles has become an issue in our lifetimes and will continue to grow as countries become more prosperous. Although 20 different cross-sectional studies have used individual data to cross-sectionally compare long working hours and mental health, only a few found an association with the two variables (19). This suggests there are in fact various confounding factors like SES and family history that factor into overall health. This is another indication that with the delay in global health efforts, healthcare management on the individual scale becomes of relevance. Making healthcare an individual science promotes the best results for individuals, but unfortunately this is impractical to achieve. Therefore, analyzing the causes of chronic diseases such as depression and anxiety is a more sustainable solution.

Urbanization in coordination with acceleration of daily tasks has placed increased performance burdens on individuals with pressure to produce results (20). Concurrent with these lifestyle changes and the burden of daily life are physiological consequences such as unhealthy diets and physical inactivity. Lower income countries are doubly disadvantaged as they not only face these types of burdens but also the burden of infection and NCDs.

Addressing the next generation of citizens, students with lower SES need more interventions due to reasons provided by David Card and Jesse Rothstein in their paper titled *Racial Segregation and the Black-White Test Score Gap.* Taking a specified look at minority students can help to assess a large portion of low-SES students. Primarily, direct exposure effects explain how different students have exposure to different resources and people. This phenomenon is explained by minority students' exposure to peers with lower achievement expectations and aspirations than their non-minority counterparts. More simply stated, minority students may not have successful role models that challenge them intellectually and push them to pursue success. Here, the quality of peers and peer interaction is poorer for minority students. Being surrounded by strong-willed people and leaders creates mentorship and facilitation of personal goals for students. Furthermore, students with lower SES tend to have poorer family structures such as single-parent homes and other instabilities, which affects their performance in school. Thus, evaluating the social realities of low SES students, demonstrates the need for more funding for these students. Reform would start to equalize the educational outcomes of disadvantaged students compared to advantaged students.

Continuing the discussion, assessing the needs of school children and students raises the topic of social discrimination. A polarized economy in the United States creates a severe wealth gap. The United States faces low generational financial mobility compared to other OECD (Organization for Economic Cooperation and Development) countries. This is exemplified through the understanding that the correlation between earnings of fathers and sons is higher than in other countries. In other words, sons are expected to be just as poor as their fathers. The article *Rethinking American Poverty* by Mark Rank explains these concepts rather explicitly. Rank suggests that social structures do in fact perpetuate inequality. According to Rank, the consensus of poverty is that people bring it upon themselves and those that are inflicted with poverty are deserving of it when they are a part of the working class (21). This specific article demonstrates how people fail to consider children and students a part of classes of individuals affected by poverty. School-age children are definitively not a part of the working class and, therefore, they could not have brought poverty upon themselves. Not only is institutional reform needed, but also social reform in which people must begin to realize the implications of their judgment and inactivity. For now, however, these current practices are contributing to the increase in chronic diseases. Holistically identifying factors that are a part of overall well-being serves to provide a more encompassing answer to why chronic diseases are perpetuating.

Implications of globalization

With globalization reaching far corners of the world, anthropogenic attributes are promoting a damaged environment and ecosystem. Namely, anthropogenic activities including population growth, deforestation, pollution, and overconsumption are heightening the loss of biodiversity. This loss not only alters the physical environment but also our bodies, with increased exposure to harm. Rapid deforestation and habitat destruction are initiating a cascade of events that bring people in closer contact to animals and wildlife. Close proximities give rise to zoonotic viruses and transmissible diseases. Destroying the natural pharmacies found within forests eliminates sources of medicinal and nutrient-dense microbiomes. Moreover, potential medicinal active ingredients could be lost with the destruction of forests (22). Eliminating micro- and macronutrients from the environment has detrimental effects, still waiting to be observed through hierarchical disturbances.

Pollutants and greenhouse gases are trapping heat in the atmosphere, resulting in poorer air quality. Referring to globalization and anthropogenic effects, factories and automobiles deposit pollutants into the air with their fumes, regardless of EPA (Environmental Protection Association) standards. Forested areas and trees deposit pollutants into their vegetation while promoting air quality through the release of oxygen. Having established these natural relationships, apparent links between the environment and human health are explained. An interesting study revealed that forests, compared to cities, may increase the prevalence of natural killer cells (NKs) in the body, thus promoting immune system health (23). Human activity has also affected climate change, which promotes the presence of chronic disease. Global warming has caused reductions in food security and increases in wildfires, floods, allergens, vector diseases, and poorer air quality, among other impacts. Overall, urbanization promotes chronic illnesses such as cardiovascular diseases, cancers, and respiratory diseases as some of its epidemiological effects.

Structural drivers alongside individual causes of illness are promoting the growth of chronic illnesses. Trends in structural drivers that bolster poor nutrition include "liberalization of food trade; the escalation of foreign investment by transnational food companies; the rapid spread of cheap, ultra-processed food that affect the availability, affordability and acceptability of what people eat; and deregulation of commodities markets that cause food price volatility and dietary dependency" (24). Rapid inflation alongside demands of supply chains and consumers lessens the democratization of food. In other words, lack of affordable healthy foods contributes to poor health outcomes. Numerous studies have been conducted correlating poor nutrition with the growth in chronic diseases. For example, the World Health Organization identified that there has been an increase from 108 million adults in 1980 to 422 million adults living with diabetes today (25).

What is interesting is that intervention using a low-glycemic and low-carb regimen could reverse prediabetes and improve insulin sensitivity (26). Research on nutrition indicates prediabetes can be reversed, obesity can be controlled, and other chronic diseases can go into remission or reach anti-inflammatory statuses (26). With these outcomes possible, the question of how to implement a multidimensional approach to treat chronic diseases remains.

Result of pandemics

Coronaviruses are positive-sense RNA viruses, which means they are immediately translated by the host (due to its similarities with mRNA) and therefore directly cause infection. Positive-sense single-stranded RNA viruses are also known as the sense strand. Anti-sense, or negative-sense, viral RNA on the other hand needs RNA polymerase to be transcribed into positive-sense RNA. These statements preface the fact that coronaviruses are extremely contagious through respiratory droplets, and they proliferate in the body very efficiently. According to cell theory, viruses are not alive; in fact, life is the most basic property of a cell, a quality that viruses alone do not possess without a host organism. To discuss this further, coronaviruses have a symbiotic relationship with their hosts; they are intracellular obligate parasites and cannot replicate outside of host machinery. For these reasons, they must arrest the normal cell cycle. Hijacking of normal regulatory cell cycle process allows the proliferation (or the growing in numbers) of viral infections. Within the last few decades (within our lifetimes), we have seen coronavirus outbreaks. Many people are hearing about coronaviruses for the first time, when in fact they have been around for quite some time now, and previous research has been a large driver in why vaccines were created so quickly. The first notable coronavirus outbreak was the SARS-CoV outbreak in 2003. Researchers lacked a comprehensive basis of the mechanism of virus then. Animal models of SARS-CoV infection could not describe the disease progression in humans because of an incongruent pathogenesis. Again, animals are quite different than humans, so this is understandable. And once the SARS epidemic ended, human trials were not possible. Although several candidate vaccines against SARS-CoV have been produced and tested, at present, unfortunately, there is no FDA-approved vaccine against SARS (27). The second coronavirus outbreak was the Middle East respiratory syndrome coronavirus (MERS-CoV) outbreak in 2012. What is known is that dromedary camel populations of East Africa and the Middle East were reservoirs for the Middle East Respiratory Syndrome-Coronavirus (MERS-CoV). Both outbreaks in 2003 *and* 2012 were infectious. The question must be asked, why didn't we have a global pandemic then in 2003, and in 2012? The answer is due to the lack of medications and vaccines, the options for early intervention were limited to public health measures. There was rigorous infection control, including isolating camels, animals, and people. Relevant to the current state, notably, we are most familiar with Covid-19. Many studies have been done suggesting why it failed to be contained, but for the purposes of this book, there is simply discussion of pandemics and their implication with chronic diseases.

The focus is on global health, and therefore identifying fallacies in the US healthcare system serves to elaborate on the broken systems of developed countries. Looking at global health trends in coordination with the Covid-19 pandemic, the emergence of a new set of obstacles further intensifies the global health crisis. Emergency rooms and makeshift hospitals are generally crowded and functioning with limited means. The global pandemic had filled up the capacity of hospital beds with Covid; what we failed to realize was that all other illnesses and hospital visits did not subside. They were still present. Looking at the data, we can really start to understand the depth of legitimate claims that current healthcare infrastructure is at capacity. Now more than before intervention is a necessity, with chronic diseases showing no sign of reversion.

Global pandemics, a force of extreme transmission, have made chronic disease prevention of utmost priority. As discussed earlier, not only do viruses arrest normal cell cycles efficiently but they do so quietly at first, signaling an asymptomatic progression. During this time, before the onset of symptoms, people remain contagious and are oblivious of the attack on their bodies. Not only are people at greater risk of contracting viruses and other foreign pathogens when they have chronic illnesses, but they also are at higher risk of poor recovery and death (28). Therefore, pandemics have made it harder for people with disabilities to receive proper care. When resources are being allocated for those in need of immediate care, people with disabilities are at higher risk for complications. The nature of pandemics reveals fallacies in healthcare systems, as those with disabilities are unproportionately impacted due to their preexisting conditions. As a result, chronic illnesses perpetuate and worsen during these times. On the flip side of the argument, because of pandemics, healthcare infrastructure and global health measures are progressed. Notably, for example, the American Medical Association (AMA) and the Centers for Disease Control and Prevention (CDC) have started to collaborate on efforts to reduce prediabetes (28). This is merely one example of how pandemics have pushed organizations to pivot healthcare for people with disabilities. Thus, fewer people are susceptible for noncommunicable and communicable diseases. It is often when time becomes of the essence that productive movements occur. Unfortunately, however, progress is slow, and the number of chronic diseases outweighs current efforts to combat them.

Age of medicalization and the sick role

Four chronic conditions have been highlighted as priority, including cardiovascular disease, common cancers, chronic respiratory disease, and diabetes. These NCDs share common risk factors, including poor diet, inactivity or sedentary lifestyle, smoking, and harmful alcohol consumption, which have been pronounced on global health agendas (29). Medicalization in its essence is treating nonmedical problems as the subject of medical conditions and thus creating a diagnosis that must be treated. In modern times, newly medicalized illnesses include alcoholism, ADHD, PTSD, eating disorders, infertility, and sleep disorders, for example (30). What is being observed is an increase in the number of people medically defined as sick. With less social stigma regarding illness, it becomes a normal social phenomenon to report illnesses, and resulting health discrimination subsides. Medicalization places the encumbrance of illness outside the control of the individual, promoting the social construction of illness (31). Individuals start to then place expectations on healthcare systems and governments to share the responsibility of their healthcare management. Feelings of less autonomy over one's body and well-being allow more people to assume the sociological sick role.

The sociologist Talcott Parsons identified the sick role as the cause of increased chronic illnesses, particularly in capitalist societies. Since the 1970s, there has been a transition from acute illnesses to chronic illnesses, as well as new privileges given to sick individuals. Talcott Parsons' theory of the sick role denotes those sick as a form of deviant behavior in terms of physiological incapacity. The theory places expectations on those who are sick and what dependency behaviors the sick exhibit. Those assuming the sick role have a legitimization for their deviance from society and must not be blamed for their condition. As another tenant of the sick role, those who are unwell must seek to become better as an indication of a healthy mind. They must want professional help and cooperate with the help offered (32). What Parsons failed to consider is the achieved normality of chronic illness in current global health. If an overwhelming number of people participate in deviant behavior, the state of being sick, that is, the behavior, then becomes the norm and not deviant behavior. Contradictory to what Parsons previously postulated, today sickness is not just a temporary state. Global medicalization has caused the prolonging, and to some extent the permanence, of illness. The individual has a responsibility to their own well-being, supplemented by available means in society.

Other contributors to the increase of medicalization include doctors, pharmacists, and other prominent market stakeholders. Prescription drug spending has become a global problem that perpetuates the state of illness. The *American Journal of Public Health* states that prescription drugs account for the largest portion of a person's healthcare expenditure (33). These findings are substantial when the cycle of overprescribing is considered. Doctors are prescribing medications for the side effects of other mediations that a patient takes. As a result, increased drug dependence, organ failure, behavioral changes, and microbial resistance are observed. People with chronic diseases in particular share the greatest drug expenditure and are most responsive to changes in drug prices (33).

Conclusion

The introduction to the rise in chronic diseases is accompanied by justification from external and internal forces. Overall, the burden of chronic diseases is far too severe for inaction. Discussing comorbidities associated with chronic diseases, many of which are due to side effects of prescription medications, a third of adults suffer from multiple chronic conditions. As societies become advanced and industrious, the world witnesses aging populations that are outliving the generations prior. This shift in demographics indicates that as people live longer, the chances of developing chronic illnesses increases (1). In terms of the global cost of healthcare, the World Health Organization's 2020 analysis showed that global healthcare costs reached 10% of the world's GDP, with out-of-pocket expenses from low- and middle-income countries comprising over 40% in 2018 (7). DALY stats indicate illnesses such as COPD, ILD, cardiovascular disease, stroke, and overnutrition are risk factors with severe growth. Epidemiological transitions demonstrate not only aging populations but also ground observations of morbidity shifting from communicable to noncommunicable diseases.

The UN World Summit in 2011 addressed with extreme priority that intervention must be taken on NCDs that result from tobacco product usage, unhealthy diets, use of alcohol, and physical inactivity (17). From a sociological perspective, sleep deprivation has increasingly become more prominent over time as people are working longer hours. And with more women in the workforce, there are more people living goal-driven lives. The emergence of hyper-productivity lifestyles will continue to grow, as countries become more prosperous. Evaluating the social realities of low-SES students demonstrates the need for more funding for these students. Reform would start to equalize the educational outcomes of disadvantaged students compared to advantaged students. In other words, sons are expected to be just as poor as their fathers. With more opportunity given to children, it becomes more probable that they will receive better quality healthcare and thus more access to chronic disease preventative measures. Holistically identifying the factors that are a part of overall well-being serves to provide a more encompassing answer to why chronic diseases are perpetuating. Anthropogenic activities like population growth, deforestation, pollution, and overconsumption are heightening the loss of biodiversity. This loss not only alters the physical environment but also our bodies, with increased exposure to harm. This urbanization promotes chronic illnesses such as cardiovascular disease, cancer, and respiratory disease as some of its epidemiological effects. Lack of affordable healthy food contributes to poor health outcomes. Numerous studies have been conducted correlating poor nutrition with a growth in chronic diseases. The World Health Organization identified that there has been an increase from 108 million adults in 1980 to 422 million adults living with diabetes (25).

Global pandemics, a force of extreme transmission, have made chronic disease prevention of utmost priority. The nature of pandemics reveals fallacies in healthcare systems, as those with disabilities are unproportionately impacted due to their preexisting conditions. As a result, chronic illnesses perpetuate and worsen during these times. In modern times, newly medicalized conditions, including alcoholism, ADHD, PTSD, eating disorders, infertility, menopause, and sleep disorders, for example (30), increase the number of people medically defined as sick. Talcott Parsons' theory of the sick role denotes those sick as a form of deviant behavior in terms of physiological incapacity. Contradictory to what Parsons previously postulated, today sickness is not just a temporary state. Global medicalization has caused the prolonging, and to some extent the permanence, of illness. Other contributors in the increase of medicalization include doctors, pharmacists, and other prominent market stakeholders. Doctors are prescribing medications for the side effects of other medications that patients take. As a result, increased drug dependence, organ failure, behavioral changes, and microbial resistance are observed. An investigation into functional foods and nutraceuticals aims to diversify strategies to combat the growth of chronic diseases.

References

1. Divo, M. J., Martinez, C. H., & Mannino, D. M. (2014). Ageing and the epidemiology of multimorbidity. *European Respiratory Journal, 44*(4), 1055–1068. https://doi.org/10.1183/09031936.00059814
2. Mclaughlin, M., & Rank, M. (2017). *Estimating the economic cost of childhood poverty in the United States*. St. Louis, MO: George Warren Brown School of Social Work Washington University.
3. WHO. (2013). *Noncommunicable diseases*. Geneva, Switzerland: World Health Organization.
4. Nugent, R., & Feigl, A. B. (2010). Where have all the donors gone? Scarce funding for non-communicable disease. Washington, DC: Center for Global Development. Working paper 228.
5. Holman, H. R. (2020). The relation of the chronic disease epidemic to the health care crisis. *ACR Open Rheumatology, 2*(3), 167–173. https://doi.org/10.1002/acr2.11114
6. Hackbarth, G., & Boccuti, C. (2011). Transforming graduate medical education to improve health care value. *New England Journal of Medicine, 364*(8), 693–695. https://doi.org/10.1056/nejmp1012691
7. World Health Organization. (n.d.). Global spending on health: Weathering the storm. World Health Organization. Retrieved November 21, 2021, from https://www.who.int/publications/i/item/9789240017788.
8. Global population at a glance: 2002 and beyond – census. (n.d.). Retrieved November 21, 2021, from https://www.census.gov/library/publications/2004/demo/wp02-1.html.
9. McKeown, R. E. (2009). The epidemiologic transition: Changing patterns of mortality and population dynamics. *American Journal of Lifestyle Medicine, 3*(Suppl. 1), 19S–26S.
10. Lozano, R., Naghavi, M., Foreman, K., Lim, S., Shibuya, K., Aboyans, V., Abraham, J., Adair, T., Aggarwal, R., Ahn, S. Y., Alvarado, M., Anderson, H. R., Anderson, L. M., Andrews, K. G., Atkinson, C., Baddour, L. M., Barker-Collo, S., Bartels, D. H., Bell, M. L., …, Murray, J. L. C. (2012). Global and regional mortality from 235 causes of death for 20 Age groups in 1990 and 2010: A systematic analysis for the global burden of disease study 2010. *The Lancet, 380*(9859), 2095–2128. https://doi.org/10.1016/s0140-6736(12)61728-0
11. Centers for Disease Control and Prevention. (2019, December 19). Frequently asked questions. Centers for Disease Control and Prevention. Retrieved November 21, 2021, from https://www.cdc.gov/nchhstp/socialdeterminants/faq.html#:~:text=Health%20is%20influenced%20by%20many,medical%20care%20and%20social%20factors.
12. St John, P. D., Tyas, S. L., Menec, V., & Tate, R. (2014, May). Multimorbidity, disability, and mortality in community-dwelling older adults. *Canadian Family Physician Medecin de Famille Canadien*. Retrieved November 21, 2021, from https://www.ncbi.nlm.nih.gov/pmc/articles/PMC4020665/.
13. Hajat, C., & Stein, E. (2018). The global burden of multiple chronic conditions: A narrative review. *Preventive Medicine Reports, 12*, 284–293. https://doi.org/10.1016/j.pmedr.2018.10.008
14. Khanam, M. A., Streatfield, P. K., Kabir, Z. N., Qiu, C., Cornelius, C., & Wahlin, Å. (2011). Prevalence and patterns of multimorbidity among elderly people in rural Bangladesh: A cross-sectional study. *Journal of Health, Population and Nutrition, 29*(4). https://doi.org/10.3329/jhpn.v29i4.8458
15. Blakely, T., Hales, S., Kieft, C., Wilson, N., & Woodward, A. (n.d.). The global distribution of risk factors by poverty level. *Bulletin of the World Health Organization*. Retrieved November 21, 2021, from https://pubmed.ncbi.nlm.nih.gov/15744404/.
16. Barnett, K., Mercer, S. W., Norbury, M., Watt, G., Wyke, S., & Guthrie, B. (2012). Epidemiology of multimorbidity and implications for health care, research, and Medical Education: A cross-sectional study. *The Lancet, 380*(9836), 37–43. https://doi.org/10.1016/s0140-6736(12)60240-2

17. United Nations. (2011). *Political declaration of the high-level meeting of the General Assembly on the prevention and control of non-communicable diseases (Document A/RES/66.2)*. High Level Meeting on Prevention and Control of Non-communicable Diseases, New York, NY: United Nations.
18. Centers for Disease Control and Prevention. (2018, August 8). CDC – sleep and chronic disease – sleep and sleep disorders. Centers for Disease Control and Prevention. Retrieved November 21, 2021, from https://www.cdc.gov/sleep/about_sleep/chronic_disease.html
19. Virtanen, M., & Kivimäki, M. (2012). Saved by the bell: Does working too much increase the likelihood of depression? *Expert Review of Neurotherapeutics, 12*(5), 497–499. https://doi.org/10.1586/ern.12.29
20. Centers for Disease Control and Prevention. (n.d.). The New Global Health – volume 19, number 8-August 2013 – emerging infectious diseases journal – CDC. Centers for Disease Control and Prevention. Retrieved November 21, 2021, from https://wwwnc.cdc.gov/eid/article/19/8/13-0121_article.
21. Rank, M. (2011). "Rethinking American poverty". *American Sociological Association*, 10, 16–21.
22. Karjalainen, E., Sarjala, T., & Raitio, H. (2009). Promoting human health through forests: Overview and major challenges. *Environmental Health and Preventive Medicine, 15*(1), 1–8. https://doi.org/10.1007/s12199-008-0069-2
23. Li, Q., Morimoto, K., Kobayashi, M., Inagaki, H., Katsumata, M., Hirata, Y., Hirata, K., Suzuki, H., Li, Y. J., Wakayama, Y., Kawada, T., Park, B. J., Ohira, T., Matsui, N., Kagawa, T., Miyazaki, Y., & Krensky, A. M. (2008). Visiting a forest, but not a city, increases human natural killer activity and expression of anti-cancer proteins. *International Journal of Immunopathology and Pharmacology, 21*(1), 117–127. https://doi.org/10.1177/039463200802100113
24. Clark, J. (2014). Medicalization of global health 3: The medicalization of the non-communicable diseases agenda. *Global Health Action*, *7*(1), 24002. https://doi.org/10.3402/gha.v7.24002
25. WHO. (n.d.). Global report on diabetes.Retrieved November 21, 2021, from https://www.who.int/news-room/fact-sheets/detail/diabetes
26. Ojo, O. (2019). Nutrition and chronic conditions. *Nutrients, 11*(2), 459. https://doi.org/10.3390/nu11020459
27. Yang, Y., Peng, F., Wang, R., Guan, K., Jiang, T., Xu, G., Sun, J., & Chang, C. (2020). The deadly coronaviruses: The 2003 SARS pandemic and the 2020 novel coronavirus epidemic in China. *Journal of Autoimmunity, 109*, 102434. https://doi.org/10.1016/j.jaut.2020.102434
28. Kmetik, K. S., Skoufalos, A., & Nash, D. B. (2021). Pandemic makes chronic disease prevention a priority. *Population Health Management, 24*(1), 1–2. https://doi.org/10.1089/pop.2020.0126
29. WHO, UNICEF, Government of Sweden, & Government of Botswana. Health in the post-2015 agenda. Report of the Global Thematic Consultation on Health.
30. Conrad, P., Mackie, T., & Mehrotra, A. (2010). Estimating the costs of medicalization. *Social Science and Medicine, 70*(12), 1943–1947.
31. Puhl, R.M., & Heuer, C.A. (2009). The stigma of obesity: A review and update. *Obesity (Silver Spring)*, 17(5), 941–964.
32. Varul, M. Z. (2010). Talcott Parsons, the sick role and chronic illness. *Body & Society, 16*(2), 72–94. https://doi.org/10.1177/1357034x10364766
33. Mueller, C., Schur, C., & O'Connell, J. (1997). Prescription drug spending: The impact of age and chronic disease status. *American Journal of Public Health, 87*(10), 1626–1629. https://doi.org/10.2105/ajph.87.10.1626

2

Importance of Integrative Health Sciences in the Therapy of Chronic Diseases

Ghazaleh Eslamian, Arman Ghorbani, and Morvarid Noormohammadi

Contents

Chronic diseases 17
Integrative medicine 19
Different types of complementary treatments 20
  Natural treatments 20
    Some kinds of diet 20
    Nutritional therapy 21
  Herbal supplements 22
    Probiotics 23
  Mind-body practices 23
  Bodywork 24
Management of chronic diseases with integrative health sciences 25
References 29

Chronic diseases

Chronic diseases prevail among the top reasons for mortality in the majority of industrialized and developing countries (1). World Health Organization (WHO) describes a chronic disease as one that lasts for an extended period, progresses gradually, and is not transmitted between individuals (2). Chronic diseases are described generically as problems that endure for one year or more and need continuing medical treatment, restrict day-to-day life activities, or both (3). They are classified following the biomedical disease categorization, and include conditions such as diabetes, asthma, and depression, among others (3).

Chronic diseases are on the rise, according to the latest data. About one in every three individuals has more than one or numerous chronic conditions and bears extreme health and financial burdens resulting from their conditions (4). Chronic diseases have a significant financial effect, with worldwide

DOI: 10.1201/9781003220053-3

economic losses of US$7 trillion anticipated between 2011 and 2025 as a result of chronic diseases (5). Chronic diseases have hurt many people's health and quality of life throughout the world (6) and have grown more noticeable as the speed of death and the burden of disease have increased (7). According to recent tendencies in population increase, age variations, and disorder dynamics, the prevalence of chronic diseases, further chronic disorders, and varieties of chronic conditions will increase in the years to come. These tendencies, which include the fast-expanding number of seniors, life expectancy improvement linked to advancements in public health and medical care, and the consistently increased prevalence of particular risk factors, pose a danger to public health and financial health (8).

Healthcare systems, which should manage chronic diseases globally, have evolved primarily to cope with acute periodical treatment rather than supply structured treatment for individuals with long-lasting disorders (2). With chronic diseases bearing a significant economic and social impact, and an increasing prevalence of these disorders, new and novel solutions must be developed to prevent and treat chronic diseases (9).

Cardiovascular disease, cancer, chronic lung disorders, and diabetes are the leading causes of mortality globally, accounting for three out of every five fatalities (10). Medical science has achieved significant lifetime extensions at the expense of tremendous investigation expenditures in many nations during the previous 50 years, but they have mainly been nullified by the inability to address the chronic disease issue (11). There is widespread agreement that significant reforms in the health system will be required to face the difficulties of the rising prevalence of chronic diseases and the consequences for the health system and quality of life (12). The primary preventive measures to prevent disease onset and the secondary preventive measures for those at risk for disease onset or in the early stages of the disease are rarely implemented; as a result, individuals with chronic diseases are underdiagnosed and undertreated (13). Therapy compliance, which is a public health concern, is particularly important in the treatment of chronic diseases (14). Patient compliance with recommended therapy is poor (7). Insufficient compliance to therapy may impair the effectiveness of the recommended medical care, leading to treatment results that are less than intended (7). In the United States, where efforts to turn primary care practices into Patient-Centered Medical Homes are becoming more widespread, they have shown only modest increases in quality of treatment, indicating that these treatments need additional refining. The methods and results of chronic care delivery must be transformed, and according to the investigation results, multicomponent interventions are necessary for this transformation (13).

People at risk for chronic diseases must be identified, and the correct information about them must be available. Actionable insights about these patients must be generated, and these patients must be coached through better lifestyle options. These aspects are essential for community health and chronic disease managing programs to be successful (6).

Integrative medicine

Approximately three-quarters of all healthcare expenses in the United States are attributed to five of the most expensive disorders. A large amount of that 75 percent is currently avoidable by dietary and lifestyle modifications (11). Individuals' requirements for complementary therapies, their usage of complementary and conventional treatments, the problems of chronic disease control, and a growing emphasis on disease prevention and health promotion have contributed to the emergence of integrative healthcare practices as a method of providing health-promoting services. In turn, more integrative healthcare-specific research activity, the creation and performance of integrative healthcare practices and programs, and health professional training programming in this field have all increased in recent years (15). The merging of Western and alternative medicine brings a holistic strategy to patient care that may culturally benefit various populations. The term "integrative medicine" has been used to describe a notion that goes beyond just mixing various systems of medicine and authorizes for a more individualized strategy to patient treatment that involves the mind, spirituality, and a feeling of community as well as the body (16).

Integrative medicine is defined in different ways, and the terms used to describe it are as diverse (17). An evolution from conventional medicine, integrative medicine represents a new growth step, representing a transition from specialization to integration in the medical system. It is not a reversal but rather a step forward. It aims to achieve the following objectives: the integration of biological factors with those of psychological, social, and environmental factors; the integration of the recent medical findings in all life-related areas with that of the most practical clinical experience in all medical specialties (18).

Integration of conventional and complementary medicine is a holistic approach to well-being that incorporates:

Conventional medicine – medicine, surgery, and lifestyle modifications.

Complementary medicine – not considered part of standard medicine. It refers to a range of therapeutic and diagnostic disciplines that operate mostly outside of the institutions where conventional healthcare is both taught and provided. Chiropractic therapy, yoga, meditation, and other forms of treatment may be used.

Complementary medicine is becoming more prevalent in healthcare training, but there is still a great deal of ambiguity regarding what it is and what place the disciplines covered under this term should have in respect to traditional medicine (19).

When these disciplines were first introduced in the 1970s and 1980s, they were primarily used as an alternative to traditional healthcare, and as a result, they were known together as "alternative medicine." As the two approaches started to be utilized in conjunction with (and therefore "complementing") one another, the term "complementary medicine" was coined.

Complementary medicine has evolved through time from characterizing the relationship between unconventional healthcare disciplines and traditional treatment to defining the whole range of fields. Some experts use the terms "unconventional medicine" and "alternative medicine" interchangeably. This shifting and overlapping vocabulary may help to explain some of the ambiguity that exists in the discussion of the issue (19).

Different types of complementary treatments

Natural treatments, mind-body practices, and bodywork are just a few of the alternative therapies that are available.

Natural treatments

Natural chemicals, which may be found in nature, assist to optimize the amount of nutrients available. Medical nutrition therapy, herbal supplements, probiotics, and other natural medicines are examples of natural remedies (9).

Medical nutrition therapy is a method of managing chronic health issues, particularly in the context of weight loss and weight management (20). When you follow a specific diet and meal plan, you may help lessen the symptoms of some diseases and avoid subsequent health concerns (20). There are hundreds of different diets to choose from. Weight loss is a primary goal for some, but increasing weight, decreasing cholesterol, living a long and healthy life, and a variety of other reasons are also important (20).

Some kinds of diet

The Atkins diet is a low-carbohydrate diet that is designed to help you lose weight by lowering insulin levels in your body. Consuming significant quantities of refined carbs causes insulin levels to increase and decrease quickly in the body. The body responds to rising insulin levels by storing energy from the food that has been ingested, decreasing the likelihood that the body will utilize stored fat as a source of energy. As a result, those who follow the Atkins diet avoid carbs while eating as much protein and fat as they like (21).

The Zone diet strives for a nutritional balance of 40 percent carbs, 30 percent fats, and 30 percent protein in each meal. The emphasis is also on managing insulin levels, which may result in more effective weight reduction and body weight management than other techniques. The Zone diet supports the eating of high-quality carbs – that is, unprocessed carbohydrates – and fats such as olive oil, avocado, and almonds (22).

The Ketogenic diet is high in fat and low in carbohydrates. It has been used as a therapy for epilepsy for decades, and is now being researched for potential use in other fields. It may seem counterintuitive, but it permits the body to use fat as fuel rather than carbs, which is opposed to conventional sense. Ketogenic diets are characterized by a significant reduction in carbohydrate intake (usually to less than 50 g/day) and a corresponding increase in the relative proportions of protein and fat – usually extremely high percentages of fat because it is difficult to increase protein intake beyond a certain point on the ketogenic diet. In order to maintain an overall focus on fat in the diet, dietary fats such as those found in avocados, coconuts, Brazil nuts, seeds, fatty salmon, and olive oil are freely added to the diet (23).

Vegetarians may be classified into many categories: lacto-vegetarians, fruitarian vegetarians, lacto-ovo vegetarians, vegetarians on a living food diet, ovo-vegetarians, pesco-vegetarians, and semi-vegetarians. Lacto-ovo vegetarianism is the most common kind of vegetarianism. This means that no animal-based items other than eggs, dairy products, and honey are consumed. Studies conducted in the previous several years have shown that vegetarians have a lower body weight, suffer from fewer ailments, and have a higher life expectancy than persons who consume meat (24).

The Mediterranean diet involves a high intake of plant foods, including fresh fruits for dessert, legumes, nuts, whole grains, seeds, and olive oil as the primary source of dietary fats, among other things. The most common dairy products are cheese and yogurt. Fish and poultry are also included in the diet in moderate proportions, with up to four eggs per week being the maximum. Small amounts of red meat are included, as is low to moderate alcohol use. Fat accounts for up to one-third of the calories consumed in the Mediterranean diet, with saturated fats accounting for no more than 8 percent of total calories.

It is the most widely researched diet to date, with trustworthy evidence supporting its usage for increasing a person's quality of life while also decreasing illness risk in both humans and animals (25).

Nutritional therapy

Nutritional therapy is also concerned with the manner in which meals are produced and given to the consumer to be consumed (26). In order for meals to be therapeutically useful, they must have the right micro- and macronutrients and be supplied in a nutritionally rich style that is free from pollutants (26). Despite the fact that plants are the primary source of micronutrients, the nutritional content of plants may be changed by incorrect management. Fruits and vegetables selected before they reach their full maturity contain less phytochemicals than those gathered later in the season (26). In addition to protecting plants against a variety of pests and diseases, phytochemicals also provide protection to the human body when they are ingested (26). Cooking and processing may cause phytochemicals and antioxidants to lose even more of their beneficial characteristics (26).

There are several scientific findings suggesting that nutrition is a significant modifiable factor that may have an impact on the occurrence of a wide range of chronic illnesses (20). Food, in reality, includes a variety of biologically active components that are beneficial to one's health.

In recent years, advances in -omics technology have shed light on the molecular and cellular impacts of nutrition on a person (20). In turn, this has enabled the individualization of dietary treatments as well as the manufacture of custom-made meals that are based on an individual's physiological, genetic, ethnic, cultural, and economic background (20, 27).

Herbal supplements

Herbal supplements are produced from plants and/or their oils, roots, seeds, berries, or flowers, as well as their fruits and leaves.

Herbal supplements have been utilized for hundreds of years to treat a variety of ailments. There are several herbal supplements available, each of which has a variety of applications.

There are many various methods and forms in which herbs and plants may be processed and consumed, including entire herbs, teas, syrup, essential oils, ointments, salves, rubs, capsules, and tablets that contain a pulverized or powdered version of a raw herb or a dried extract of a raw herb (28).

Herbal supplements include the following, which are some of the most popular:

Aloe vera: Topically used on burns, psoriasis, and osteoarthritis, among other things. In the oral form, it is used to treat digestive problems such as gastritis and constipation.

Black cohosh: Hot flashes, nocturnal sweats, vaginal dryness, and other menopausal symptoms are treated with this supplement.

Chamomile: Sleeplessness, anxiety, upset stomach, gas, and diarrhea are all treated with this supplement. It is also used topically to alleviate skin disorders. People who are allergic to ragweed should use caution.

Echinacea: used to treat the symptoms of the common cold and flu.

Flaxseed: a supplement that lowers cholesterol levels. Fiber and omega-3 fatty acids are abundant in these seeds.

Ginko: Tinnitus and memory issues are two conditions that may be treated with this supplement. Those on antidepressant drugs and experiencing adverse effects may benefit from ginko used in conjunction with selective serotonin reuptake inhibitors to increase sexual desire and performance.

Peppermint oil: used to treat digestive issues such as nausea, indigestion, stomach troubles, and bowel disorders.

Soy: Menopausal symptoms, cognitive issues, and excessive cholesterol levels are all treated with this supplement.

St. John's Wort: Depression, anxiety, and sleep difficulties are all treated with this supplement. There are several more medication and herb interactions with this plant.

Tea tree oil: Acne, athlete's foot, nail fungus, wounds, infections, lice, oral yeast infection, cold sores, and dandruff are just a few of the ailments that may be treated topically using tea tree oil.

Probiotics

Probiotics are beneficial microorganisms that help to maintain healthy digestive and immune systems. Probiotics are made up of bacteria and yeast, and they are beneficial to the body. Lactobacillus and Bifidobacterium are two common types of probiotic bacteria. Saccharomyces boulardii is the yeast that is most often seen in probiotics. Probiotics and prebiotics are microbiota-management tools that may be used to improve human health once they have been shown to be useful for maintaining a healthy population of gut microbiota and normal bowel function in animals (29).

Mind-body practices

Mind-body activities make use of the link that exists between the mental and physical aspects of wellness. Biofeedback, hypnosis, meditation, Reiki, yoga, and tai chi are some of the techniques available (30).

Biofeedback is a technique that helps a person become more aware of particular bodily processes. Biofeedback is a method that is used to supplement the regular sensory input and improve control over bodily activities that are often considered involuntary, such as heart rate and breathing. The overall technique begins by monitoring a bodily parameter, which may be either physiologic or biomechanical in nature, and then converting that measurement into a visual, audio, or tactile signal for presentation. The person learns to manipulate the signal by adjusting the body parameter.

Because of the positive feedback modifications, the actions become more deeply embedded in memory, providing for an enhanced degree of control over a previously automatic function (31).

Hypnotherapy helps a person become more aware of their surroundings and more receptive to suggestions. Although hypnosis is not a treatment in and of itself, it may be used to enhance the delivery of therapy in the same way that a syringe promotes the administration of medications (32). Hypnosis does not make the impossible feasible; rather, it may assist patients in believing and experiencing what they feel is possible for them to attain via hypnosis (32).

Meditation is a technique that employs mental concentration to calm unwanted thoughts and sensations. Meditation incorporates strategies such as paying attention to the breath, repeating a mantra, or disconnecting from the cognitive process in order to concentrate the mind and achieve self-awareness and inner tranquility (33).

Reiki (energy healing) is a technique that helps to repair the body's energy fields in order to facilitate healing. This biofield energy treatment has been studied with community-dwelling older persons, as well as with particular illness states such as cancer, chronic tiredness, diabetic neuropathy, and surgical patients, among other groups (34).

Yoga and tai chi are both forms of exercise that combine particular postures and motions with breathing to help harmonize body and mind (35). Wellness movement techniques such as yoga, tai chi, and qigong, which are usually referred to as complementary and alternative medicine (CAM) are becoming more popular. They are used in conjunction with other complementary and alternative medical methods, such as acupuncture, herbal treatments, chiropractic, and homoeopathy, to provide a comprehensive treatment plan (35). Among supplementary and alternative weight-loss methods reported in the United States, yoga was the most frequently mentioned (36).

Bodywork

Bodywork treatments are those that work with the body to improve health and well-being. Acupuncture, chiropractic adjustments, reflexology, and therapeutic massage are some of the types of bodywork that are available (37).

Acupuncture is a medical therapy and philosophy that is founded on the notion of putting tiny needles or pressure on particular places on the body in order to alleviate pain and other symptoms. The roots of this therapeutic approach may be traced back to traditional Chinese medicine, with underlying philosophical beliefs linked with Confucianism and Taoism serving as guiding principles (38).

Chiropractic adjustments are performed using the hands of a healer in order to restore joints to their normal position. Specifically, chiropractic is defined as "a health-care profession that is concerned with the diagnosis, treatment, and prevention of problems of the neuromusculoskeletal system, as well as the implications of these disorders on overall health." Chiropractic treatment processes have a strong emphasis on manual adjustments (39).

Reflexology is a systematic procedure in which applying mild pressure to certain areas on the feet and hands has an influence on the health of other regions of the body that are connected to the feet and hands (40).

Therapeutic massage is the manipulation of the soft tissue of entire body areas in order to bring about generalized improvements in health, such as relaxation or improved sleep, or specific physical benefits, such as the relief of muscular aches and pains. Therapeutic massage is a type of bodywork that is becoming increasingly popular (41).

Management of chronic diseases with integrative health sciences

A first step toward comprehending the phenomena of integrative healthcare, according to Stumpf, Shapiro, and Hardy, is to define integrative healthcare as something changeable. However, it is more probable that a description will appear from the fundamental concerns influencing the future of integrated medicine. That is, bilateral integrative medicine is more similar to the assimilation of complementary and alternative medicine by biomedicine than actual acculturation in the traditional sense (42).

Over the last decades, several different techniques and models of care to control chronic diseases have been developed (43). Although the research implies that components of these approaches may be helpful in improving personal and systemic results (44–47), most of these approaches for therapy have just a tiny amount of data to support their efficacy (46, 48).

While a medico-centric/conventional biomedical method may be effective to treat acute diseases, especially life-threatening ones, it may not be consistent with the concepts of chronic disease control (49). As stated by the Pan American Health Organization, chronic disease control is recommended to be patient centered, aggressive, and prevention focused, instead of the conventional biomedical practice, which emphasizes disease-centered, reactive, and treatment-focused care. Additionally, the measure of care must be coordinated and integrated (50). The absence of proof in traditional, complementary, and integrative medicine (TCIM) is being criticized daily (51). Since 2002, the WHO's Traditional Medicine Strategy has promoted and supported the integration, credit, and usefulness of TCIM products and physicians in national health systems at all grades, including primary healthcare, specialized care, and hospital care (52).

According to a *2019 Global Report on Traditional and Complementary Medicine* by WHO, a "shortage of investigation" in the field is the most significant obstacle to growing regulatory procedures to incorporate TCIM in health strategies. This view is mainly founded on a shortage of available evidence, obstacles to access (e.g., the language of publishing, payment), problems in analyzing findings, and the study specificities (53). According to recent research, integrative healthcare is a style of treatment that looks to be more in line with the concepts of chronic disease control and less medico-centric than other chronic disease models of care. "A client-centered measure of care supplied by a group of biomedical, allied, and complementary health experts who work together and respectfully to provide available, holistic, evidence-based care that is personalized and corresponded with a focus on disease prevention, recovery, and health promotion" is what this approach represents (54).

For example, interest in integrative healthcare has grown significantly over the last decades, in contrast to other forms of treatment. A variety of manifestations of this interest have occurred, like an increase in the number of skilled groups, clinics, publications, and academic chairs devoted to integrative healthcare. In light of this, an integrated healthcare paradigm may have some extent of ecological validity. As with chronic disease models of treatment, there is a great deal of variation in how integrative healthcare is performed worldwide, which is no different (54, 55).

In a real-world healthcare system, observational treatment result studies are able to supply valuable data on the clinical substances of integrating philosophically additional strategies to human health and recovery, like the kinds of patients attending for integrative medical treatment, whether personalized, integrative strategies to patient care are associated with enhanced health-related quality of life (HRQoL) (56). Greeson et al. have shown that integrative medical care in a university-based center is related to considerable improvements in HRQoL in a medically heterogeneous group with severe comorbidity and practical boundaries (57). Baker et al. did not find any contrast between integrative healthcare and standard medical treatment in terms of quality of life (i.e., The Functional Assessment of Cancer Therapy [FACT-G] subscale) scores at three and six months among women with breast cancer (58).

The findings of Sundberg et al. revealed no statistically significant contrast between integrative healthcare and conventional healthcare in any 36-Item Short Form Survey (SF-36) health dimension among patients with nonspecific lower back or neck pain after 16 weeks of treatment (59, 60).

According to the findings of Chen et al., individuals with Parkinson's disease who received integrative healthcare showed a statistically significant increase in World Health Organization Quality of Life (WHOQOL) in contrast to those who received just Western medical therapy after 18 weeks (61).

According to Krucoff et al. (62, 63), the findings of two Monitoring and Actualisation of Noetic Trainings (MANTRA) studies conducted by them were not statistically significant for any therapy result comparison. Nevertheless, when comparing individuals treated with any Noetic treatments to those treated with routine care, a reduction was seen (25–30 percent) in the total incidence of principal adverse cardiovascular occasions or negative clinical occurrences. In addition, in a pilot analysis of four separate Noetic interventions, off-site prayer was related with the lowest absolute mortality both in-hospital and six months after discharge (63).

After conducting a similar study to determine the advantages of Noetic therapies on mood assessed by the Visual Analogue Scale (VAS) before percutaneous coronary intervention for unstable coronary syndromes, they discovered that only one of the eight VAS scales on mood [the VAS for worry] showed statistically significant differences in the stress control, imagery, and touch therapy groups compared to the routine therapy group (64).

Using multidimensional intervention concepts, Edelman et al. conducted a randomized controlled trial on integrative healthcare to reduce cardiovascular risk (65). The study found that when a relationship-centered, mind-body approach (including mindfulness meditation, relaxation training, stress management, motivational techniques, and health education and coaching) was used to support behavior transformation, significant advancements in ten-year cardiovascular risk as calculated by the Framingham risk score (FRS) were observed when compared to routine care (65).

Neither the HbA1c nor the blood pressure nor the lipid levels (i.e., total cholesterol, LDL-C, HDL-C, or triglycerides) between integrative healthcare and routine care were found to be statistically significant at 6 or 12 months in patients with diabetes in a study conducted by Bradley et al. (66) Two non randomized trials by Carlsson et al., compared the quality of life of women with breast cancer eather treated with anthroposophic therapy (ABCW) after operation for breast cancer or received conventional care (67, 68).

Carlsson et al. (67) performed their first research to assess the acceptability and feasibility of an anthroposophic treatment (ABCW) following operation for breast cancer when compared to standard care. They examined the disparities in subjective quality of life/life pleasure and coping methods between the women who received ABCW at the time of enrollment in the trial and those who received just regular care. For the purposes of this study, three primary result instruments were used:

- The European Organization for Research and Treatment of Cancer quality-of-life questionnaire core 30 (EORTC QLQ-C30)
- The life satisfaction questionnaire (LSQ)
- The differences in coping were assessed using the mental adjustment to cancer scale

The findings revealed that women who chose ABCW significantly expressed a worse quality of life and increased psychological distress on all measures than women who got standard treatment at the start of the research. The ABCW group had a dropout rate of around 13 percent, whereas the standard care group had a dropout rate of approximately 7 percent. The authors investigated the idea that the women in the ABCW group, who opted for individualistic care, would have had more autonomous personalities in their decision making as compared to the women in the standard care group when it came to their decision making about care. According to follow-up research, the women in the ABCW group had greater passive and nervous coping on admission, but this reduced over time as the study progressed (69). This might explain why the ABCW had a greater dropout rate than the general population.

According to the findings of a second study conducted by the same team (2004), women with breast cancer who experienced a six-month Conventional Care Treatment (CCT) and one-year follow-up declared a significant higher quality-of-life/life satisfaction (EORTC QLQ-C30 and LSQ) after ABCW treatment than women who received only standard care (68). According to follow-up research conducted on the same group of women, improvements in overall quality of life and emotional and social functioning were seen after five years when comparing the women who chose ABCW to those who were admitted. Progress was seen in these women between the time of admission and one year, but not beyond that (70).

Physical, social, and emotional side effects of cancer treatment include fatigue, pain, neuropathy, lymphedema, difficulty sleeping, weight gain, cognitive dysfunction, sexual dysfunction, psychological distress, and fear of recurrence (71).

According to Baker et al. (58), integrative healthcare was found to be statistically significantly more practical than standard medical care in enhancing the Multidimensional Fatigue Inventory (MFI)-mental fatigue subscale at six months, but not at three months or for any of the other MFI subscale scores in patients with breast cancer. Integrative healthcare was shown to be statistically significantly distinct from standard medical treatment when it came to endocrine-related symptom subscale scores and FACT physical composite scores after six months, with the standard medical care being significantly better.

At three and six months, there were no statistically significant differences between the groups in terms of NK cell activity, PBMC arginase activity, or symptoms of stress questionnaire subscale scores, according to the findings.

An individual's physical, emotional, and spiritual health is enhanced via the integration of complementary treatments (such as physical exercise, food, dietary supplements, mind-body therapies, acupuncture, and massage therapy) with conventional therapy in the practice of integrative medicine (72).

According to Arvidsdotter and colleagues (73), integrative healthcare and therapeutic acupuncture were significantly more practical than conventional treatment in enhancing mean HAD-anxiety, HAD-depression, MCS, and SOC-13 scores at 8 to 12 weeks compared to conventional treatment in a study of patients with psychological distress. At 24 weeks, the only statistically significant contrast among the groups was in mean HAM-A scores in favor of integrated healthcare compared to therapeutic acupuncture.

According to Bradley et al., statistically significant progress in mood (PHQ-8), self-efficacy (SES), problem areas in diabetes (PAID – a measure of emotional distress), and motivation for changing lifestyle were found in patients with diabetes who received integrative healthcare at 6 and 12 months (i.e., adjunctive naturopathic care). It should be noted that there was no statistically significant difference between the composite PSS scores in the integrative healthcare and control groups at any time point. There were no differences between integrative healthcare and standard treatment (66).

In patients with Parkinson's disease, Chen et al. (61) discovered that integrative healthcare was considerably more successful than Western medicine control in improving BDI-II scores after 18 weeks of observation. A statistically significant difference in BAI between the groups was not found after 18 weeks of testing.

In women with breast cancer, Baker et al. (58) observed no statistically significant difference between integrative healthcare and standard medical treatment in mean depressive, emotional, and cognitive subscores (measured by the SOSI) at three and six months after diagnosis, indicating that routine medical care was superior.

According to the findings of Eisenberg et al., when compared to usual care, patients with low back pain who received integrative healthcare demonstrated statistically significant enhancements in mean bothersomeness of worst symptom scores, pain scores, and difficulty with the worst of three self-reported activities scores at 12 weeks, except for bothersomeness (51).

The differences in SF-12 physical subdomain scores between integrative healthcare and standard treatment did not achieve statistical significance at 12 weeks but were statistically significant after 26 weeks, indicating that integrative healthcare was superior to standard care (74)

Sundberg et al. (59, 60) found no statistically significant difference between integrative healthcare and conventional therapy for any SF-36 dimensions after 16 weeks of follow-up in patients with nonspecific back pain.

Recent research has established the efficacy of many integrative medicine modalities, albeit there have been some restrictions in the techniques and sample sizes used in the studies (75). More strong randomized controlled trials are necessary to establish whether integrative medicine strategies are helpful in treating chronic illnesses, given the predicted demand for symptomatic therapies (75). Numerous research articles have been published in the previous decade that have examined the feasibility, safety, and effectiveness of integrative medicine, like several unique methodologies, among other things. Nevertheless, there seems to be a common trend throughout many of these researches: the small sample size and varied methodology. More significant numbers of randomized controlled studies are needed to determine whether integrative medicine strategies are actually helpful in chronic disorders. Such research should be aimed at identifying the most suitable individuals and applications for these treatments in order for them to be most successfully applied in clinical practice. When counseling patients, it is essential to remember that combinations of health-related interventions (diet, exercise, stress reduction, social support) will have the most significant impact on disease outcomes, and integrative medicine is invaluable in helping to promote and support positive, holistic health-related behavior change (75, 76).

References

1. Davis RM, Wagner EG, Groves T. Advances in managing chronic disease. Research, performance measurement, and quality improvement are key. BMJ (Clinical Research Ed). 2000;320(7234):525–6.
2. Reynolds R, Dennis S, Hasan I, Slewa J, Chen W, Tian D, et al. A systematic review of chronic disease management interventions in primary care. BMC Family Practice. 2018;19(1):1–13.

3. Martin CM. Chronic disease and illness care: adding principles of family medicine to address ongoing health system redesign. Canadian Family Physician. 2007;53(12):2086–91.
4. Hajat C, Stein E. The global burden of multiple chronic conditions: a narrative review. Preventive Medicine Reports. 2018;12:284–93.
5. World Health Organization. Global status report on noncommunicable diseases 2014. World Health Organization; 2014.
6. Raghupathi W, Raghupathi V. An empirical study of chronic diseases in the United States: a visual analytics approach. International Journal of Environmental Research and Public Health. 2018;15(3):431.
7. Danielson E, Melin-Johansson C, Modanloo M. Adherence to treatment in patients with chronic diseases: from alertness to persistence. International Journal of Community Based Nursing and Midwifery. 2019;7(4):248–57.
8. Goodman RA, Posner SF, Huang ES, Parekh AK, Koh HK. Defining and measuring chronic conditions: imperatives for research, policy, program, and practice. Preventing Chronic Disease. 2013;10:E66.
9. Yacobucci KL. Natural medicines. Journal of the Medical Library Association. 2016;104(4):371–4.
10. GBD 2015 Mortality and Causes of Death Collaborators. Global, regional, and national life expectancy, all-cause mortality, and cause-specific mortality for 249 causes of death, 1980-2015: a systematic analysis for the Global Burden of Disease Study 2015. Lancet (London, England). 2016;388(10053):1459–544. https://pubmed.ncbi.nlm.nih.gov/27733281/
11. Roy R. Integrative medicine to tackle the problem of chronic diseases. Journal of Ayurveda and Integrative Medicine. 2010;1(1):18–21.
12. Ahmed S, Ware P, Visca R, Bareil C, Chouinard MC, Desforges J, et al. The prevention and management of chronic disease in primary care: recommendations from a knowledge translation meeting. BMC Research Notes. 2015;8:571.
13. Cramm JM, Nieboer AP. Disease management: the need for a focus on broader self-management abilities and quality of life. Population Health Management. 2015;18(4):246–55.
14. Fernandez-Lazaro CI, García-González JM, Adams DP, Fernandez-Lazaro D, Mielgo-Ayuso J, Caballero-Garcia A, et al. Adherence to treatment and related factors among patients with chronic conditions in primary care: a cross-sectional study. BMC Family Practice. 2019;20(1):132.
15. Kania-Richmond A, Metcalfe A. Integrative health care – what are the relevant health outcomes from a practice perspective? A survey. BMC Complementary and Alternative Medicine. 2017;17(1):548.
16. Gannotta R, Malik S, Chan AY, Urgun K, Hsu F, Vadera S. Integrative medicine as a vital component of patient care. Cureus. 2018;10(8):e3098.
17. Boon H, Verhoef M, O'Hara D, Findlay B, Majid N. Integrative healthcare: arriving at a working definition. Alternative Therapies in Health and Medicine. 2004;10(5):48–56.
18. Fan D, Wang N. Holistic integrative medicine. Integrative Ophthalmology. 2020;3:3–14.
19. Zollman C, Vickers A. What is complementary medicine? BMJ (Clinical Research Ed). 1999;319(7211):693–6.
20. Di Renzo L, Gualtieri P, Romano L, Marrone G, Noce A, Pujia A, et al. Role of personalized nutrition in chronic-degenerative diseases. Nutrients. 2019;11(8):1707.
21. Mahdi GS. The Atkin's diet controversy. Annals of Saudi Medicine. 2006;26(3):244–5.
22. Cheuvront SN. The Zone Diet phenomenon: a closer look at the science behind the claims. Journal of the American College of Nutrition. 2003;22(1):9–17.
23. Shilpa J, Mohan V. Ketogenic diets: boon or bane? The Indian Journal of Medical Research. 2018;148(3):251–3.
24. Medawar E, Huhn S, Villringer A, Veronica Witte A. The effects of plant-based diets on the body and the brain: a systematic review. Translational Psychiatry. 2019;9(1):226.

25. Schwingshackl L, Morze J, Hoffmann G. Mediterranean diet and health status: active ingredients and pharmacological mechanisms. British Journal of Pharmacology. 2020;177(6):1241–57.
26. Koithan M, Devika J. New approaches to nutritional therapy. The Journal for Nurse Practitioners. 2010;6(10):805–6.
27. German JB, Roberts MA, Watkins SM. Genomics and metabolomics as markers for the interaction of diet and health: lessons from lipids. The Journal of Nutrition. 2003; 133(6 Suppl 1):2078s–83s.
28. Mehta DH, Gardiner PM, Phillips RS, McCarthy EP. Herbal and dietary supplement disclosure to health care providers by individuals with chronic conditions. Journal of Alternative and Complementary Medicine (New York, NY). 2008;14(10):1263–9.
29. Fijan S. Microorganisms with claimed probiotic properties: an overview of recent literature. International Journal of Environmental Research and Public Health. 2014;11(5):4745–67.
30. Taylor SL, Hoggatt KJ, Kligler B. Complementary and integrated health approaches: what do veterans use and want. Journal of General Internal Medicine. 2019;34(7):1192–9.
31. Kos A, Tomažič S, Umek A. Suitability of smartphone inertial sensors for real-time biofeedback applications. Sensors (Basel, Switzerland). 2016;16(3):301.
32. Williamson A. What is hypnosis and how might it work? Palliative Care. 2019;12: 1178224219826581.
33. Canter PH. The therapeutic effects of meditation. BMJ (Clinical Research Ed). 2003; 326(7398):1049–50.
34. Singg S. Use of Reiki as a biofield therapy: an adjunct to conventional medical care. Clinical Case Reports and Reviews. 2015;1(3):54–60.
35. Vergeer I. Trends in Yoga, Tai Chi, and Qigong Use: differentiations between practices and the need for dialogue and diffusion. American Journal of Public Health. 2019;109(5):662–3.
36. Sharpe PA, Blanck HM, Williams JE, Ainsworth BE, Conway JM. Use of complementary and alternative medicine for weight control in the United States. Journal of Alternative and Complementary Medicine (New York, NY). 2007;13(2):217–22.
37. McPartland J, Miller B. Bodywork therapy systems. Physical Medicine and Rehabilitation Clinics of North America. 1999;10(3):583–602, viii.
38. Van Hal M, Dydyk AM, Green MS. Acupuncture. In: StatPearls [Internet]. Treasure Island (FL): StatPearls Publishing; 2021.
39. Salehi A, Hashemi N, Imanieh MH, Saber M. Chiropractic: is it efficient in treatment of diseases? Review of systematic reviews. International Journal of Community Based Nursing and Midwifery. 2015;3(4):244–54.
40. Embong NH, Soh YC, Ming LC, Wong TW. Perspectives on reflexology: a qualitative approach. Journal of Traditional and Complementary Medicine. 2017;7(3):327–31.
41. Vickers A, Zollman C, Reinish JT. Massage therapies. Western Journal of Medicine. 2001;175(3):202–4.
42. Stumpf SH, Shapiro SJ, Hardy ML. Divining integrative medicine. Evidence-Based Complementary and Alternative Medicine. 2008;5(4):409–13.
43. Grover A, Joshi A. An overview of chronic disease models: a systematic literature review. Global Journal of Health Science. 2014;7(2):210–27.
44. Clark NM, Gong M, Kaciroti N. A model of self-regulation for control of chronic disease. Health Education & Behavior: The Official Publication of the Society for Public Health Education. 2014;41(5):499–508.
45. Leventhal H, Phillips LA, Burns E. The Common-Sense Model of Self-Regulation (CSM): a dynamic framework for understanding illness self-management. Journal of Behavioral Medicine. 2016;39(6):935–46.
46. Davy C, Bleasel J, Liu H, Tchan M, Ponniah S, Brown A. Effectiveness of chronic care models: opportunities for improving healthcare practice and health outcomes: a systematic review. BMC Health Services Research. 2015;15:194.

47. Davy C, Bleasel J, Liu H, Tchan M, Ponniah S, Brown A. Factors influencing the implementation of chronic care models: a systematic literature review. BMC Family Practice. 2015;16:102.
48. Chan RJ, Crawford-Williams F, Crichton M, Joseph R, Hart NH, Milley K, et al. Effectiveness and implementation of models of cancer survivorship care: an overview of systematic reviews. Journal of Cancer Survivorship: Research and Practice. 2021:1–25.
49. Leach MJ, Eaton H, Agnew T, Thakkar M, Wiese M. The effectiveness of integrative healthcare for chronic disease: a systematic review. International Journal of Clinical Practice. 2019;73(4):e13321.
50. Pan American Health Organization (PAHO). Innovative care for chronic conditions: Organizing and delivering high quality care for chronic noncommunicable diseases in the Americas. Washington, DC: PAHO; 2013.
51. Schveitzer MC, Abdala CVM, Portella CFS, Ghelman R. Traditional, complementary, and integrative medicine evidence map: a methodology to an overflowing field of data and noise. Revista panamericana de salud publica (Pan American Journal of Public Health). 2021;45:e48.
52. World Health Organization. WHO traditional medicine strategy: 2014–2023. Geneva, Switzerland: World Health Organization; 2013.
53. World Health Organization. WHO global report on traditional and complementary medicine 2019. Geneva, Switzerland: World Health Organization; 2019.
54. Leach MJ, Wiese M, Thakkar M, Agnew T. Integrative health care – Toward a common understanding: a mixed method study. Complementary Therapies in Clinical Practice. 2018;30:50–7.
55. Mann D, Gaylord S, Norton S. Moving toward integrative care: rationales, models, and steps for conventional-care providers. Complementary Health Practice Review. 2004;9(3):155–72.
56. Walach H, Jonas WB, Lewith GT. The role of outcomes research in evaluating complementary and alternative medicine. Alternative Therapies in Health and Medicine. 2002;8(3):88–95.
57. Greeson JM, Rosenzweig S, Halbert SC, Cantor IS, Keener MT, Brainard GC. Integrative medicine research at an academic medical center: patient characteristics and health-related quality-of-life outcomes. Journal of Alternative and Complementary Medicine (New York, NY). 2008;14(6):763–7.
58. Baker BS, Harrington JE, Choi BS, Kropf P, Muller I, Hoffman CJ. A randomised controlled pilot feasibility study of the physical and psychological effects of an integrated support programme in breast cancer. Complementary Therapies in Clinical Practice. 2012;18(3):182–9.
59. Sundberg T, Petzold M, Wändell P, Rydén A, Falkenberg T. Exploring integrative medicine for back and neck pain – a pragmatic randomised clinical pilot trial. BMC Complementary and Alternative Medicine. 2009;9:33.
60. Sundberg T, Hagberg L, Zethraeus N, Wändell P, Falkenberg T. Integrative medicine for back and neck pain: exploring cost-effectiveness alongside a randomized clinical pilot trial. European Journal of Integrative Medicine. 2014;6(1):29–38.
61. Chen FP, Chang CM, Shiu JH, Chiu JH, Wu TP, Yang JL, et al. A clinical study of integrating acupuncture and Western medicine in treating patients with Parkinson's disease. The American Journal of Chinese Medicine. 2015;43(3):407–23.
62. Krucoff MW, Crater SW, Gallup D, Blankenship JC, Cuffe M, Guarneri M, et al. Music, imagery, touch, and prayer as adjuncts to interventional cardiac care: the Monitoring and Actualisation of Noetic Trainings (MANTRA) II randomised study. Lancet (London, England). 2005;366(9481):211–7.
63. Krucoff MW, Crater SW, Green CL, Maas AC, Seskevich JE, Lane JD, et al. Integrative noetic therapies as adjuncts to percutaneous intervention during unstable coronary syndromes: monitoring and actualization of noetic training (MANTRA) feasibility pilot. American Heart Journal. 2001;142(5):760–9.

64. Seskevich JE, Crater SW, Lane JD, Krucof MW. Beneficial effects of noetic therapies on mood before percutaneous intervention for unstable coronary syndromes. Nursing Research. 2004;53(2):116–21.
65. Edelman D, Oddone EZ, Liebowitz RS, Yancy WS, Jr., Olsen MK, Jeffreys AS, et al. A multidimensional integrative medicine intervention to improve cardiovascular risk. Journal of General Internal Medicine. 2006;21(7):728–34.
66. Bradley R, Sherman KJ, Catz S, Calabrese C, Oberg EB, Jordan L, et al. Adjunctive naturopathic care for type 2 diabetes: patient-reported and clinical outcomes after one year. BMC Complementary and Alternative Medicine. 2012;12:44.
67. Carlsson M, Arman M, Backman M, Hamrin E. Perceived quality of life and coping for Swedish women with breast cancer who choose complementary medicine. Cancer Nursing. 2001;24(5):395–401.
68. Carlsson M, Arman M, Backman M, Flatters U, Hatschek T, Hamrin E. Evaluation of quality of life/life satisfaction in women with breast cancer in complementary and conventional care. Acta Oncologica (Stockholm, Sweden). 2004;43(1):27–34.
69. Carlsson M, Arman M, Backman M, Hamrin E. Coping in women with breast cancer in complementary and conventional care over 5 years measured by the mental adjustment to cancer scale. Journal of Alternative and Complementary Medicine (New York, NY). 2005;11(3):441–7.
70. Carlsson M, Arman M, Backman M, Flatters U, Hatschek T, Hamrin E. A five-year follow-up of quality of life in women with breast cancer in anthroposophic and conventional care. Evidence-Based Complementary and Alternative Medicine. 2006;3(4): 523–31.
71. Gegechkori N, Haines L, Lin JJ. Long-term and latent side effects of specific cancer types. Medical Clinics of North America. 2017;101(6):1053–73.
72. Deng GE, Frenkel M, Cohen L, Cassileth BR, Abrams DI, Capodice JL, et al. Evidence-based clinical practice guidelines for integrative oncology: complementary therapies and botanicals. Journal of the Society for Integrative Oncology. 2009;7(3): 85–120.
73. Arvidsdotter T, Marklund B, Taft C. Six-month effects of integrative treatment, therapeutic acupuncture and conventional treatment in alleviating psychological distress in primary care patients–follow up from an open, pragmatic randomized controlled trial. BMC Complementary and Alternative Medicine. 2014;14:210.
74. Eisenberg DM, Buring JE, Hrbek AL, Davis RB, Connelly MT, Cherkin DC, et al. A model of integrative care for low-back pain. Journal of Alternative and Complementary Medicine (New York, NY). 2012;18(4):354–62.
75. Viscuse PV, Price K, Millstine D, Bhagra A, Bauer B, Ruddy KJ. Integrative medicine in cancer survivors. Current Opinion in Oncology. 2017;29(4):235–42.
76. Ornish D, Lin J, Chan JM, Epel E, Kemp C, Weidner G, et al. Effect of comprehensive lifestyle changes on telomerase activity and telomere length in men with biopsy-proven low-risk prostate cancer: 5-year follow-up of a descriptive pilot study. The Lancet Oncology. 2013;14(11):1112–20.

II

Effect of Functional Foods in Treating Chronic Diseases

3

Safety and Efficacy Determination in Functional Foods and Nutraceuticals

Arun Soni, Tulsi Patil, Sanjeev Acharya, and Niyati Acharya

Contents

Introduction 38
Nutraceuticals 38
  Classification of nutraceuticals 39
  Mode of action of nutraceuticals 39
Functional food 39
  Health benefits of functional foods 41
Safety of functional foods and nutraceuticals 42
  Evaluation of safety 42
    Long-term repeat-dose toxicity 44
    Acute toxicity 44
    Reproductive toxicology 44
Efficacy of functional food and nutraceuticals 45
Natural bio-constituents 46
  Soy Isoflavones 46
    Soy-derived isoflavones 46
    Effect on antioxidant and lipid status 46
    Soy as a functional ingredient 47
  Plant sterols and stanols 48
    Efficacy for reducing cholesterol levels 48
    Dose required for efficacy 49
    Timing for efficacy 49
    Plant sterols and stanols as functional ingredients 50
  Fiber and its various components 51
    β-glucan and inulin 51
    β-glucan 51

DOI: 10.1201/9781003220053-5

Regulatory framework for functional food and nutraceuticals 55
Dietary Supplement Ingredient Advisory List 56
Reference body weight...56
Recommended Dietary Allowance (RDA)...56
Fats and oils 57
Fiber 57
Carbohydrate..58
Minerals..58
Conclusion 60
References..61

Introduction

Food, clothing, and shelter are the three fundamental and necessary prerequisites for human survival. Food is the most important requirement among these. "Functional foods" is a well-known term in scientific and social media. To foresee the relationship between food for health and its specific therapeutic effects, Hippocrates remarked, "Let food be your medicine, and medicine be your food." Food manufacturers have committed resources to the production of processed foods that may give consumers additional functional benefits (Hasler, 2002). The phrase "functional foods" was coined in Japan in the 1980s, but its meaning is frequently misconstrued because functional foods are regulated but not legally recognized in the majority of countries, resulting in no formal definition. Functional foods can be defined as synthetically processed or natural foods that, when taken at adequate amounts in a diverse diet, have potentially beneficial influence on health beyond the basic nutrition (Cencic & Chingwaru, 2010). Functional foods and nutraceuticals, due to their nutritional content, may promote optimal health and lower the risk of one or more noninfectious diseases. In terms of functional foods and nutraceuticals, quality, safety, and efficacy are important. Validation of the safety and efficacy of functional foods and nutraceuticals is required for their nutritional value to be maximized. Consumers nowadays want foods that are safe, pure, fresh, and natural, as well as having enriched formulas. Functional food design, like all newly produced food products, is costly, demanding, and arduous (Hilliam, 2000).

Nutraceuticals

Dr. Stephen De Felice (Chairman of the Foundation for Innovation in Medicine) invented the term "nutraceutical" in 1989 as a mixture of two terms: "nutrition" and "pharmaceutical." A nutraceutical is any substance that can be regarded as a component of food that aids in the maintenance of people's health. A nutraceutical is a functional food that is meant to help prevent and treat illnesses other than deficient conditions such as anemia. The distinction between terminology such as nutraceuticals, medicines, dietary supplements, functional foods, and so on is very fine, and these phrases are sometimes used interchangeably (Table 3.1). Nutraceuticals are sometimes referred to as "functional foods," however the distinction between food and medicine is not always clear.

Table 3.1 Difference between Food Ingredients, Pharmaceuticals, and Nutraceuticals

Curative Effect	Traditional medicine	Pharmaceutical
	Food ingredients	Nutraceuticals
	⟶	
Preventive Effect	Natural origin	Artificial or man-made origin
	Source of origin	

Classification of nutraceuticals

Nutraceuticals are categorized based on their chemical constituents, food ingredients, or whether they are conventional or not and also on the basis of their mechanism of action.

1. On the basis of chemical ingredients such as alkaloids, phenols, terpenes, carbohydrates and fibers, fatty acids and other lipids, stilbenes, organosulfur, and so on.
2. On the basis of standard and unconventional sources: fruits, vegetables, and juices from plant sources, animal sources, minerals, and microbial sources.
3. On the basis of category of food items: vitamins, minerals, dairy products, nondairy alternatives.
4. On the basis of mechanism of action: anticancer, antihyperlipidemic, anti-inflammatory, antioxidant, and so on.

Mode of action of nutraceuticals

Nutraceuticals are said to work by enhancing the supply of vital building blocks to the body. These necessary building pieces can be obtained in two ways:

1. By lowering illness symptoms as a buffering agent for alleviation and
2. By immediately benefiting persons' health

Functional food

The phrase "functional food" originated in Japan (Kubomara, 1998). When food is prepared using "scientific intelligence," it is referred to as a "functional food." Functional food contains vitamins, vital amino acids, lipids, proteins, and carbohydrates that are necessary for an individual's health. It is a food, whether natural or formulated or fortified, that will help enhance physiological functions beyond the traditional value of the food.

The Japanese Ministry of Health and Welfare defines foods for specified health uses (FOSHU) as:

- Foods with essential components that are believed to provide special health advantages.
- Foods whose safety and efficacy have been scientifically tested as a result of such a removal or addition.
- Permission has been allowed to make claims about the potential health benefits that can be assumed from their usage.

Foods that are categorized as FOSHU must also offer evidence that the final food product, rather than isolated individual components, meets the requirements. When ingested as part of a normal diet, it also has a health and physiological effect. Last but not least, FOSHU products should be available in the types of ordinary foods (i.e., not pills or capsules).

In 1989, the term "nutraceutical" was coined to separate functional or therapeutic foods from medicines. Nutraceuticals are not pharmaceuticals, which are pharmacologically active chemicals capable of enhancing, suppressing, or otherwise altering any physiological or metabolic function. The popularity of functional meals among consumers, rising healthcare expenditures, new regulations, and scientific discoveries are major drivers of increased interest in functional food (Milner, 2000).

The US Food and Drug Administration (FDA) regulates both finished dietary supplement products and dietary ingredients. FDA regulates dietary supplements under a different set of regulations than those covering "conventional" foods and drug products. Under the Dietary Supplement Health and Education Act of 1994 (DSHEA): Manufacturers and distributors of dietary supplements and dietary ingredients are prohibited from marketing products that are adulterated or misbranded. That means that these firms are responsible for evaluating the safety and labeling of their products before marketing to ensure that they meet all the requirements of DSHEA and FDA regulations. FDA is responsible for taking action against any adulterated or misbranded dietary supplement product after it reaches the market (Kannan et al., 2020).

Under existing law, including the DSHEA passed by Congress in 1994, the FDA can take action to remove products from the market, but the agency must first establish that such products are adulterated (i.e., that the product is unsafe) or misbranded (i.e., that the labeling is false or misleading). Safety concerns and related information on the various products and ingredients like *Acacia rigidula*, BMPEA, Cesium Chloride, DMAA, DMBA, DMHA, Methylsynephrine, Phenibut, Picamilon, Pure Powdered Caffeine,

Tianeptine, and Vinpocetine are available on the FDA website (Jenkins et al., 1978).

Health benefits of functional foods

The average daily dietary intake level sufficient to meet the nutrient requirements of nearly all (97–98 percent) healthy individuals in a particular life stage and gender group is represented as recommended dietary intake (RDI). Many of the RDI values of functional ingredients are established, but some are still under research and their RDI/RDA (recommended daily allowance) values have yet to be established.

Table 3.2 lists some of the functional bioactive ingredients of food, their biological functions, and common sources along with the types of evidence-based observations. Functional components usually occur in food in multiple forms such as glycosylated, esterified, thiolated, and hydroxylated materials. They also have multiple metabolic activities, allowing for beneficial effects on several diseases.

Table 3.2 Health Benefits of Functional Foods with Evidence-Based Research in US Market

Functional Food	Bioactive Ingredients	Health Benefits	Evidence Type	Recommended Intake
Garlic	Organosulfur compounds	Reduce LDL and total cholesterol level	Clinical trials	600–900 mg/d
Fortified margarines	Plant sterol and stanol esters	Reduce LDL and total cholesterol level	Clinical trials	1.3 g/d for sterols and 1.7 g/d for stanols
Psyllium	Soluble fiber	Reduce LDL and total cholesterol level	Clinical trials	1 g/d
Soy	Protein	Reduce LDL and total cholesterol level	Clinical trials	25 g/d
Green tea	Catechins	Reduce risk of some types of cancer	Epidemiological	Unknown
Whole oat products	Beta-Glucan	Reduce total and LDL cholesterol	Clinical trials	3 g/d
Spinach	Lutein	Reduce risk of macular degeneration	Epidemiological	6 mg/d
Cranberry juice	Proanthocyanins	Reduce risk of urinary tract infections	Clinical trials	300 mL/d
Tomatoes	Lycopene	Reduce risk of prostate cancer	Epidemiological	–
Fermented dairy products	Probiotics	Boost gastrointestinal health and immunity	*In vivo* and *in vitro* studies	–
Fatty fish	Fatty acids	Reduce triglycerides level, reduce risk of heart disease and myocardial infarction	Clinical trials; epidemiological	–
Cruciferous, vegetables	Glycosylates, indoles	Reduce risk of some types of cancer	Epidemiological	–

Source: Earl and Thomas (1994) and Baigent et al. (2005).

Safety of functional foods and nutraceuticals

Functional foods are designed to play an important part in modern diets by enhancing health and minimizing the risk of many long-term diseases. Functional foods must adhere to all food safety assessment criteria. Furthermore, for this type of diet, the concept of long-term benefits versus risk must be researched, developed, and proven. Protocols must be improved for pre- and post-marketing nutrition studies on functional foods, considering the hazards, such as increased contamination, negative metabolic effects, excessive immune system stimulation, and probable gene transfer.

Manufacturers must conduct placebo-controlled clinical investigations and analyze their results in four phases, according to the Food and Agriculture Organization (FAO) standards for functional food safety: Safety, Efficiency, Effectiveness, and Surveillance (Nowicka & Naruszewicz, 2004).

A dietitian, nutritionist, food science specialist, or other health professional must be able to answer several important questions in order to assess the safety of a functional meal (Hasler et al., 2001):

- What is(are) the name(s) of the active ingredient(s)?
- What is(are) the ingredient(s) of the functional food(s)?
- What is an ordinary portion size of a serving?
- How often is functional food eaten?
- Is the company aware of any animal or human safety research on the ingredients?
- Is there peer-reviewed research on the component that has been published?
- Will the manufacturer give safety information or context for past research studies?
- Is there any interaction between the functional ingredient and prescribed medications?

Evaluation of safety

Fundamental concepts guide the demonstration of safety for functional ingredient as follows (Kruger & Mann, 2003):

1. At varied intake quantities, functional components are physiologically active and can have a variety of outcomes, ranging from poor physiological activity to therapeutic impact to outright toxicity. Understanding the mechanisms of pharmacological action and toxicological potential is essential for forecasting the effects of different exposure dose levels.
2. Functional ingredients are a broad group of compounds that can be described by single component ingredients, complex herbal extracts, or compounds collected from novel sites or novel processes. The

compositional analysis for each of these types of products is an important factor for the method used to determine the ingredient's safety. Each of these objects is connected with unique safety issues that must be addressed independently.

3. Based on the compound, historical exposure, and experimental data (animal toxics, absorption, distribution, metabolism, excretion (ADME), clinical studies), this safety level may be measured by analyzing the intended use and the potential risk to a functional ingredient – the margin of safety between the desired uptake level and the potential risk to a functional ingredient.
4. There is the chances of contraindication with food and drug interaction, similar to medicines to access potential hazards if possible. These should be considered as an important aspect.

The ability of functional components to cause detrimental consequences not only of toxicity, but also of physiological activity, complicates safety assurance. Furthermore, a functional component can be the "active primary ingredient" in a complicated and uncharacterized mixture.

Interactions and impurities of food ingredients of a mixture must also be assessed for their potential to induce toxicological reactions. To use animal toxicity studies for functional ingredients, the toxicological endpoints of concern, such as long-term repeated dose toxicity with objective organ assessment, mutagenicity testing, carcinogenicity testing, developmental toxicity testing, reproductive toxicity testing, immunotoxicity testing, and neurotoxicity testing, should be evaluated as suitable, as should the input of appropriate endpoints for human trials.

Furthermore, because of some limitations in animal testing in terms of characteristics like pharmacokinetic and pharmacological properties of drug therapy, human clinical trials must provide the primary evidence of safety. When assessing the safety of functional components, the difficulties included in transferring from animal models to humans must be addressed, and clinical proof of safety is therefore crucial. Specific aspects of the animal testing design, such as species, dose level, length, administration method, and control groups, must be determined on a case-by-case basis in order to achieve and understand whether results from the animal study may be extended to humans (Kruger & Mann, 2003).

A technique for establishing the safety of a functional element should be based on two factors: the constituents' exposure and structural analysis. The intensity of toxicological analysis is essential for a supplement's characterization by both components (active as well as its impurities) and the preexisting safety data base that could be used to assess the possibility of negative health advantages resulting from ingestion at the recommended dose ranges with reasonable certainty. An animal testing model system can be used to identify the target organs and the consequences of toxicity.

To determine a safe level of exposure without additional toxicology testing, the following criteria must be met:

1. Active components and related chemicals have been thoroughly defined, and there is a clear knowledge of the absence of toxicity potential at dose level of humans advised by literature review findings.
2. Impurities must be clearly described based on existing knowledge from the literature, and with a solid comprehension of the absence of toxicity potential.
3. The manufacturing procedure must be standardized and repeatable (Kruger & Mann, 2003).

Long-term repeat-dose toxicity

The primary goal of long-term repeat-dose toxicity testing in an animal's system is to detect organs that are more vulnerable to toxicity as well as toxic substances with threshold effects. In terms of describing NOELs (no observable effect levels), NOAELs (no observable adverse effect levels), and MTD (maximum tolerable dose), the target organs of toxicity, estimating the human dose, and probable prodromal indicators that may be utilized to monitor the beginning of adverse reactions in man, repeat-dose toxicity studies are the backbone of the preclinical drug development procedure. They also define the multiple dose-response relationship for any reported toxicities, as well as the dose at which they emerge for the first time (threshold dose). The possibility of reversing a negative effect should also be investigated.

Acute toxicity

Acute toxicity studies with a single dose sometimes couldn't fulfill the claim that many administrations of doses are non-toxic, as they only gives data about parameters (pathological, biochemical, and histological) or to account for the effects of more frequent dosing because they are not designed to monitor the standard toxicity criteria (Kruger & Mann, 2003). ADME investigations can reveal therapeutic efficacy and aid in dose-response comparisons between animals and humans in the development of a particular molecular component of a drug.

Reproductive toxicology

Reproductive toxicology investigations, which include fertility or reproductive testing, teratogenicity, and prenatal or perinatal development in animals, give data on the possible toxicity of any pharmaceutical substance during the various stages of reproductive and developmental or discovery processes.

Carcinogenicity studies: Depending on the time period of intended treatment or any particular cause of concern, carcinogenicity testing may or may not be required or performed. Screening for mode of action and sites of action to evaluate the pharmacological activity of a component on bodies or systems is quite beneficial. These investigations, when appropriate and relevant, can provide explanations for hazardous findings in animal experiments. Due to limitations in experimentation in animal models, such as pharmacokinetic and pharmacological parameters of the compounds, human clinical studies must provide the primary evidence of safety for functional components. When assessing the safety of functional components, the challenges of transfer from animal models to human trials should be addressed, and clinical supporting evidence of safety is therefore required. However, it is crucial to investigate the quality of clinical evidence, not simply the numbers but overall parameters, to support the parameters of safety of the functional component (Kruger & Mann, 2003).

Efficacy of functional food and nutraceuticals

Except for a few exceptions, plants are the source of the vast majority of bioactive chemicals categorized as nutraceuticals. Bioactive components derived from legumes, cereals, fruits, and vegetables have been shown to be useful in decreasing lipid and cholesterol levels, increasing bone mineral density, and improving antioxidant status, as well as having anticancer potential (Eskin & Tamir, 2005). However, few of the many plant-derived nutraceuticals discovered have been integrated into common foods for routine consumption.

Some factors that influence the appropriateness of identifying bioactive components for inclusion as constituents in prepared foods follows:

1. The needed dose range to achieve effect may render the usage of a nutraceutical an ineffective strategy of disease prohibition. There are some studies that demonstrate efficacy for individual phytochemicals also use highly pure doses of plant extracts on cell or animal models (Espín et al., 2007). As a result, incorporating effective amounts of identified nutraceuticals into foods that humans may take as a part of the typical dietary regimen might be tough.
2. Nutraceuticals based on plant origin may alter the sensory properties of foods by imparting an unpleasant taste, odor, or texture. Extracts of legume give a 'bean'-like flavor that some people may find unpleasant (Fogliano & Vitaglione, 2005).
3. The food structure employed to transport the bioactive may impact the solubility, bioavailability, and stability of the nutraceutical components in a formulation matrix (Fogliano & Vitaglione, 2005). As a result of that, the potential of observing any effect on illness risk is reduced.

Natural bio-constituents

Taking all the aforementioned factors into account, there are some plant-based nutraceutical compounds that have easily made their path into various food matrix structures whereby effective dosages have been possible to develop. This chapter now focuses on plant-based bioactive substances and their constituents, such as soy proteins and its isoflavones, plant sterols, and fibers, as prototypes for establishing criteria, in general or whenever applicable, that must be considered when designing functional foods employing plant-based nutraceutical items.

Soy Isoflavones

Soy extracts consist of two components, one is from the isoflavone category and the other one is soy-derived proteins. Isoflavones have been studied regularly as an effective measure for reducing some risk factors that are related to heart disease and cancer via decreased circulating lipids like LDL (low-density lipids) and cholesterol levels, and also increased antioxidant properties. Nonetheless, efficacy results from human trials are inconsistent, raising issues about whether other soy-derived components are the true biologically active substances present in soy items or not (Xiao et al., 2008).

Soy-derived isoflavones

The soybean (isoflavones) exists largely as glucosides. Genistein and daidzein are the two primary glucosides present in soy. Following ingestion, genistein and daidzein are digested with the help of enzymes and native intestine glucosidases, resulting in the conversion of aglycones part genistein and daidzein. Aglycones are taken either directly or after additional metabolism into the daughter molecules, whereas glucosides must be translated to their equivalent aglycone part before absorption. There really is no agreement on which molecules are more bioavailable, glycosides or aglycones (Cassidy, 2006; Nielsen & Williamson, 2007).

Effect on antioxidant and lipid status

The effect of soy isoflavones as antioxidants is inconsistently observed. There was no change in low-density lipoprotein cholesterol values, ferric ion reducing, and O_2 radical absorbance activity when participants were fed protein shakes consisting of 107 mg of soy-based isoflavones (Heneman et al., 2007). Goldin et al. discovered similar results when hyperlipidemic males were fed meals containing either minimal or large doses of isoflavones (50 mg/1,000 kcal) originating from animal or soy sources. Although overall antioxidant potential improved by 10 percent when patients consumed soy protein either without isoflavones or with it, capacity against antioxidant measures such as protein carbonyls, malondialdehyde, F2-isoprostanes, and oxidized low-density lipoprotein did not differ between treatments (Goldin et al., 2005). On the other side, certain studies on the antioxidant capacity of soy-based isoflavones have proved to be beneficial.

Furthermore, males and females who took 50 or 100 mg of soy-based isoflavones each day for three weeks showed substantial reductions in DNA damage through oxidation (Djuric et al., 2001). The research just cited provides some amount of evidence that soy isoflavones can improve antioxidant status.

Soy as a functional ingredient

The investigation of soy-based ingredients as efficacious and biologically active is sufficient for food manufacturing industries to take benefits of the nutraceutical market share and incorporate soy-based extracts in prepared food products. According to the literature on the efficacy of soy isoflavones in food matrix structure, soy-based isoflavones are effective in improving antioxidant potential *in vivo*. Simultaneously, foods that contain soy proteins have strong lipid-lowering benefits. When introduced into diverse food matrices form, human *in vivo* experiments using soy-based isoflavones have convincingly revealed enhancements in antioxidant capacity and inflammatory indicators (Tikkanen et al., 1998; Wiseman et al., 2000; Hall et al., 2005; Hallund et al., 2006).

When researchers incorporated soy-derived doses of 57 mg/day aglycone into cereal bars, the lag time for low-density lipoprotein oxidation mediated through copper was significantly raised (Tikkanen et al., 1998). The cereal bars used in the study contained soy protein, whose fraction may have also contained bioactive that would improve antioxidant status. However, when Wiseman et al. examined the effects of soy burgers containing low dose 1.9 mg/kg or high dose 56 mg/kg levels of soy aglycones, only isoflavone-enriched burgers significantly raised the low-density lipoprotein oxidation (Wiseman et al., 2000). Marked increase in brachial flow dilation and decreases in CRP (C-reactive protein), both of which are indicators of vascular health and inflammation, were also found. The observed efficacy was also linked to glucosidases found naturally in semolina wheat, according to the researchers. Soy glucosides should be hydrolyzed into their corresponding aglycones throughout the production process, resulting in isoflavones that are rapidly absorbed because they do not require hydrolysis by intestinal enzymes or bacteria (Clerici et al., 2007). Soy protein-enriched bread and fortified beverages have been demonstrated to reduce total cholesterol (TC) values by 10 percent, low-density lipoprotein cholesterol by 12.5 percent, high-density lipoprotein cholesterol values by 12 percent, and circulating testosterone by 19 percent (Ridges et al., 2001). The later has been linked to the prevention of cancer. Despite this, no decreases in circulating lipid levels were detected when hyperlipidemic males and females ingested cereals having a concentration of soy isoflavones 168 mg/day. Jenkins et al. (2000) discovered a 9.2 percent decrease in low-density lipoprotein-affiliated conjugated dienes in the same trial, perhaps indicating an increase in antioxidant capacity. It is yet unknown whether soy protein, isoflavones derived from soy, or a mix of the two best induce the functional qualities of soy. Finally, production techniques may explain why the outcomes of research involving soy extracts in food products are inconsistent.

Plant sterols and stanols

Plant sterols can be found in virtually all plant materials and serve as structural components of cell walls. The biochemical structure of plant sterols is very similar to that of cholesterol with the addition of either a methyl or an ethyl functional group. The most common plant sterols found in nature are sitosterol, stigmasterol, and campesterol (Ntanios et al., 2002).

Plant sterols have been shown efficacious in reducing circulating cholesterol levels in humans since 1954 (Varady et al., 2007). Since then, an increasing number of studies and reviews have been published highlighting plant sterols and their efficacy in producing *in vivo* reductions in circulating tumor cells and low-density lipoprotein cholesterol levels of up to 9 percent and 10 percent, respectively, in humans (Lau et al., 2005; Alhassan et al., 2006; Hallikainen et al., 2006). Most studies regarding the use of plant sterols as nutraceuticals have used plant sterols as food ingredients rather than purified supplements. Reasons for using food as a vehicle for supplementation stem from the observation that free and esterified plant sterols are more soluble in the gastrointestinal tract as an emulsion rather than a crystalline form. When properly solubilized, free and esterified plant sterols possess similar cholesterol-lowering efficacy (Hallikainen et al., 2006; Chan et al., 2007).

Efficacy for reducing cholesterol levels

Hundreds of studies have investigated several aspects of the clinical efficacy of plant sterols for lowering low-density lipoprotein cholesterol (LDL-C), including chemical form (sterol vs. stanol), food matrix, and other factors associated with delivery of these compounds (Ras et al., 2014). First, when plant sterols are compared with plant stanols, consistent evidence demonstrates that both plant sterols and plant stanols lower LDL-C levels by 7.5 percent to 12 percent at intakes of 1.5 to 3 g/d (Talati et al., 2010). At intakes of up to 3 g/d, which is the current recommended range of intake in most countries, plant sterols and plant stanols produce equal LDL-C–lowering effects. A systematic review of 14 studies showed a nonsignificant weighted mean difference in LDL-C lowering between plant sterols and plant stanols (Sairanen et al., 2007).

Various foods were chosen as vehicles for investigation of the free and/or esterified plant sterols' cholesterol-lowering characteristics, as crystalline forms of intestinal solubility may be an issue. Experiments on efficacy using plant sterols supplements commonly use soft gel formulations. Plant sterols are suspended in an oil and used (Woodgate et al., 2006). Tablets studies add lecithin as an emulsifier (McPherson et al., 2005; Goldberg et al., 2006). Effective food matrices useful for the management of plant sterols include lipid-based propagations (Ntanios et al., 2002; Lau et al., 2005; Alhassan et al., 2006; Hallikainen et al., 2006; Chan et al., 2007; Varady et al., 2007; Clifton et al., 2008), milk or dairy products (Clifton et al., 2004), yogurt, and mayonnaise (Volpe et al., 2001). Plant cereals and breads that are rich in sterol have also proven effective in reducing cholesterol (Clifton et al., 2004). Collectively, a substantial number of clinical data is available to show that plant sterols is a functional element for cholesterol-reducing effectiveness. However, dose, timing, plant sterols content of food lipid, and safety considerations need to be defined so that the producers may manufacture plant sterols–enriched meals that result in significant cholesterol reductions if they are eaten outside of the healthcare setting.

Dose required for efficacy

Studies indicate that plant sterols are efficacious over a different range of doses 0.3 to 8.0 g/day (Abumweis et al., 2008). However, some dose ranges may be more beneficial than others, depending on the cholesterol-reducing ability of individuals. It is therefore crucial that the maximum dose should be established so that most persons who hope to gain benefits from a supplement of plant sterol have the highest chance of preventing cholesterol with plant sterols foods.

Males and females ingested increased dosages of margarine plant sterols at 0.0, 0.8, 1.6, 2.3, and 3.0 g/day for four weeks in research (Hallikainen et al., 2000). The low-density lipoprotein dropped sharply to 1.6, 7.0, 10.6, and 11.5 percent. While low-density lipoprotein decomposition increased with every further higher plant sterol dose, there was no significant impact than 1.6 g/day (Hallikainen et al., 2000; Abumweis et al., 2008). Additional investigations found that the maximal dose response with plant sterol complementation was 1.6 g/day for cholesterol-reducing efficacy (Hendriks et al., 1999; Clifton et al., 2008).

Current recommendations propose that people who have a high level of cholesterol can take 2.0–2.5 g/day plant sterol to decrease the LDL-C circulation (Abumweis et al., 2008). When producing functional foods, emphasis must be given to the ability to integrate effective doses of nutraceuticals in new foods.

Timing for efficacy

Physiological factors may make a nutraceutical more beneficial if consumed at a given point in the day. For example, whether you swallow a single dose in the morning with prolonged efficacy or in a lesser dosage all day long, a bioactive may be most beneficial. Plant sterols are an excellent example of the "timing phenomenon" as a functional food ingredient as a nutraceutical. One way for plant sterols to induce a cholesterol-reducing impact is cholesterol absorption, leading to an increase in the production of liver cholesterol and its uptake (Marinangeli et al., 2006). Because hepatic cholesterol synthesis has been demonstrated to follow a diurnal cycle, plant sterols intake would be ideal when liver synthesis is at its lowest level of the day, forcing the hepatic to receive cholesterol from circulating low-density lipoprotein values. According to research on circadian rhythms, the highest rates of endogenous cholesterol biosynthesis occur at night, while the lowest rates occur in the day, mainly in the afternoon (McNamara et al., 1980; Parker et al., 1982; Cella et al., 1995). As a result, taking plant sterols in the afternoon when cholesterol biosynthesis is lowest may be more effective, enabling the liver greater flexibility to get cholesterol from circulating lipoprotein. According to Abumweis et al., a single morning dose of plant sterols failed to lower the levels of low-density lipoprotein cholesterol (Abumweis et al., 2008). Other studies using single morning bolus doses of plant sterols report significant but moderate decreases in TC values from non-significant to 8 percent and low-density lipoprotein from 4.4 to 9 percent (Kassis et al., 2008). Smaller dosages of plant sterols distributed throughout the day results in larger cholesterol reduction, ranging from 8.2 to 13.1 percent for total cholesterol and 8.8 to 13.4 percent for low-density lipoprotein, respectively (Abuajah et al., 2015). In light of the findings just discussed, plant sterols have been studied for efficacy in a variety of food-based vehicles, giving consumers options and convenience when incorporating them into their dietary intake regimen.

Plant sterols and stanols as functional ingredients

The efficacy and safety of statin therapy was explored in a prospective meta-analysis of data from over 90,000 individuals in 14 randomized trials. The study concluded that, on average, a reduction of 1 mmol/L (38.7 mg/dL) in low-density lipoprotein cholesterol levels by statin therapy yields a consistent 23 percent reduction in the risk of major coronary events over five years (Sabatine et al., 2015). In this regard, the recent development of PCSK9 (proprotein convertase subtilsin/kexin type 9) inhibitors is also of note. These agents reduce the degradation of low-density lipoprotein receptors, thereby enhancing low-density lipoprotein cholesterol clearance from the circulation and reducing low-density lipoprotein cholesterol levels by as much as 60 percent (Momtazi et al., 2017). Definitive clinical trials with these agents are ongoing. While the data are still too preliminary to be conclusive, it appears that plant sterol/stanol supplements efficiently decrease plasma low-density lipoprotein cholesterol levels and intestinal cholesterol absorption by influencing PCSK9 expression (Sahebkar et al., 2016).

Plant sterols are a proven positive story in the field of plant-derived therapeutic ingredients with substantial evidence-based studies. Numerous human trials utilizing a variety of food matrices have reported their usefulness as cholesterol-reducing agents. Nonetheless, plant sterols are brilliant examples of the several additional elements that must be taken into account before designing a viable functional diet that would be effective outside of the clinical context. Using food matrices capable of providing the proper amount while taking into account the time of day, how a product should be taken – single or repeated doses throughout the day – and safety are all obvious considerations. Furthermore, the source of active ingredients may influence whether a food product becomes functional or not. Plant sterols have proven to be a success story as a functional food ingredient due to substantial study around plant sterols supplementation under a variety of scenarios.

Fiber and its various components

β-glucan and inulin

Fibers are defined as plant cell wall material that has been separated from plants. They are classified as indigestible carbohydrates that are fermented by bacteria. Fibers are classified into three types: water-insoluble fibers, water-soluble fibers, and resistant starch. Celluloses are water-insoluble fibers, whereas gums, pectin, β-glucan, psyllium, and oligosaccharides are water-soluble fibers. The American population is aware that a fiber-rich diet is beneficial for better gut health. As a result, fiber-based food claims are very common in North American supermarkets. Although most people identify fiber with bowel regularity, in recent decades, certain components of water-soluble fibers have been linked to additional physiological advantages such as lowering the circulating lipids and cholesterol levels. However, including fiber as a useful element into foods can be challenging due to diversified physico-chemical properties. Because different fibers have distinct physical qualities, such as water retention, sweetness, and viscosity, adding them to marketable food products to increase fiber content can be troublesome on the part of manufacturing processes. As a result, it is a challenge for food firms to not only fortify food products with fiber but also to choose the proper type of fiber to produce the aforementioned favorable physiological changes. Soluble fibers like β-glucan and fructo-oligosaccharides like inulin have been effectively incorporated into marketable functional food products evidenced by a range of products available in the market.

β-glucan

β-glucan is among the most researched fiber-derived functional ingredients primarily extracted from oats and barley (Lazaridou & Biliaderis, 2007). Researchers have started to assess the health-promoting benefits of β-glucan as a functional component in cookies, crackers (Casiraghi et al., 2006), juices (Keenan et al., 2007), pastas (Bourdon et al., 1999), breads (Davidson et al., 1991), and cereals (Davidson et al., 1991). Significant reductions in cardiovascular- and diabetes-related risk variables, including total cholesterol, low-density lipoprotein, satiety (Nilsson et al., 2008), postprandial glucose (Casiraghi et al., 2006), and postprandial insulin (Casiraghi et al., 2006), have been assigned to the capacity for β-glucan to give a highly viscous mass in the gastrointestinal tract, and slow gastric emptying.

Similarly to soybean isoflavones, some research findings on β-glucan as a functional food component, has not shown health-related efficacy (Biörklund et al., 2005). When adding β-glucan to foods as bioactive, a number of things must be considered, including food preparation, the food matrix for β-glucan delivery, β-glucan extraction, processing, and final dose or serving size.

β-glucan efficacy The food matrix's influencers like cooking temperature, baking time, pressure, and starch gelatinization have all been shown to affect the postprandial glycemic reactions of various foods (Törrönen et al., 1992; García-Alonso & Goni, 2000). In a study that evaluated the postprandial glycemic reaction of cookies and crackers prepared with 3.5 g of barley β-glucan, β-glucan crackers had a 33 percent and 32 percent drop in glycemic and insulin reactions, correspondingly, when compared to crackers made with whole wheat flour. On the other hand, cookies significantly reduced postprandial glucose and insulin responses 60 percent and 31 percent, respectively, compared to whole wheat cookies. Casiraghi et al. suggested that differences in dough moisture, cooking temperature, pressure, as well as the lower levels of simple sugars in β-glucan cookies may have accounted for better postprandial glucose responses compared to β-glucan crackers (Casiraghi et al., 2006). Törrönen et al. attributed the ineffectiveness of 11 g/day of oat bran-derived β-glucan at lowering lipid levels in hypercholesterolemic males to the poor solubility of β-glucan in the bread matrix employed to give the treatment. A study comparing high-carbohydrate diets and beverages fortified with β-glucan yielded similar outcomes. Only high-carbohydrate products supplemented with β-glucan significantly reduced postprandial glycemic reactions, according to the findings (Törrönen et al., 1992).

The inadequacy of the β-glucan-based beverages to reduce postprandial glycemia was attributed to two reasons. First, the beverage delivered glucose to the gastrointestinal cells quicker than the whole foods, which could be attributed to rapid stomach emptying. Second, the beverage matrix did not include the necessary environment for β-glucan to form a very viscous mass (Jenkins et al., 1978). In another trial, a beverage containing a single 5 g dosage of oat-derived β-glucan reduced total cholesterol and postprandial glucose rates compared to the same beverage containing 10 g β-glucan. Researchers hypothesized that because the solubility of β-glucan decreases with storage time, the process is accelerated in highly concentrated solutions, limiting its efficacy (Biörklund et al., 2005). Overall, the food matrix can have a strong influence on β-glucan's potential to render health benefits.

The effect of extraction and processing When it comes to the ability of glucans to convert an ordinary food into a functional food, food processing and preparation can make all the difference. The ability of β-glucan fractions to form a viscous mass in the digestive tract is ultimately determined by their solubility.

Varied β-glucan proportions or molecular weights impart different solubilities and influence the overall viscosity (Aaman & Graham, 1987; Redgwell & Fischer, 2005). The molecular weight of β-glucan can vary according to the extraction technique, resulting in either a high-molecular-weight extract with higher viscosity or a low-molecular-weight product with reduced viscosity but higher palatability (Davidson et al., 1991). Törrönen et al. used a large dose of β-glucan, 11 g, in comparison to other trials that have reported efficacy. And with as little as 3.5 g/day, β-glucan has been found to be an effective cholesterol-reducing food element (Törrönen et al., 1992; Keenan et al., 2007).

Volunteers in research who replaced whole wheat flour with barley (8 g/day) in pasta, biscuits, bread, and cereal had seen a 6 percent to 7 percent reduction in total cholesterol and low-density lipoprotein levels, respectively. According to the authors, the quality of β-glucan used in food products is more significant than the amount, which is why their investigation and studies using even lower concentrations demonstrated better health advantages when β-glucan is incorporated as a bioactive (Davidson et al., 1991). Researchers discovered that only an 84 g/day dose of oatmeal caused a substantial drop in low-density lipoprotein level of 10.1 percent over six weeks in a dose-dependent trial comparing the effects of a 28, 56, or 84 g/day dose of oatmeal and oat bran on lipid levels. However, with 56 and 84 g/day dosages of oat bran, respectively, effective low-density lipoprotein level reductions of 15.9 percent and 11.5 percent were noted. There was no difference in fat content across treatments. As a result, differences in efficacy between the two treatments were linked to oat bran having larger quantities of β-glucan than oatmeal (Wood et al., 2000), which could be attributed to how oat products are processed.

In terms of β-glucan molecular weight, a direct relationship has been found between molecular weight, efficacy, and viscosity (Roberfroid, 2005). Wood et al. used molecular weight and viscosity as dependent variables in a retrospective study of previously published data on the effect of β-glucan on glucose tolerance. The results showed that viscosity and molecular weight both play a substantial role in β-glucan's capacity to lower postprandial glucose and insulin levels (Roberfroid, 2005). The authors proposed that the low-molecular-weight β-glucan utilized in this study had even more promise as a nutraceutical ingredient since its molecular structure gives better sensory qualities than equal doses of high-molecular-weight β-glucan. Regardless of the aforesaid findings, it would be premature to attribute β-glucan's cholesterol and glycemic efficacy exclusively to its molecular weight. The general structure of β-glucan changes substantially between oats and barley. Thus, starting with a highly variable material, it can be hypothesized that different extraction procedures would break or even reform chemical bonds and produce highly variable structures with considerable diversity regarding structural branching.

Fructo-oligosaccharides: inulin The word "fructo-oligosaccharide" refers to any indigestible oligosaccharide that contains fructose in its molecular structure.

Because fructo-oligosaccharides are not absorbed, many of their health benefits are thought to be derived from prebiotic activity in the colon. Fructo-oligosaccharides are fermented in the intestine by resident microflora, trying to promote the growth and proliferation of beneficial bacteria while inhibiting the growth of harmful bacteria. According to research, maintaining a healthy milieu of particular strains of colonic microflora can enhance both local colonic and systemic health. Inulin is the most common fructo-oligosaccharide that has found its way into popular food items as a useful component.

Efficacy of inulin as a prebiotic Inulin is widely extracted from Jerusalem artichoke and chicory (Menne et al., 2000). There are two isoforms of inulin available. A d-glucopyranose molecule is connected to repeated Soy Isoflavones units with a terminal d-fructofuranose (GpyFn) molecule in the most prevalent form of inulin. The second, and rarer, inulin isoform, has a d-fructopyranose in place of the glucopyranose (FpyFn).

According to the findings reported on the animal research, eating inulin may also stimulate the formation of favorable *clostridium* strains. Both isoforms of inulin (GpyFn and FpyFn) have been found to be effective as prebiotics when consumed orally (Grizard & Barthomeuf, 1999). As scientists continue to research prebiotics, colonic microbiota, and health, a clear symbiotic relationship between humans and resident bacteria emerges. Inadvertently using substances like inulin to encourage the growth of specific microflora reduces levels of pathogenic and putrefactive bacteria while increasing colonic absorption of dietary calcium and magnesium (Grizard & Barthomeuf, 1999). Furthermore, reviews by Sauer et al., Kanauchi et al., and Pereira & Gibson (Pereira & Gibson, 2002; Kanauchi et al., 2003; Sauer et al., 2007) recommend that short-chain fatty acids produced by some of these colonic bacteria whose growth is aided by fructo-oligosaccharides, respectively, can prohibit colon cancer and irritable bowel syndrome and lowers the circulating lipid profile. According to Kolida et al., fermentation of inulin in human colonic microflora produced short-chain fatty acids that greatly enhanced glutathione S-transferases, which are utilized to detoxify carcinogens (Kolida et al., 2007).

Efficacious dose The amount of inulin required per day to modify intestinal bacterial populations has yet to be determined. A study comparing low and high doses of inulin in females and males found that 5 and 8 g/day inulin for two weeks can dramatically boost *bifidobacteria* populations in the colon (Ghoddusi et al., 2007). According to the researchers, alterations in *bifidobacteria* are not always dose-dependent. Rather, modifications are determined by the initial amounts of *bifidobacteria*. That is, volunteers with lower levels of colonic *bifidobacteria* at the start of the trial show bigger alterations than those with a well-established colony (Ghoddusi et al., 2007). Adding more inulin to functional meals, on the other hand, may increase the likelihood of finding a prebiotic impact. Higher levels of fermentable carbohydrate in the colon, on the other hand, may result in greater discomfort due to excess gas and bloating (Aryana et al., 2007), a big worry that food companies should take into consideration when developing functional foods containing fructo-oligosaccharides or other fermentable fibers. Inulin incorporation into functional food products is an achievement. Inulin has been successfully included into a variety of foods, including yogurt (Kleessen et al., 2007), snack bars (Seidel et al., 2007), bread (Sairanen et al., 2007), and milk (Su et al., 2007). Inulin has also been added to commercial infant formula in order to promote good intestinal flora colonization (Su et al., 2007).

Recently, research has begun to look into the use of prebiotic inulin in conjunction with probiotics as a means of enhancing the survival and subsequent colonization of probiotic bacteria in the colon. In an animal study, inulin extended the lifespan of the probiotic *Lactobacillus casei* from two to six days (Arai, 1996). A similar study in humans found that adding 4 g inulin to probiotic milk did not significantly boost the colon colonization of *lactobacilli* and *bifidobacteria* compared to probiotic only. Inulin coupled with probiotics, on the other hand, considerably increases gastrointestinal side effects in research volunteers. Overall, inulin is an effective functional component with a prebiotic impact in a variety of dietary matrices. However, unfavorable gastrointestinal side effects may prevent their use if the dose and effects when paired with probiotics are not taken into account (Kim et al., 2007).

Regulatory framework for functional food and nutraceuticals

It has been already known that functional food and nutraceuticals are not a distinct category/class of foods and they both have no regulatory definition or meaning. As a result, functional food and nutraceuticals are regulated by the FDA, as all other food items. In 2005, the National Academy Institute of Medicine and the National Research Council formed a Blue-Ribbon Association to offer the FDA with an easy way to analyze dietary supplements without having to divide nutraceuticals into their own category. The government approves only medications, but both pharmaceuticals and functional foods can heal and prevent disease (Sabatine et al., 2015).

The DSHEA, 1994, is the major set of guidelines governing the functional food and nutraceuticals market. This rule prohibits the FDA from classifying a new product as a "drug" or "food additive" if it meets the definition of a "dietary supplement," which includes, among other things, any feasible dietary component as well as concentrations, constituent, extract, or metabolite of these components. The act's other significant component shifts the burden of safety. Instead of the manufacturer showing the safety of a substance, the FDA must now have to prove it is hazardous or not. The DSHEA regulations do not apply to animal nutraceuticals. In a summation, the federal government has cited variations in substance metabolism between humans and animals, as well as potential safety issues with nutraceuticals used in food-producing animals, as reasons to exempt animals from DSHEA regulations. As a result, expressed or implicit statements pertaining to the use of a product with the treatment or prevention of the disease, or with an influence on the structure or function of the body in a way other than what would generally be attributed to "food."

On the other hand, nutraceuticals are defined in India under Clause 22 of the Food Safety and Standards Act (FSSA) of 2006. Under the FSSAI, total of around nine different categories, functional food and such supplements are categorized and relevant guidance is available for each. However, many issues still remain unresolved at the regulatory level in India. Labeling standards and health claims are one of these difficulties. Due to a lack of a regulatory framework, substandard enterprises join the market and compromise product quality.

Dietary Supplement Ingredient Advisory List

As a result of the FDA Dietary Supplement Ingredient Advisory List, the public will be swiftly alerted when the FDA identifies substances that do not appear to be legitimately included in goods promoted as supplements (Table 3.3).

Table 3.3 Dietary Supplement Products That Have Been the Subject of FDA Action

1	1,4-DMAA (1,4 dimethylamylamine or 1,4 dimethylpentylamine)
2	5-Alpha-Hydroxy-Laxogenin (5-Alpha-Hydroxy-Laxogenin or 5a-Hydroxy-Laxogenin)
3	Andarine (GTx 007 or SARM S-4)
4	Bismuth nitrate (Bismuth(III) nitrate or Bismuth trinitrate)
5	Higenamine(Norcoclaurine)
6	Hordenine (anhaline or eremursine)
7	N-Methyltyramine (Methyl-4-tyramine)
8	Octopamine (Analet or Norden)
9	Sodium tetrachloroaurate (EINECS 239-241-3 or Gold chloride sodium)

Reference body weight

Earlier an expert committee on RDA used data generated during 1989 on body weights and heights of well-to-do Indian children and adolescents, which was based only on a segment of the Indian population and did not have an all-India character. The reference weights for men and women were 60 kg and 50 kg, respectively.

Recommended Dietary Allowance (RDA)

The Indian Council of Medical Research (ICMR) and National Institute of Nutrition (NIN) have published an updated information about the RDA and estimated average requirements (EAR) (Momtazi et al., 2017). The summary of this recommendation is given here:

Reference body weight: as per the modified definition, reference Indian adult men and women are individuals between the ages of 20 and 39 years with body weight of 65 kg and 55 kg, respectively.

Energy: 10 percent and 9 percent reduction in body mass index (BMI) for males and females, respectively.

Table 3.4 Protein Requirements for All Age Groups

Men	54.0 g/d
Women	46.0 g/d
Pregnant women	+9.5 (2nd trimester) +22.0 (3rd trimester)
Lactating women (0–6m)	+17.0 g/d
Lactating women (7–12m)	+13.0 g/d
Infants (0–6m)	8.0 g/d
Infants (6–12m)	10.5 g/d
Children (1–3y)	12.5 g/d
Children (4–6y)	16.0 g/d
Children (7–9 y)	23.0 g/d
Boys (10–12y)	32.0 g/d
Boys (13–15y)	45.0 g/d
Boys (16–18y)	55.0 g/d
Girls (10–12y)	33.0 g/d
Girls (13–15y)	43.0 g/d
Girls (16–18y)	46.0 g/d.

Fats and oils

WHO recommendations are used to meet the requirements for optimal fetal and infant growth and development, maternal health, and combating chronic energy deficiency (children and adults) and diet-related noncommunicable diseases (DR-NCD) in adults. The recommended intake of visible fat for sedentary, moderate, and heavy activity is 25, 30, and 40 g/d for adult men and 20, 25, and 30 g/d for adult women as against the single level recommended earlier.

Fiber

The amount of fiber is recommended based on energy intake. Fiber intake of 30 g/2000 kcal is considered to be safe.

Carbohydrate

Dietary intake of carbohydrate is considered for the first time in RDA. A minimum intake of 100–130 g/day is recommended to ensure brain glucose utilization for ages one year and above.

Minerals

Calcium and phosphorus Calcium recommendation for adult men and adult women is 1.5 times the earlier recommendation. For all age groups except infants, the recommended value of phosphorus is in the ratio of 1:1. For infants, it is 1.5 times the value recommended for calcium (Table 3.5).

Table 3.5 Calcium and Phosphorus Recommendations for All Age Groups

Men	1000 mg/d
Women	1000 mg/d
Pregnant women	1000 mg/d
Lactating women	1200 mg/d
Infants	300 mg/d
Children (1–3y)	500 mg/d
Children (4–6y)	550 mg/d
Children (7–9y)	650 mg/d
Boys/girls (10–12y)	850 mg/d
Boys/girls (13–15y)	1,000 mg/d
Boys/girls (16–18y)	1,050 mg/d

Magnesium

Table 3.6 Magnesium Recommendation for All Age Groups

Men	440 mg/d
Women	370 mg/d
Pregnant women	440 mg/d
Lactating women	400 mg/d
Infants	40 mg/d
Children (1–3y)	90 mg/d
Children (4–6y)	125 mg/d
Children (7–9y)	175 mg/d
Boys/girls (10–12y)	240 mg/d
Boys/girls (13–15y)	345 mg/d
Boys/girls (16–18y)	440 mg/d

Sodium and potassium

Recommendations for sodium and potassium have been made according to recommendation (Table 3.7).

Table 3.7 Sodium and Potassium Recommendations

Recommended sodium	2000 mg/day
Recommended potassium	3510 mg/day

Iron Eight percent absorption for men, women, and adolescents, and 6 percent for children is considered by the present committee (Table 3.8). It is recommended that the density of ascorbic acid in the daily diet should be at least 20 mg/1000 kcal for improved iron absorption.

Table 3.8 Iron Recommendations for All Age Groups

Men	19 mg/d
Women	29 mg/d
Pregnant women	27 mg/d
Lactating women	23 mg/d
Infants	3 mg/d
Children (1–3y)	8 mg/d
Children (4–6y)	11 mg/d
Children (7–9y)	15 mg/d
Boys/Girls (10–12y)	16 mg/d
Boys/Girls (13–15y)	22 mg/d
Boys (16–18y)	26 mg/d

Zinc RDA for zinc is recommended after considering all processes of zinc loss for better bioavailability (Table 3.9).

Table 3.9 Zinc Recommendations for All Age Groups

Men	17 mg/d
Women	13 mg/d
Pregnant women	14.5 mg/d
Lactating women	14 mg/d
Infants	2.5 mg/d
Children (1–3y)	3.3 mg/d
Children (4–6y)	4.5 mg/d
Children (7–9y)	5.9 mg/d
Boys/Girls (10–12y)	8.5 mg/d
Boys/Girls (13–15y)	14.3 mg/d
Boys (16–18y)	17.6 mg/d

Copper, chromium, and manganese Due to the nutritional importance, RDA for copper, chromium, and manganese have been considered separately (Table 3.10).

Table 3.10 Amount of Copper, Chromium, and Manganese Required per Day

Copper	2 mg/day
Chromium	50 μg/day
Manganese	4 mg/day

Selenium Recommended adequate intake of selenium is 40 μg/day.

Iodine The recommendation for iodine is based on iodine intake in the diet through food and fortified salt (Table 3.11).

Table 3.11 Iodine Recommendations for All Age Groups

Men	150 μg/d
Women	150 μg/d
Pregnant women	250 μg/d
Lactating women	280 μg/d
Infants	100 μg/d
Children (1–3y)	100 μg/d
Children (4–6y)	120 μg/d
Children (7–9y)	120 μg/d
Boys/Girls (10–12y)	150 μg/d
Boys/Girls (13–15y)	150 μg/d
Boys (16–18y)	150 μg/d

Conclusion

Functional foods are foods that include a variety of biologically active substances that, when taken as part of a balanced diet, help consumers maintain their healthy lifestyles with the purpose to lower the risk of certain health issues. Antioxidants are found in abundant quantities in most of the functional foods. These molecules aid in the neutralization of dangerous substances known as free radicals, which helps prevent cell damage and chronic diseases and maintain good health. Although many functional foods may hold promise for public health, there are concerns that the promotion of functional foods and structure/function claims may not rest on sufficiently strong scientific evidence. Confusion also exists about claims applied to foods and those applied to dietary supplements. With the addition to foods of ingredients usually found only in dietary supplements, such confusion has increased. Although claims about the potential health benefits from functional foods or food ingredients must be communicated effectively to consumers, the differences between health claims and structure-function claims must also be more widely addressed to allow consumers to understand the differences in the scientific bases of such claims. Any health benefits attributed to functional foods should be based on sound and accurate scientific criteria, including rigorous studies of safety and efficacy.

There are no good and bad "foods," only good and bad dietary patterns. Thus, consumers should be wary of many of the promoted or implied benefits of these foods, and must realize that there is no consistent regulation or enforcement of existing regulations in the functional foods area. Diet is only one aspect of a comprehensive lifestyle approach to good health, which should also include regular exercise, tobacco avoidance, stress reduction, maintenance of healthy body weight, and other positive health practices. Only when all of these issues are addressed can functional foods become part of an effective strategy to maximize health and reduce disease risk.

References

Aaman P, Graham H. Analysis of total and insoluble mixed-linked (1 → 3)(1 → 4)-.beta.-D-glucans in barley and oats. Journal of Agricultural and Food Chemistry. 1987 Sep; 35(5):704–9.

Abuajah CI, Ogbonna AC, Osuji CM. Functional components and medicinal properties of food: a review. Journal of Food Science and Technology. 2015 May;52(5):2522–9.

Abumweis S, Barake R, Jones P. Plant sterols/stanols as cholesterol lowering agents: a meta-analysis of randomized controlled trials. Food & Nutrition Research. 2008 Jan 1;52(1):1811.

Alhassan S, Reese KA, Mahurin J, Plaisance EP, Hilson BD, Garner JC, Wee SO, Grandjean PW. Blood lipid responses to plant stanol ester supplementation and aerobic exercise training. Metabolism. 2006 Apr 1;55(4):541–9.

Arai S. Studies on functional foods in Japan—state of the art. Bioscience, Biotechnology, and Biochemistry. 1996 Jan 1;60(1):9–15.

Aryana KJ, Plauche S, Rao RM, McGrew P, Shah NP. Fat-free plain yogurt manufactured with inulins of various chain lengths and Lactobacillus acidophilus. Journal of Food Science. 2007 Apr;72(3):M79–84.

Baigent C, Keech A, Kearney PM, Blackwell L, Buck G, Pollicino C, Kirby A, Sourjina T, Peto R, Collins R, Simes R, Cholesterol Treatment Trialists'(CTT) Collaborators. Efficacy and safety of cholesterol-lowering treatment: prospective meta-analysis of data from 90,056 participants in 14 randomised trials of statins. Lancet. 2005;366:1267–78.

Biörklund M, Van Rees A, Mensink RP, Önning G. Changes in serum lipids and postprandial glucose and insulin concentrations after consumption of beverages with β-glucans from oats or barley: a randomised dose-controlled trial. European Journal of Clinical Nutrition. 2005 Nov;59(11):1272–81.

Bourdon I, Yokoyama W, Davis P, Hudson C, Backus R, Richter D, Knuckles B, Schneeman BO. Postprandial lipid, glucose, insulin, and cholecystokinin responses in men fed barley pasta enriched with β-glucan. The American Journal of Clinical Nutrition. 1999 Jan 1;69(1):55–63.

Casiraghi MC, Garsetti M, Testolin G, Brighenti F. Post-prandial responses to cereal products enriched with barley β-glucan. Journal of the American College of Nutrition. 2006 Aug 1;25(4):313–20.
Cassidy A. Factors affecting the bioavailability of soy isoflavones in humans. Journal of AOAC International. 2006 Jul 1;89(4):1182–8.
Cella LK, Van Cauter E, Schoeller DA. Diurnal rhythmicity of human cholesterol synthesis: normal pattern and adaptation to simulated "jet lag". American Journal of Physiology-Endocrinology and Metabolism. 1995 Sep 1;269(3):E489–98.
Cencic A, Chingwaru W. The role of functional foods, nutraceuticals, and food supplements in intestinal health. Nutrients. 2010 Jun;2(6):611–25.
Chan YM, Demonty I, Pelled D, Jones PJ. Olive oil containing olive oil fatty acid esters of plant sterols and dietary diacylglycerol reduces low-density lipoprotein cholesterol and decreases the tendency for peroxidation in hypercholesterolaemic subjects. British Journal of Nutrition. 2007 Sep;98(3):563–70.
Clerici C, Setchell KD, Battezzati PM, Pirro M, Giuliano V, Asciutti S, Castellani D, Nardi E, Sabatino G, Orlandi S, Baldoni M, Morelli O, Mannarino E, Morelli A. Pasta naturally enriched with isoflavone aglycons from soy germ reduces serum lipids and improves markers of cardiovascular risk. The Journal of Nutrition. 2007 Oct;137(10):2270–8.
Clifton PM, Mano M, Duchateau GS, Van der Knaap HC, Trautwein EA. Dose-response effects of different plant sterol sources in fat spreads on serum lipids and C-reactive protein and on the kinetic behavior of serum plant sterols. European Journal of Clinical Nutrition. 2008 Aug;62(8):968–77.
Clifton PM, Noakes M, Ross D, Fassoulakis A, Cehun M, Nestel P. High dietary intake of phytosterol esters decreases carotenoids and increases plasma plant sterol levels with no additional cholesterol lowering. Journal of Lipid Research. 2004 Aug 1;45(8):1493–9.
Davidson MH, Dugan LD, Burns JH, Bova J, Story K, Drennan KB. The hypocholesterolemic effects of β-glucan in oatmeal and oat bran: a dose-controlled study. JAMA. 1991 Apr 10;265(14):1833–9.
Djuric Z, Chen G, Doerge DR, Heilbrun LK, Kucuk O. Effect of soy isoflavone supplementation on markers of oxidative stress in men and women. Cancer Letters. 2001 Oct 22;172(1):1–6.
Earl R, Thomas PR, editors. Opportunities in the nutrition and food sciences: research challenges and the next generation of investigators. National Academies Press; 1994 Feb 1.
Eskin M, Tamir S. Dictionary of nutraceuticals and functional foods. CRC Press; 2005 Dec 19.
Espín JC, García-Conesa MT, Tomás-Barberán FA. Nutraceuticals: facts and fiction. Phytochemistry. 2007 Nov 1;68(22–24):2986–3008.
Fogliano V, Vitaglione P. Functional foods: planning and development. Molecular Nutrition and Food Research. 2005 Mar;49(3):256–62.
García-Alonso A, Goni I. Effect of processing on potato starch: in vitro availability and glycaemic index. Food/Nahrung. 2000 Jan 1;44(1):19–22.
Ghoddusi HB, Grandison MA, Grandison AS, Tuohy KM. In vitro study on gas generation and prebiotic effects of some carbohydrates and their mixtures. Anaerobe. 2007 Oct 1;13(5–6):193–9.
Goldberg AC, Ostlund Jr RE, Bateman JH, Schimmoeller L, McPherson TB, Spilburg CA. Effect of plant stanol tablets on low-density lipoprotein cholesterol lowering in patients on statin drugs. The American Journal of Cardiology. 2006 Feb 1;97(3):376–9.
Goldin BR, Brauner E, Adlercreutz H, Ausman LM, Lichtenstein AH. Hormonal response to diets high in soy or animal protein without and with isoflavones in moderately hypercholesterolemic subjects. Nutrition and Cancer. 2005 Jan 1;51(1):1–6.
Grizard D, Barthomeuf C. Non-digestible oligosaccharides used as prebiotic agents: mode of production and beneficial effects on animal and human health. Reproduction Nutrition Development. 1999;39(5–6):563–88.

Hall WL, Vafeiadou K, Hallund J, Bügel S, Koebnick C, Reimann M, Ferrari M, Branca F, Talbot D, Dadd T, Nilsson M, Dahlman-Wright K, Gustafsson J-A, Minihane A-M, Williams CM. Soy-isoflavone-enriched foods and inflammatory biomarkers of cardiovascular disease risk in postmenopausal women: interactions with genotype and equol production. The American Journal of Clinical Nutrition. 2005 Dec 1;82(6):1260–8.
Hallikainen M, Lyyra-Laitinen T, Laitinen T, Ågren JJ, Pihlajamäki J, Rauramaa R, Miettinen TA, Gylling H. Endothelial function in hypercholesterolemic subjects: effects of plant stanol and sterol esters. Atherosclerosis. 2006 Oct 1;188(2):425–32.
Hallikainen MA, Sarkkinen ES, Uusitupa MI. Plant stanol esters affect serum cholesterol concentrations of hypercholesterolemic men and women in a dose-dependent manner. The Journal of Nutrition. 2000 Apr 1;130(4):767–76.
Hallund J, Bügel S, Tholstrup T, Ferrari M, Talbot D, Hall WL, Reimann M, Williams CM, Wiinberg N. Soya isoflavone-enriched cereal bars affect markers of endothelial function in postmenopausal women. British Journal of Nutrition. 2006 Jun;95(6):1120–6.
Hasler C, Moag-Stahlberg A, Webb D, Hudnall M. How to evaluate the safety, efficacy, and quality of functional foods and their ingredients. Journal of the American Dietetic Association. 2001 Jul 1;101(7):733–6.
Hasler CM. Functional foods: benefits, concerns and challenges—a position paper from the American Council on Science and Health. The Journal of Nutrition. 2002 Dec 1;132(12):3772–81.
Hendriks HF, Weststrate JA, Van Vliet T, Meijer GW. Spreads enriched with three different levels of vegetable oil sterols and the degree of cholesterol lowering in normocholesterolaemic and mildly hypercholesterolaemic subjects. European Journal of Clinical Nutrition. 1999 Apr;53(4):319–27.
Heneman KM, Chang HC, Prior RL, Steinberg FM. Soy protein with and without isoflavones fails to substantially increase postprandial antioxidant capacity. The Journal of Nutritional Biochemistry. 2007 Jan 1;18(1):46–53.
Hilliam M. Functional food: how big is the market. The World of Food Ingredients. 2000 Dec;12(50–52).
Jenkins DJ, Kendall CW, Vidgen E, Vuksan V, Jackson CJ, Augustin LS, Lee B, Garsetti M, Agarwal S, Rao AV, Cagampang GB, Fulgoni 3rd V. Effect of soy-based breakfast cereal on blood lipids and oxidized low-density lipoprotein. Metabolism: Clinical and Experimental. 2000 Nov 1;49(11):1496–500.
Jenkins DJ, Wolever TM, Leeds AR, Gassull MA, Haisman P, Dilawari J, Goff DV, Metz GL, Alberti KG. Dietary fibres, fibre analogues, and glucose tolerance: importance of viscosity. British Medical Journal. 1978 May 27;1(6124):1392–4.
Kanauchi O, Mitsuyama K, Araki Y, Andoh A. Modification of intestinal flora in the treatment of inflammatory bowel disease. Current Pharmaceutical Design. 2003 Feb 1;9(4):333–46.
Kannan S, Naha A, Singh RR, Bansal P, Nayak VC, Goud S, Rani U. Role of dietary supplements in sports performance. Research Journal of Pharmacy and Technology. 2020 Dec 1;13(12):6259–65.
Kassis AN, Vanstone CA, AbuMweis SS, Jones PJ. Efficacy of plant sterols is not influenced by dietary cholesterol intake in hypercholesterolemic individuals. Metabolism. 2008 Mar 1;57(3):339–46.
Keenan JM, Goulson M, Shamliyan T, Knutson N, Kolberg L, Curry L. The effects of concentrated barley β-glucan on blood lipids in a population of hypercholesterolaemic men and women. British Journal of Nutrition. 2007 Jun;97(6):1162–8.
Kleessen B, Schwarz S, Boehm A, Fuhrmann H, Richter A, Henle T, Krueger M. Jerusalem artichoke and chicory inulin in bakery products affect faecal microbiota of healthy volunteers. British Journal of Nutrition. 2007 Sep;98(3):540–9.
Kolida S, Meyer D, Gibson GR. A double-blind placebo-controlled study to establish the bifidogenic dose of inulin in healthy humans. European Journal of Clinical Nutrition. 2007 Oct;61(10):1189–95.

Kruger CL, Mann SW. Safety evaluation of functional ingredients. Food and Chemical Toxicology. 2003 Jun 1;41(6):793–805.
Kubomara K. Japan redefines functional foods. Prepared Foods. 1998;167(5):129–32.
Lau VW, Journoud M, Jones PJ. Plant sterols are efficacious in lowering plasma LDL and non-HDL cholesterol in hypercholesterolemic type 2 diabetic and nondiabetic persons. The American Journal of Clinical Nutrition. 2005 Jun 1;81(6):1351–8.
Lazaridou A, Biliaderis CG. Molecular aspects of cereal β-glucan functionality: Physical properties, technological applications and physiological effects. Journal of Cereal Science. 2007 Sep 1;46(2):101–18.
Marinangeli CP, Varady KA, Jones PJ. Plant sterols combined with exercise for the treatment of hypercholesterolemia: overview of independent and synergistic mechanisms of action. The Journal of Nutritional Biochemistry. 2006 Apr 1;17(4):217–24.
McIntosh GH, Whyte J, McArthur R, Nestel PJ. Barley and wheat foods: influence on plasma cholesterol concentrations in hypercholesterolemic men. The American Journal of Clinical Nutrition. 1991 May 1;53(5):1205–9.
McNamara DJ, Davidson NO, Fernandez S. In vitro cholesterol synthesis in freshly isolated mononuclear cells of human blood: effect of in vivo administration of clofibrate and/or cholestyramine. Journal of Lipid Research. 1980 Jan 1;21(1):65–71.
McPherson TB, Ostlund RE, Goldberg AC, Bateman JH, Schimmoeller L, Spilburg CA. Phytostanol tablets reduce human LDL-cholesterol. Journal of Pharmacy and Pharmacology. 2005 Jul;57(7):889–96.
Menne E, Guggenbuhl N, Roberfroid M. Fn-type chicory inulin hydrolysate has a prebiotic effect in humans. The Journal of Nutrition. 2000 May 1;130(5):1197–9.
Milner JA. Functional foods: the US perspective. The American Journal of Clinical Nutrition. 2000 Jun 1;71(6):1654S–9S.
Momtazi AA, Banach M, Pirro M, Katsiki N, Sahebkar A. Regulation of PCSK9 by nutraceuticals. Pharmacological Research. 2017 Jun 1;120:157–69.
Nielsen IL, Williamson G. Review of the factors affecting bioavailability of soy isoflavones in humans. Nutrition and Cancer. 2007 May 4;57(1):1–10.
Nilsson AC, Ostman EM, Holst JJ, Björck IM. Including indigestible carbohydrates in the evening meal of healthy subjects improves glucose tolerance, lowers inflammatory markers, and increases satiety after a subsequent standardized breakfast. The Journal of Nutrition. 2008 Apr 1;138(4):732–9.
Nowicka G, Naruszewicz M. Assessing health claims for functional foods. In Functional Foods, Cardiovascular Disease and Diabetes; 2004 Jan 1 (pp. 10–17). Woodhead Publishing.
Ntanios FY, Homma Y, Ushiro S. A spread enriched with plant sterol-esters lowers blood cholesterol and lipoproteins without affecting vitamins A and E in normal and hypercholesterolemic Japanese men and women. The Journal of Nutrition. 2002 Dec 1;132(12):3650–5.
Parker TS, McNamara DJ, Brown C, Garrigan O, Kolb R, Batwin H, Ahrens EH. Mevalonic acid in human plasma: relationship of concentration and circadian rhythm to cholesterol synthesis rates in man. Proceedings of the National Academy of Sciences. 1982 May 1;79(9):3037–41.
Pereira DI, Gibson GR. Effects of consumption of probiotics and prebiotics on serum lipid levels in humans. Critical Reviews in Biochemistry and Molecular Biology. 2002 Jan 1;37(4):259–81.
Ras RT, Geleijnse JM, Trautwein EA. LDL-cholesterol-lowering effect of plant sterols and stanols across different dose ranges: a meta-analysis of randomised controlled studies. British Journal of Nutrition. 2014 Jul;112(2):214–9.
Redgwell RJ, Fischer M. Dietary fiber as a versatile food component: an industrial perspective. Molecular Nutrition & Food Research. 2005 Jun;49(6):521–35.
Ridges L, Sunderland R, Moerman K, Meyer B, Astheimer L, Howe P. Cholesterol lowering benefits of soy and linseed enriched foods. Asia Pacific Journal of Clinical Nutrition. 2001 Sep 27;10(3):204–11.

Roberfroid MB. Introducing inulin-type fructans. British Journal of Nutrition. 2005 Apr;93(S1):S13–25.

Sabatine MS, Giugliano RP, Wiviott SD, Raal FJ, Blom DJ, Robinson J, Ballantyne CM, Somaratne R, Legg J, Wasserman SM, Scott R, Koren MJ, Stein EA, Open-Label Study of Long-Term Evaluation against LDL Cholesterol (OSLER) Investigators. Efficacy and safety of evolocumab in reducing lipids and cardiovascular events. New England Journal of Medicine. 2015 Apr 16;372(16):1500–9.

Sahebkar A, Serban MC, Gluba-Brzózka A, Mikhailidis DP, Cicero AF, Rysz J, Banach M. Lipid-modifying effects of nutraceuticals: an evidence-based approach. Nutrition. 2016 Nov 1;32(11–12):1179–92.

Sairanen U, Piirainen L, Gråsten S, Tompuri T, Mättö J, Saarela M, Korpela R. The effect of probiotic fermented milk and inulin on the functions and microecology of the intestine. Journal of Dairy Research. 2007 Aug;74(3):367–73.

Sauer J, Richter KK, Pool-Zobel BL. Products formed during fermentation of the prebiotic inulin with human gut flora enhance expression of biotransformation genes in human primary colon cells. British Journal of Nutrition. 2007 May;97(5):928–37.

Seidel C, Boehm V, Vogelsang H, Wagner A, Persin C, Glei M, Pool-Zobel BL, Jahreis G. Influence of prebiotics and antioxidants in bread on the immune system, antioxidative status and antioxidative capacity in male smokers and non-smokers. British Journal of Nutrition. 2007 Feb;97(2):349–56.

Sook-He K, Lee D, Meyer D. Supplementation of infant formula with native inulin has a prebiotic effect in formula-fed babies. Asia Pacific Journal of Clinical Nutrition. 2007 Mar 1;16(1):172.

Su P, Henriksson A, Mitchell H. Prebiotics enhance survival and prolong the retention period of specific probiotic inocula in an in vivo murine model. Journal of Applied Microbiology. 2007 Dec;103(6):2392–400.

Talati R, Sobieraj DM, Makanji SS, Phung OJ, Coleman CI. The comparative efficacy of plant sterols and stanols on serum lipids: a systematic review and meta-analysis. Journal of the American Dietetic Association. 2010 May 1;110(5):719–26.

Tikkanen MJ, Wähälä K, Ojala S, Vihma V, Adlercreutz H. Effect of soybean phytoestrogen intake on low density lipoprotein oxidation resistance. Proceedings of the National Academy of Sciences. 1998 Mar 17;95(6):3106–10.

Törrönen R, Kansanen L, Uusitupa M, Hänninen O, Myllymäki O, Härkönen H, Mälkki Y. Effects of an oat bran concentrate on serum lipids in free-living men with mild to moderate hypercholesterolaemia. European Journal of Clinical Nutrition. 1992 Sep 1;46(9):621–7.

Varady KA, Houweling AH, Jones PJ. Effect of plant sterols and exercise training on cholesterol absorption and synthesis in previously sedentary hypercholesterolemic subjects. Translational Research. 2007 Jan 1;149(1):22–30.

Volpe R, Niittynen L, Korpela R, Sirtori C, Bucci A, Fraone N, Pazzucconi F. Effects of yoghurt enriched with plant sterols on serum lipids in patients with moderate hypercholesterolaemia. British Journal of Nutrition. 2001 Aug;86(2):233–9.

Wiseman H, O'Reilly JD, Adlercreutz H, Mallet AI, Bowey EA, Rowland IR, Sanders TA. Isoflavone phytoestrogens consumed in soy decrease F2-isoprostane concentrations and increase resistance of low-density lipoprotein to oxidation in humans. The American Journal of Clinical Nutrition. 2000 Aug 1;72(2):395–400.

Wood P, Beer MU, Butler G. Evaluation of role of concentration and molecular weight of oat β-glucan in determining effect of viscosity on plasma glucose and insulin following an oral glucose load. British Journal of Nutrition. 2000 Jul;84(1):19–23.

Woodgate D, Chan CH, Conquer JA. Cholesterol-lowering ability of a phytostanol softgel supplement in adults with mild to moderate hypercholesterolemia. Lipids. 2006 Feb;41(2):127–32.

Xiao CW, Mei J, Wood CM. Effect of soy proteins and isoflavones on lipid metabolism and involved gene expression. Frontiers in Bioscience. 2008 Jan 1;13:2660–73.

Global Burden of Disease, the Heavy Cost of Sun Deprivation

Implications for Mass Food Fortification with Vitamin D

Zahra Yari, Bahareh Nikooyeh, Samira Ebrahimof, and Tirang R. Neyestani

Contents

Introduction 68
Global Burden of Disease (GBD): Main contributors 69
Vitamin D biology 70
Metabolism 70
Vitamin D functions 72
Assessment methods 73
Deficiency 73
Requirement 73
Contribution of VDD to GBD 74
Obesity 74
Etiology, pathophysiology, health consequences 74
Vitamin D and obesity: evidence from experimental, observational and clinical trial studies 74
Contribution of obesity to GBD: does vitamin D play a role? 75
Diabetes mellitus 75
Etiology, pathophysiology and health consequences 75
Vitamin D and diabetes mellitus: evidence from experimental, observational and clinical trial studies 76
Contribution of diabetes to GBD: the possible role of vitamin D 78
Vitamin D and cardiovascular disease 78
Etiology, pathophysiology and health consequences 78
Vitamin D and cardiovascular disease: evidence from experimental, observational and clinical trial studies 78
Contribution of CVD to GBD: the role of vitamin D is still debatable 79

DOI: 10.1201/9781003220053-6

Vitamin D and bone health 80
Etiology, pathophysiology and health consequences 80
Vitamin D and MBDs: evidence from experimental, observational and clinical trial studies 81
Contribution of metabolic bone disorders to GBD: the crucial role of vitamin D 81
Multiple sclerosis, rheumatoid arthritis, other autoimmune disorders and vitamin D 82
Etiology, pathophysiology and health consequences 82
Vitamin D and autoimmune disorders: evidence from experimental, observational and clinical trial studies 82
Contribution of autoimmune disorders to GBD: vitamin D may reduce the burden 84
Vitamin D and cancers 84
Etiology, pathophysiology and health consequences 84
Vitamin D and cancers: evidence from experimental, observational and clinical trial studies 85
Contribution of cancers to GBD: improvement of vitamin D status is worth investment 86
Depression 86
Etiology, pathophysiology and health consequences 86
Vitamin D and MDD: evidence from experimental, observational and clinical trial studies 87
Contribution of MDD to GBD: where is the place of vitamin D? 88
Vitamin D and Covid-19 infection 88
Etiology, pathophysiology and health consequences 88
Vitamin D and Covid-19: evidence from experimental, observational and clinical trial studies 89
Contribution of Covid-19 to GBD: does vitamin D status matter? 90
Strategies to combat VDD 90
Vitamin D supplementation 90
Vitamin D fortification 91
Concluding remarks 93
References 94

Introduction

The ultimate goal of public health nutrition services is to establish justice through health and nutrition equity as a human right. To do this, we need to have a clear picture of health loss due to diverse human diseases and their related risk factors so that by using these data nutrition and health systems can be improved to minimize and ultimately eliminate disparities. The Global

Burden of Disease (GBD) is a tool for providing this picture. Thousands of researchers from 145 countries from all around the world are contributing to GBD studies via evaluating mortality, morbidity and disability rates (1). Disability is usually expressed as disability-adjusted life year (DALY), which is a measure of overall burden of disease resulting from years lost because of disease, disability or premature death. DALY is used to compare the health condition and life expectancy among different countries (2). Data from different studies indicate that a great proportion of GBD is attributable to modifiable risk factors including nutrition and body weight (3, 4). Among nutritional factors, micronutrient deficiencies (MND), the so-called hidden hunger, are the most widespread nutritional problem and have a strong link with overall morbidity and mortality whereby contributing to GBD (5–7). Though deficiencies of iron, zinc, vitamin A and iodine are the most common MND around the world (8), a huge body of evidence indicates that vitamin D deficiency (VDD) is the most prevalent MND globally with no exception between different subpopulations (9–11). Severe VDD can bring about serious metabolic bone disorders including rickets in children and osteomalacia as well as osteoporosis in adults. But the story of adverse effects of VDD does not end to the bone disorders. A growing body of evidence indicates diverse roles for vitamin D beyond bones, the so-called non-calcemic functions. While suboptimal vitamin D status may not represent with any clinical symptoms, it may predispose for many human diseases, notably cardiovascular disease (CVD), diabetes, autoimmune disorders including multiple sclerosis (MS) and rheumatoid arthritis (RA) and also certain malignancies such as colorectal cancer (12, 13). Therefore, it seems plausible that improvement of vitamin D status of the whole community can contribute to reducing the burden of several human diseases.

Global Burden of Disease (GBD): Main contributors

Until about three decades ago, there was no reliable source of information regarding GBD and risk factors. In 1991, in a collaborative effort, World Bank and the World Health Organization (WHO) inaugurated the GBD study (14). In GBD studies, consistent methods are used to critically evaluate and make comparable the available information on each condition using standardized metrics (1).

According to the latest GBD report, in 2019 ischemic heart disease and stroke were the main causes of DALYs in age 50–74 years and older subgroups (15). During 2015 to 2020, heart diseases, cancers, Covid-19 and diabetes were among the top-ranked causes of death in the United States (16). As for diabetes, it has been estimated that the global prevalence of type 2 diabetes (T2D) will reach 7079 per 100,000 by 2030, with an accelerated rate in both developed and developing countries (17).

It is noteworthy that in 2017, noncommunicable diseases (NCDs) contributed over 60% of GBD, while infections and maternal, neonatal and nutritional disease were responsible for 28% of it (18).

It has been estimated that 71% of all deaths (41 million) annually are due to NCDs globally. It is unfortunate that about 77% of NCD deaths occur in low- and middle-income countries, where more children suffer from undernutrition (19).

CVDs are responsible for the most deaths due to NCDs (17.9 million a year), followed by cancer (9.3 million), respiratory disease (4.1 million) and diabetes (1.5 million) (19), all of which are predisposed by poor nutrition and low physical activity (20).

It is worthwhile to note that with the emergence of the coronavirus infectious disease (Covid-19) pandemic, with higher mortality rates in countries with high prevalence of undernutrition (21), calculation of the impact of Covid-19 in future will probably put it among the top ten conditions of GBD (22). In the following sections, after a brief review of certain biological, epidemiological and nutritional aspects of vitamin D, the association of this nutrient with the occurrence of human diseases, with an emphasis on the major contributors of GBD, will be discussed.

Vitamin D biology

Several decades after the discovery of vitamin D, there is still a lot to learn about it. Vitamin D is commonly known as a hormone; nonetheless the only hormone that has a recommended dietary intake (DRI) amount. This implies that vitamin D is also known as a nutrient. Naturally, vitamin D is synthesized in the body following direct exposure to the sun. Hunter-gatherer humans, who were almost naked, could obtain enough vitamin D through their skin. However, cultural, environmental and social factors have made this natural connection of human and sun fade away, so that the role of vitamin D as a nutrient has become more prominent, indicating that it must be obtained through diet, including fortified foods and supplements.

Metabolism

Vitamin D is a generic term for at least two compounds, that is, ergocalciferol (D_2) and cholecalciferol (D_3), and their several metabolites found in the body. Ergocalciferol has a plant origin, whereas cholecalciferol is synthesized in animals and humans. Under the influence of solar ultraviolet (UV) beams in the wavelength range of 280–320 nm, 7-dehydrocholesterol is transformed to cholecalciferol, which is a biologically inactive pre-hormone. To be fully activated, cholecalciferol needs to undergo two steps of hydroxylation, the first in the liver and the second in the kidney, thus producing 25-hydroxycholecalciferol ($25(OH)D_3$) or calcidiol and 1, 25-dihydroxycholecalciferol

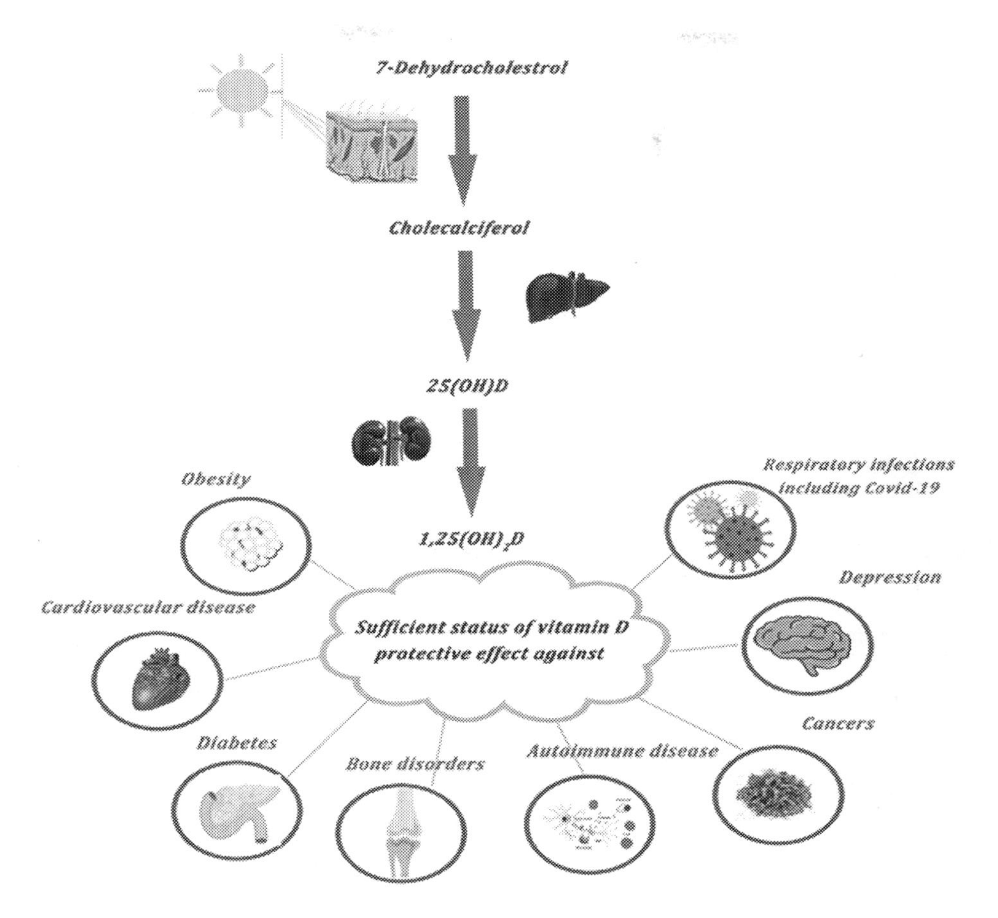

Figure 4.1 *Endogenous biosynthesis pathway of vitamin D and some of its beneficial health effects.*

(1, 25$(OH)_2D_3$) or calcitriol, respectively (Figure 4.1). Hepatic hydroxylation occurs at C-25 under the influence of 25-hydroxylase (CYP2R1), whereas nephrotic 1-α-hydroxylase (CYP27B1), located in the proximal convoluted tubules (PCT), hydroxylates 25$(OH)D_3$ at C-1 of the A ring thus producing the biologically active form of the vitamin, 1, 25$(OH)_2D_3$. In PCT, at least two endocytic receptors, megalin and cubilin, contribute to DBP-bound 25(OH)D uptake for further hydroxylation step. In some disease states, including diabetic nephropathy, the function of these receptors may be impaired, resulting in vitamin D deficiency (23). Many factors contribute to regulation of calcitriol biosynthesis, including parathyroid hormone (PTH), fibroblast growth factor (FGF)23, circulating concentrations of calcium, phosphate and calcitriol itself. FGF23, with the assistance of a signaling cofactor named klotho, suppresses renal 1-α-hydroxylase and stimulates 24-hydroxylase. Klotho gene is upregulated by calcitriol (24).

Vitamin D_2, photosynthesized from its precursor, ergosterol, in plants and mushrooms, has a double bond between C22 and C23 and a methyl group at C24 of its side chain, which attenuates its affinity for DBP. Consequently, the clearance of D_2 from the circulation is faster than D_3. However, the activation pathway of these two isoforms is similar (25).

Ingested vitamin D, both dietary and supplement, is absorbed along with fats. Vitamin D and its metabolites are transported in circulation mostly bound to a specific carrier, vitamin D binding protein (DBP) and to a lesser extent to albumin. A very small fraction of 25(OH)D is unbound or free. The sum of albumin-bound and free 25(OH)D is known as bioavailable 25(OH)D (26).

The biological effects of the active vitamin D metabolite 1, $25(OH)_2D$ is exerted through both genomic and non-genomic pathways. In the classic genomic response that may take a few hours, 1, $25(OH)_2D$ interacts with its specific nuclear receptor, vitamin D receptor (VDR), which then undergoes heterodimerization with retinoid receptor, RXR. Some evidence indicates that 9-cis retinoic acid at very high concentrations may segregate RXR as its homodimer form and thus may suppress this pathway (25). Identification of VDRs in a vast variety of cells and tissues opened a new window to the roles of vitamin D in many body functions (27).

Vitamin D-related rapid non-genomic responses that occur in seconds to minutes are mediated through another class of VDRs, which are located on plasma membrane caveolae of several cell types. Activation of the vitamin D non-genomic pathway results in recruitment of membrane calcium transport proteins from intracellular vesicles to the cell surface and thus a rapid increase in calcium absorption, which takes place before induction of calcium-binding protein (calbindin) expression. The other function of this pathway is opening of intracellular calcium channels and activation of protein kinase C and mitogen-activated protein kinases (MAP kinases), which results in inhibition of cellular proliferation and induction of cellular differentiation (28).

Vitamin D functions

Biological actions of vitamin D can be categorized to classical or calcemic and novel or non-calcemic. The calcemic effect of vitamin D, mainly on bone health, was recognized even long before its exploration. Vitamin D has a crucial role in calcium and phosphate homeostasis and musculoskeletal health (29). However, a growing body of evidence indicates several non-calcemic functions for vitamin D, including immunomodulation, antioxidant, cardiovascular health, protection against diabetes and several types of malignancies, mental health, hematopoiesis, fertility, as well as pregnancy and birth outcomes (30–37). However, the effectiveness of vitamin D supplementation in prevention of some non-calcemic function-related disorders, including cardiovascular disease, is still debatable (38, 39).

Assessment methods

The main isoform assayed to assess vitamin D status is circulating 25(OH)D. As 1, $25(OH)_2D$ is under tight homeostatic regulation, its circulating concentration is not beneficial for the assessment (40). Some investigators believe that free 25(OH)D must also be assayed to assess vitamin D status (41). One of the major drawbacks of evaluation of free 25(OH)D is that, due to its very low concentration, it is commonly calculated. Some evidence shows that there might be discrepancies between directly measured and calculated values of free 25(OH)D (42). Furthermore, measurement of free 25(OH)D may not confer an additional advantage, at least in healthy adults (41, 43). Notwithstanding, the relationship between total and free 25(OH)D may be influenced by health conditions and body mass index. The clinical importance of free 25(OH)D measurement still needs to be clarified (44, 45).

There are several methods to assay circulating 25(OH)D, including radioimmunoassay (RIA), enzyme immunoassay (EIA), competitive chemiluminescent immunoassay (CLIA), high performance liquid chromatography (HPLC) and mass spectrometric methods, whose results may not have agreement since each method has its own advantages and limitations (46). Therefore, efforts have been made to standardize or harmonize the results obtained from different methods (47, 48).

There is no global consensus on the optimal circulating 25(OH)D concentration (49). However, it is almost commonly agreed that a concentration below 50 nmol/L (20 ng/mL) is undesirable, apart from the name given to it, insufficiency or deficiency (50). However, it has been recommended to maintain circulating 25(OH)D concentrations above 75 nmol/L (30 ng/mL) to take advantage of both its calcemic and non-calcemic health effects (51).

Deficiency

VDD is a worldwide public health problem. It is estimated that over one billion people from all age, sex and ethnicity groups globally have undesirable circulating 25(OH)D concentrations (10). Many social, environmental, geographical, nutritional and cultural factors contribute to this high prevalence of VDD, among them are limited natural dietary sources of vitamin D, living in high latitudes, air pollution, dark skin, using sunscreens and low outdoor activities (52). As 25(OH)D may be sequestered in body fat, obesity is also considered one of the predisposing factors for suboptimal vitamin D status (53).

Requirement

There is no universal consensus on the adequate amount of vitamin D intake. The recommended dietary intake (RDA) for subjects aged 4–70 years is 600 IU (15 μg) a day. However, many experts believe that higher amounts (1000–2000 IU/day) are needed to maximize the health effects of vitamin D. Evidence indicates that daily intake of up to 5000 IU and 10,000 IU are safe for children and adults, respectively (54).

Contribution of VDD to GBD

Obesity

Etiology, pathophysiology, health consequences

Obesity, defined as excess body weight as fat, is considered one of the most serious public health concerns. Although not being an accurate measure of adiposity, body mass index (BMI) (weight in kg/height in m^2) is simple to use in screening and epidemiological investigations. BMI, as a major indicator of obesity, is categorized as: underweight (below 18.5 kg/m2), normal (18.5–25 kg/m^2), overweight (25–29.9 kg/m^2) and obese (> 30 kg/m^2) (55).

The obesity results from more energy intake than energy expenditure. Notwithstanding, the etiology of obesity is complex, with a spectrum of environmental, sociocultural, physiological, medical, behavioral, genetic, epigenetic and numerous other factors contributing to causation (56). Excess energy intake and subsequent storage in adipose tissue leads to adipocyte hyperplasia and hypertrophy. Enlarged fat cells contribute to the clinical problems of obesity due to increased secretion of free fatty acids, multiple peptides, proinflammatory cytokines and hormones. Hence, adipose tissue can be viewed as an endocrine organ that has metabolic activity. The consequence of these events is an increase of obesity-associated morbidity and mortality. Numerous chronic diseases including diabetes mellitus, hypertension, dyslipidemia, CVD, osteoarthritis and some forms of cancer are attributed to obesity (57).

Vitamin D and obesity: evidence from experimental, observational and clinical trial studies

Evaluation of the dose-response relationship between vitamin D and BMI showed a quadratic curve indicating a rate-limiting mechanism to prevent excessive formation of $1,25(OH)_2D_3$ to toxic levels (58). In confirmation of the relationship between weight and vitamin D, many studies have reported changes in vitamin D status following weight changes (59, 60). Vitamin D status has even been reported to be linked with visceral or subcutaneous adipose tissue (61, 62). It has also been shown that although increased circulating 25(OH)D concentrations may not help regulate weight, VDD could contribute to the adverse health effects associated with obesity (63). According to a meta-analysis, the prevalence of VDD in obese, as compared with normal weight, people is more than three times (pooled OR (95% CI) was 3.43 (2.33–5.06)) (64). Several mechanisms for the low status of vitamin D in obesity have been proposed, including lower sun exposure in obese people than in lean people (65); sequestration of calciferol in the larger body pool of fat (66); inadequate vitamin D intake through food and supplements due to unhealthy food choices (67); decreased hepatic production of 25(OH)D due to steatosis, which is common in obesity (68); increased production of $1,25(OH)_2D_3$ and consequent decreased 25(OH)D synthesis through negative feedback, which is reported more in obese compared with non-obese subjects (69), though more recently larger studies showed the opposite is true (70, 71), that is decreased circulating concentrations of 25(OH)D due to volumetric dilution (72); and finally increased metabolic clearance of 25(OH)D in obese people, possibly due to increased uptake by adipose tissue (73).

Alternatively, the results of experimental studies indicate that VDD can increase lipogenesis and lead to obesity by enhancing secretion of PTH and calcium overflow into adipocytes (74). A murine experience showed that targeted expression of human VDRs in adipocytes was accompanied by reduced energy expenditure and increased adiposity (75). On the other hand, vitamin D is an essential factor in the synthesis of leptin, the satiety hormone. VDD can, therefore, lead to increased appetite and obesity by disrupting leptin production (62, 76). Leptin exerts its stimulatory effects on lipolysis and inhibitory effects on lipogenesis in adipocytes by interacting with the vitamin D receptor, and so it exerts its modulating effects on adipocyte lipid metabolism (77). Elevated leptin levels following vitamin D supplementation have been shown in several clinical trials (78, 79). However, a meta-analysis of nine randomized controlled trials (RCT) failed to show a significant effect of vitamin D supplementation on serum leptin concentrations (80). The effect of vitamin D on adipose tissue and body weight still needs more elucidation (81, 82).

Contribution of obesity to GBD: does vitamin D play a role?

According to the GBD Study and the World Health Organization (WHO) reports, obesity is the key contributor to many chronic diseases, disabilities and all-cause mortality worldwide (83, 84). The effect of vitamin D status on body weight and more specifically on fat mass is controversial. Though some studies have reported a significant reduction in adiposity following improvement of vitamin D status (31, 85, 86), several meta-analytical studies failed to support this finding (87–89). Be that as it may, vitamin D status has been shown to have an association with several obesity comorbidities (90–92). It is, therefore, highly plausible that raising vitamin D status of the community may contribute to reducing GBD through decreasing the occurrence and severity of obesity comorbidities (93).

Diabetes mellitus

Etiology, pathophysiology and health consequences

Diabetes mellitus (DM) means the body's inability to regulate blood sugar. Various hormones are involved in regulating blood glucose, the most important of which is insulin. Insulin is secreted from the β-cells of the Langerhans islets of the pancreas and allows glucose to move out of the blood into muscle and fat cells, where it is used as a fuel (94).

The main problem in type 1 diabetes (T1D) is insufficient insulin production and secretion, whereas in type 2 diabetes (T2D), insulin resistance and inability of the body to use insulin properly are the initial causes. Notwithstanding, T2D patients may later develop insulin deficiency. Inefficient cellular uptake and use of glucose causes hyperglycemia (high blood sugar) and consequent cell starvation and tissue damage (94). Other categories of diabetes include maturity-onset diabetes of the young (MODY), gestational diabetes (GDM), neonatal diabetes and diabetes due to secondary causes such as steroid use and certain endocrinopathies (95). Although both T1D and T2D have genetic predispositions, T2D involves a more complex interaction of genetics and lifestyle and has a stronger hereditary basis as compared with T1D (96). The pathophysiology of diabetes is multifactorial and has not yet been fully elucidated. In fact, there is a vicious cycle between hyperglycemia and impaired insulin secretion due to β-cell dysfunction, which ultimately leads to metabolic abnormalities (97).

Complications of diabetes range from acute, life-threatening conditions such as severe hypoglycemia and ketoacidosis to chronic, debilitating complications affecting multiple organ systems. Nonetheless, the chronic pathologic hallmark of diabetes involves the vasculature leading to both microvascular damages (such as retinopathy, nephropathy and neuropathy) and macrovascular complications (including coronary artery disease, cerebrovascular disease and peripheral vascular disease) (98). The duration of diabetes and the degree of glycemic control are two important determinants of the complications of diabetes (99). This section mostly focuses on T2D.

Vitamin D and diabetes mellitus: evidence from experimental, observational and clinical trial studies

As previously mentioned, the first step in T2D onset is insulin resistance. At first, β-cells overcome insulin resistance by increasing insulin secretion and thus prevent hyperglycemia. This compensatory activity increases the exposure of β-cells to excessive calcium and reactive oxygen species (ROS), which eventually leads to cell death, hyperglycemia and the onset of diabetes (100). Diabetes is not just an inability to control blood sugar properly, it is an inflammatory disease as well (101). VDD contributes to this process, from insulin resistance to β-cell death and thus modifies the risk of diabetes (100).

A consistent, significant inverse association between VDD and development of insulin resistance and T2D is well supported by observational, longitudinal and cross-sectional studies (102, 103). Adequate serum concentrations of 25(OH)D (> 30 ng/mL) were associated with a 20–50% reduction of the risk of T2D development (102, 103). VDD predisposes individuals to develop diabetes, but once diabetes has developed, vitamin D does not appear to make much change to the course of the disease. However, VDD may lead to suboptimal responses to therapy in the affected patients and consequently vitamin D supplementation during the course of the disease may reduce the complications of the disease (104–106). Some evidence shows that improvement of vitamin D status of adult subjects with T2D might not only help better glycemic control but also suppress systemic inflammation and oxidative stress (31–33), both of which are major contributors to diabetic complications (107). The beneficial effects of vitamin D on glycemic as well as inflammatory status may partly be mediated through sirtuins, a family of highly conserved nicotinamide adenine dinucleotide (NAD+)-dependent enzymes that revamp certain proteins like histones post-translationally (108). Nevertheless, the overall findings of clinical trials have been controversial to date (109–111), part of which might be due to VDR polymorphisms of different populations (112).

The link between vitamin D and diabetes may follow several pathways. Vitamin D conserves the normal resting levels of calcium and does not allow ROS to rise excessively in the β-cells during diabetes (100). Also, due to its anti-inflammatory properties, many health benefits are attributed to vitamin D (113, 114). Exploration of VDRs in β-cells and the vitamin D–dependent calcium-binding proteins (CaBP) in the pancreas has provided ample evidence for the role of vitamin D in insulin secretion (115). The role of vitamin D has been perceived in various aspects, including glucose-induced insulin secretion, improving islet cell functions, decreasing insulin resistance and protecting against inflammation that indirectly improves pancreatic β-cell function (116, 117). Vitamin D also mediates regulation of extracellular ionized calcium levels and calcium influx to β-cells, making it a potential modulator of depolarization-induced secretion of insulin (118, 119). Vitamin D is thus involved in the action of calcium-dependent endopeptidases in β-cells and the conversion of proinsulin to insulin (120).

Another finding that can partially explain the link between vitamin D and diabetes is the expression of 1-α-hydroxylase in pancreatic β cells, the key enzyme that regulates the conversion of 25(OH)D into its active form, $1,25(OH)_2D_3$ (121). As previously mentioned, VDRs are abundantly expressed in β cells and its response element is also found in the promoter of the insulin receptor gene. Hence, $1,25(OH)_2D_3$, through binding to its receptor, affects pancreatic β cell function and insulin secretion (122).

VDD typically increases circulating PTH, which in turn impairs glucose tolerance and decreases insulin sensitivity by altering the intracellular free calcium concentration in target cells (123). PTH may be further involved in adipocyte metabolism and promote obesity by stimulating lipogenesis and inhibiting lipolysis (124). Findings from a recent meta-analytical study support the role of vitamin D supplementation in prevention of diabetes in the subjects with prediabetes (125).

Contribution of diabetes to GBD: the possible role of vitamin D

The rising burden of T2D is a major concern in healthcare globally. T2D accounts for more than 90% of cases of diabetes and its prevalence is estimated to increase to 7079 individuals per 100,000 by 2030 globally, including lower-income countries (126).

The financial burden of diabetes is also enormous. In 2021, the International Diabetes Federation (IDF) reported that 10% of global health expenditure is spent on diabetes (127). Prevention and treatment of diabetes and its complications and comorbidities require a comprehensive assessment with multisectoral efforts (128). Interventions at the population level should be concentrated on the obesogenic and diabetogenic environment and lifestyle modifications (129). Current evidence indicates that rising circulating 25(OH)D concentrations of the population to a sufficient level can contribute to the prevention of diabetes and its further complications (109–111, 125).

Vitamin D and cardiovascular disease

Etiology, pathophysiology and health consequences

CVD is the number one cause of illness, death and rising healthcare costs worldwide (130, 131). CVDs are a group of disorders of the heart and blood vessels. They include coronary heart disease, cerebrovascular disease, peripheral arterial disease, rheumatic heart disease, congenital heart disease and deep vein thrombosis and pulmonary embolism. The major drivers of CVD include tobacco use, high blood pressure, high blood glucose, lipid abnormalities, obesity and physical inactivity (130).

Vitamin D and cardiovascular disease: evidence from experimental, observational and clinical trial studies

It has been suggested that suboptimal vitamin D status may increase CVD risk (132). Recognition of VDRs in cardiomyocytes, vascular smooth muscle cells and endothelial cells indicated a direct role of vitamin D in cardiovascular function (133, 134). The role of vitamin D in cardiovascular health is either by affecting the disease risk factors or by affecting the cardiovascular system directly (135).

Observational studies, as well as recent meta-analyses, have indicated the association of circulating 25(OH)D with cardiometabolic risk factors including hypertension (136–143) and lipid profile (144–149). The association has been shown for atherosclerosis (150–152), heart failure (153–156) and CVD mortality (156–159), as well. Results from a meta-analysis showed that the association between circulating 25(OH)D and CVD is not linear, with the highest risk at 25(OH)D levels below 50 nmol/L and no decrease in the risk of CVD at higher levels up to 137.5 nmol/L (160). Two large observational studies added that the risk of CVD is again increased at levels of 90 to 100 nmol/L (161, 162). It is suggested that such J-shaped associations could be due to hypovitaminosis D-related diseases masked with self-supplementation before onset (163, 164). The observational data cannot show causality because they are prone to several confounders. Therefore, data from Mendelian randomization studies and randomized control trials could be more reliable.

About 7.5% of variations in circulating 25(OH)D is due to single nucleotide polymorphisms (SNPs), and a few Mendelian randomization studies have investigated the effects of SNPs on the relationship of circulating 25(OH)D and CVD outcomes. Because of the lower effect of genes on circulating 25(OH)D compared with sun exposure and oral supplementation, the effect of genetically determined vitamin D status on CVD events is not clear (165). From five studies (166–170) using the Mendelian randomization approach to investigate the association of CVD outcomes, such as hypertension, stroke, coronary artery disease, myocardial infarction and cardiovascular mortality with circulating 25(OH)D, only one study reported that a 10% increase in genetically determined 25(OH)D concentration is associated with a significant blood pressure reduction of 0.3 mm Hg and a significant reduction in the risk of hypertension (169). A number of meta-analyses have indicated the association of VDRs and CYP24A1 gene polymorphisms with CVD risk (171–176) and only one meta-analysis reported no association of gene variants with coronary artery disease (171). However, the *p* value reported in none of the aforementioned analyses was lower than 10^{-7}, which is the threshold for reliable associations of gene polymorphisms with a disease outcome (165).

Data from large meta-analyses indicated a positive effect of vitamin D supplementation on lipid profile through reducing triglycerides, total cholesterol and low-density lipoprotein-cholesterol (149, 177, 178). The effects were more pronounced if the supplementation period was longer than six months (177) or if circulating 25(OH)D was less than 50 nmol/l (178). However, data on the beneficial effects of vitamin D supplementation on CVD outcomes are not convincing. Results of three large RCTs, in line with other earlier RCTs and meta-analyses, did not support the positive effect of vitamin D supplementation on CVD outcome.

Contribution of CVD to GBD: the role of vitamin D is still debatable

At the beginning of the 20th century, fewer than 10% of all deaths worldwide were attributed to CVD. However, by 2010 that value raised to 30% of all deaths, more than all communicable, maternal, neonatal and nutritional disorders combined (179). The total number of DALYs due to CVD has risen steadily between 1990 and 2019, reaching 6.55 million deaths for stroke and 9.14 million deaths for ischemic heart disease in the year 2019 (131). Though effectiveness of vitamin D in the prevention of CVD has been documented (180), this finding was not supported by a large clinical trial using monthly high-dose vitamin D supplementation (181). Notwithstanding, the effect of improvement of vitamin D status mostly through daily intake of physiologic doses in prevention of CVD needs to be addressed by well-designed prospective studies. Considering the inverse association between circulating 25(OH)D concentrations and atherogenic profile of blood lipids (148), investment in raising vitamin D status to the adequate level can be a part of the cost-effective public health programs for combating CVDs.

Vitamin D and bone health

Etiology, pathophysiology and health consequences

Metabolic bone diseases (MBD) encompass a wide spectrum of different clinical diseases that lead to bone abnormalities. MBD are usually characterized by significant clinical manifestations that are almost reversible after treatment of the underlying defect. Rickets/osteomalacia and osteoporosis are the two important MBDs. Rickets, the clinical outcome of VDD, is usually characterized by various bone deformities. Though rickets is far less common than osteoporosis, some studies reported that the prevalence of rickets is on the rise in some areas (182, 183).

Osteoporosis is the most important and common MBD, which is caused by a reduction in bone mass and a change in its microarchitecture. It subsequently increases the susceptibility to fracture, and since osteoporosis is primarily found in the elderly, it can be an essential cause of senile frailty and falls. Bone mass loss occurs due to an imbalance between bone formation by osteoblasts and bone resorption by osteoclasts, a dynamic process that ensues continuously throughout the skeleton to achieve its functions (184).

Osteoporosis is often perceived as an inevitable consequence of age-related hormonal changes associated with aging. However, evidence suggests that inadequate physical activity, sarcopenia, inappropriate nutrition and other unhealthy lifestyle factors should not be overlooked in the etiology of osteoporosis (185, 186). Moreover, genetic, sex, lack of sex hormones such as postmenopausal estrogen depletion, diseases such as hyperparathyroidism, and widely prevalent VDD, as the key determinants of postmenopausal osteoporosis are among the factors that reduce bone mass and thus may contribute to the etiology of osteoporosis (187). Adequate and constant supply of nutrients is required for bone formation and metabolic processes related to bone in general, of which vitamin D is perhaps the most important (188). The expression of VDRs on the surface of osteoblasts and immature cells of osteoclast precursor confirms the direct effects of vitamin D on bones (189). Main devastating health consequences of osteoporotic fractures include an increased mortality and morbidity, which causes disability and imposes a considerable burden to the healthcare system (190).

Vitamin D and MBDs: evidence from experimental, observational and clinical trial studies

Both animal and human studies indicate an axial rôle for vitamin D in bone health (191, 192). Although large-scale epidemiological studies have revealed an association between circulating 25(OH)D concentrations and bone density in both women and men, the results of the effectiveness of vitamin D supplementation in preventing bone loss and fractures in the elderly are inconsistent. Some studies have shown the effectiveness of vitamin D in reducing the risk of osteoporosis and subsequent fractures (193); some studies have failed to show these effects (194); others have even reported the harms of vitamin D supplementation.

A Cochrane review of 45 RCTs in postmenopausal women or men over 65 years failed to show the effects of vitamin D supplement alone on reducing the risk of fractures. In this review, vitamin D supplementation was compared with placebo, calcium supplementation, or no intervention (195). In one study, only in patients with insufficient serum levels of 25(OH)D and institutionalized patients, vitamin D together with calcium supplementation showed some benefits on primary fracture prevention (196). In adults over the age of 50, both men and women, combining vitamin D with calcium, rather than vitamin D alone, prevented 5–30% of fractures. It is noteworthy that vitamin D alone or in combination with calcium was not effective in preventing secondary fractures, and also the active form of vitamin D was not recommended for the treatment of osteoporosis (189).

Altogether, though even high doses of vitamin D may not be beneficial to bone density (197), VDD is a crucial determinant of development of osteopenia, osteoporosis and hence predisposition to fractures (198).

Contribution of metabolic bone disorders to GBD: the crucial role of vitamin D

Among musculoskeletal diseases, which are the second leading contributor to disability in the world, bone diseases have the highest health burden. Around 50% of women and 20% of men over the age of 50 may suffer from osteoporosis-related fracture and comorbidities (199). On average, an osteoporotic fracture occurs every three seconds (200). It is estimated that more than 37 billion Euro are spent annually on hospitalizations, with serious consequences in terms of morbidity and mortality in the European Union (201). The annual cost to the United States is estimated at around $17.9 billion (202). Owing to the increasing trend of population aging in the world, osteoporosis as a silent epidemic is on the rise (203) and the global costs of osteoporotic fractures are projected to increase by 25% during the years 2010–2025 (201). Improvement of vitamin D status through supplementation and food fortification is a highly cost-effective strategy to reduce the burden of MBDs including osteoporosis, falls and consequent fractures in the elderly (204, 205) and also rickets in pregnant women and young children (206).

Multiple sclerosis, rheumatoid arthritis, other autoimmune disorders and vitamin D

Etiology, pathophysiology and health consequences

Autoimmune disorders are conditions in which the immune system is not able to distinguish between healthy tissue (self) and disease-causing cells (non-self) (207). Some common autoimmune diseases include celiac disease, type 1 diabetes (T1D), rheumatoid arthritis (RA), multiple sclerosis (MS), systemic lupus erythematosus (SLE), thyroid disease (Hashimoto thyroiditis and Graves' disease), psoriasis and inflammatory bowel disease (IBD) (207).

Vitamin D and autoimmune disorders: evidence from experimental, observational and clinical trial studies

The presence of VDRs and expression of 1-α-hydroxylase enzyme in several immune regulatory cells propose a role for vitamin D in the pathogenesis of autoimmune diseases including T1D, autoimmune thyroid disease, MS, IBD, SLE and RA (208–215). Autoimmune diseases result from the organism's reaction against self-components and involve several mechanisms, including increased inflammatory response, oxidative stress, pro-fibrotic effects and glucocorticoid response. VDD has been demonstrated to affect the regulation of dendritic cells, regulatory T-lymphocytes and T helper (Th)1 cells (212). Recently, several studies demonstrated an association between VDR polymorphism and autoimmune diseases including BsmI or TaqI polymorphisms with autoimmune thyroid disease; BsmI and FokI polymorphisms with SLE; FokI polymorphism with diabetic nephropathy; and finally ApaI, BsmI and TaqI polymorphisms with RA (216–219). Notably, VDR activation has a central role in modulating the immunological responses and its polymorphisms may lead to functional changes such as reducing vitamin D regulatory effects on the immune response (220).

RA is a common autoimmune disease characterized by peripheral joint inflammation. Pathogenesis of RA is complex and along with genetic susceptibility, some environmental factors including vitamin D status could affect the occurrence and progression of the disease (215). Vitamin D modulates the pathogenesis of RA through suppression of the proliferation and activity of Th1 and Th17 and enhancing the activity of T regulatory (Treg) cells (221). Results of observational and cross-sectional studies on the relationship of vitamin D status and RA are inconsistent. Although many studies have reported an inverse relationship between 25(OH)D and RA (222–228), some others did not support this link (229–236). Heterogeneity in the environmental factors, including ethnicity, ultraviolet B (UVB) exposure, physical activity and dietary intake, as well as severity and treatment regimen of the disease, may explain the inconsistency in the results (215, 220, 237). The results of clinical trials on the effects of vitamin D supplementation on RA are also conflicting.

The beneficial effects of vitamin D supplementation on pain relief are limited to the patients with VDD (238, 239) but not those with sufficient 25(OH)D at the baseline (240). Some studies failed to show significant effects of vitamin D supplementation on the recurrence of the RA after 6- to 24-month follow-up of the patients with initial VDD (241, 242). Meta-analyses also indicated no positive effect of vitamin D supplementation on pain relief (241, 243, 244). However, the combination of vitamin D with anti-rheumatic drugs has been shown to have a significantly more positive impact on pain and joint inflammation (245, 246).

The discrepancy in these observations could partly be explained by differences in patient characteristics, small sample sizes and hence limited power of the trials, dosage of vitamin D, the severity of the disease and the duration of treatment. More large-scale trials are needed to conclude the treatment dose of vitamin D and effective 25(OH)D concentration in patients with RA.

Multiple sclerosis (MS), one of the leading causes of neurological disability in adults worldwide, is an autoimmune disease characterized by chronic inflammation and degeneration of the central nervous system (247). Like other immune disorders, interplay of genes and environmental factors determines the risk of developing the disease (248, 249). The possible role of vitamin D in the pathogenesis and progression of MS has been widely investigated in animal as well as human studies (250, 251). It has been suggested that circulating 25(OH)D concentration is not only correlated with the onset of MS (252–255) but also to activity and progression of the disease, especially lower relapse rate (256–258) and fewer new MRI lesions (252, 259).

It is noteworthy that Mendelian randomization studies have confirmed the causal relationship between vitamin D and MS risk and reported that the presence of some SNPs is associated with higher 25(OH)D concentration (260–263). However, the possible positive effects of vitamin D supplementation in patients with MS have not been confirmed yet (264). A significant reduction in disease activity and severity has been reported in some studies (265–268), while others have not confirmed this effect (269–271).

The association of increased serum 25(OH)D concentration with a modest decrease in relapse rate and radiological inflammatory activities in patients with MS has been reported (272). Nevertheless, the effect of vitamin D on disability progression is still questionable (264, 273, 274). A significant reduction in disease activity and severity has been reported in some studies (265–268) while others have not confirmed this effect (269–271). Clinical trials and meta-analyses investigating effects of various doses of vitamin D supplementation (ranging from 10,000 to 40,000 international units [IU]/day) have indicated no therapeutic effect on disability or relapse rate (264, 273, 274). High-dose vitamin D supplementation also failed to show a positive effect despite the reasonable increase in circulating concentrations of 25(OH)D (275–278).

Recently, it was hypothesized that autoimmune diseases may result from vitamin D resistance and hence must be treated using high-dose supplementation (279). Along the same line, a recent intention toward using megadoses of vitamin D in the range of 50,000–300,000 IU/day in the subjects with MS has been reported that is not based on study evidence and can cause life-threatening complications due to vitamin D toxicity. Vitamin D toxicity may manifest as fatigue, muscle weakness, or urinary dysfunction, which may mimic the natural course of progressive MS (264).

Though current evidence is not sufficient yet to recommend routine vitamin D supplementation, beyond the correction of hypovitaminosis D, to reduce the incidence and severity of autoimmune disorders, emerging data may reveal new therapeutic implications for it (220).

Contribution of autoimmune disorders to GBD: vitamin D may reduce the burden

In 2010, the American Autoimmune Related Diseases Association (AARDA) estimated that autoimmune disorders affected 50 million Americans, and this number appears to be growing (280). Autoimmune disease has many direct and indirect impacts on rising healthcare costs, as well as the quality of life for individuals with these diseases (281). It has been widely hypothesized that VDD acts as an environmental trigger for the induction of autoimmunity, and that high-dose vitamin D supplementation could be preventive (282). For unknown reasons, most autoimmune diseases affect women 75% more often than men (207). Interestingly, VDD is also commonly more prevalent in women than in men (283). The prophylactic effect of raising vitamin D status of the general population against autoimmune disorders warrants well-designed prospective studies. Nonetheless, this approach may help lessen the burden of diseases at least through decreasing the severity and the resulting disabilities of these disorders (215).

Vitamin D and cancers

Etiology, pathophysiology and health consequences

There are more than a hundred distinct types of cancer that develop across time and involve the unregulated proliferation of the body's cells. Since the development of malignant cells is a complex multistep process, various factors may affect the chance that cancer will develop. However, many factors, including radiation, chemicals and viruses, have been found to cause cancer in both experimental animals and humans. Cancer is the second-leading cause of death in the world. Nevertheless, survival rates are improving for many types of cancers, thanks to improvements in cancer screening, treatment and prevention (284).

Vitamin D and cancers: evidence from experimental, observational and clinical trial studies

Since 40 years ago when the UVB–vitamin D–cancer hypothesis was proposed for the first time, over 25,000 articles have been published on the role of UVB exposure and vitamin D in the risk of cancer incidence, progression, survival and mortality (285, 286). However, the definite proof of the hypothesis has yet to be determined.

Vitamin D is a cell growth and differentiation regulator in a broad range of cell types including cancer cells (287). A large body of evidence supports the inhibitory role of vitamin D in cancer cell proliferation, angiogenesis and metastasis, as well as its role in apoptosis and differentiation promotion (288, 289).

Ecological studies have demonstrated that higher 25(OH)D levels are negatively associated with incidence and mortality rates of about 20 cancer types (285, 286). Considering the confounding factors, several observational studies, including prospective cohort studies and case-control studies in line with meta-analyses, have indicated inverse correlations between 25(OH)D concentration and site-specific cancer outcomes including incidence and mortality. Observational studies showed the reverse correlations for brain (glioma), cervical, esophageal squamous-cell carcinoma, gastric adenocarcinoma, head and neck, larynx and hypopharynx, liver, oral cavity and gum, ovarian and pancreatic cancers. Meta-analyses indicated the same correlation for bladder, breast, colorectal, kidney and lung cancer incidence (197, 285, 286). The strongest associations were reported for breast and colorectal malignancies (289–291). Regarding mortality, meta-analyses on prospective studies on participants with or without cancer at baseline indicated an adverse association of 25(OH)D levels and mortality of various cancer sites including breast, colorectal and lymphoma (292, 293). Dose-response meta-analyses reported inverse trends for liver, lung, colorectal, colon and breast cancers; 10–40 nmol/L increments of 25(OH)D levels were correlated with a 2–8% decrease in both the cancer risk and mortality (197, 294–296). Although an inverse association between 25(OH)D and cancer incidence has been reported for many cancer sites, prostate cancer is an exception. It is reported that high solar UVB exposure and higher circulating 25(OH)D concentrations are related to a higher incidence rate of prostate cancer (297, 298). This observation may be due to the role of vitamin D in dietary calcium absorption, which has been suggested as a potential risk factor for prostate cancer (286, 299, 300).

In contrast to the aforementioned observational studies that indicated the protective effect of 25(OH)D on cancer incidence and mortality, nearly all meta-analyses showed that vitamin D supplementation has a stronger beneficial impact on cancer mortality than cancer incidence (293, 301–303). One meta-analysis showed no overall effect of vitamin D supplementation (304). In this meta-analysis, the number of participants and events were apparently lower compared with the other meta-analyses, mainly because it missed three large studies because one study co-administered vitamin D and calcium (305) and two others were published after it (306, 307).

A recent report of the secondary analysis of the Vitamin D and Omega-3 Trial (VITAL) indicated that the primary 17% reduction in advanced cancer risk was augmented when BMI stratification was used (308). A significant 38% risk reduction was reported in normal-weight individuals but not in overweight and obese individuals. Such findings suggest that individuals do not benefit the same from the supplementation and more attention should be paid to interindividual differences while interpreting the study results (309).

Contribution of cancers to GBD: improvement of vitamin D status is worth investment

Cancers are among the leading causes of death worldwide. In accordance with estimates from WHO in 2019, cancers are the first or second leading cause of death before the age of 70 in 112 of 183 countries and ranks third or fourth in a further 23 countries (310). A series of advances in our understanding of the associations between vitamin D and cancer in recent years have been accompanied by increasing potentials for cancer prevention. This is of specific importance for public health in view of the high prevalence of hypovitaminosis D around the world (13). Considering the benefits of vitamin D in reducing risk and mortality of many types of cancers, in our opinion, increasing serum 25(OH)D levels of at least 75 nmol/L (30 ng/mL) in the general population is a cost-effective way to reduce global mortality rates.

Depression

Etiology, pathophysiology and health consequences

Major depressive disorder (MDD) is among the most debilitating and burdensome disorders across the globe, with an estimated 3.8% of the population affected and a higher prevalence among the elderly (311, 312). MDD severely decreases people's quality of life and daily functioning. It was reported that more than 80% of MDD patients have difficulty with work and other daily activities; subsequently, MDD contributes considerably to the global disability (313–315).

MDD is commonly characterized by cognition impairments and emotional dysregulation, with memory malfunctioning and neurocognitive symptoms (316). The etiology is still unclear. Nonetheless, some plausible mechanism have been proposed that appear to have a role both in the etiology and progression of the disorder. Inflammation (317), hypothalamic–pituitary (HPA) axis abnormalities (318), neutrophic growth disorder (319) and VDD (320) are among the most important proposed mechanisms that have substantial support in the literature.

Like many diseases and disorders, complex interaction of biological, genetic, epigenetic, psychological and social factors contribute to MDD (321). Depression can, in turn, lead to several severe disabilities, which may lead to secondary disabilities because of interrelationships between depression and physical health (316).

Vitamin D and MDD: evidence from experimental, observational and clinical trial studies

The expression of VDRs in the nuclei of neurons in several regions involved in mood and cognition, including the cingulate cortex and hippocampus and production of the active form of vitamin D in the brain made it biologically plausible that vitamin D might be associated with depression (322, 323). Besides, DBPs have been discovered in the central nervous system (CNS), particularly in areas associated with mood and depression (90, 102). In addition, various processes in the brain, including neuroimmunomodulation, regulation of neurotrophic factors, neuroprotection, neuroplasticity and brain development, require vitamin D (324). Seasonal variations in the prevalence and incidence of depression can also be attributed in part to vitamin D (325, 326).

Evidence from cellular studies on the possible contribution of vitamin D to depression is limited but altogether has proposed a possible link between vitamin D, HPA and depression. Notwithstanding, further studies are still needed (327).

The huge body of human investigations dealing with the possible roles of vitamin D in MDD are cross-sectional and cohort studies that agree on the inverse association between serum 25(OH)D concentration and depressive symptoms and/or MDD risk (320, 328–333). Albeit some clinical literature failed to replicate the finding of significant associations and have led to controversial results regarding the effects of vitamin D supplementation on depression severity (334–338). Heterogeneity of participants in terms of differences in age, clinical severity, circulating 25(OH)D concentration at the beginning of the study, different assessment measures, different dose of vitamin D supplementation and duration of intervention, and different potential confounders are among the causes listed for these conflicting outcomes. Another possible explanation for the ineffectiveness of vitamin D supplementation in improving mood in clinical trials is that the dose of vitamin D was not sufficient to compensate for hypovitaminosis D and consequently could not overcome the severity of depression (339). Nevertheless, in a clinical trial, vitamin D supplementation with 50,000 IU biweekly for eight weeks resulted in a significant improvement of depression severity, as compared with the control group. It is noteworthy that most of the patients (64% and 43% in the intervention and placebo groups, respectively) had initial circulating 25(OH)D concentrations above 75 nmol/L (340).

Several mechanisms have been proposed for the effect of vitamin D in improving the moods of the subjects with MDD, among them are a regulatory role in the expression of brain neurotransmitters and neurotrophic factors as well as having a cross-talk with glucocorticoids in hippocampal cells (341, 342); maintaining normal serotonin levels and enhancing mood status by inducing the expression of tryptophan hydroxylase-2, the rate-limiting enzyme in serotonin biosynthesis pathway (343); downregulation of tryptophan hydroxylase-1 gene in the gut, which subsequently inhibits the production of serotonin in extra cranial tissues, where when found in excess promotes inflammation (344); reducing inflammation and oxidative stress that can prevent or delay the onset or progression of the disease and contribution to neuroimmunomodulation and neuroplasticity (345).

Contribution of MDD to GBD: where is the place of vitamin D?

According to a recent WHO report, the number of people suffering from MDD increased by 12.9% between 2010 and 2018 (346). Given that MDD is the most prevalent 21th-century mood disorder with undesirable health implications, it is not unexpected that it imposes a heavy personal and economic burden on societies. In line with this fact, and according to the GBD Study, depression has the heaviest burden compared with other mental and behavioral health disorders, as about 2.7 million DALYs in 2016 were attributed to MDD (347). Across the same period, the incremental economic burden of MDD increased by about 38%, of which 35% was related to direct costs, 4% to suicide-related costs and 61% to workplace costs (346).

Despite the large body of data, the causal relationship between vitamin D and depression is still uncertain. On one hand, the role of VDD in the pathophysiology of depression is discussed, and on the other hand, VDD can be a consequence of depression and the resulting behavioral and dietary changes, such as reduced sun exposure and decreased outdoor activity (341, 348). Also, though serotonin levels increase following vitamin D supplementation, it is unclear whether this approach is successful in treating depression (349, 350). Overall, although research on vitamin D and depression is still in its infancy, there is growing evidence proposing a potential role for this micronutrient in mental health, especially depression, which provides incentive for further research.

Vitamin D and Covid-19 infection

Etiology, pathophysiology and health consequences

The year 2019 did not end happily due to the emergence of a highly contagious and occasionally deadly coronavirus disease, hence named Covid-19. The resulting pandemic has brought about millions of cases of morbidity and mortality globally (351) with a heavy economic impact (352).

Covid-19 is caused by the severe acute respiratory syndrome (SARS-CoV)-2 virus. Clinical symptoms of the disease vary from common cold-like symptoms to acute respiratory distress syndrome (ARDS) and septic shock. SARS-CoV-2, as compared with seasonal flu, tends to have a higher pulmonary pathogenicity and hence more severe respiratory complications and mortality (353). A growing body of evidence suggests a pathogenic role for a sudden release of extra amounts of pro-inflammatory cytokines, including interleukin (IL)-1, IL-6 and tumor necrosis factor (TNF)-α, into the circulation, the so-called cytokine storm due to coronavirus infection (354). Several factors have also been reported to affect the severity and outcomes of Covid-19, among them are concomitant infections, comorbidities including diabetes and obesity, as well as race and age (355). Malnutrition and especially micronutrient deficiency can also adversely affect Covid-19 outcomes (356, 357).

Vitamin D and Covid-19: evidence from experimental, observational and clinical trial studies

Some observational as well as meta-analytical studies have reported a significant association between poor vitamin D status and severity of Covid-19 outcomes (357–361). Nevertheless, the protective effect of vitamin D against coronavirus remained controversial, necessitating solid evidence from clinical trials (362, 363).

Findings from vitamin D supplementation studies in Covid-19 have been controversial to date. A study on 8297 adults who had records of Covid-19 test results using UK Biobank showed that those who consumed vitamin D supplements had a lower risk of Covid-19 infection (364). Based on these findings, a single 300,000 IU dose of vitamin D was proposed as a means to prevent and treat Covid-19 (365). However, high dose parenteral vitamin D supplementation to 175 Covid-19 patients with vitamin D deficiency (serum 25(OH)D < 12 ng/mL) who were hospitalized in the ICU ward did not affect disease severity and outcomes (366). It is noteworthy that vitamin D supplementation with 300–2000 IU/day, as compared with large doses with long intervals (100,000–200,000 IU in a month or every three months) has been more effective against viral infections (367).

The active form of vitamin D, 1, $25(OH)_2D$, contributes in innate antimicrobial immunity via upregulation of cathelicidin antimicrobial peptide (CAMP) and defensin β-4 (DEFB4) genes. Vitamin D also regulates angiotensin converting enzyme (ACE)-2, which is used by the coronavirus as an entry receptor.

Vitamin D also has an immunomodulatory function whereby it downregulates pro-inflammatory cytokines, including TNF-α, interferon (IFN)-γ, IL-6 and IL-12, which may critically contribute in the clinical outcomes of Covid-19 patients (368). Furthermore, reduced amounts of circulating adiponectin have been reported in patients with Covid-19-induced acute respiratory failure (369) and dietary approaches have been proposed to suppress pro-inflammatory but enhance anti-inflammatory cytokines including adiponectin (370). A recent meta-analysis reported that vitamin D might be considered as an adiponectin secretagogue, especially in subjects with diabetes (371).

Contribution of Covid-19 to GBD: does vitamin D status matter?

Several studies have estimated the heavy burden of Covid-19 (372, 373), which can be heavier in the communities with higher proportions of elderly (372) and nutritional deficiencies (356). There is no conclusive evidence indicating the protective effect of vitamin D against Covid-19 and hence lowering the related disease burden. Notwithstanding, improvement of vitamin status of the communities with high prevalence rates of suboptimal circulating 25(OH) D concentrations through supplementation and food fortification strategies, apart from its beneficial effects on bone health and metabolism, could boost immunity against microbial pathogens including SARS-CoV-2 (374, 375).

Strategies to combat VDD

VDD is a common micronutrient deficiency (MND) worldwide that warrants immediate action. As a whole, micronutrient supplementation, fortification, as well as nutrition-sensitive interventions are the major strategies to combat MND (376). Though sun is the natural source of vitamin D, some studies have shown that sun exposure may not be efficient against VDD in certain subpopulations (377, 378). Considering the potential harmful effect of sun exposure on skin including malignancies (379), this strategy is less considered to improve vitamin D status at the population level. Both supplementation and food fortification have been employed by most countries with different degrees of success (380).

Vitamin D supplementation

Generally, nutritional supplementation is a feasible intervention for many subgroups, including children, pregnant women and elderly people, especially those residing in institutions (381). There are several guidelines for vitamin D supplementation, some consider bone health and others weight up the pleiotropic effects of vitamin D on general health (382). Vitamin D supplements may be present in a preparation either as a single nutrient, in combination with calcium, or as a constituent of a multi-nutrient compound. Also, different forms of vitamin D may be used for supplementation, including calciferol, calcidiol and calcitriol (383).

Different regimens of vitamin D supplementation may be applied, including daily physiologic dose or high dose with weekly or monthly intervals. However, some evidence indicates that high-dose, long interval strategy may not be as effective as physiologic daily dose despite reaching the desired circulating 25(OH)D concentrations (383), and even may have detrimental effects like increased falls and fractures in elderly (384, 385).

In both supplementation and fortification, desirable a circulating concentration of 25(OH)D is a very critical issue. Though there is no global consensus on the optimal circulating concentration of 25(OH)D and hence the optimal amount of vitamin D supplementation (9, 49), there is a universal consensus that 25(OH)D concentration below 50 nmol/L (20 ng/mL) is not desirable (382). However, this cutoff point has been set based on the calcemic effect of vitamin D (386) and a study using histomorphometric analysis of iliac crest bone biopsies showed that this level of 25(OH)D may not actually be adequate for bone health, and circulating 25(OH)D concentrations above 75 nmol/L (30 ng/mL) may be more appropriate (387).

To determine the desirable vitamin D status, a very critical issue is the method employed to assay 25(OH)D concentration. Several studies have reported remarkable disagreement among different assay methods (388–390). This issue can dramatically affect the judgment of vitamin D status in clinical settings and in community studies, as well. Efforts have been made to standardize the assay methods and/or harmonize the results obtained from different methods (47, 48).

Body weight, and more specifically body fat, is a crucial determinant of response to vitamin D supplementation. It has been shown that the rise of circulating 25(OH)D in supplemented obese children and adults is lower than in their normal weight counterparts (391, 392). Therefore, body weight must be taken into consideration for any recommendation for vitamin D supplementation (382).

Safety issues of vitamin D supplementation have always been a great concern (393, 394). Vitamin D supplementation with daily doses of 400, 4000 and 10,000 IU in adult subjects for three years resulted in no significant adverse effect but hypercalciuria and mild, transient hypercalcemia, both of which were more frequent with higher doses (395). Though some evidence indicated that vitamin D intake as much as 10,000 IU/day is not associated with any adverse effect (386, 395, 396), a number of studies reported a J-shaped association between circulating concentration of 25(OH)D and CVD risk or mortality (397, 398). However, this J-shaped association has been argued by a meta-analytical study (399).

By and large, vitamin D supplementation can be useful for treatment and prevention purposes in a subgroup of population like children and pregnant women. However, when the prevalence of deficiency is high, supplementation may not be an effective strategy, as supplement use is affected by several socioeconomic factors and may not have good coverage at the community level (400–402).

Vitamin D fortification

Wherever the prevalence of a nutrient deficiency is high, affecting various subpopulations, supplementation may not be a cost-effective, wide coverage and, above all, sustainable strategy (403). Under these situations, food fortification may be a preferred approach to tackle VDD at the community level (404).

For an efficient vitamin D fortification program, a safe amount of the fortificant must be determined, as excessive intake of nutrients has been, and continues to be, a great concern of food fortification (405). Though several approaches have been developed (406), the amount of the fortificant must be determined by considering prevalence of VDD in the target community, mean circulating concentrations of 25(OH)D in the subgroups, seasonal fluctuations of 25(OH)D, as well as its target concentration following fortification.

Food fortification can be performed as either mandatory (mass fortification) or discretionary (market driven). In market-driven fortification, usually the minimum amount of the allowed level of the fortificant is added to the vehicle, which is not necessarily consumed by the majority of the population. Nevertheless, consumption of such fortified foods can contribute to the overall intake of the added nutrient. A mass fortification program, on the other hand, is implemented whenever the prevalence of deficiency is high and other strategies, including supplementation, cannot be carried out with a desirable coverage and sustainability. Recent evidence indicates that food fortification is the only environment-friendly strategy to improve vitamin D status of the general population without extra carbon emission to the environment (407).

In a mass fortification program, several important issues must be precisely taken into consideration, among them are selection of a suitable vehicle and the amount of the fortificant. Ideally, several foods must be fortified so that the effectiveness will increase and the risk of toxicity will remarkably decrease. However, fortification of several foods under a mandatory program is absolutely difficult. Therefore, fortification of a staple food, which has a significant contribution to the usual diet of the target population, is highly recommended (408). Obviously, the stability of the added nutrient during processing and storage as well as its bioavailability in the vehicle (final product) must be determined.

We conducted several efficacy trials on vitamin D-fortified food items using various vehicles, including milk and orange juice for targeted food fortification in primary schools (409), Persian yogurt drink (*doogh*) (31) and cooking oil (86) for voluntary fortification, and bakery's wheat flour mostly for a mass (mandatory) fortification program (85). It is noteworthy that the added vitamin D may be lost to some extent during the cooking process, especially in cooking oils (up to 30%) and flat breads made from vitamin D-fortified wheat flour (about 50%) (85, 410). This lost amount of vitamin D during cooking must be taken into consideration on calculation of the amount of the fortificant in the final product. Some evidence indicates interindividual difference in response of circulating 25(OH)D to the consumption of vitamin D-fortified foods according to the different variants of VDRs (112). However, an average increase of 2 nmol/L and 0.7 nmol/L in circulating 25(OH)D for every 100 IUs of vitamin D intake through fortified foods is expected in general adult and children populations, respectively (411, 412).

Concluding remarks

Suboptimal vitamin D status is a global problem with various adverse health impacts (9). A huge body of evidence indicates a role for VDD in increased risk, progression, severity or adverse outcomes of various human diseases (13), including Covid-19 (413). Notwithstanding, the results obtained from clinical trials have been inconsistent to date. Though vitamin D is not a panacea, current evidence strongly indicates that raising the vitamin D status of the general population to a sufficient level can help the burden of several prevalent diseases including cancers, CVD, MS and respiratory infections (Figure 4.1) (12, 414, 415). Even if this effect is minimal, the overall impact can be appreciable due to the heavy burden of those diseases (Figure 4.2). Just for osteoporosis, for instance, it has been estimated that direct annual cost only in the United States, Canada and Europe is US$5000–6500 billion and this will be much more when considering the costs of the resulting disability and loss of productivity (284).

Though well-designed cellular, experimental and clinical trial studies are still needed to elucidate the association of vitamin D with various aspects of human diseases, future studies must also be directed toward prospective community trials to evaluate the effectiveness of improvement of vitamin D status of the general population, through food fortification and supplementation, on overall health of the community and burden of diseases.

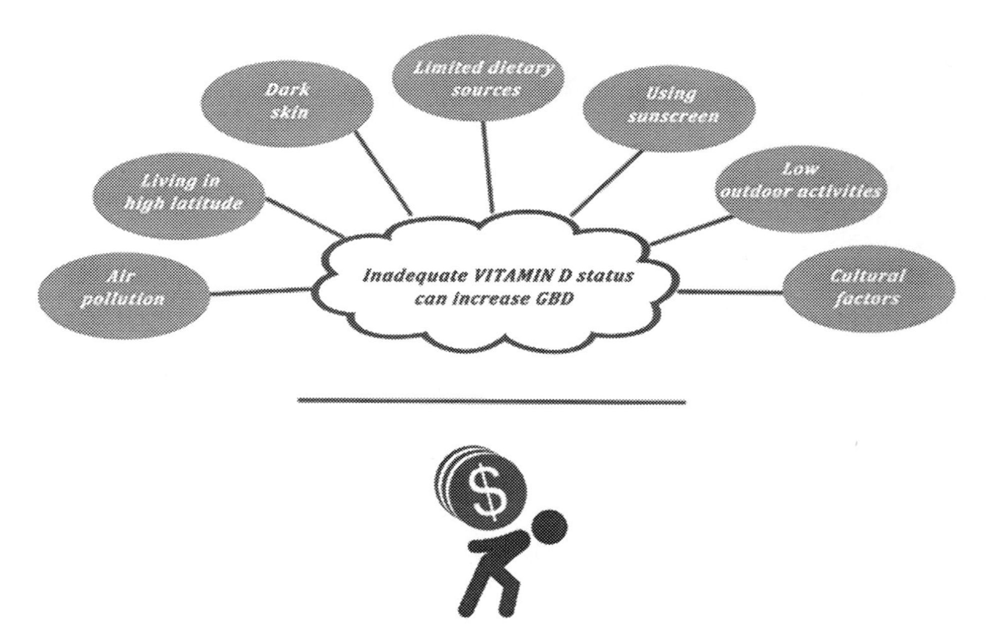

Figure 4.2 *Inadequate vitamin D status of the general population can increase burden of disease.*

References

1. Murray CJ, Lopez AD. Measuring the global burden of disease. New England Journal of Medicine. 2013;369(5):448–57.
2. Kyu HH, Abate D, Abate KH, Abay SM, Abbafati C, Abbasi N, et al. Global, regional, and national disability-adjusted life-years (DALYs) for 359 diseases and injuries and healthy life expectancy (HALE) for 195 countries and territories, 1990–2017: a systematic analysis for the Global Burden of Disease Study 2017. The Lancet. 2018;392(10159):1859–922.
3. Dai H, Alsalhe TA, Chalghaf N, Riccò M, Bragazzi NL, Wu J. The global burden of disease attributable to high body mass index in 195 countries and territories, 1990-2017: an analysis of the Global Burden of Disease Study. PLoS Medicine. 2020;17(7):e1003198.
4. Feigin VL, Roth GA, Naghavi M, Parmar P, Krishnamurthi R, Chugh S, et al. Global burden of stroke and risk factors in 188 countries, during 1990–2013: a systematic analysis for the Global Burden of Disease Study 2013. The Lancet Neurology. 2016;15(9):913–24.
5. Black RE. Global distribution and disease burden related to micronutrient deficiencies. International Nutrition: Achieving Millennium Goals and Beyond. Karger Publishers. 2014;78:21–8.
6. Hoeft B, Weber P, Eggersdorfer M. Micronutrients—A global perspective on intake, health benefits and economics. International Journal of Vitamin and Nutrition Research. 2012;82(5):316–20.
7. Gödecke T, Stein AJ, Qaim M. The global burden of chronic and hidden hunger: Trends and determinants. Global Food Security. 2018;17:21–9.
8. Bhutta ZA, Salam RA, Das JK. Meeting the challenges of micronutrient malnutrition in the developing world. British Medical Bulletin. 2013;106(1):7–17.
9. Amrein K, Scherkl M, Hoffmann M, Neuwersch-Sommeregger S, Köstenberger M, Tmava Berisha A, et al. Vitamin D deficiency 2.0: an update on the current status worldwide. European Journal of Clinical Nutrition. 2020;74(11):1498–513.
10. Palacios C, Gonzalez L. Is vitamin D deficiency a major global public health problem? The Journal of Steroid Biochemistry and Molecular Biology. 2014;144(Pt A):138–45.
11. Van Schoor N, Lips P. Global overview of vitamin D status. Endocrinology and Metabolism Clinics. 2017;46(4):845–70.
12. Holick MF. Health benefits of vitamin D and sunlight: a D-bate. Nature Reviews Endocrinology. 2011;7(2):73–5.
13. Charoenngam N, Holick MF. Immunologic effects of vitamin D on human health and disease. Nutrients. 2020;12(7):2097.
14. Bank W. World Development Report 1993: Investing in Health, Volume 1: The World Bank; 1993.
15. Vos T, Lim SS, Abbafati C, Abbas KM, Abbasi M, Abbasifard M, et al. Global burden of 369 diseases and injuries in 204 countries and territories, 1990–2019: a systematic analysis for the Global Burden of Disease Study 2019. The Lancet. 2020; 396(10258):1204–22.
16. Ahmad FB, Anderson RN. The leading causes of death in the US for 2020. JAMA. 2021.
17. Khan MAB, Hashim MJ, King JK, Govender RD, Mustafa H, Al Kaabi J. Epidemiology of type 2 diabetes–global burden of disease and forecasted trends. Journal of Epidemiology and Global Health. 2020;10(1):107.
18. Our World in Data. https://ourworldindata.org/burden-of-disease#:~:text=At%20a%20global%20level%2C%20in,over%2010%20percent%20from%20injuries. [Internet]. Accessed in 9 May 2021.
19. Noncommunicable Disease. https://www.who.int/news-room/fact-sheets/detail/noncommunicable-diseases#:~:text=Cardiovascular%20diseases%20account%20for%20most,of%20all%20premature%20NCD%20deaths. [Internet]. World Health Organization. Accessed in 9 May 2021.

20. Peters R, Ee N, Peters J, Beckett N, Booth A, Rockwood K, et al. Common risk factors for major noncommunicable disease, a systematic overview of reviews and commentary: the implied potential for targeted risk reduction. Therapeutic Advances in Chronic Disease. 2019;10:2040622319880392.
21. Mertens E, Peñalvo JL. The burden of malnutrition and fatal COVID-19: a global burden of disease analysis. Frontiers in Nutrition. 2021;7:351.
22. Wyper G, Assunção R, Colzani E, Grant I, Haagsma JA, Lagerweij G, et al. Burden of disease methods: a guide to calculate COVID-19 disability-adjusted life years. International Journal of Public Health. 2021;66.
23. Kaseda R, Hosojima M, Sato H, Saito A. Role of megalin and cubilin in the metabolism of vitamin D(3). Therapeutic Apheresis and Dialysis: Official Peer-Reviewed Journal of the International Society for Apheresis, the Japanese Society for Apheresis, the Japanese Society for Dialysis Therapy. 2011;15(Suppl 1):14–7.
24. Haussler MR, Whitfield GK, Kaneko I, Haussler CA, Hsieh D, Hsieh JC, et al. Molecular mechanisms of vitamin D action. Calcified Tissue International. 2013;92(2):77–98
25. Saponaro F, Saba A, Zucchi R. An update on vitamin D metabolism. International Journal of Molecular Sciences. 2020;21(18):6573.
26. Peris P, Filella X, Monegal A, Guañabens N, Foj L, Bonet M, et al. Comparison of total, free and bioavailable 25-OH vitamin D determinations to evaluate its biological activity in healthy adults: the LabOscat study. Osteoporosis International. 2017;28(8):2457–64.
27. Wang Y, Zhu J, DeLuca HF. Where is the vitamin D receptor? Archives of Biochemistry and Biophysics. 2012;523(1):123–33.
28. Nikooyeh B, Anari R, Neyestani TR. Vitamin D, oxidative stress, and diabetes: crossroads for new therapeutic approaches. Diabetes: Elsevier; 2020. p. 385–95.
29. Lieben L, Carmeliet G, Masuyama R. Calcemic actions of vitamin D: effects on the intestine, kidney and bone. Best Practice & Research Clinical Endocrinology & Metabolism. 2011;25(4):561–72.
30. Pludowski P, Holick MF, Pilz S, Wagner CL, Hollis BW, Grant WB, et al. Vitamin D effects on musculoskeletal health, immunity, autoimmunity, cardiovascular disease, cancer, fertility, pregnancy, dementia and mortality—a review of recent evidence. Autoimmunity Reviews. 2013;12(10):976–89.
31. Nikooyeh B, Neyestani TR, Farvid M, Alavi-Majd H, Houshiarrad A, Kalayi A, et al. Daily consumption of vitamin D–or vitamin D+ calcium–fortified yogurt drink improved glycemic control in patients with type 2 diabetes: a randomized clinical trial. The American Journal of Clinical Nutrition. 2011;93(4):764–71.
32. Neyestani TR, Nikooyeh B, Alavi-Majd H, Shariatzadeh N, Kalayi A, Tayebinejad N, et al. Improvement of vitamin D status via daily intake of fortified yogurt drink either with or without extra calcium ameliorates systemic inflammatory biomarkers, including adipokines, in the subjects with type 2 diabetes. The Journal of Clinical Endocrinology & Metabolism. 2012;97(6):2005–11.
33. Nikooyeh B, Neyestani T, Tayebinejad N, Alavi-Majd H, Shariatzadeh N, Kalayi A, et al. Daily intake of vitamin D- or calcium+vitamin D-fortified Persian yogurt drink (doogh) attenuates diabetes-induced oxidative stress: evidence for antioxidative properties of vitamin D. Journal of Human Nutrition and Dietetics. 2014;27:276–83.
34. Motamed S, Nikooyeh B, Kashanian M, Hollis BW, Neyestani TR. Efficacy of two different doses of oral vitamin D supplementation on inflammatory biomarkers and maternal and neonatal outcomes. Maternal & child nutrition. 2019;15(4):e12867.
35. Nikooyeh B, Neyestani TR. Poor vitamin D status increases the risk of anemia in school children: national food and nutrition surveillance. Nutrition. 2018;47:69–74.
36. Adams JS, Hewison M. Update in vitamin D. The Journal of Clinical Endocrinology & Metabolism. 2010;95(2):471–8.
37. Zittermann A, Gummert JF. Nonclassical vitamin D actions. Nutrients. 2010;2(4):408–25.

38. Manson JE, Bassuk SS, Cook NR, Lee I-M, Mora S, Albert CM, et al. Vitamin D, marine n-3 fatty acids, and primary prevention of cardiovascular disease current evidence. Circulation Research. 2020;126(1):112–28.
39. Kouvari M, Panagiotakos D, Chrysohoou C, Yannakoulia M, Georgousopoulou E, Tousoulis D, et al. Dietary vitamin D intake, cardiovascular disease and cardiometabolic risk factors: a sex-based analysis from the ATTICA cohort study. Journal of Human Nutrition and Dietetics. 2020;33(5):708–17.
40. Shah I, Akhtar MK, Hisaindee S, Rauf MA, Sadig M, Ashraf SS. Clinical diagnostic tools for vitamin D assessment. The Journal of Steroid Biochemistry and Molecular Biology. 2018;180:105–17.
41. Tsuprykov O, Chen X, Hocher CF, Skoblo R, Lianghong Y, Hocher B. Why should we measure free 25(OH) vitamin D? The Journal of Steroid Biochemistry and Molecular Biology. 2018;180:87–104.
42. Schwartz JB, Lai J, Lizaola B, Kane L, Markova S, Weyland P, et al. A comparison of measured and calculated free 25(OH) vitamin D levels in clinical populations. The Journal of Clinical Endocrinology and Metabolism. 2014;99(5):1631–7.
43. Thambiah SC, Wong TH, Gupta ED, Radhakrishnan AK, Gun SC, Chembalingam G, et al. Calculation of free and bioavailable vitamin D and its association with bone mineral density in Malaysian women. The Malaysian Journal of Pathology. 2018;40(3):287–94.
44. Schwartz JB, Gallagher JC, Jorde R, Berg V, Walsh J, Eastell R, et al. Determination of free 25(OH)D concentrations and their relationships to Total 25(OH)D in multiple clinical populations. The Journal of Clinical Endocrinology & Metabolism. 2018;103(9):3278–88.
45. Bouillon R. Free or total 25OHD as marker for vitamin D status? Journal of Bone and Mineral Research. 2016;31(6):1124–7.
46. Altieri B, Cavalier E, Bhattoa HP, Pérez-López FR, López-Baena MT, Pérez-Roncero GR, et al. Vitamin D testing: advantages and limits of the current assays. European Journal of Clinical Nutrition. 2020;74(2):231–47.
47. Stokes CS, Lammert F, Volmer DA. Analytical methods for quantification of Vitamin D and implications for research and clinical practice. Anticancer Research. 2018;38(2):1137–44.
48. Nikooyeh B, Samiee SM, Farzami MR, Alavimajd H, Zahedirad M, Kalayi A, et al. Harmonization of serum 25-hydroxycalciferol assay results from high-performance liquid chromatography, enzyme immunoassay, radioimmunoassay, and immunochemiluminescence systems: A multicenter study. Journal of Clinical Laboratory Analysis. 2017;31(6):e22117.
49. Nikooyeh B, Neyestani TR. What is the definition of "vitamin D deficiency" and who is considered "vitamin D deficient"? Urgent need for a national consensus. Nutrition and Food Sciences Research. 2017;4(2):1–5.
50. Bischoff-Ferrari HA. Optimal serum 25-hydroxyvitamin D levels for multiple health outcomes. Advances in Experimental Medicine and Biology. 2014;810:500–25.
51. Holick MF. Vitamin D status: measurement, interpretation, and clinical application. Annals of Epidemiology. 2009;19(2):73–8.
52. Cashman KD. Vitamin D deficiency: defining, prevalence, causes, and strategies of addressing. Calcified Tissue International. 2020;106(1):14–29
53. Oliai Araghi S, van Dijk SC, Ham AC, Brouwer-Brolsma EM, Enneman AW, Sohl E, et al. BMI and body fat mass is inversely associated with vitamin D levels in older individuals. The Journal of Nutrition, Health & Aging. 2015;19(10):980–5.
54. Holick MF. Vitamin D and health: evolution, biologic functions, and recommended dietary intakes for vitamin D. Vitamin D: Springer; 2010. p. 3–33.
55. Collaborators GO. Health effects of overweight and obesity in 195 countries over 25 years. New England Journal of Medicine. 2017;377(1):13–27.

56. Heymsfield SB, Wadden TA. Mechanisms, pathophysiology, and management of obesity. New England Journal of Medicine. 2017;376(3):254–66.
57. Bray GA. Medical consequences of obesity. The Journal of Clinical Endocrinology & Metabolism. 2004;89(6):2583–9.
58. Wamberg L, Christiansen T, Paulsen S, Fisker S, Rask P, Rejnmark L, et al. Expression of vitamin D-metabolizing enzymes in human adipose tissue—the effect of obesity and diet-induced weight loss. International Journal of Obesity. 2013;37(5):651–7.
59. Metrics IfH, Evaluation. The State of US Health: Innovations, Insights, and Recommendations from the Global Burden of Disease Study: IHME; 2013.
60. Ceglia L, Nelson J, Ware J, Alysandratos K-D, Bray GA, Garganta C, et al. Association between body weight and composition and plasma 25-hydroxyvitamin D level in the Diabetes Prevention Program. European Journal of Nutrition. 2017;56(1):161–70.
61. Hannemann A, Thuesen BH, Friedrich N, Völzke H, Steveling A, Ittermann T, et al. Adiposity measures and vitamin D concentrations in Northeast Germany and Denmark. Nutrition & Metabolism. 2015;12(1):1–9.
62. Kremer R, Campbell PP, Reinhardt T, Gilsanz V. Vitamin D status and its relationship to body fat, final height, and peak bone mass in young women. The Journal of Clinical Endocrinology & Metabolism. 2009;94(1):67–73.
63. Vimaleswaran KS, Berry DJ, Lu C, Tikkanen E, Pilz S, Hiraki LT, et al. Causal relationship between obesity and vitamin D status: bi-directional Mendelian randomization analysis of multiple cohorts. PLoS Medicine. 2013;10(2):e1001383.
64. Yao Y, Zhu L, He L, Duan Y, Liang W, Nie Z, et al. A meta-analysis of the relationship between vitamin D deficiency and obesity. International Journal of Clinical and Experimental Medicine. 2015;8(9):14977.
65. Compston JE, Vedi S, Ledger JE, Webb A, Gazet J-C, Pilkington T. Vitamin D status and bone histomorphometry in gross obesity. The American Journal of Clinical Nutrition. 1981;34(11):2359–63.
66. Wortsman J, Matsuoka LY, Chen TC, Lu Z, Holick MF. Decreased bioavailability of vitamin D in obesity. The American Journal of Clinical Nutrition. 2000;72(3):690–3.
67. Hyppönen E, Power C. Hypovitaminosis D in British adults at age 45 y: nationwide cohort study of dietary and lifestyle predictors. The American Journal of Clinical Nutrition. 2007;85(3):860–8.
68. Stein E, Strain G, Sinha N, Ortiz D, Pomp A, Dakin G, et al. Vitamin D insufficiency prior to bariatric surgery: risk factors and a pilot treatment study. Clinical Endocrinology. 2009;71(2):176–83.
69. Bell NH, Epstein S, Greene A, Shary J, Oexmann MJ, Shaw S. Evidence for alteration of the vitamin D-endocrine system in obese subjects. The Journal of Clinical Investigation. 1985;76(1):370–3.
70. Parikh SJ, Edelman M, Uwaifo GI, Freedman RJ, Semega-Janneh M, Reynolds J, et al. The relationship between obesity and serum 1, 25-dihydroxy vitamin D concentrations in healthy adults. The Journal of Clinical Endocrinology & Metabolism. 2004;89(3):1196–9.
71. Lagunova Z, Porojnicu AC, Vieth R, Lindberg FA, Hexeberg S, Moan J. Serum 25-hydroxyvitamin D is a predictor of serum 1, 25-dihydroxyvitamin D in overweight and obese patients. The Journal of Nutrition. 2011;141(1):112–7.
72. Pourshahidi LK. Vitamin D and obesity: current perspectives and future directions. The Proceedings of the Nutrition Society. 2015;74(2):115–24.
73. Liel Y, Ulmer E, Shary J, Hollis BW, Bell NH. Low circulating vitamin D in obesity. Calcified Tissue International. 1988;43(4):199–201.
74. Hjelmesæth J, Hofsø D, Aasheim ET, Jenssen T, Moan J, Hager H, et al. Parathyroid hormone, but not vitamin D, is associated with the metabolic syndrome in morbidly obese women and men: a cross-sectional study. Cardiovascular Diabetology. 2009;8(1):1–7.

75. Wong KE, Kong J, Zhang W, Szeto FL, Ye H, Deb DK, et al. Targeted expression of human vitamin D receptor in adipocytes decreases energy expenditure and induces obesity in mice. The Journal of Biological Chemistry. 2011;286(39):33804–10.
76. Kong J, Chen Y, Zhu G, Zhao Q, Li YC. 1, 25-Dihydroxyvitamin D3 upregulates leptin expression in mouse adipose tissue. Journal of Endocrinology. 2013;216(2):265–71.
77. Abbas MA. Physiological functions of Vitamin D in adipose tissue. The Journal of Steroid Biochemistry and Molecular Biology. 2017;165:369–81.
78. Ghavamzadeh S, Mobasseri M, Mahdavi R. The effect of vitamin D supplementation on adiposity, blood glycated hemoglobin, serum leptin and tumor necrosis factor-α in type 2 diabetic patients. International Journal of Preventive Medicine. 2014;5(9):1091.
79. Maggi S, Siviero P, Brocco E, Albertin M, Romanato G, Crepaldi G. Vitamin D deficiency, serum leptin and osteoprotegerin levels in older diabetic patients: an input to new research avenues. Acta Diabetologica. 2014;51(3):461–9.
80. Gallagher JC, Sai A, Templin T, Smith L. Dose response to vitamin D supplementation in postmenopausal women: a randomized trial. Annals of Internal Medicine. 2012;156(6):425–37.
81. Koszowska AU, Nowak J, Dittfeld A, Brończyk-Puzoń A, Kulpok A, Zubelewicz-Szkodzińska B. Obesity, adipose tissue function and the role of vitamin D. Central-European Journal of Immunology. 2014;39(2):260.
82. Dinca M, Serban M-C, Sahebkar A, Mikhailidis DP, Toth PP, Martin SS, et al. Does vitamin D supplementation alter plasma adipokines concentrations? A systematic review and meta-analysis of randomized controlled trials. Pharmacological Research. 2016;107:360–71.
83. Lim SS, Vos T, Flaxman AD, Danaei G, Shibuya K, Adair-Rohani H, et al. A comparative risk assessment of burden of disease and injury attributable to 67 risk factors and risk factor clusters in 21 regions, 1990–2010: a systematic analysis for the global burden of disease study 2010. The Lancet. 2012;380(9859):2224–60.
84. Organization WH. Global Health Risks: Mortality and Burden of Disease Attributable to Selected Major Risks: World Health Organization; 2009.
85. Nikooyeh B, Neyestani TR, Zahedirad M, Mohammadi M, Hosseini SH, Abdollahi Z, et al. Vitamin D-fortified bread is as effective as supplement in improving vitamin D status: a randomized clinical trial. The Journal of Clinical Endocrinology & Metabolism. 2016;101(6):2511–9.
86. Nikooyeh B, Zargaraan A, Kalayi A, Shariatzadeh N, Zahedirad M, Jamali A, et al. Vitamin D-fortified cooking oil is an effective way to improve vitamin D status: an institutional efficacy trial. European Journal of Nutrition. 2020;59(6):2547–55.
87. Mallard SR, Howe AS, Houghton LA. Vitamin D status and weight loss: a systematic review and meta-analysis of randomized and nonrandomized controlled weight-loss trials. The American Journal of Clinical Nutrition. 2016;104(4):1151–9.
88. Pathak K, Soares MJ, Calton EK, Zhao Y, Hallett J. Vitamin D supplementation and body weight status: a systematic review and meta-analysis of randomized controlled trials. Obesity Reviews. 2014;15(6):528–37.
89. Duan L, Han L, Liu Q, Zhao Y, Wang L, Wang Y. Effects of Vitamin D supplementation on general and central obesity: results from 20 randomized controlled trials involving apparently healthy populations. Annals of Nutrition & Metabolism. 2020;76(3): 153–64.
90. Williams R, Novick M, Lehman E. Prevalence of Hypovitaminosis D and its association with comorbidities of childhood obesity. The Permanente Journal. 2014;18(4):32.
91. Al Zarooni AAR, Al Marzouqi FI, Al Darmaki SH, Prinsloo EAM, Nagelkerke N. Prevalence of Vitamin D deficiency and associated comorbidities among Abu Dhabi Emirates population. BMC Research Notes. 2019;12(1):1–6.
92. Guasch A, Bulló M, Rabassa A, Bonada A, Del Castillo D, Sabench F, et al. Plasma vitamin D and parathormone are associated with obesity and atherogenic dyslipidemia: a cross-sectional study. Cardiovascular Diabetology. 2012;11(1):1–11.

93. Wamberg L, Pedersen SB, Rejnmark L, Richelsen B. Causes of vitamin D deficiency and effect of vitamin D supplementation on metabolic complications in obesity: a review. Current Obesity Reports. 2015;4(4):429–40.
94. Siddiqui AA, Siddiqui SA, Ahmad S, Siddiqui S, Ahsan I, Sahu K. Diabetes: Mechanism, pathophysiology and management-A review. International Journal of Drug Development and Research. 2013;5(2):1–23.
95. Sapra A, Bhandari P. Diabetes mellitus. 2022. In: StatPearls [Internet]. Treasure Island (FL): StatPearls Publishing; PMID: 31855345.
96. Klein BE, Klein R, Moss SE, Cruickshanks KJ. Parental history of diabetes in a population-based study. Diabetes Care. 1996;19(8):827–30.
97. Unger RH, Orci L. Paracrinology of islets and the paracrinopathy of diabetes. Proceedings of the National Academy of Sciences. 2010;107(37):16009–12.
98. Orasanu G, Plutzky J. The pathologic continuum of diabetic vascular disease. Journal of the American College of Cardiology. 2009;53(5S):S35–S42.
99. Zhang Z-Y, Miao L-F, Qian L-L, Wang N, Qi M-M, Zhang Y-M, et al. Molecular mechanisms of glucose fluctuations on diabetic complications. Frontiers in Endocrinology. 2019;10(640).
100. Berridge MJ. Vitamin D deficiency and diabetes. Biochemical Journal. 2017; 474(8):1321–32.
101. Dandona P, Aljada A, Bandyopadhyay A. Inflammation: the link between insulin resistance, obesity and diabetes. Trends in Immunology. 2004;25(1):4–7.
102. Pittas AG, Nelson J, Mitri J, Hillmann W, Garganta C, Nathan DM, et al. Plasma 25-Hydroxyvitamin D and progression to diabetes in patients at risk for diabetes: an ancillary analysis in the Diabetes Prevention Program. Diabetes Care. 2012;35(3):565–73.
103. González-Molero I, Rojo-Martínez G, Morcillo S, Gutiérrez-Repiso C, Rubio-Martín E, Almaraz MC, et al. Vitamin D and incidence of diabetes: a prospective cohort study. Clinical Nutrition. 2012;31(4):571–3.
104. Gupta AK, Brashear MM, Johnson WD. Prediabetes and prehypertension in healthy adults are associated with low Vitamin D levels. Diabetes Care. 2011;34(3):658–60.
105. Littorin B, Blom P, Schölin A, Arnqvist H, Blohme G, Bolinder J, et al. Lower levels of plasma 25-Hydroxyvitamin D among young adults at diagnosis of autoimmune type 1 diabetes compared with control subjects: results from the nationwide Diabetes Incidence Study in Sweden (DISS). Diabetologia. 2006;49(12):2847–52.
106. Oh J, Weng S, Felton SK, Bhandare S, Riek A, Butler B, et al. 1, 25 (OH) 2 vitamin D inhibits foam cell formation and suppresses macrophage cholesterol uptake in patients with type 2 diabetes mellitus. Circulation. 2009;120(8):687–98.
107. Oguntibeju OO. Type 2 diabetes mellitus, oxidative stress and inflammation: examining the links. International Journal of Physiology, Pathophysiology and Pharmacology. 2019;11(3):45–63.
108. Nikooyeh B, Hollis BW, Neyestani TR. The effect of daily intake of Vitamin D-fortified yogurt drink, with and without added calcium, on serum adiponectin and sirtuins 1 and 6 in adult subjects with type 2 diabetes. Nutrition & Diabetes. 2021;11(1):26.
109. Hu Z, Chen J, Sun X, Wang L, Wang A. Efficacy of Vitamin D supplementation on glycemic control in type 2 diabetes patients: A meta-analysis of interventional studies. Medicine (Baltimore). 2019;98(14):e14970.
110. Pramono A, Jocken JW, Blaak EE, van Baak MA. The effect of vitamin D supplementation on insulin sensitivity: a systematic review and meta-analysis. Diabetes Care. 2020;43(7):1659–69.
111. Mirhosseini N, Vatanparast H, Mazidi M, Kimball SM. The effect of improved serum 25-Hydroxyvitamin D status on glycemic control in diabetic patients: a meta-analysis. The Journal of Clinical Endocrinology & Metabolism. 2017;102(9):3097–110.
112. Neyestani TR, Djazayery A, Shab-Bidar S, Eshraghian MR, Kalayi A, Shariátzadeh N, et al. Vitamin D Receptor Fok-I polymorphism modulates diabetic host response to vitamin D intake: need for a nutrigenetic approach. Diabetes Care. 2013;36(3):550–6.

113. Adorini L, Amuchastegui S, Corsiero E, Laverny G, Le Meur T, Penna G. Vitamin D receptor agonists as anti-inflammatory agents. Expert Review of Clinical Immunology. 2007;3(4):477–89.
114. Chagas CEA, Borges MC, Martini LA, Rogero MM. Focus on vitamin D, inflammation and type 2 diabetes. Nutrients. 2012;4(1):52–67.
115. John AN, Jiang F-X. An overview of type 2 diabetes and importance of vitamin D3-vitamin D receptor interaction in pancreatic β-cells. Journal of Diabetes and Its Complications. 2018;32(4):429–43.
116. Pilz S, Kienreich K, Rutters F, de Jongh R, van Ballegooijen AJ, Grübler M, et al. Role of vitamin D in the development of insulin resistance and type 2 diabetes. Current Diabetes Reports. 2013;13(2):261–70.
117. Sung C-C, Liao M-T, Lu K-C, Wu C-C. Role of vitamin D in insulin resistance. Journal of Biomedicine and Biotechnology. 2012;2012:634195.
118. Schöttker B, Herder C, Rothenbacher D, Perna L, Müller H, Brenner H. Serum 25-hydroxyvitamin D levels and incident diabetes mellitus type 2: a competing risk analysis in a large population-based cohort of older adults. European Journal of Epidemiology. 2013;28(3):267–75.
119. Gagnon C, Lu ZX, Magliano DJ, Dunstan DW, Shaw JE, Zimmet PZ, et al. Serum 25-hydroxyvitamin D, calcium intake, and risk of type 2 diabetes after 5 years: results from a national, population-based prospective study (the Australian Diabetes, Obesity and Lifestyle study). Diabetes Care. 2011;34(5):1133–8.
120. Chiu KC, Chu A, Go VLW, Saad MF. Hypovitaminosis D is associated with insulin resistance and β cell dysfunction. The American Journal of Clinical Nutrition. 2004;79(5):820–5.
121. Bland R, Markovic D, Hills CE, Hughes SV, Chan SL, Squires PE, et al. Expression of 25-hydroxyvitamin D3-1α-hydroxylase in pancreatic islets. The Journal of Steroid Biochemistry and Molecular Biology. 2004;89:121–5.
122. Maestro B, Dávila N, Carranza MC, Calle C. Identification of a Vitamin D response element in the human insulin receptor gene promoter. The Journal of Steroid Biochemistry and Molecular Biology. 2003;84(2–3):223–30.
123. Alvarez JA, Ashraf AP, Hunter GR, Gower BA. Serum 25-hydroxyvitamin D and parathyroid hormone are independent determinants of whole-body insulin sensitivity in women and may contribute to lower insulin sensitivity in African Americans. The American Journal of Clinical Nutrition. 2010;92(6):1344–9.
124. Xue B, Greenberg AG, Kraemer FB, Zemel MB. Mechanism of intracellular calcium ([Ca^2+] i) inhibition of lipolysis in human adipocytes. The FASEB Journal. 2001;15(13):2527–9.
125. Zhang Y, Tan H, Tang J, Li J, Chong W, Hai Y, et al. Effects of vitamin D supplementation on prevention of type 2 diabetes in patients with prediabetes: a systematic review and meta-analysis. Diabetes Care. 2020;43(7):1650–8.
126. Khan MAB, Hashim MJ, King JK, Govender RD, Mustafa H, Al Kaabi J. Epidemiology of type 2 diabetes - global burden of disease and forecasted trends. Journal of Epidemiology and Global Health. 2020;10(1):107–11.
127. Federation ID. IDF diabetes atlas http://www.diabetesatlas.org.2021 [cited 29 September 2021].
128. Arokiasamy P, Salvi S, Selvamani Y. Global burden of diabetes mellitus. Handbook of Global Health: Springer; 2020, pp. 1–44.
129. Sauder KA, Ritchie ND. Reducing intergenerational obesity and diabetes risk. Diabetologia. 2021:1–10.
130. Gaziano T, Reddy KS, Paccaud F, et al. Cardiovascular disease. In: Jamison DT, Breman JG, Measham AR, et al. (eds.). Disease Control Priorities in Developing Countries. 2nd edition. The International Bank for Reconstruction and Development/The World Bank; 2006. Chapter 33. Available from: https://www.ncbi.nlm.nih.gov/books/NBK11767/ Co-published by Oxford University Press, New York.

131. Roth GA, Mensah GA, Johnson CO, Addolorato G, Ammirati E, Baddour LM, et al. Global burden of cardiovascular diseases and risk factors, 1990–2019: update from the GBD 2019 study. Journal of the American College of Cardiology. 2020;76(25):2982–3021.
132. Altieri B, Muscogiuri G, Barrea L, Mathieu C, Vallone CV, Mascitelli L, et al. Does vitamin D play a role in autoimmune endocrine disorders? A proof of concept. Reviews in Endocrine & Metabolic Disorders. 2017;18(3):335–46.
133. Nibbelink KA, Tishkoff DX, Hershey SD, Rahman A, Simpson RU. 1,25(OH)2-vitamin D3 actions on cell proliferation, size, gene expression, and receptor localization, in the HL-1 cardiac myocyte. The Journal of Steroid Biochemistry and Molecular Biology. 2007;103(3):533–7.
134. Tishkoff DX, Nibbelink KA, Holmberg KH, Dandu L, Simpson RU. Functional vitamin D receptor (VDR) in the t-tubules of cardiac myocytes: VDR knockout cardiomyocyte contractility. Endocrinology. 2008;149(2):558–64.
135. Latic N, Erben RG. Vitamin D and cardiovascular disease, with emphasis on hypertension, atherosclerosis, and heart failure. International Journal of Molecular Sciences. 2020;21(18).
136. Rostand SG. Ultraviolet light may contribute to geographic and racial blood pressure differences. Hypertension. 1997;30(2 Pt 1):150–6.
137. Kunutsor SK, Apekey TA, Steur M. Vitamin D and risk of future hypertension: meta-analysis of 283,537 participants. European Journal of Epidemiology. 2013;28(3):205–21.
138. Ke L, Graubard BI, Albanes D, Fraser DR, Weinstein SJ, Virtamo J, et al. Hypertension, pulse, and other cardiovascular risk factors and vitamin D status in Finnish men. American Journal of Hypertension. 2013;26(8):951–6.
139. Forman JP, Giovannucci E, Holmes MD, Bischoff-Ferrari HA, Tworoger SS, Willett WC, et al. Plasma 25-hydroxyvitamin D levels and risk of incident hypertension. Hypertension. 2007;49(5):1063–9.
140. Scragg R, Sowers M, Bell C. Serum 25-hydroxyvitamin D, ethnicity, and blood pressure in the Third National Health and Nutrition Examination Survey. American Journal of Hypertension. 2007;20(7):713–9.
141. Judd SE, Nanes MS, Ziegler TR, Wilson PW, Tangpricha V. Optimal vitamin D status attenuates the age-associated increase in systolic blood pressure in white Americans: results from the third National Health and Nutrition Examination Survey. The American Journal of Clinical Nutrition. 2008;87(1):136–41.
142. Zhang D, Cheng C, Wang Y, Sun H, Yu S, Xue Y, et al. Effect of vitamin d on blood pressure and hypertension in the general population: an update meta-analysis of cohort studies and randomized controlled trials. Preventing Chronic Disease. 2020;17:E03.
143. Qi D, Nie XL, Wu S, Cai J. Vitamin D and hypertension: prospective study and meta-analysis. PloS One. 2017;12(3):e0174298.
144. Lupton JR, Faridi KF, Martin SS, Sharma S, Kulkarni K, Jones SR, et al. Deficient serum 25-hydroxyvitamin D is associated with an atherogenic lipid profile: the very large database of lipids (VLDL-3) study. Journal of Clinical Lipidology. 2016;10(1):72–81.e1.
145. Nikooyeh B, Abdollahi Z, Hajifaraji M, Alavi-Majd H, Salehi F, Yarparvar AH, et al. Vitamin D status and cardiometabolic risk factors across latitudinal gradient in Iranian adults: National food and nutrition surveillance. Nutrition and Health. 2017;23(2):87–94.
146. Nikooyeh B, Abdollahi Z, Hajifaraji M, Alavi-Majd H, Salehi F, Yarparvar AH, et al. Healthy changes in some cardiometabolic risk factors accompany the higher summertime serum 25-hydroxyvitamin D concentrations in Iranian children: National Food and Nutrition Surveillance. Public Health Nutrition. 2018;21(11):2013–21.
147. Nikooyeh B, Neyestani TR. Contribution of vitamin D status as a determinant of cardiometabolic risk factors: a structural equation model, national food and nutrition surveillance. BMC Public Health. 2021;21(1):1–7.
148. Wang Y, Si S, Liu J, Wang Z, Jia H, Feng K, et al. The associations of serum lipids with vitamin D status. PLoS One. 2016;11(10):e0165157.

149. Li Y, Tong CH, Rowland CM, Radcliff J, Bare LA, McPhaul MJ, et al. Association of changes in lipid levels with changes in vitamin D levels in a real-world setting. Scientific Reports. 2021;11(1):1–7.
150. Melamed ML, Muntner P, Michos ED, Uribarri J, Weber C, Sharma J, et al. Serum 25-hydroxyvitamin D levels and the prevalence of peripheral arterial disease: results from NHANES 2001 to 2004. Arteriosclerosis, Thrombosis, and Vascular Biology. 2008;28(6):1179–85.
151. Fahrleitner A, Dobnig H, Obernosterer A, Pilger E, Leb G, Weber K, et al. Vitamin D deficiency and secondary hyperparathyroidism are common complications in patients with peripheral arterial disease. Journal of General Internal Medicine. 2002;17(9):663–9.
152. Fahrleitner-Pammer A, Obernosterer A, Pilger E, Dobnig H, Dimai HP, Leb G, et al. Hypovitaminosis D, impaired bone turnover and low bone mass are common in patients with peripheral arterial disease. Osteoporosis International: A Journal Established as Result of Cooperation between the European Foundation for Osteoporosis and the National Osteoporosis Foundation of the USA. 2005;16(3):319–24.
153. Kim DH, Sabour S, Sagar UN, Adams S, Whellan DJ. Prevalence of hypovitaminosis D in cardiovascular diseases (from the National Health and nutrition examination survey 2001 to 2004). The American Journal of Cardiology. 2008;102(11):1540–4.
154. Zittermann A, Schleithoff SS, Koerfer R. Vitamin D insufficiency in congestive heart failure: why and what to do about it? Heart Failure Reviews. 2006;11(1):25–33.
155. Gotsman I, Shauer A, Zwas DR, Hellman Y, Keren A, Lotan C, et al. Vitamin D deficiency is a predictor of reduced survival in patients with heart failure; vitamin D supplementation improves outcome. European Journal of Heart Failure. 2012;14(4):357–66.
156. Brøndum-Jacobsen P, Benn M, Jensen GB, Nordestgaard BG. 25-hydroxyvitamin d levels and risk of ischemic heart disease, myocardial infarction, and early death: population-based study and meta-analyses of 18 and 17 studies. Arteriosclerosis, Thrombosis, and Vascular Biology. 2012;32(11):2794–802.
157. Cubbon RM, Lowry JE, Drozd M, Hall M, Gierula J, Paton MF, et al. Vitamin D deficiency is an independent predictor of mortality in patients with chronic heart failure. European Journal of Nutrition. 2019;58(6):2535–43.
158. Melamed ML, Michos ED, Post W, Astor B. 25-hydroxyvitamin D levels and the risk of mortality in the general population. Archives of Internal Medicine. 2008; 168(15):1629–37.
159. Pilz S, Tomaschitz A, März W, Drechsler C, Ritz E, Zittermann A, et al. Vitamin D, cardiovascular disease and mortality. Clinical Endocrinology. 2011;75(5):575–84.
160. Zhang R, Li B, Gao X, Tian R, Pan Y, Jiang Y, et al. Serum 25-hydroxyvitamin D and the risk of cardiovascular disease: dose-response meta-analysis of prospective studies. The American Journal of Clinical Nutrition. 2017;105(4):810–9.
161. Dror Y, Giveon SM, Hoshen M, Feldhamer I, Balicer RD, Feldman BS. Vitamin D levels for preventing acute coronary syndrome and mortality: evidence of a nonlinear association. The Journal of Clinical Endocrinology and Metabolism. 2013;98(5):2160–7.
162. Durup D, Jorgensen HL, Christensen J, Tjonneland A, Olsen A, Halkjaer J, et al. A reverse j-shaped association between serum 25-hydroxyvitamin D and cardiovascular disease mortality: The CopD study. The Journal of Clinical Endocrinology and Metabolism. 2015;100(6):2339–46.
163. Grant WB, Karras SN, Bischoff-Ferrari HA, Annweiler C, Boucher BJ, Juzeniene A, et al. Do studies reporting 'U'-shaped serum 25-hydroxyvitamin D-health outcome relationships reflect adverse effects? Dermatoendocrinol. 2016;8(1):e1187349–e.
164. Zittermann A. Vitamin D status, supplementation and cardiovascular disease. Anticancer Research. 2018;38(2):1179–86.
165. Zittermann A, Trummer C, Theiler-Schwetz V, Lerchbaum E, März W, Pilz S. Vitamin D and cardiovascular disease: an updated narrative review. International Journal of Molecular Sciences. 2021;22(6).

166. Brøndum-Jacobsen P, Benn M, Afzal S, Nordestgaard BG. No evidence that genetically reduced 25-hydroxyvitamin D is associated with increased risk of ischaemic heart disease or myocardial infarction: a Mendelian randomization study. International Journal of Epidemiology. 2015;44(2):651–61.
167. Leong A, Rehman W, Dastani Z, Greenwood C, Timpson N, Langsetmo L, et al. The causal effect of vitamin D binding protein (DBP) levels on calcemic and cardiometabolic diseases: a Mendelian randomization study. PLoS Medicine. 2014;11(10):e1001751.
168. Afzal S, Brondum-Jacobsen P, Bojesen SE, Nordestgaard BG. Genetically low vitamin D concentrations and increased mortality: Mendelian randomisation analysis in three large cohorts. BMJ (Clinical Research ed). 2014;349:g6330.
169. Vimaleswaran KS, Cavadino A, Berry DJ, LifeLines Cohort Study i, Jorde R, Dieffenbach AK, et al. Association of vitamin D status with arterial blood pressure and hypertension risk: a Mendelian randomisation study. The Lancet Diabetes & Endocrinology. 2014;2(9):719–29.
170. Manousaki D, Mokry LE, Ross S, Goltzman D, Richards JB. Mendelian randomization studies do not support a role for Vitamin D in coronary artery disease. Circulation Cardiovascular Genetics. 2016;9(4):349–56.
171. Alizadeh S, Djafarian K, Alizadeh H, Mohseni R, Shab-Bidar S. Common variants of Vitamin D receptor gene polymorphisms and susceptibility to coronary artery disease: a systematic review and meta-analysis. Journal of Nutrigenetics and Nutrigenomics. 2017;10(1–2):9–18.
172. Liu Z, Liu L, Chen X, He W, Yu X. Associations study of vitamin D receptor gene polymorphisms with diabetic microvascular complications: a meta-analysis. Gene. 2014;546(1):6–10.
173. Lu S, Guo S, Hu F, Guo Y, Yan L, Ma W, et al. The associations between the polymorphisms of Vitamin D receptor and coronary artery disease: a systematic review and meta-analysis. Medicine. 2016;95(21):e3467.
174. Shen H, Bielak LF, Ferguson JF, Streeten EA, Yerges-Armstrong LM, Liu J, et al. Association of the vitamin D metabolism gene CYP24A1 with coronary artery calcification. Arteriosclerosis, Thrombosis, and Vascular Biology. 2010;30(12):2648–54.
175. Song N, Yang S, Wang YY, Tang SQ, Zhu YQ, Dai Q, et al. The impact of Vitamin D receptor gene polymorphisms on the susceptibility of diabetic vascular complications: a meta-analysis. Genetic Testing and Molecular Biomarkers. 2019;23(8):533–56.
176. Zhang D, Wang L, Zhang R, Li S. Association of Vitamin D receptor gene polymorphisms and the risk of multiple sclerosis: a meta analysis. Archives of Medical Research. 2019;50(6):350–61.
177. Mirhosseini N, Rainsbury J, Kimball SM. Vitamin D supplementation, serum 25(OH) D concentrations and cardiovascular disease risk factors: a systematic review and meta-analysis. Frontiers in Cardiovascular Medicine. 2018;5:87.
178. Dibaba DT. Effect of vitamin D supplementation on serum lipid profiles: a systematic review and meta-analysis. Nutrition Reviews. 2019;77(12):890–902.
179. Nabel EG. Cardiovascular disease. New England Journal of Medicine. 2003;349(1): 60–72.
180. Wang L, Manson JE, Song Y, Sesso HD. Systematic review: vitamin D and calcium supplementation in prevention of cardiovascular events. Annals of Internal Medicine. 2010;152(5):315–23.
181. Scragg R, Stewart AW, Waayer D, Lawes CM, Toop L, Sluyter J, et al. Effect of monthly high-dose vitamin D supplementation on cardiovascular disease in the vitamin D assessment study: a randomized clinical trial. JAMA Cardiology. 2017;2(6):608–16.
182. Thacher TD, Fischer PR, Tebben PJ, Singh RJ, Cha SS, Maxson JA, et al. Increasing incidence of nutritional rickets: a population-based study in Olmsted County, Minnesota. Mayo Clinic Proceedings. 2013;88(2):176–83.
183. Mahmoud AO, Ahmed AY, Aly H-TAM. The prevalence of active nutritional rickets in Egyptian infants in Cairo. Egyptian Pediatric Association Gazette. 2016;64(3):105–10.

184. El Demellawy D, Davila J, Shaw A, Nasr Y. Brief review on metabolic bone disease. Academic Forensic Pathology. 2018;8(3):611–40.
185. Fausto-Sterling A. The bare bones of race. Social Studies of Science. 2008;38(5):657–94.
186. Bonjour J-P, Chevalley T, Ferrari S, Rizzoli R. The importance and relevance of peak bone mass in the prevalence of osteoporosis. Salud publica de Mexico. 2009;51:s5–s17.
187. Clarke BL, Khosla S. Physiology of bone loss. Radiologic Clinics. 2010;48(3):483–95.
188. Palacios C. The role of nutrients in bone health, from A to Z. Critical Reviews in Food Science and Nutrition. 2006;46(8):621–8.
189. Ebeling PR, Eisman JA. Vitamin D and osteoporosis. Vitamin D: Elsevier; 2018. p. 203–20.
190. Cauley JA. Public health impact of osteoporosis. Journals of Gerontology Series A: Biomedical Sciences and Medical Sciences. 2013;68(10):1243–51.
191. Fischer V, Haffner-Luntzer M, Prystaz K, vom Scheidt A, Busse B, Schinke T, et al. Calcium and vitamin-D deficiency marginally impairs fracture healing but aggravates posttraumatic bone loss in osteoporotic mice. Scientific Reports. 2017;7(1):7223.
192. Turner AG, Anderson PH, Morris HA. Vitamin D and bone health. Scandinavian Journal of Clinical and Laboratory Investigation. 2012;72(sup243):65–72.
193. Thanapluetiwong S, Chewcharat A, Takkavatakarn K, Praditpornsilpa K, Eiam-Ong S, Susantitaphong P. Vitamin D supplement on prevention of fall and fracture: a meta-analysis of randomized controlled trials. Medicine (Baltimore). 2020;99(34):e21506.
194. Zhao J-G, Zeng X-T, Wang J, Liu L. Association between calcium or vitamin d supplementation and fracture incidence in community-dwelling older adults: a systematic review and meta-analysis. JAMA. 2017;318(24):2466–82.
195. Avenell A, Gillespie WJ, Gillespie LD, O'Connell D. Vitamin D and vitamin D analogues for preventing fractures associated with involutional and post-menopausal osteoporosis. Cochrane Database Syst Rev. 2009 Apr 15;(2):CD000227. doi: 10.1002/14651858.CD000227.pub3. Update in: Cochrane Database Syst Rev. 2014;4:CD000227. PMID: 19370554.
196. Jackson RD, LaCroix AZ, Gass M, Wallace RB, Robbins J, Lewis CE, et al. Calcium plus vitamin D supplementation and the risk of fractures. New England Journal of Medicine. 2006;354(7):669–83.
197. Burt LA, Billington EO, Rose MS, Raymond DA, Hanley DA, Boyd SK. Effect of high-dose vitamin D supplementation on volumetric bone density and bone strength: a randomized clinical trial. JAMA. 2019;322(8):736–45.
198. De Martinis M, Allegra A, Sirufo MM, Tonacci A, Pioggia G, Raggiunti M, et al. Vitamin D deficiency, osteoporosis and effect on autoimmune diseases and hematopoiesis: a review. International Journal of Molecular Sciences. 2021;22(16):8855.
199. Harvey N, Dennison E, Cooper C. Epidemiology of osteoporotic fractures. Primer on the Metabolic Bone Diseases and Disorders of Mineral Metabolism. 2008;6:244–8.
200. Johnell O, Kanis J. An estimate of the worldwide prevalence and disability associated with osteoporotic fractures. Osteoporosis International. 2006;17(12):1726–33.
201. Hernlund E, Svedbom A, Ivergård M, Compston J, Cooper C, Stenmark J, et al. Osteoporosis in the European Union: medical management, epidemiology and economic burden. Archives of Osteoporosis. 2013;8(1):1–115.
202. Foundation NO. Annual Report https://cdn.nof.org/wp-content/uploads/Annual-Report-2019.pdf [cited 9 October 2021].
203. Drake MT, Cremers S, Russell RG, Bilezikian JP. Drugs for the treatment of metabolic bone diseases. British Journal of Clinical Pharmacology. 2019;85(6):1049.
204. Patil R, Kolu P, Raitanen J, Valvanne J, Kannus P, Karinkanta S, et al. Cost-effectiveness of vitamin D supplementation and exercise in preventing injurious falls among older home-dwelling women: findings from an RCT. Osteoporosis International. 2016; 27(1):193–201.
205. Weaver CM, Bischoff-Ferrari HA, Shanahan CJ. Cost-benefit analysis of calcium and vitamin D supplements. Archives of Osteoporosis. 2019;14(1):50.
206. Floreskul V, Juma FZ, Daniel AB, Zamir I, Rawdin A, Stevenson M, et al. Cost-effectiveness of Vitamin D supplementation in pregnant women and young children in preventing rickets: a modeling study. Frontiers in Public Health. 2020;8:439.

207. Julian MK. Autoimmune disease: Cost-effective care. Nursing Management. 2014; 45(11):24–9.
208. Anaya JM, Restrepo-Jimenez P, Ramirez-Santana C. The autoimmune ecology: an update. Current Opinion in Rheumatology. 2018;30(4):350–60.
209. Mathieu C. Vitamin D and diabetes: Where do we stand? Diabetes Research and Clinical Practice. 2015;108(2):201–9.
210. Bizzaro G, Shoenfeld Y. Vitamin D and autoimmune thyroid diseases: facts and unresolved questions. Immunologic Research. 2015;61(1–2):46–52.
211. Czaja AJ, Montan Loza AJ. Evolving role of Vitamin D in immune-mediated disease and its implications in autoimmune hepatitis. Digestive Diseases and Sciences. 2018;64:324–44.
212. Alswailmi FK, Shah SIA, Nawaz H. Immunomodulatory role of Vitamin D: clinical implications in infections and autoimmune disorders. Gomal Journal of Medical Sciences. 2020;18:132–8.
213. Cantorna MT. Vitamin D, multiple sclerosis and inflammatory bowel disease. Archives of Biochemistry and Biophysics. 2012;523(1):103–6.
214. Khairallah MK, Makarem YS, Dahpy MA. Vitamin D in active systemic lupus erythematosus and lupus nephritis: a forgotten player. The Egyptian Journal of Internal Medicine. 2020;32(1):16.
215. Harrison SR, Li D, Jeffery LE, Raza K, Hewison M. Vitamin D, autoimmune disease and rheumatoid arthritis. Calcified Tissue International. 2020;106(1):58–75.
216. Carvalho C, Marinho A, Leal B, Bettencourt A, Boleixa D, Almeida I, et al. Association between vitamin D receptor (VDR) gene polymorphisms and systemic lupus erythematosus in Portuguese patients. Lupus. 2015;24(8):846–53.
217. Giovinazzo S, Vicchio TM, Certo R, Alibrandi A, Palmieri O, Campennì A, et al. Vitamin D receptor gene polymorphisms/haplotypes and serum 25(OH)D3 levels in Hashimoto's thyroiditis. Endocrine. 2017;55(2):599–606.
218. Bettencourt A, Boleixa D, Guimarães AL, Leal B, Carvalho C, Brás S, et al. The vitamin D receptor gene FokI polymorphism and multiple sclerosis in a northern portuguese population. Journal of Neuroimmunology. 2017;309:34–7.
219. Gallone G, Haerty W, Disanto G, Ramagopalan SV, Ponting CP, Berlanga-Taylor AJ. Identification of genetic variants affecting vitamin D receptor binding and associations with autoimmune disease. Human Molecular Genetics. 2017;26(11): 2164–76.
220. Murdaca G, Tonacci A, Negrini S, Greco M, Borro M, Puppo F, et al. Emerging role of vitamin D in autoimmune diseases: An update on evidence and therapeutic implications. Autoimmunity Reviews. 2019;18(9):102350.
221. Aslam MM, John P, Bhatti A, Jahangir S, Kamboh MI. Vitamin D as a principal factor in mediating rheumatoid arthritis-derived immune response. BioMed Research International. 2019;2019:3494937.
222. Lin J, Liu J, Davies ML, Chen W. Serum Vitamin D level and rheumatoid arthritis disease activity: review and meta-analysis. PloS One. 2016;11(1):e0146351.
223. Bragazzi NL, Watad A, Neumann SG, Simon M, Brown SB, Abu Much A, et al. Vitamin D and rheumatoid arthritis: an ongoing mystery. Current Opinion in Rheumatology. 2017;29(4):378–88.
224. Di Franco M, Barchetta I, Iannuccelli C, Gerardi MC, Frisenda S, Ceccarelli F, et al. Hypovitaminosis D in recent onset rheumatoid arthritis is predictive of reduced response to treatment and increased disease activity: a 12 month follow-up study. BMC Musculoskeletal Disorders. 2015;16:53.
225. Lee YH, Bae SC. Vitamin D level in rheumatoid arthritis and its correlation with the disease activity: a meta-analysis. Clinical and Experimental Rheumatology. 2016;34(5):827–33.
226. Zakeri Z, Sandoughi M, Mashhadi MA, Raeesi V, Shahbakhsh S. Serum vitamin D level and disease activity in patients with recent onset rheumatoid arthritis. International Journal of Rheumatic Diseases. 2016;19(4):343–7.

227. Hong Q, Xu J, Xu S, Lian L, Zhang M, Ding C. Associations between serum 25-hydroxyvitamin D and disease activity, inflammatory cytokines and bone loss in patients with rheumatoid arthritis. Rheumatology (Oxford, England). 2014;53(11):1994–2001.
228. Abourazzak FE, Talbi S, Aradoini N, Berrada K, Keita S, Hazry T. 25-Hydroxy vitamin D and its relationship with clinical and laboratory parameters in patients with rheumatoid arthritis. Clinical Rheumatology. 2015;34(2):353–7.
229. Raczkiewicz A, Kisiel B, Kulig M, Tłustochowicz W. Vitamin D status and its association with quality of life, physical activity, and disease activity in rheumatoid arthritis patients. Journal of Clinical Rheumatology: Practical Reports on Rheumatic & Musculoskeletal Diseases. 2015;21(3):126–30.
230. Pakchotanon R, Chaiamnuay S, Narongroeknawin P, Asavatanabodee P. The association between serum vitamin D Level and disease activity in Thai rheumatoid arthritis patients. International Journal of Rheumatic Diseases. 2016;19(4):355–61.
231. Haque UJ, Bartlett SJ. Relationships among vitamin D, disease activity, pain and disability in rheumatoid arthritis. Clinical and Experimental Rheumatology. 2010;28(5):745–7.
232. Baykal T, Senel K, Alp F, Erdal A, Ugur M. Is there an association between serum 25-hydroxyvitamin D concentrations and disease activity in rheumatoid arthritis? Bratislavske Lekarske Listy. 2012;113(10):610–1.
233. Rajaee E, Ghorbani A, Mowla K, Zakerkish M, Mohebi M, Dargahi-Malamir M. The relationship between serum level of vitamin D3 and the severity of new onset rheumatoid arthritis activity. Journal of Clinical and Diagnostic Research: JCDR. 2017;11(3):Oc28–oc30.
234. Matsumoto Y, Sugioka Y, Tada M, Okano T, Mamoto K, Inui K, et al. Relationships between serum 25-hydroxycalciferol, vitamin D intake and disease activity in patients with rheumatoid arthritis–tomorrow study. Modern Rheumatology. 2015;25(2):246–50.
235. Baker JF, Baker DG, Toedter G, Shults J, Von Feldt JM, Leonard MB. Associations between vitamin D, disease activity, and clinical response to therapy in rheumatoid arthritis. Clinical and Experimental Rheumatology. 2012;30(5):658–64.
236. Haga HJ, Schmedes A, Naderi Y, Moreno AM, Peen E. Severe deficiency of 25-hydroxyvitamin D_3 (25-OH-D_3) is associated with high disease activity of rheumatoid arthritis. Clinical Rheumatology. 2013;32(5):629–33.
237. Heidari B, Hajian-Tilaki K, Babaei M. Vitamin D deficiency and rheumatoid arthritis: epidemiological, immunological, clinical and therapeutic aspects. Mediterranean Journal of Rheumatology. 2019;30(2):94–102.
238. Gendelman O, Itzhaki D, Makarov S, Bennun M, Amital H. A randomized double-blind placebo-controlled study adding high dose vitamin D to analgesic regimens in patients with musculoskeletal pain. Lupus. 2015;24(4–5):483–9.
239. Buondonno I, Rovera G, Sassi F, Rigoni MM, Lomater C, Parisi S, et al. Vitamin D and immunomodulation in early rheumatoid arthritis: A randomized double-blind placebo-controlled study. Plos One. 2017;12(6):e0178463.
240. Rozmus D, Ciesielska A, Plominski J, Grzybowski R, Fiedorowicz E, Kordulewska N, et al. Vitamin D Binding Protein (VDBP) and Its Gene Polymorphisms-The Risk of Malignant Tumors and Other Diseases. International Journal of Molecular Sciences. 2020;21(21).
241. Yang J, Liu L, Zhang Q, Li M, Wang J. Effect of vitamin D on the recurrence rate of rheumatoid arthritis. Experimental and Therapeutic Medicine. 2015;10(5):1812–6.
242. Dehghan A, Rahimpour S, Soleymani-Salehabadi H, Owlia MB. Role of vitamin D in flare ups of rheumatoid arthritis. Zeitschrift fur Rheumatologie. 2014;73(5):461–4.
243. Guan Y, Hao Y, Guan Y, Bu H, Wang H. The effect of Vitamin D supplementation on rheumatoid arthritis patients: a systematic review and meta-analysis. Frontiers in Medicine. 2020;7:596007.
244. Nguyen Y, Sigaux J, Letarouilly JG, Sanchez P, Czernichow S, Flipo RM, et al. Efficacy of oral vitamin supplementation in inflammatory rheumatic disorders: a systematic review and meta-analysis of randomized controlled trials. Nutrients. 2020;13(1).

245. Li C, Yin S, Yin H, Cao L, Zhang T, Wang Y. Efficacy and safety of 22-oxa-calcitriol in patients with rheumatoid arthritis: a phase ii trial. Medical Science Monitor: International Medical Journal of Experimental and Clinical Research. 2018; 24: 9127–35.
246. Gopinath K, Danda D. Supplementation of 1,25 dihydroxy vitamin D3 in patients with treatment naive early rheumatoid arthritis: a randomised controlled trial. International Journal of Rheumatic Diseases. 2011;14(4):332–9.
247. Thompson AJ, Baranzini SE, Geurts J, Hemmer B, Ciccarelli O. Multiple sclerosis. Lancet. 2018;391(10130):1622–36.
248. Waubant E, Lucas R, Mowry E, Graves J, Olsson T, Alfredsson L, et al. Environmental and genetic risk factors for MS: an integrated review. Annals of Clinical and Translational Neurology. 2019;6(9):1905–22.
249. Belbasis L, Bellou V, Evangelou E, Ioannidis JP, Tzoulaki I. Environmental risk factors and multiple sclerosis: an umbrella review of systematic reviews and meta-analyses. Lancet Neurol. 2015;14(3):263–73.
250. Neyestani TR. Is multiple sclerosis a sun deprivation disease?: Lessons from the past for the future path. Bioactive Nutraceuticals and Dietary Supplements in Neurological and Brain Disease: Elsevier; 2015. p. 481–94.
251. Miclea A, Bagnoud M, Chan A, Hoepner R. A brief review of the effects of vitamin D on multiple sclerosis. Frontiers in Immunology. 2020;11(781).
252. Fitzgerald KC, Munger KL, Kochert K, Arnason BG, Comi G, Cook S, et al. Association of vitamin D levels with multiple sclerosis activity and progression in patients receiving interferon beta-1b. JAMA Neurology. 2015;72(12):1458–65.
253. Bagur MJ, Murcia MA, Jimenez-Monreal AM, Tur JA, Bibiloni MM, Alonso GL, et al. Influence of diet in multiple sclerosis: a systematic review. Advances in Nutrition. 2017;8(3):463–72.
254. Ascherio A, Munger KL, White R, Kochert K, Simon KC, Polman CH, et al. Vitamin D as an early predictor of multiple sclerosis activity and progression. JAMA Neurology. 2014;71(3):306–14.
255. Munger KL, Levin LI, Hollis BW, Howard NS, Ascherio A. Serum 25-hydroxyvitamin D levels and risk of multiple sclerosis. JAMA. 2006;296(23):2832–8.
256. Runia TF, Hop WC, de Rijke YB, Buljevac D, Hintzen RQ. Lower serum vitamin D levels are associated with a higher relapse risk in multiple sclerosis. Neurology. 2012;79(3):261–6.
257. Simpson S, Jr., Taylor B, Blizzard L, Ponsonby AL, Pittas F, Tremlett H, et al. Higher 25-hydroxyvitamin D is associated with lower relapse risk in multiple sclerosis. Annals of Neurology. 2010;68(2):193–203.
258. Graves JS, Barcellos LF, Krupp L, Belman A, Shao X, Quach H, et al. Vitamin D genes influence MS relapses in children. Multiple Sclerosis. 2020;26(8):894–901.
259. Mowry EM, Waubant E, McCulloch CE, Okuda DT, Evangelista AA, Lincoln RR, et al. Vitamin D status predicts new brain magnetic resonance imaging activity in multiple sclerosis. Annals of Neurology. 2012;72(2):234–40.
260. Yeh WZ, Gresle M, Jokubaitis V, Stankovich J, van der Walt A, Butzkueven H. Immunoregulatory effects and therapeutic potential of vitamin D in multiple sclerosis. British Journal of Pharmacology. 2020;177(18):4113–33.
261. Mokry LE, Ross S, Ahmad OS, Forgetta V, Smith GD, Goltzman D, et al. Vitamin D and risk of multiple sclerosis: a Mendelian randomization study. PLoS Medicine. 2015;12(8):e1001866.
262. Jacobs BM, Noyce AJ, Giovannoni G, Dobson R. BMI and low vitamin D are causal factors for multiple sclerosis: a Mendelian randomization study. Neurology(R) Neuroimmunology & Neuroinflammation. 2020;7(2).
263. Rhead B, Baarnhielm M, Gianfrancesco M, Mok A, Shao X, Quach H, et al. Mendelian randomization shows a causal effect of low vitamin D on multiple sclerosis risk. Neurology Genetics. 2016;2(5):e97.

264. Feige J, Moser T. Vitamin D supplementation in multiple sclerosis: a critical analysis of potentials and threats. Nutrients.2020;12(3):783.
265. Laursen JH, Søndergaard HB, Sørensen PS, Sellebjerg F, Oturai AB. Vitamin D supplementation reduces relapse rate in relapsing-remitting multiple sclerosis patients treated with natalizumab. Multiple Sclerosis and Related Disorders. 2016;10:169–73.
266. Etemadifar M, Janghorbani M. Efficacy of high-dose vitamin D3 supplementation in vitamin D deficient pregnant women with multiple sclerosis: Preliminary findings of a randomized-controlled trial. Iranian Journal of Neurology. 2015;14(2):67–73.
267. Soilu-Hänninen M, Aivo J, Lindström BM, Elovaara I, Sumelahti ML, Färkkilä M, et al. A randomised, double blind, placebo controlled trial with vitamin D3 as an add on treatment to interferon β-1b in patients with multiple sclerosis. Journal of Neurology, Neurosurgery, and Psychiatry. 2012;83(5):565–71.
268. Jelinek GA, Marck CH, Weiland TJ, Pereira N, van der Meer DM, Hadgkiss EJ. Latitude, sun exposure and vitamin D supplementation: associations with quality of life and disease outcomes in a large international cohort of people with multiple sclerosis. BMC Neurology. 2015;15:132.
269. Shaygannejad V, Janghorbani M, Ashtari F, Dehghan H. Effects of adjunct low-dose vitamin d on relapsing-remitting multiple sclerosis progression: preliminary findings of a randomized placebo-controlled trial. Multiple Sclerosis International. 2012;2012:452541.
270. Stein MS, Liu Y, Gray OM, Baker JE, Kolbe SC, Ditchfield MR, et al. A randomized trial of high-dose vitamin D2 in relapsing-remitting multiple sclerosis. Neurology. 2011;77(17):1611–8.
271. Burton JM, Kimball S, Vieth R, Bar-Or A, Dosch HM, Cheung R, et al. A phase I/II dose-escalation trial of vitamin D3 and calcium in multiple sclerosis. Neurology. 2010; 74(23):1852–9.
272. Martínez-Lapiscina EH, Mahatanan R, Lee CH, Charoenpong P, Hong JP. Associations of serum 25(OH) vitamin D levels with clinical and radiological outcomes in multiple sclerosis, a systematic review and meta-analysis. Journal of the Neurological Sciences. 2020;411:116668.
273. Zheng C, He L, Liu L, Zhu J, Jin T. The efficacy of vitamin D in multiple sclerosis: a meta-analysis. Multiple Sclerosis and Related Disorders. 2018;23:56–61.
274. Doosti-Irani A, Tamtaji OR, Mansournia MA, Ghayour-Mobarhan M, Ferns G, Daneshvar Kakhaki R, et al. The effects of vitamin D supplementation on expanded disability status scale in people with multiple sclerosis: a critical, systematic review and metaanalysis of randomized controlled trials. Clinical Neurology and Neurosurgery. 2019;187:105564.
275. Hupperts R, Smolders J. Randomized trial of daily high-dose vitamin D(3) in patients with RRMS receiving subcutaneous interferon β-1a. Neurology. 2019; 93(20):e1906–e16.
276. Smolders J, Mimpen M, Oechtering J, Damoiseaux J, van den Ouweland J, Hupperts R, et al. Vitamin D3 supplementation and neurofilament light chain in multiple sclerosis. Acta Neurologica Scandinavica. 2020;141(1):77–80.
277. Camu W, Lehert P, Pierrot-Deseilligny C, Hautecoeur P, Besserve A, Jean Deleglise AS, et al. Cholecalciferol in relapsing-remitting MS: a randomized clinical trial (CHOLINE). Neurology(R) Neuroimmunology & Neuroinflammation. 2019;6(5).
278. Dorr J, Backer-Koduah P, Wernecke KD, Becker E, Hoffmann F, Faiss J, et al. High-dose vitamin D supplementation in multiple sclerosis – results from the randomized EVIDIMS (efficacy of vitamin D supplementation in multiple sclerosis) trial. Multiple Sclerosis Journal - Experimental, Translational and Clinical. 2020; 6(1): 2055217320903474.
279. Lemke D, Klement RJ, Schweiger F, Schweiger B, Spitz J. Vitamin D resistance as a possible cause of autoimmune diseases: a hypothesis confirmed by a therapeutic high-dose Vitamin D protocol. Frontiers in Immunology. 2021;12(1110).

280. Tobias L. A briefing report on autoimmune diseases and AARDA: past, present, and future. Eastpointe: American Autoimmune Related Diseases Association (AARDA). 2010.
281. Ramos-Casals M, Brito-Zerón P, Kostov B, Sisó-Almirall A, Bosch X, Buss D, et al. Google-driven search for big data in autoimmune geoepidemiology: analysis of 394,827 patients with systemic autoimmune diseases. Autoimmunity Reviews. 2015; 14(8):670–9.
282. Kriegel MA, Manson JE, Costenbader KH, editors. Does vitamin D affect risk of developing autoimmune disease?: a systematic review. Seminars in Arthritis and Rheumatism; 2011: Elsevier.
283. Johnson LK, Hofsø D, Aasheim ET, Tanbo T, Holven KB, Andersen LF, et al. Impact of gender on vitamin D deficiency in morbidly obese patients: a cross-sectional study. European Journal of Clinical Nutrition. 2012;66(1):83–90.
284. Rashki Kemmak A, Rezapour A. Economic burden of osteoporosis in the world: a systematic review. Medical Journal of the Islamic Republic of Iran. 2020;34:154.
285. Grant WB. A Review of the evidence supporting the Vitamin D-cancer prevention hypothesis in 2017. Anticancer Research. 2018;38(2):1121–36.
286. Grant WB. Review of recent advances in understanding the role of Vitamin D in reducing cancer risk: breast, colorectal, prostate, and overall cancer. Anticancer Research. 2020;40(1):491–9.
287. Hansen CM, Binderup L, Hamberg KJ, Carlberg C. Vitamin D and cancer: effects of 1,25(OH)2D3 and its analogs on growth control and tumorigenesis. Frontiers in Bioscience: A Journal and Virtual Library. 2001;6:D820–48.
288. Feldman D, Krishnan AV, Swami S, Giovannucci E, Feldman BJ. The role of vitamin D in reducing cancer risk and progression. Nature Reviews Cancer. 2014;14(5):342–57.
289. Sluyter JD, Manson JE, Scragg R. Vitamin D and clinical cancer outcomes: a review of meta-analyses. JBMR Plus. 2021;5(1):e10420.
290. Hossain S, Beydoun MA, Beydoun HA, Chen X, Zonderman AB, Wood RJ. Vitamin D and breast cancer: a systematic review and meta-analysis of observational studies. Clinical Nutrition ESPEN. 2019;30:170–84.
291. Akutsu T, Kitamura H, Himeiwa S, Kitada S, Akasu T, Urashima M. Vitamin D and cancer survival: does Vitamin D supplementation improve the survival of patients with cancer? Current Oncology Reports. 2020;22(6):62.
292. Young MRI, Xiong Y. Influence of vitamin D on cancer risk and treatment: Why the variability? Trends in Cancer Research. 2018;13:43–53.
293. Keum N, Lee DH, Greenwood DC, Manson JE, Giovannucci E. Vitamin D supplementation and total cancer incidence and mortality: a meta-analysis of randomized controlled trials. Annals of Oncology: Official Journal of the European Society for Medical Oncology. 2019;30(5):733–43.
294. Guo XF, Zhao T, Han JM, Li S, Li D. Vitamin D and liver cancer risk: a meta-analysis of prospective studies. Asia Pacific Journal of Clinical Nutrition. 2020;29(1):175–82.
295. Feng Q, Zhang H, Dong Z, Zhou Y, Ma J. Circulating 25-hydroxyvitamin D and lung cancer risk and survival: a dose-response meta-analysis of prospective cohort studies. Medicine. 2017;96(45):e8613.
296. Ma Y, Zhang P, Wang F, Yang J, Liu Z, Qin H. association between vitamin D and risk of colorectal cancer: a systematic review of prospective studies. Journal of Clinical Oncology. 2011;29(28):3775–82.
297. Nair-Shalliker V, Smith DP, Egger S, Hughes AM, Kaldor JM, Clements M, et al. Sun exposure may increase risk of prostate cancer in the high UV environment of New South Wales, Australia: a case-control study. International Journal of Cancer. 2012;131(5):E726–32.
298. Gao J, Wei W, Wang G, Zhou H, Fu Y, Liu N. Circulating vitamin D concentration and risk of prostate cancer: a dose-response meta-analysis of prospective studies. Therapeutics and Clinical Risk Management. 2018;14:95–104.

299. Batai K, Murphy AB, Ruden M, Newsome J, Shah E, Dixon MA, et al. Race and BMI modify associations of calcium and vitamin D intake with prostate cancer. BMC Cancer. 2017;17(1):64.
300. Capiod T, Barry Delongchamps N, Pigat N, Souberbielle JC, Goffin V. Do dietary calcium and vitamin D matter in men with prostate cancer? Nature Reviews Urology. 2018;15(7):453–61.
301. Bjelakovic G, Gluud LL, Nikolova D, Whitfield K, Krstic G, Wetterslev J, et al. Vitamin D supplementation for prevention of cancer in adults. The Cochrane Database of Systematic Reviews. 2014;(6):CD007469.
302. Keum N, Giovannucci E. Vitamin D supplements and cancer incidence and mortality: a meta-analysis. British Journal of Cancer. 2014;111(5):976–80.
303. Zhang Y, Fang F, Tang J, Jia L, Feng Y, Xu P, et al. Association between vitamin D supplementation and mortality: systematic review and meta-analysis. BMJ (Clinical Research ed). 2019;366:l4673.
304. Goulão B, Stewart F, Ford JA, MacLennan G, Avenell A. Cancer and vitamin D supplementation: a systematic review and meta-analysis. The American Journal of Clinical Nutrition. 2018;107(4):652–63.
305. Wactawski-Wende J, Kotchen JM, Anderson GL, Assaf AR, Brunner RL, O'Sullivan MJ, et al. Calcium plus vitamin D supplementation and the risk of colorectal cancer. The New England Journal of Medicine. 2006;354(7):684–96.
306. Scragg R, Khaw KT, Toop L, Sluyter J, Lawes CMM, Waayer D, et al. Monthly high-dose Vitamin D supplementation and cancer risk: a post hoc analysis of the vitamin D assessment randomized clinical trial. JAMA Oncology. 2018;4(11):e182178.
307. Manson JE, Cook NR, Lee IM, Christen W, Bassuk SS, Mora S, et al. Vitamin D supplements and prevention of cancer and cardiovascular disease. The New England Journal of Medicine. 2019;380(1):33–44.
308. Chandler PD, Chen WY, Ajala ON, Hazra A, Cook N, Bubes V, et al. Effect of Vitamin D3 supplements on development of advanced cancer: a secondary analysis of the VITAL randomized clinical trial. JAMA Network Open. 2020;3(11):e2025850–e.
309. Zgaga L. Heterogeneity of the effect of Vitamin D supplementation in randomized controlled trials on cancer prevention. JAMA Network Open. 2020;3(11):e2027176.
310. Sung H, Ferlay J, Siegel RL, Laversanne M, Soerjomataram I, Jemal A, et al. Global cancer statistics 2020: GLOBOCAN estimates of incidence and mortality worldwide for 36 cancers in 185 countries. CA: A Cancer Journal for Clinicians. 2021; 71(3): 209–49.
311. Smith K. Mental health: a world of depression. Nature News. 2014;515(7526):180.
312. Evaluation. IoHMa. Global Health Data Exchange (GHDx) http://ghdx.healthdata.org/gbd-results-tool?params=gbd-api-2019-permalink/d780dffbe8a381b25e1416884959e88b [cited 2 October 2021].
313. Friedrich MJ. Depression is the leading cause of disability around the world. JAMA. 2017;317(15):1517.
314. Liu Q, He H, Yang J, Feng X, Zhao F, Lyu J. Changes in the global burden of depression from 1990 to 2017: findings from the global burden of disease study. Journal of Psychiatric Research. 2020;126:134–40.
315. Brody DJ, Pratt LA, Hughes JP. Prevalence of depression among adults aged 20 and over: United States, 2013–2016. NCHS Data Brief. 2018;303:1–8.
316. WHO Depression Fact Sheet. https://www.who.int/news-room/fact-sheets/detail/depression: World Health Organization (WHO). 2021 [updated 13 September 2021; cited 2 October 2021].
317. Vogelzangs N, Duivis HE, Beekman AT, Kluft C, Neuteboom J, Hoogendijk W, et al. Association of depressive disorders, depression characteristics and antidepressant medication with inflammation. Translational Psychiatry. 2012;2(2):e79–e.
318. Vreeburg SA, Hoogendijk WJ, van Pelt J, DeRijk RH, Verhagen JC, van Dyck R, et al. Major depressive disorder and hypothalamic-pituitary-adrenal axis activity: results from a large cohort study. Archives of General Psychiatry. 2009;66(6):617–26.

319. Molendijk ML, Bus BA, Spinhoven P, Penninx BW, Kenis G, Prickaerts J, et al. Serum levels of brain-derived neurotrophic factor in major depressive disorder: state–trait issues, clinical features and pharmacological treatment. Molecular Psychiatry. 2011; 16(11):1088–95.
320. Milaneschi Y, Hoogendijk W, Lips P, Heijboer A, Schoevers R, Van Hemert A, et al. The association between low vitamin D and depressive disorders. Molecular Psychiatry. 2014;19(4):444–51.
321. Dean J, Keshavan M. The neurobiology of depression: an integrated view. Asian Journal of Psychiatry. 2017;27:101–11.
322. Eyles DW, Smith S, Kinobe R, Hewison M, McGrath JJ. Distribution of the vitamin D receptor and 1α-hydroxylase in human brain. Journal of Chemical Neuroanatomy. 2005;29(1):21–30.
323. Kesby JP, Eyles DW, Burne TH, McGrath JJ. The effects of vitamin D on brain development and adult brain function. Molecular and Cellular Endocrinology. 2011;347(1–2):121–7.
324. de Abreu DF, Eyles D, Feron F. Vitamin D, a neuro-immunomodulator: implications for neurodegenerative and autoimmune diseases. Psychoneuroendocrinology. 2009;34:S265–S77.
325. Gu Y, Luan X, Ren W, Zhu L, He J. Impact of seasons on stroke-related depression, mediated by vitamin D status. BMC Psychiatry. 2018;18(1):1–7.
326. Gu Y, Zhu Z, Luan X, He J. Vitamin D status and its association with season, depression in stroke. Neuroscience Letters. 2019;690:99–105.
327. Berridge MJ. Vitamin D and depression: cellular and regulatory mechanisms. Pharmacological Reviews. 2017;69(2):80–92.
328. Jääskeläinen T, Knekt P, Suvisaari J, Männistö S, Partonen T, Sääksjärvi K, et al. Higher serum 25-hydroxyvitamin D concentrations are related to a reduced risk of depression. British Journal of Nutrition. 2015;113(9):1418–26.
329. Briggs R, McCarroll K, O'Halloran A, Healy M, Kenny RA, Laird E. Vitamin D deficiency is associated with an increased likelihood of incident depression in community-dwelling older adults. Journal of the American Medical Directors Association. 2019; 20(5):517–23.
330. Hoogendijk WJ, Lips P, Dik MG, Deeg DJ, Beekman AT, Penninx BW. Depression is associated with decreased 25-hydroxyvitamin D and increased parathyroid hormone levels in older adults. Archives of General Psychiatry. 2008;65(5):508–12.
331. Lee DM, Tajar A, O'Neill TW, O'Connor DB, Bartfai G, Boonen S, et al. Lower vitamin D levels are associated with depression among community-dwelling European men. Journal of Psychopharmacology. 2011;25(10):1320–8.
332. Song BM, Kim HC, Rhee Y, Youm Y, Kim CO. Association between serum 25-hydroxyvitamin D concentrations and depressive symptoms in an older Korean population: a cross-sectional study. Journal of Affective Disorders. 2016;189:357–64.
333. Pu D, Luo J, Wang Y, Ju B, Lv X, Fan P, et al. Prevalence of depression and anxiety in rheumatoid arthritis patients and their associations with serum vitamin D level. Clinical Rheumatology. 2018;37(1):179–84.
334. Zhao G, Ford ES, Li C, Balluz LS. No associations between serum concentrations of 25-hydroxyvitamin D and parathyroid hormone and depression among US adults. British Journal of Nutrition. 2010;104(11):1696–702.
335. Michaëlsson K, Melhus H, Larsson SC. Serum 25-hydroxyvitamin D concentrations and major depression: a Mendelian randomization study. Nutrients. 2018;10(12):1987.
336. Dean AJ, Bellgrove MA, Hall T, Phan WMJ, Eyles DW, Kvaskoff D, et al. Effects of vitamin D supplementation on cognitive and emotional functioning in young adults–a randomised controlled trial. PloS One. 2011;6(11):e25966.
337. Kjærgaard M, Wang CE, Almås B, Figenschau Y, Hutchinson MS, Svartberg J, et al. Effect of vitamin D supplement on depression scores in people with low levels of serum 25-hydroxyvitamin D: nested case—control study and randomised clinical trial. The British Journal of Psychiatry. 2012;201(5):360–8.

338. Yalamanchili V, Gallagher JC. Dose ranging effects of vitamin D3 on the geriatric depression score: A clinical trial. The Journal of Steroid Biochemistry and Molecular Biology. 2018;178:60–4.
339. Gustafson C. Bruce Ames, PhD, and Rhonda Patrick, PhD: discussing the triage concept and the Vitamin D-serotonin connection. Integrative Medicine: A Clinician's Journal. 2014;13(6):34.
340. Kaviani M, Nikooyeh B, Zand H, Yaghmaei P, Neyestani TR. Effects of vitamin D supplementation on depression and some involved neurotransmitters. Journal of Affective Disorders. 2020;269:28–35.
341. Eyles DW, Burne TH, McGrath JJ. Vitamin D, effects on brain development, adult brain function and the links between low levels of vitamin D and neuropsychiatric disease. Frontiers in Neuroendocrinology. 2013;34(1):47–64.
342. Obradovic D, Gronemeyer H, Lutz B, Rein T. Cross-talk of vitamin D and glucocorticoids in hippocampal cells. Journal of Neurochemistry. 2006;96(2):500–9.
343. Patrick RP, Ames BN. Vitamin D and the omega-3 fatty acids control serotonin synthesis and action, part 2: Relevance for ADHD, bipolar disorder, schizophrenia, and impulsive behavior. The FASEB Journal. 2015;29(6):2207–22.
344. Kuhn DM, Hasegawa H. Tryptophan hydroxylase and serotonin synthesis regulation. In: Huston JP, Steiner H (eds.). Handbook of Behavioral Neuroscience, Volume 31: Elsevier; 2020. pp. 239–56.
345. Berridge MJ. Vitamin D, reactive oxygen species and calcium signalling in ageing and disease. Philosophical Transactions of the Royal Society B: Biological Sciences. 2016;371(1700):20150434.
346. Greenberg PE, Fournier A-A, Sisitsky T, Simes M, Berman R, Koenigsberg SH, et al. The economic burden of adults with major depressive disorder in the United States (2010 and 2018). Pharmacoeconomics. 2021;39(6):653–65.
347. Mokdad AH, Ballestros K, Echko M, Glenn S, Olsen HE, Mullany E, et al. The state of US health, 1990-2016: burden of diseases, injuries, and risk factors among US states. JAMA. 2018;319(14):1444–72.
348. McCann JC, Ames BN. Is there convincing biological or behavioral evidence linking vitamin D deficiency to brain dysfunction? The FASEB Journal. 2008;22(4):982–1001.
349. Muss C, Mosgoeller W, Endler T. Mood improving potential of a vitamin trace element composition—a randomized, double blind, placebo controlled clinical study with healthy volunteers. Neuroendocrinology Letters. 2016;37(1):18–28.
350. Kim HK, Kim SH, Jang CS, Kim SI, Kweon CO, Kim BW, et al. The combined effects of yogurt and exercise in healthy adults: Implications for biomarkers of depression and cardiovascular diseases. Food science & nutrition. 2018;6(7):1968–74.
351. Matta S, Chopra K, Arora V. Morbidity and mortality trends of Covid 19 in top 10 countries. Indian Journal of Tuberculosis. 2020.
352. McKibbin W, Fernando R. The economic impact of COVID-19. Economics in the Time of COVID-19. 2020;45–51(10.1162).
353. Piroth L, Cottenet J, Mariet A-S, Bonniaud P, Blot M, Tubert-Bitter P, et al. Comparison of the characteristics, morbidity, and mortality of COVID-19 and seasonal influenza: a nationwide, population-based retrospective cohort study. The Lancet Respiratory Medicine. 2021;9(3):251–9.
354. Tang L, Yin Z, Hu Y, Mei H. Controlling cytokine storm is vital in COVID-19. Frontiers in Immunology. 2020;11:3158.
355. Zhang X, Tan Y, Ling Y, Lu G, Liu F, Yi Z, et al. Viral and host factors related to the clinical outcome of COVID-19. Nature. 2020;583(7816):437–40.
356. Mertens E, Peñalvo JL. The burden of malnutrition and fatal COVID-19: a global burden of disease analysis. Frontiers in Nutrition. 2021;7(351).
357. Damayanthi H, Prabani KIP. Nutritional determinants and COVID-19 outcomes of older patients with COVID-19: a systematic review. Archives of Gerontology and Geriatrics. 2021;95:104411.

358. Liu N, Sun J, Wang X, Zhang T, Zhao M, Li H. Low vitamin D status is associated with coronavirus disease 2019 outcomes: a systematic review and meta-analysis. International Journal of Infectious Diseases. 2021;104:58–64.
359. Meltzer DO, Best TJ, Zhang H, Vokes T, Arora V, Solway J. Association of vitamin D status and other clinical characteristics with COVID-19 test results. JAMA Network Open. 2020;3(9):e2019722–e.
360. Kaya MO, Pamukçu E, Yakar B. The role of vitamin D deficiency on the Covid-19: a systematic review and meta-analysis of observational studies. Epidemiology and Health. 2021:e2021074.
361. Carpagnano GE, Di Lecce V, Quaranta VN, Zito A, Buonamico E, Capozza E, et al. Vitamin D deficiency as a predictor of poor prognosis in patients with acute respiratory failure due to COVID-19. Journal of Endocrinological Investigation. 2021;44(4):765–71.
362. Li X, van Geffen J, van Weele M, Zhang X, He Y, Meng X, et al. An observational and Mendelian randomisation study on vitamin D and COVID-19 risk in UK Biobank. Scientific Reports. 2021;11(1):18262.
363. Pizzini A, Aichner M, Sahanic S, Böhm A, Egger A, Hoermann G, et al. Impact of vitamin D deficiency on COVID-19—a prospective analysis from the CovILD Registry. Nutrients. 2020;12(9):2775.
364. Ma H, Zhou T, Heianza Y, Qi L. Habitual use of vitamin D supplements and risk of coronavirus disease 2019 (COVID-19) infection: a prospective study in UK Biobank. The American Journal of Clinical Nutrition. 2021;113(5):1275–81.
365. Liu G, Hong T, Yang J. A single large dose of vitamin D could be used as a means of coronavirus disease 2019 prevention and treatment. Drug Design, Development and Therapy. 2020;14:3429–34.
366. Güven M, Gültekin H. The effect of high-dose parenteral vitamin D3 on COVID-19-related inhospital mortality in critical COVID-19 patients during intensive care unit admission: an observational cohort study. European Journal of Clinical Nutrition. 2021;75(9):1383–8.
367. Sharma SK, Mudgal SK, Pai VS, Chaturvedi J, Gaur R. Vitamin D: a cheap yet effective bullet against coronavirus disease-19–Are we convinced yet? National Journal of Physiology, Pharmacy and Pharmacology. 2020;10(7):511–8.
368. Verdoia M, De Luca G. Potential role of hypovitaminosis D and vitamin D supplementation during COVID-19 pandemic. QJM: An International Journal of Medicine. 2020;114(1):3–10.
369. Kearns SM, Ahern KW, Patrie JT, Horton WB, Harris TE, Kadl A. Reduced adiponectin levels in patients with COVID-19 acute respiratory failure: a case-control study. Physiological Reports. 2021;9(7):e14843.
370. Messina G, Polito R, Monda V, Cipolloni L, Di Nunno N, Di Mizio G, et al. Functional role of dietary intervention to improve the outcome of COVID-19: a hypothesis of work. International Journal of Molecular Sciences. 2020;21(9):3104.
371. Nikooyeh B, Neyestani T. Can vitamin D be considered an adiponectin secretagogue? A systematic review and meta-analysis. The Journal of Steroid Biochemistry and Molecular Biology. 2021:105925.
372. Wyper GM, Assunção R, Cuschieri S, Devleesschauwer B, Fletcher E, Haagsma JA, et al. Population vulnerability to COVID-19 in Europe: a burden of disease analysis. Archives of Public Health. 2020;78:1–8.
373. Miller IF, Becker AD, Grenfell BT, Metcalf CJE. Disease and healthcare burden of COVID-19 in the United States. Nature Medicine. 2020;26(8):1212–7.
374. Martineau AR, Forouhi NG. Vitamin D for COVID-19: a case to answer? The Lancet Diabetes & Endocrinology. 2020;8(9):735–6.
375. Ranaei V, Pilevar Z, Neyestani TR. Can raising vitamin D status slow down Covid-19 waves? Nutrition and Food Sciences Research. 2021;8(1):1–3.

376. Lawrence M, Wingrove K, Naude C, Durao S. Evidence synthesis and translation for nutrition interventions to combat micronutrient deficiencies with particular focus on food fortification. Nutrients. 2016;8(9):555.
377. Kift R, Rhodes LE. Is sunlight exposure enough to avoid wintertime vitamin D deficiency in United Kingdom population groups? International Journal of Environmental Research and Public Health. 2018;15(8):1624.
378. Webb AR, Kazantzidis A, Kift RC, Farrar MD, Wilkinson J. Colour counts: sunlight and skin type as drivers of vitamin D deficiency at UK latitudes. Nutrients. 2018; 10(4):457.
379. Moan J, Grigalavicius M, Baturaite Z, Dahlback A, Juzeniene A. The relationship between UV exposure and incidence of skin cancer. Photodermatology, Photoimmunology & Photomedicine. 2015;31(1):26–35.
380. Aguiar M, Andronis L, Pallan M, Högler W, Frew E. Preventing vitamin D deficiency (VDD): a systematic review of economic evaluations. European Journal of Public Health. 2017;27(2):292–301.
381. Payette H, Boutier V, Coulombe C, Gray-Donald K. Benefits of nutritional supplementation in free-living, frail, undernourished elderly people: a prospective randomized community trial. Journal of the American Dietetic Association. 2002;102(8):1088–95.
382. Pludowski P, Holick MF, Grant WB, Konstantynowicz J, Mascarenhas MR, Haq A, et al. Vitamin D supplementation guidelines. The Journal of Steroid Biochemistry and Molecular Biology. 2018;175:125–35.
383. Vieth R. Vitamin D supplementation: cholecalciferol, calcifediol, and calcitriol. European Journal of Clinical Nutrition. 2020;74(11):1493–7.
384. Sanders KM, Stuart AL, Williamson EJ, Simpson JA, Kotowicz MA, Young D, et al. Annual high-dose oral vitamin D and falls and fractures in older women: a randomized controlled trial. JAMA. 2010;303(18):1815–22.
385. Bischoff-Ferrari HA, Dawson-Hughes B, Orav EJ, Staehelin HB, Meyer OW, Theiler R, et al. Monthly high-dose vitamin D treatment for the prevention of functional decline: a randomized clinical trial. JAMA Internal Medicine. 2016;176(2):175–83.
386. Ross AC, Manson JE, Abrams SA, Aloia JF, Brannon PM, Clinton SK, et al. The 2011 report on dietary reference intakes for calcium and vitamin D from the Institute of Medicine: what clinicians need to know. The Journal of Clinical Endocrinology & Metabolism. 2011;96(1):53–8.
387. Priemel M, von Domarus C, Klatte TO, Kessler S, Schlie J, Meier S, et al. Bone mineralization defects and vitamin D deficiency: histomorphometric analysis of iliac crest bone biopsies and circulating 25-hydroxyvitamin D in 675 patients. Journal of Bone and Mineral Research. 2010;25(2):305–12.
388. Lee JH, Choi J-H, Kweon OJ, Park AJ. Discrepancy between vitamin D total immunoassays due to various cross-reactivities. Journal of Bone Metabolism. 2015;22(3):107–12.
389. Snellman G, Melhus H, Gedeborg R, Byberg L, Berglund L, Wernroth L, et al. Determining vitamin D status: a comparison between commercially available assays. PloS One. 2010;5(7):e11555.
390. Fraser WD, Tang JCY, Dutton JJ, Schoenmakers I. Vitamin D measurement, the debates continue, new analytes have emerged, developments have variable outcomes. Calcified Tissue International. 2020;106(1):3–13.
391. Aguirre Castaneda R, Nader N, Weaver A, Singh R, Kumar S. Response to Vitamin D_3 supplementation in obese and non-obese caucasian adolescents. Hormone Research in Paediatrics. 2012;78(4):226–31.
392. de Oliveira LF, de Azevedo LG, da Mota Santana J, de Sales LPC, Pereira-Santos M. Obesity and overweight decreases the effect of vitamin D supplementation in adults: systematic review and meta-analysis of randomized controlled trials. Reviews in Endocrine and Metabolic Disorders. 2020;21(1):67–76.
393. Heaney RP. Vitamin D: criteria for safety and efficacy. Nutrition Reviews. 2008; 66(suppl_2):S178–S81.

394. Rizzoli R. Vitamin D supplementation: upper limit for safety revisited? Aging Clinical and Experimental Research. 2021;33(1):19–24.
395. Billington EO, Burt LA, Rose MS, Davison EM, Gaudet S, Kan M, et al. Safety of high-dose Vitamin D supplementation: secondary analysis of a randomized controlled trial. The Journal of Clinical Endocrinology & Metabolism. 2019;105(4):1261–73.
396. Holick MF, Binkley NC, Bischoff-Ferrari HA, Gordon CM, Hanley DA, Heaney RP, et al. Evaluation, treatment, and prevention of vitamin D deficiency: an endocrine society clinical practice guideline. The Journal of Clinical Endocrinology & Metabolism. 2011;96(7):1911–30.
397. Sempos CT, Durazo-Arvizu RA, Dawson-Hughes B, Yetley EA, Looker AC, Schleicher RL, et al. Is there a reverse J-shaped association between 25-hydroxyvitamin D and all-cause mortality? Results from the US nationally representative NHANES. The Journal of Clinical Endocrinology & Metabolism. 2013;98(7):3001–9.
398. Durup D, Jørgensen HL, Christensen J, Tjønneland A, Olsen A, Halkjær J, et al. A reverse J-shaped association between serum 25-hydroxyvitamin D and cardiovascular disease mortality: the CopD study. The Journal of Clinical Endocrinology & Metabolism. 2015;100(6):2339–46.
399. Garland CF, Kim JJ, Mohr SB, Gorham ED, Grant WB, Giovannucci EL, et al. Meta-analysis of all-cause mortality according to serum 25-hydroxyvitamin D. American Journal of Public Health. 2014;104(8):e43–e50.
400. Lee J-S, Kim J. Factors affecting the use of dietary supplements by Korean adults: data from the Korean National Health and Nutrition Examination Survey III. Journal of the American Dietetic Association. 2009;109(9):1599–605.
401. Tessema J, Jefferds ME, Cogswell M, Carlton E. Motivators and barriers to prenatal supplement use among minority women in the United States. Journal of the American Dietetic Association. 2009;109(1):102–8.
402. Cashman KD. Vitamin D: dietary requirements and food fortification as a means of helping achieve adequate vitamin D status. The Journal of Steroid Biochemistry and Molecular Biology. 2015;148:19–26.
403. de Lourdes Samaniego-Vaesken M, Alonso-Aperte E, Varela-Moreiras G. Vitamin food fortification today. Food & Nutrition Research. 2012;56(1):5459.
404. Hayes A, Cashman KD. Food-based solutions for vitamin D deficiency: putting policy into practice and the key role for research. The Proceedings of the Nutrition Society. 2017;76(1):54–63.
405. Sacco JE, Dodd KW, Kirkpatrick SI, Tarasuk V. Voluntary food fortification in the United States: potential for excessive intakes. European Journal of Clinical Nutrition. 2013;67(6):592–7.
406. Kloosterman J, Fransen HP, de Stoppelaar J, Verhagen H, Rompelberg C. Safe addition of vitamins and minerals to foods: setting maximum levels for fortification in the Netherlands. European Journal of Nutrition. 2007;46(4):220–9.
407. Bruins MJ, Létinois U. Adequate Vitamin D intake cannot be achieved within carbon emission limits unless food is fortified: a simulation study. Nutrients. 2021;13(2):592.
408. Datta M, Vitolins MZ. Food fortification and supplement use—are there health implications? Critical Reviews in Food Science and Nutrition. 2016;56(13):2149–59.
409. Neyestani T, Hajifaraji M, Omidvar N, Nikooyeh B, Eshraghian M, Shariatzadeh N, et al. Calcium-vitamin D-fortified milk is as effective on circulating bone biomarkers as fortified juice and supplement but has less acceptance: a randomised controlled school-based trial. Journal of Human Nutrition and Dietetics. 2014;27(6):606–16.
410. Saghafi Z, Nikooyeh B, Jamali A, Mehdizadeh M, Zargaraan A. Influence of time and temperature on stability of added vitamin D3 during cooking procedure of fortified vegetable oils. Nutrition and Food Sciences Research. 2018;5(4):43–8.
411. Nikooyeh B, Neyestani T. The effects of vitamin D-fortified foods on circulating 25 (OH) D concentrations in adults: A systematic review and meta-analysis. British Journal of Nutrition. 2021:1–47.

412. Nikooyeh B, Ghodsi D, Neyestani TR. How much does serum 25 (OH) D improve by Vitamin D supplement and fortified food in children? A systematic review and meta-analysis. Journal of Pediatric Gastroenterology and Nutrition. 2021.
413. Maghbooli Z, Sahraian MA, Ebrahimi M, Pazoki M, Kafan S, Tabriz HM, et al. Vitamin D sufficiency, a serum 25-hydroxyvitamin D at least 30 ng/mL reduced risk for adverse clinical outcomes in patients with COVID-19 infection. PloS One. 2020; 15(9):e0239799.
414. Charoenngam N, Shirvani A, Holick MF. Vitamin D and its potential benefit for the COVID-19 pandemic. Endocrine Practice. 2021.
415. Bradshaw MJ, Holick MF, Stankiewicz JM. Vitamin D and multiple sclerosis. Clinical Neuroimmunology. 2020:197–212.

5

Functional Vegetables and Medicinal Uses

Cure and Curse

Sivakumar S. Moni

Contents

Introduction 117
Vegetables as a source 118
Spinach 119
Murraya koenigii 120
Moringa oleifera 120
Carrot 121
Onions 122
Garlic 122
Ginger 123
Lady's finger 124
Brinjal 125
Chili 125
Green beans 126
Cauliflower 127
Broccoli 128
References 129

Introduction

Food is our most basic requirement and has been accorded a prominent role in our lives. It is a fundamental requirement for humans because it is the source of both mental and physical energy. Other than major nutrient components, all foods have an adequate quantity of functional components that directly or indirectly benefit one's health. Nutritional elements of normal food categories that provide health advantages in addition to their nutritional worth are known as functional foods. Therefore, functional foods play an essential part in promoting human health and lowering illness risk (1, 2). The term "functional food" was coined in 1980 in Japan, when government authorities

DOI: 10.1201/9781003220053-7

began licensing meals with proven health advantages in an effort to improve the general public's health. More recently the term "nutraceuticals" is common (3). Vitamins and other extra supplements that are intended to improve one's health in some ways are included in some formulations, say for example, fortified food products with vitamins, minerals, probiotics, fiber, extracts of fruits, and so on (4). The dietary recommendations from health and nutrition experts have repeatedly emphasized the importance of consumption of fruits and vegetables (5). Many researchers report that naturally occurring vegetables may play an important role in maintaining cardiac health. For example, consumption of β carotene and lycopene are associated with a reduced risk of heart disease (6, 7). Despite repeated dietary guidelines emphasizing the significance of boosting fruit and vegetable consumption, global intakes are below recommended levels. Because of this, many people may be consuming a diet that is deficient in the nutrients and phytonutrients that are found in a diet that is rich in a variety of fruits and vegetables (5). The Food and Agriculture Organization of the United Nations and the World Health Organization have both issued warnings about chronic diseases caused by poor lifestyle choices and unhealthy diets (8, 9). Disease prevention has always been regarded more highly than disease treatment, and nutritious health-promoting components have long been a part of the traditional foods of many different cultures. According to several scientific studies, the consumption of specific vegetables, fruits, and cereals can help prevent or delay the onset of specific diseases and are regarded as important for the benefit of human society. Several previous studies have demonstrated a decreased risk of cardiovascular disease, cancer, chronic obstructive pulmonary disease, diabetes mellitus, and stroke when specific fruits and vegetables are consumed, as these are well-known sources of minerals, vitamins, dietary fiber, and other phytonutrients (10, 11). The advancement of human civilization and the adoption of an ultramodern living model have led to an increase in the number of industries producing valuable goods. Due to metals, elements, acids, and alkalis, industrialization has transformed human society and caused environmental contamination that impacts food materials. Moreover, consuming chemically processed foods frequently results in the consumption of refined materials and artificial substances such as chemical flavoring agents, coloring agents, and sweeteners, which are become unavoidable in this fast-paced ultra-modern world. These characteristics have an impact on the critical features of food components, as well as on the normal health of people and other animals (12, 13). The present chapter describes the cure and the curse of functional foods.

Vegetables as a source

Vegetables are beneficial to human health because they contain vitamins, minerals, phytochemicals, and dietary fiber. Vitamins A, C, and E, as well as dietary fiber, play critical roles in human health. An adequate vegetable diet has been shown to be protective against some chronic diseases, including diabetes, cancer, obesity, metabolic syndrome, and cardiovascular diseases, as well as improving risk factors associated with these conditions (14). With the increased interest in the effect of foods on a healthy life, people have begun to change their eating habits in recent years (15). The most essential aspect of these diets is their high vegetable consumption, which provides fiber, vitamins, minerals, flavonoids, phytoestrogens, sulfur compounds, and phenolic compounds such as monoterpenes and bioactive peptides, all of which are beneficial to health (15).

Spinach

Figure 5.1

Spinach is a native Persian, leafy green vegetable that is a rich source of calcium, iron, vitamin K, vitamin A, carotenoids, lutein, zeaxanthin, manganese, magnesium, folic acid, vitamin C, vitamin B2, potassium, dietary fiber, vitamin B6, vitamin E, and omega-3 fatty acids, which have antioxidant properties, and are essential for the maintenance, improvement, and regulation of the function of human tissues. Studies show that consumption of spinach could protect against cancer of the mouth, esophagus, and stomach (16). Glycoglycerolipid such as sulfoquinovosyl diacylglycerol (SQDG) and monogalactosyl diacylglycerol (MGDG) are present in spinach (16). Although eating spinach has many medicinal values, consuming an excessive amount of spinach can impair the body's ability to absorb nutrients. Spinach contains oxalic acid, which binds to zinc, magnesium, and calcium, preventing the body from absorbing adequate nutrients, resulting in mineral deficiencies. Purines and oxalates in high concentrations can cause kidney stones and gout. Gout or gouty arthritis can be made worse by the high purine content in spinach, which can cause joint discomfort, edema, and inflammation (17). Spinach is abundant in vitamin K, which may interact with anticoagulants and alter their action on other coagulating substances in the blood. Therefore, individuals who are on anticoagulant therapy such as warfarin should avoid eating spinach. Eating too much spinach can produce gas, bloating, cramps, diarrhea, and stomach pain (17).

Murraya koenigii

Figure 5.2

Murraya koenigii is a member of the Rutaceae family and is often known as the curry leaf plant. The leaf of the plant is highly regarded for its therapeutic properties as well as its characteristic aroma (18, 19). The extracts of *Murraya koenigii* leaves were reported to exhibit antioxidant, antibacterial, antifungal, anticaries, and antidiabetic properties, and for being hepatoprotective (20–21). A recent research has shown that methanolic leaf extracts contain unique bioactive compounds that describe the presence of antibacterial biomolecules in the plant (22). An earlier study suggested that methanolic extract of leaves of *Murraya koenigii* exhibited toxicity above the dosage of 50 mg/kg/day in an acute toxicity study (23). In contrast, Sakarkar et al., 2017 demonstrated 28 days toxicity study of the ethanolic extract of *Murraya koenigii* (24). The study showed that ingestion up to a dose level of 500 mg/kg is safe and does not result in any structural damage to the internal organs. An earlier study suggested that the methanolic extract of *Murraya koenigii* leaves exhibited prominent cytotoxic properties against Caco-2 human colon adenocarcinoma and minimal toxic effects against Hep2G human liver cells (25).

Moringa oleifera

Figure 5.3

Moringa oleifera is a tree that is frequently referred to as the drumstick tree, miracle tree, ben oil tree, or horseradish tree. The Moringaceae plant is a major crop in Asia and Africa, where it is grown as a root crop. *Moringa oleifera* leaves have been investigated extensively, and they have been demonstrated to be effective in a variety of chronic illnesses, including hypercholesterolemia, hypertension, type 2 diabetes, insulin resistance, nonalcoholic liver disease, cancer, and general inflammation (26). In 2009, Sreelatha and Padma reported the antioxidant properties of the *Moringa oleifera* leaf extracts (27). They stated the aqueous extract of *Moringa oleifera* has a strong scavenging impact on the free radicals 2, 2-diphenyl-2-picryl hydrazyl (DPPH), superoxide, and nitric oxide radicals, as well as prevention of lipid per oxidation. In line with this finding, the methanolic extract of leaves showed the highest free radical scavenging activity with IC50 value of 49.30 μg/mL in DPPH assay and 11.73 μg/mL in 2,2'-azinobis-(3-ethylbenzothiazoline-6-sulfonate (ABTS) assay (28). *M. oleifera* leaves are a better natural supply of antioxidants and anti-inflammatory compounds, and they have a lot of potential for being developed into health-promoting dietary supplements (29). The methanolic extract of *M. oleifera* leaves was exhibiting a good spectrum of antibacterial activity (30). The aqueous leaf extract of *M. oleifera* was exhibiting genotoxicities in human peripheral blood mononuclear cells. The study suggests that the *M. oleifera* leaf supplementation induces genotoxic effect at 3,000 mg/kg body weight. However, intake is safe at levels ≤ 1,000 mg/kg body weight (31).

Carrot

Figure 5.4

Carrots are a root vegetable that can be consumed raw or cooked (fried or steamed). Carrots are high in phytochemicals such as carotenoids, flavonoids, polyacetylenes, vitamins, and minerals, all of which have significant nutritional and health benefits (32, 33). Carrots have a wealth of antioxidants and offer many health benefits: good for eyes, lowers the risk of cancer and cardiac complications, improves the immune system, controls diabetes, and relieves constipation (34).

Onions

Figure 5.5

The onion, *Allium cepa*, is a native of central Asia and is one of the world's oldest cultivated plants, having been grown for almost 4,000 years. It is extensively used around the world for a variety of reasons, including food preparation and as a flavoring agent (35). Onions have excellent antioxidant, antibacterial, antifungal, antiparasitic, and antiviral properties; they have cardioprotective effects, and are anti-inflammatory, antihistaminic, anticancer, and so on. Onions are one of the most abundant sources of dietary flavonoids, which have been shown to lower the risk of cancer and inflammatory illnesses in several studies. A rich source of vitamin C, onions are also a good source of dietary fiber, which is crucial for maintaining healthy digestion. The ingestion of organosulfur compounds such as those present in onions on a regular basis may also help to prevent the development of cardiovascular disease (36). They are extremely high in fructans, which are beneficial soluble fibers. Fructans are prebiotic fibers, which means they provide food for the healthy bacteria in the gut. Onions are an important source for vitamin C and anthocyanins, pigments that found in red or purple onions, which are powerful antioxidants. Onions are high in flavonoid quercetin, a powerful antioxidant that may help decrease blood pressure and enhance heart health (37, 38). Onions are high in sulfur compounds such as sulfides and polysulfides, which may have anticancer properties (39). Furthermore, onions are rich in folate (B9), which is necessary for cell growth and metabolism, and is especially crucial for pregnant women. Onions are high in potassium, an essential element that lowers blood pressure and is beneficial to heart health. The sulfur-containing chemicals thiosulfinates, which are abundant in onions, may limit the growth of dangerous germs and prevent blood clot formation (40, 41). The use of onions may also result in bloating, cramps, and gas, despite their numerous health benefits.

Garlic

Garlic, *Allium sativum*, is a perennial herb that originates in western Asia and around the Mediterranean coast. It is frequently used as a flavoring in cooking,

Figure 5.6

but it has also been used medicinally throughout history, both ancient and modern, to prevent and treat a broad variety of ailments and diseases (42, 43). Garlic, either fresh or aged, has been reported to have potent antioxidant and antibacterial effects. However, aged garlic exhibited better antioxidant and antibacterial activity when compared to fresh garlic (44). Traditionally, research on garlic phytochemicals has concentrated on their cancer–chemo preventive qualities; however, there has been little published evidence demonstrating their therapeutic potential in cancer treatment (45, 46). Garlic has been shown to have immunomodulating properties and to be effective in the prevention of cancer, according to an earlier report (47). Garlic has been touted in alternative medicine as a potentially effective treatment for high blood pressure, coronary artery disease (hardened arteries), stomach cancer, colon cancer, or rectal cancer; to control high cholesterol levels; as well as a preventative for tick bites and for other ailments. Garlic applied topically to the skin may also be beneficial in the treatment of fungal skin infections such as ringworm, jock itch, and athlete's foot. Raw garlic has a number of side effects, the most common of which are unpleasant breath or body odor; heartburn, burning in your mouth or throat; nausea, vomiting, gas formation, and diarrhea, among other things.

Ginger

Figure 5.7

Zingiber officinale Roscoe (ginger) is a well-known herbaceous plant that has been widely used as a flavoring agent and herbal medicine for hundreds of years in various cultures. The consumption of the ginger rhizome, on the other hand, is a typical traditional remedy for the relief of common health problems such as pain, inflammation, nausea, and vomiting, as well as for the prevention and treatment of bacterial and viral infections and cancer (48, 49). Despite the fact that ginger is found to contain more than 400 compounds, the spice is primarily composed of carbohydrate, fat, and volatile oils. Ginger's fragrance and flavor are derived from its volatile oils, which are composed primarily of zingerone, shogaols, and gingerols, with 6-gingerol serving as the most pungent compound (48). Ginger supplementation has been shown to inhibit liver carcinogenesis by scavenging free radical formation and decreasing lipid peroxidation in the liver. NF-kB may be inactivated by ginger, which suppresses the pro-inflammatory tumor necrosis factor-a. Ginger may have anticancer and anti-inflammatory properties by inhibiting the expression of NF=kB. Ginger has been reported to be anticancer, and to prevent cardiovascular diseases, obesity, neurodegeneration, respiratory disorders, and diabetes mellitus (52–57). An increased bleeding tendency, gastrointestinal pain, cardiac arrhythmias, depression, dermatitis, and mouth and throat irritation are all side effects of ginger.

Lady's finger

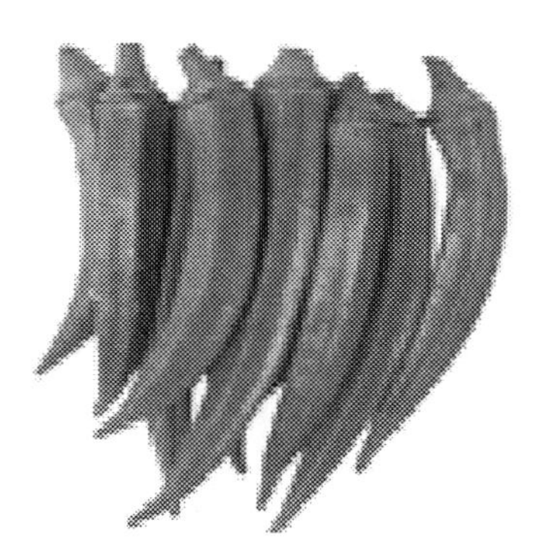

Figure 5.8 *Here*

Abelmoschus esculentus (L.), possesses remarkable biological features that confer a wide range of therapeutic benefits. This significant vegetable crop is farmed in tropical, subtropical, and warm temperate climes in countries ranging from Africa through Asia, southern Europe, and North America (58–60). The medicinal values of *Abelmoschus esculentus* are reported as antidiabetic effects, immunomodulatory activity, anticancer, anti-ulcer, hepatoprotective activity, and antioxidant (61–65). The chemical solanine is found in okra, and it is a poisonous molecule that can cause joint discomfort, arthritis, and persistent inflammation in some individuals. The high vitamin K content of okra may have an adverse effect on people who take blood-thinning medications such as warfarin or Coumadin.

Brinjal

Figure 5.9

Brinjal (*Solanum melongena* Linnaeus), known as eggplant, is a type of vegetable that belongs to the Solanaceae crop family. Eggplant is high in anthocyanins, which is a type of pigment found in eggplant that has powerful antioxidant and anticancer properties (66–68). Eggplant also reduces the risk of heart diseases by lowering the levels of both LDL cholesterol and triglycerides (69, 70). Since it contains polyphenols, eggplant helps to reduce blood glucose levels and protects against the development of diabetes mellitus (71). Eggplant helps in body weight management since it is high in fiber and low in calories. Despite the numerous health benefits that eggplant provides, eating this vegetable in large quantities can have negative consequences for your overall health. Nasunin, which can be found in eggplant, is a phytochemical that has the ability to bind with iron and remove it from the body's cells. The oxalates in this vegetable have the potential to cause kidney stones. Eggplant also causes allergy symptoms such as itchiness, coughing, diarrhea, wheezing, shortness of breath, dizziness, nausea, and vomiting (72).

Chili

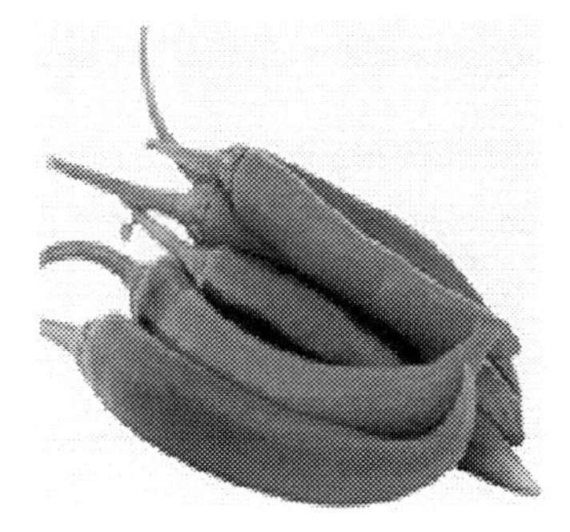

Figure 5.10

Capsicum annuum and *Capsicum frutescens* (Solanaceae) are medicinal spices utilized in several Indian traditional systems of medicine and have been shown to treat a variety of diseases. Green chilies are high in vitamin C, which helps to improve the immune system and prevent cardiac diseases such as atherosclerosis. Green chilies are a natural source of iron that can also aid in the absorption of iron through the body's digestive system. The key ingredient in green chilies, capsaicin, aids in the stimulation of blood flow through the membranes and causes mucus discharge to thin. More than 200 constituents of chili have been identified, and some of these active constituents have been shown to have numerous beneficial effects in the treatment of gastrointestinal disorders, including the stimulation of digestion and gastro mucosal defense, the reduction of gastroesophageal reflux disease (GERD) symptoms, the inhibition of gastrointestinal pathogens and cancer, the regulation of gastrointestinal secretions, and the absorption of nutrients. Green chilies have also been reported for anticancer activity and antibacterial activity (73–77). *Capsicum frutescens* is a type of hot pepper that is widely cultivated in many regions of the world, including India, and is known as the red chili. Red chilies are beneficial to one's digestive health since they enhance the secretion of digestive fluids, which helps to avoid problems such as constipation and gas from occurring. Capsaicin, found in red chili powder, is believed to enhance fat burning while simultaneously decreasing appetite. It has been shown to lower food cravings and enhance metabolism, both of which are good for weight loss. Capsaicin has also been shown to be thermogenic (78, 79).

The active ingredient, capsaicin, has been shown to lower triglycerides, cholesterol, and platelet aggregation. It may also aid in the dissolution of fibrin in the body, which may assist to avoid the formation of blood clots. This has the potential to contribute to overall better heart health. The potential of red chili powder to reduce nasal congestion by removing mucus from the nose is one of the many health advantages of this spice. Its antibacterial qualities also aid in the treatment of persistent sinus infections, as it has the ability to cause vasoconstriction in the blood vessels of the nose. Capsaicin is believed to reduce inflammation and act as a natural pain reliever. It forms a chemical bond with pain-sensing nerve endings, so alleviating the sensation of pain. Also beneficial for the lower back are capsaicin's anti-inflammatory properties, which help relieve discomfort and inflammation in the muscles and joints. The carotenoid capsanthin is the most abundant carotenoid in red chili peppers, accounting for up to 50% of the total carotenoid content. Capsanthin is responsible for the red color of the peppers. Its potent antioxidant qualities may help to prevent cancer (80–82). Although it has many beneficial features, it can also cause side effects such as ulcers, stomach pain, and diarrhea.

Green beans

Green beans, *Phaseolus vulgaris,* are a nutritional and economic staple crop in every region of the planet. Apart from supplying nutrients such as complex

Figure 5.11

carbs, increased protein, dietary fiber, minerals, and vitamins. Green beans also include a diverse array of polyphenolic chemicals with potential health advantages. Green beans are rich in vitamin C, which is a potent antioxidant. They are also rich in vitamin A, which enriches immune health. Green beans are high in soluble fiber, which helps to improve the heart health by lowering LDL cholesterol levels. It is crucial to note that green beans are a good source of manganese, which is essential for cellular metabolism, and they have antioxidant properties. Green beans are abundant in fiber, which helps to maintain a healthy digestive system. The high fiber and protein content in green beans helps to reduce body weight and reduce glycemic index, which is significant in managing diabetes. Green beans help regulate insulin and triglyceride levels, which also helps to dramatically reduce the risk of developing diabetes (83–88). Regardless of the health benefits of green beans, consuming fava beans can cause a condition known as favism, which is a neurological disorder. Anemia can be caused by fascism, which destroys red blood cells. Beans also contain phytic acid, which can interfere with the absorption of minerals by the body (89, 90). Beans can cause flatulence, stomach pain, or bloating due to indigestion (86).

Cauliflower

Figure 5.12

Cauliflower, a cruciferous vegetable, is a healthy food that includes unique plant components that may lessen the risk of a variety of ailments, including cancer and heart disease. Cauliflower is an very good source of vitamins and minerals and is rich in fiber, which reduces inflammatory bowel disease. Cauliflower is an excellent source of antioxidants, such as glucosinolates and isothiocyanates, which help to prevent the development of cancer (91–94).

The antioxidants carotenoid and flavonoid found in cauliflower have anticancer properties and may lower the chance of developing a variety of other ailments, including heart disease. Cauliflower includes a high concentration of vitamin C, which has antioxidant properties. It is well-known for its anti-inflammatory properties, which may help to improve immunological function while also lowering the risk of heart disease and cancer (95, 96). Sulforaphane, found in cauliflower, has been shown to decrease cancer growth by reducing the activity of enzymes that are involved in cancer development (96). According to a research report, sulforaphane may also help lower high blood pressure and maintain healthy arteries, both of which are important factors in the prevention of heart disease (97). Cauliflower contains a high level of fiber, which can lead to bloating and flatulence. Cauliflower also contains vitamin K, which might create issues for people who take blood thinners (98).

Broccoli

Figure 5.13

Broccoli, *Brassica oleracea*, is an edible green vegetable that is a fast-growing annual plant that is a member of the Brassicaceae family. Broccoli's medicinal significance has been well documented as an anticancer, immunomodulator, antidiabetic, antibacterial, hepatoprotective, cardioprotective, antiamnesic, and antioxidant (99–104). Broccoli has significant levels of glucoraphanin, which is transformed into sulforaphane during digestion, and sulforaphane is a powerful antioxidant (105). Besides sulforaphane, broccoli includes trace levels of the antioxidants lutein and zeaxanthin, which have been shown to protect against oxidative stress and cellular damage in the eyes (106). Broccoli contains

Kaempferol, which is a flavonoid component that has been shown to have anti-inflammatory properties (107). A number of studies have demonstrated that broccoli has anticancer qualities can lower the incidence of breast cancer, gastric cancer, colorectal cancer, renal cancer, prostate cancer, and bladder cancer, among other cancers (108–113). Broccoli has been shown to lower bad LDL cholesterol and triglyceride levels while increasing good HDL cholesterol (114). Excessive consumption of broccoli will cause hyperthyroidism because of the presence of thiocyanates.

References

1. Galanakis CM. Functionality of Food Components and Emerging Technologies. Foods. 2021;10(1):128. https://doi.org/10.3390/foods10010128
2. Tahseen FM. Functional Food - A Review. European Academic Research. 2016; 4(6):5695–5702.
3. Lang T. Functional Foods. BMJ. 2007;334(7602):1015–1016. https://doi.org/10.1136/bmj.39212.592477.BE
4. Esther E. Functional Foods. eatright.org. Academy of Nutrition and Dietetics, January 4, 2021. https://www.eatright.org/food/nutrition/healthy-eating/functional-foods
5. Mary MM, Leila MB, Judith HS, Dena RH, Keith R. Global Assessment of Select Phytonutrient Intakes by Level of Fruit and Vegetable Consumption. The British Journal of Nutrition. 2014;112:1004–1018. https://doi.org/10.1017/S000711451400193
6. Mente A, de Koning L, Shannon HS, Anand SS. A Systematic Review of the Evidence Supporting a Causal Link between Dietary Factors and Coronary Heart Disease. 2009;169(7):659–669. https://doi.org/10.1001/archinternmed.2009.38
7. Liu RH. Dietary Bioactive Compounds and Their Health Implications. Journal of Food Science. 2013;78(S1):A18–A25. https://doi.org/10.1111/1750-3841.12101
8. Day L, Seymour RB, Pitts KF, Konczak I, Lundin L. Incorporation of Functional Ingredients into Foods. Trends in Food Science & Technology. 2009;20: 388–395. https://doi.org/10.1016/j.tifs.2008.05.002
9. WHO. Diet, Nutrition and the Prevention of Chronic Diseases. WHO; Geneva, Switzerland 2002. (WHO Technical Report Series 916). www.fao.org/docrep/005/AC911E/AC911E00.HTM#Contents
10. Jayaprakash S, Sivakumar SM, Hafiz AM,Mohammed IA, Shyam S P, Rahimullah S, Mohamed EE. Antibacterial Potential of Ethanolic Extract of Broccoli (*Brassica oleracea* var. Italica) Against Humanpathogenic Bacteria. International Journal of Pharmaceutical Research. 2018;10(2):143–146.
11. Dhandevi PEM, Rajaesh J. Fruit and Vegetable Intake: Benefits and Progress of Nutrition Education Interventions–Narrative Review Article. Iranian Journal of Public Health. 2015;44(10):1309–1321.
12. Morais S, Costa FG, de Lourdes PM. Heavy Metals and Human Health. In (Ed.), Environmental Health - Emerging Issues and Practice. IntechOpen 2012: 227–246. https://doi.org/10.5772/29869.
13. Monisha J, Tenzin T, Naresh A et al. Toxicity, Mechanism and Health Effects of Some Heavy Metals. Interdisciplinary Toxicology. 2014;7(2):60–72.
14. Ülger TG, Ayşe NS, Onur C, Funda PC. Chapter 2- Role of Vegetables in Human Nutrition and Disease Prevention. Intech Open. 2018. https://www.intechopen.com/chapters/61691
15. Septembre-Malaterreb A, Remizeb F, Pouchereta P. Fruits and Vegetables, as a Source of Nutritional Compounds and Phytochemicals: Changes in Bioactive Compounds during Lactic Fermentation. Food Research International. 2018;104:86–99. https://doi.org/10.1016/j.foodres.2017.09.031

16. Naoki M, Hiromi Y, Yoshiyuki M. Chapter 26 - Spinach and Health: Anticancer Effect, Editor(s): Ronald Ross Watson, Victor R. Preedy. Bioactive Foods in Promoting Health, Academic Press, 2010: 393–405. https://doi.org/10.1016/B978-0-12-374628-3.00026-8
17. https://www.medicinenet.com/is_eating_spinach_every_day_good_for_you/article.html
18. Sivakumar SM, Muhammad HS, Hafiz AM, et al. Phytochemical, and Spectral Analysis of the Methanolic Extracts of Leaves of *Murraya koenigii* of Jazan, Saudi Arabia. Natural Product Research. 2021;35(15): 2569–2573. https://doi.org/10.1080/14786419.2019.1679137
19. Jain V, Momin M, Laddha K. *Murraya Koenigii*: An Updated Review. International Journal of Ayurvedic and Herbal Medicine. 2012;2(4):607.
20. Palwankar SM, Kale PP, Kadu PK, Prabhavalkar K. Assessment of Antidiabetic Activity of Combination of *Murraya Koenigii* Leaves Extract and *Vitis Vinifera* Seeds Extract in Alloxan-Induced Diabetic Rats. Journal of Reports in Pharmaceutical Science. 2020;9:79–85. https://www.jrpsjournal.com/text.asp?2020/9/1/79/287588
21. Tachibana Y, Kikuzaki H, Lajis NH, Nakatani N. Comparison of Anti-Oxidative Properties of Carbazole Alkaloids from *Murraya Koenigii* Leaves. Journal of Agricultural and Food Chemistry. 2003;51:6461–6467. http://dx.doi.org/10.1021/jf034700+
22. Tachibana Y, Kikuzaki H, Lajis NH, Nakatani N. Antioxidative Activity of Carbazoles from *Murraya Koenigii* Leaves. Journal of Agricultural and Food Chemistry. 2001; 49:5589–5594. http://dx.doi.org/10.1021/jf010621r
23. Marwan SA, Anil Kumar S, Ahmed GA, Abualrahman MA. Chronic LD50 vs Safest Dose for the Methanolic Extract of Curry Leaves (*Murraya Koenigii*) Cultivated in Malaysia. Journal of Applied Pharmaceutical Science. 2014;4(8):56–58. https://doi.org/10.7324/JAPS.2014.40811
24. Sakarkar DM, Tembhurne SV, More BH. 28 Days Repeated Dose Toxicity Study of Ethanolic Extract of *Murraya Koenigii* in Wistar Rats. Annals of Pharmacology and Pharmaceuticals. 2017;2(4):1047.
25. Patterson J, Verghese M. Anticancer and Toxic Effects of Curry Leaf (*Murraya Koenigii*) Extracts. Journal of Pharmacology and Toxicology. 2015;10(2):49–59. https://doi.org/10.3923/jpt.2015.49.59
26. Vergara-Jimenez M, Almatrafi MM, Fernandez ML. Bioactive Components in *Moringa oleifera* Leaves Protect against Chronic Disease. Antioxidants (Basel, Switzerland). 2017;6(4):91. https://doi.org/10.3390/antiox6040091
27. Sreelatha S, Padma PR. Antioxidant Activity and Total Phenolic Content of *Moringa oleifera* Leaves in Two Stages of Maturity. Plant Foods for Human Nutrition. 2009; 64(4):303–311. https://doi.org/10.1007/s11130-009-0141-0
28. Fitriana WD, Taslim E, Kuniyoshi S, Sri F. Antioxidant Activity of *Moringa oleifera* Extracts. Indonesian Journal of Chemistry. 2016;16(3):297–301.
29. Xu YB, Chen GL, Guo MQ. Antioxidant and Anti-Inflammatory Activities of the Crude Extracts of *Moringa oleifera* from Kenya and Their Correlations with Flavonoids. Antioxidants (Basel). 2019;8(8):296. https://doi.org/10.3390/antiox8080296
30. Abdalla M, Alwasilah HY, Rasha Al Hussein M. Evaluation of Antimicrobial Activity of *Moringa oleifera* Leaf Extracts Against Pathogenic Bacteria Isolated from Urinary Tract Infected Patients. Journal of Advanced Laboratory Research in Biology. 2016;7(2):47–51.
31. Asare GA, Gyan B, Bugyei K, et al. Toxicity Potentials of the Nutraceutical *Moringa oleifera* at Supra-Supplementation Levels. Journal of Ethnopharmacology. 2012 Jan 6;139(1): 265–272. https://doi.org/10.1016/j.jep.2011.11.009
32. Ahmad T, Cawood M, Iqbal Q, Ariño A, et al. Phytochemicals in *Daucus carota* and Their Health Benefits-Review Article. Foods. 2019;19;8(9):424. https://doi.org/10.3390/foods8090424
33. da Silva Dias JC. Nutritional and Health Benefits of Carrots and Their Seed Extracts. Food and Nutrition Sciences. 2014;5:2147–2156. https://doi.org/10.4236/fns.2014.522227

34. Liu MB, Wang W, Zhou MG. Trend Analysis on the Mortality of Cardiovascular Diseases from 2004 to 2010 in China. Chinese Journal of Endemiology. 2013;34:985–988.
35. Marrelli M, Amodeo V, Statti G, Conforti F. Biological Properties and Bioactive Components of *Allium cepa* L.: Focus on Potential Benefits in the Treatment of Obesity and Related Comorbidities. Molecules (Basel, Switzerland). 2018;24(1):119. https://doi.org/10.3390/molecules24010119
36. Kendler, B.S. Garlic (*Allium sativum*) and Onion (*Allium cepa*): A Review of Their Relationship to Cardiovascular Disease. Preventive Medicine. 1987;16:670–685.
37. Francisco PV, David BB, Federica L, et al. The Flavonoid Quercetin Induces Apoptosis and Inhibits JNK Activation in Intimal Vascular Smooth Muscle Cells. Biochemical and Biophysical Research Communications. 2006; 346(3): 919–925. https://doi.org/10.1016/j.bbrc.2006.05.198
38. Larson AJ, David S J, Thunder J. Therapeutic Potential of Quercetin to Decrease Blood Pressure: Review of Efficacy and Mechanisms. Advances in Nutrition. 2012;3(1):39–46. https://doi.org/10.3945/an.111.001271
39. Powolny AA, Singh SV. Multitargeted Prevention and Therapy of Cancer by Diallyl Trisulfide and Related *Allium* Vegetable-Derived Organosulfur Compounds. Cancer Letters. 2008;269(2):305–314. https://doi.org/10.1016/j.canlet.2008.05.027
40. Rose P, Whiteman M, Moore PK, Zhu YZ. Bioactive S-alk(en)yl Cysteine Sulfoxide Metabolites in the Genus Allium: The Chemistry of Potential Therapeutic Agents. Natural Product Reports. 2005;22(3):351–368. https://doi.org/10.1039/b417639c
41. Vazquez-Prieto MA, Miatello RM. Organosulfur Compounds and Cardiovascular Disease. Molecular Aspects of Medicine. 2010;31(6):540–545. https://doi.org/10.1016/j.mam.2010.09.009
42. Saravanan G, Prakash J. Effect of Garlic (*Allium sativum*) on Lipid Peroxidation in Experimental Myocardial Infarction in Rats. Journal of Ethnopharmacology. 2004;94(1):155–158. https://doi.org/10.1016/j.jep.2004.04.029
43. Ide N, Lau BH. Garlic Compounds Protect Vascular Endothelial Cells from Oxidized Low-Density Lipoprotein-Induced Injury. Journal of Pharmacy and Pharmacology. 1997;49(9):908–911. https://doi.org/10.1111/j.2042-7158.1997.tb06134.x
44. Jang HJ, Lee HJ, Yoon DK, et al. Antioxidant and Antimicrobial Activities of Fresh Garlic and Aged Garlic by-Products Extracted with Different Solvents. Food Science and Biotechnology. 2017;27(1):219–225. https://doi.org/10.1007/s10068-017-0246-4
45. Yan Z, Xingping L, Jun R, et al. Phytochemicals of Garlic: Promising Candidates for Cancer Therapy. Biomedicine & Pharmacotherapy. 2020;123:109730. https://doi.org/10.1016/j.biopha.2019.109730
46. Li Z, Le W, Cui Z. A Novel Therapeutic Anticancer Property of Raw Garlic Extract via Injection but not Ingestion. Cell Death Discovery. 2018;4:108. https://doi.org/10.1038/s41420-018-0122-x
47. Schäfer G, Kaschula CH. The Immunomodulation and Anti-Inflammatory Effects of Garlic Organosulfur Compounds in Cancer Chemoprevention. Anti-Cancer Agents in Medicinal Chemistry. 2014;14(2):233–240. https://doi.org/10.2174/18715206113136660370
48. Li H, Liu Y, Luo D, et al. Ginger for Health Care: An Overview of Systematic Reviews. Complementary Therapies in Medicine. 2019;45:114–123. https://doi.org/10.1016/j.ctim.2019.06.002
49. Anh NH, Kim SJ, Long NP, et al. Ginger on Human Health: A Comprehensive Systematic Review of 109 Randomized Controlled Trials. Nutrients. 2020;12(1):157. https://doi.org/10.3390/nu12010157
50. Yusof YA, Ahmad N, Das S, et al. Chemopreventive Efficacy of Ginger (*Zingiber officinale*) in Ethionine Induced Rat Hepatocarcinogenesis. African Journal of Traditional, Complementary and Alternative Medicines. 2008;6(1):87–93. https://doi.org/10.4314/ajtcam.v6i1.57078

51. Habib SH, Makpol S, Abdul Hamid NA, et al. Ginger Extract (*Zingiber officinale*) has Anti-Cancer and Anti-Inflammatory Effects on Ethionine-Induced Hepatoma Rats. Clinics. 2008;63(6):807–813. https://doi.org/10.1590/S1807-59322008000600017
52. Ho S, Chang K, Lin C. Anti-Neuroinflammatory Capacity of Fresh Ginger is Attributed Mainly to 10-Gingerol. Food Chemistry. 2013;141:3183–3191. https://doi.org/10.1016/j.foodchem.2013.06.010
53. Akinyemi AJ, Thome GR, Morsch VM, et al. Effect of Dietary Supplementation of Ginger and Turmeric Rhizomes on Angiotensin-1 Converting Enzyme (ACE) and Arginase Activities in L-NAME Induced Hypertensive Rats. Journal of Functional Foods. 2015;17:792–801. https://doi.org/10.1016/j.jff.2015.06.011
54. Suk, S, Kwon GT, Lee E, et al. Gingerenone A, a Polyphenol Present in Ginger, Suppresses Obesity and Adipose Tissue Inflammation in High-fat Diet-fed Mice. Molecular Nutrition and Food Research. 2017;61:1700139. https://doi.org/10.1002/mnfr.201700139
55. Wei C, Tsai Y, Korinek M, et al. 6-Paradol and 6-Shogaol, The Pungent Compounds of Ginger, Promote Glucose Utilization in Adipocytes and Myotubes, and 6-paradol Reduces Blood Glucose in High-fat Diet-fed Mice. International Journal of Molecular Sciences. 2017;18:168. https://doi.org/10.3390/ijms18010168
56. Walstab J, Krueger D, Stark T, et al. Ginger and its Pungent Constituents Non-Competitively Inhibit Activation of Human Recombinant and Native 5-HT3 Receptors of Enteric Neurons. Neurogastroenterology & Motility. 2013;25:439–447. https://doi.org/10.1111/nmo.12107
57. Townsend EA, Siviski ME, Zhang Y, et al. Effects of Ginger and Its Constituents on Airway Smooth Muscle Relaxation and Calcium Regulation. American Journal of Respiratory Cell and Molecular Biology. 2013;48:157–163. https://doi.org/10.1165/rcmb.2012-0231OC
58. Ndunguru J, Rajabu A. Effect of Okra Mosaic Virus Disease on the Above-Ground Morphological Yield Components of Okra in Tanzania. Scientia Horticulturae. 2004;99:225–235. https://doi.org/10.1016/S0304-4238(03)00108-0
59. Naveed A, Khan AA, Khan IA. Generation Mean Analysis of Water Stress Tolerance in Okra (*Abelmoschus esculentus* L.) Pakistan Journal of Botany. 2009;41:195–205.
60. Saifullah M, Rabbani MG. Evaluation, and Characterization of Okra (*Abelmoschus esculentus* L. Moench.) Genotypes. SAARC Journal of Agriculture. 2009;7:92–99. https://doi.org/10.52804/ijaas2020.111
61. Elkhalifa A, Alshammari E, Adnan M, et al. Okra (*Abelmoschus esculentus*) as a Potential Dietary Medicine with Nutraceutical Importance for Sustainable Health Applications. Molecules. 2021;26(3):696. https://doi.org/10.3390/molecules26030696
62. Monte LG, Santi-Gadelha T, Reis LB, et al. Lectin of *Abelmoschus esculentus* (Okra) Promotes Selective Antitumor Effects in Human Breast Cancer Cells. Biotechnology Letters. 2014;36(3):461–469. https://doi.org/10.1007/s10529-013-1382-4
63. Ortaç D, Cemek M, Karaca T, et al. In Vivo Anti-Ulcerogenic Effect of Okra (*Abelmoschus esculentus*) on Ethanol-Induced Acute Gastric Mucosal Lesions. Pharmaceutical Biology. 2018;56(1):165–175. https://doi.org/10.1080/13880209.2018.1442481
64. Wahyuningsih SPA, Sajidah ES, Atika BND, et al. Hepatoprotective Activity of Okra (*Abelmoschus esculentus* L.) in Sodium Nitrite-Induced Hepatotoxicity. Veterinary World. 2020;13(9):1815–1821. https://doi.org/10.14202/vetworld.2020.1815-1821
65. Liao Z, Zhang J, Liu B, et al. Polysaccharide from Okra (*Abelmoschus esculentus* (L.) Moench). Improves Antioxidant Capacity via PI3K/AKT Pathways and Nrf2 Translocation in a Type 2 Diabetes Model. Molecules. 2019;24(10):1906. https://doi.org/10.3390/molecules24101906
66. Abhishek TS, Dwivedi SA. Review on Integrated Management of Brinjal Shoots and Fruit Borer, *Leucinodes orbonalis* (Guenee). Journal of Entomology and Zoology Studies. 2021;9(1):181–189. https://doi.org/10.22271/j.ento.2021.v9.i1c.8143

67. Sadilova E, Stintzing FC, Carle R. Anthocyanins, Colour and Antioxidant Properties of Eggplant (*Solanum melongena* L.) and Violet Pepper (*Capsicum annuum* L.) Peel Extracts. Zeitschrift für Naturforschung C. 2006;61(7–8):527–535. https://doi.org/10.1515/znc-2006-7-810
68. Nishino H, Tokuda H, Satomi Y, et al. Cancer Prevention by Antioxidants. Biofactors. 2004;22(1–4):57–61. https://doi.org/10.1002/biof.5520220110
69. Jorge PA, Neyra LC, Osaki RM, et al. Effect of Eggplant on Plasma Lipid Levels, Lipidic Peroxidation and Reversion of Endothelial Dysfunction in Experimental Hypercholesterolemia. Arquivos Brasileiros de Cardiologia. 1998;70(2):87–91. https://doi.org/10.1590/s0066-782x1998000200004
70. Das S, Raychaudhuri U, Falchi M, et al. Cardioprotective Properties of Raw and Cooked Eggplant (*Solanum melongena* L). Food & Function. 2011;2(7):395–409. https://doi.org/10.1039/c1fo10048c
71. Hanhineva K, Törrönen R, Bondia-Pons I, et al. Impact of Dietary Polyphenols on Carbohydrate Metabolism. International Journal of Molecular Sciences. 2010; 11(4): 1365–1402. https://doi.org/10.3390/ijms11041365
72. Charles M. Health line. https://www.healthline.com/health/food-nutrition/eggplant-allergy, (2018).
73. Mann A. Biopotency Role of Culinary Spices and Herbs and Their Chemical Constituents in Health and Commonly Used Spices in Nigerian Dishes and Snacks. African Journal of Food Science. 2011;5:111–124. https://doi.org/10.5897/AJFS.9000032
74. Gantait A, Maji A, Barman T, et al. Estimation of Capsaicin Through Scanning Densitometry and Evaluation of Different Varieties of Capsicum in India. Natural Product Research. 2012;26:216–222. https://doi.org/10.1080/14786419.2010.535169
75. Khan FA, Mahmood T, Ali M, Saeed A, Maalik A. Pharmacological Importance of an Ethnobotanical Plant: *Capsicum annuum* L. Natural Product Research. 2014;28:1267–1274. https://doi.org/10.1080/14786419.2014.895723
76. Sharma J, Sharma P, Sharma B, Chaudhary P. In-Vitro Estimation of Antioxidant Activity in Green Chilli (*Capsicum annuum*) and Yellow Lantern Chilli (*Capsicum chinense*). International Journal of Research & Review. 2017;4(6):54–61.
77. Anju TR, Gopal N and Parvathy S. Antibacterial Effect of Pepper and Chilli against *Staphylococcus Aureus*: A Comparative Study. International Journal of Biochemistry & Physiology. 2019;4(3):000162. https://doi.org/10.23880/ijbp-16000162
78. Fattori V, Hohmann MS, Rossaneis AC. Capsaicin: Current Understanding of Its Mechanisms and Therapy of Pain and Other Pre-clinical and Clinical Uses. Molecules. 2016;21(7):E844. https://doi.org/10.3390/molecules21070844
79. Moscone E, Scaldaferro MA, Grabiele M, et al. The Evolution of Chili Peppers (*Capsicum* - Solanaceae): A Cytogenetic Perspective. Acta Horticulturae. 2007;745:137–170. https://doi.org/10.17660/ActaHortic.2007.745.5
80. Gómez-García Mdel R, Ochoa-Alejo N. Biochemistry and Molecular Biology of Carotenoid Biosynthesis in Chili Peppers (*Capsicum* spp.). International Journal of Molecular Sciences. 2013;14(9):19025–19053. https://doi.org/10.3390/ijms140919025
81. Kim S, Tae Youl H, In Kyeong H. Analysis, Bioavailability, and Potential Healthy Effects of Capsanthin, Natural Red Pigment from *Capsicum* spp. Food Reviews International. 2009;25(3):198–213. https://doi.org/10.1080/87559120902956141
82. Erden Y. Capsanthin Stimulates the Mitochondrial Apoptosis-Mediated Cell Death, Following DNA Damage in MCF-7 Cells. Nutrition and Cancer. 2021;73(4). https://doi.org/10.1080/01635581.2020.1819347
83. Larson AJ, Symons JD, Jalili T. Therapeutic Potential of Quercetin to Decrease Blood Pressure: Review of Efficacy and Mechanisms. Advances in Nutrition. 2012;3(1): 39–46. https://doi.org/10.3945/an.111.001271
84. Leidy HJ, Clifton PM, Astrup A, et al. The Role of Protein in Weight Loss and Maintenance. The American Journal of Clinical Nutrition. 2015;101(6):1320S–1329S. https://doi.org/10.3945/ajcn.114.084038

85. Clark MJ, Slavin JL. The Effect of Fiber on Satiety and Food Intake: A Systematic Review. Journal of American College of Nutrition. 2013;32(3):200–211. https://doi.org/10.1080/07315724.2013.791194
86. Messina V. Nutritional and Health Benefits of Dried Beans. The American Journal of Clinical Nutrition. 2014;100(Suppl 1):437S–442S. https://doi.org/10.3945/ajcn.113.071472
87. Jenkins DJ, Kendall CW, Augustin LS, et al. Effect of Legumes as Part of a Low Glycemic Index Diet on Glycemic Control and Cardiovascular Risk Factors in Type 2 Diabetes Mellitus: A Randomized Controlled Trial. Archives of Internal Medicine. 2012;172(21):1653–1660. https://doi.org/10.1001/2013.jamainternmed.70
88. Hosseinpour-Niazi S, Mirmiran P, Hedayati M, Azizi F. Substitution of Red Meat with Legumes in the Therapeutic Lifestyle Change Diet Based on Dietary Advice Improves Cardiometabolic Risk Factors in Overweight Type 2 Diabetes Patients: A Cross-Over Randomized Clinical Trial. European Journal of Clinical Nutrition. 2015;69(5):592–597. https://doi.org/10.1038/ejcn.2014.228
89. Arese P, Mannuzzu L, Turrini F. Pathophysiology of Favism. Folia Haematol Int Mag Klin Morphol Blutforsch. 1989;116(5):745–752.
90. De Flora A, Morelli A, Grasso M, et al. Alterations of Red Blood Cell Proteolysis in Favism. Biomedica Biochimica Acta. 1987;46(2–3):S184–189.
91. Abdull Razis AF, Noor NM. Cruciferous Vegetables: Dietary Phytochemicals for Cancer Prevention. Asian Pacific Journal of Cancer Prevention. 2013;14(3):1565–1570. https://doi.org/10.7314/apjcp.2013.14.3.1565
92. Higdon JV, Delage B, Williams DE, Dashwood RH. Cruciferous Vegetables and Human Cancer Risk: Epidemiologic Evidence and Mechanistic Basis. Pharmacological Research. 2007;55(3):224–236. https://doi.org/10.1016/j.phrs.2007.01.009
93. Keck AS, Finley JW. Cruciferous Vegetables: Cancer Protective Mechanisms of Glucosinolate Hydrolysis Products and Selenium. Integrative Cancer Therapies. 2004;3(1):5–12. https://doi.org/10.1177/1534735403261831
94. Ahmed FA, Ali RF. Bioactive Compounds and Antioxidant Activity of Fresh and Processed White Cauliflower. BioMed Research International. 2013;2013:367819. https://doi.org/10.1155/2013/367819
95. Chambial S, Dwivedi S, Shukla KK, et al. Vitamin C in Disease Prevention and Cure: An Overview. Indian Journal of Clinical Biochemistry. 2013;28(4):314–328. https://doi.org/10.1007/s12291-013-0375-3
96. de Figueiredo SM, Binda NS, Nogueira-Machado JA, et al. The Antioxidant Properties of Organosulfur Compounds (Sulforaphane). Recent Patent on Endocrine, Metabolic & Immune Drug Discovery. 2015;9(1):24–39. https://doi.org/10.2174/1872214809666150505164138
97. Bai Y, Wang X, Zhao S, et al. Sulforaphane Protects Against Cardiovascular Disease via Nrf2 Activation. Oxidative Medicine and Cellular Longevity. 2015;2015:407580. https://doi.org/10.1155/2015/407580
98. Ware M, 2017. https://www.medicalnewstoday.com/articles/282844
99. Pacheco-Cano RD, Salcedo-Herandez R, Lopez-Meza JE, et al. Antimicrobial Activity of Broccoli (*Brassica oleracea* Var. Italica) Cultivar Avenger Against Pathogenic Bacteria, Phytopathogenic Filamentous Fungi and Yeast. Journal of Applied Microbiology. 2017; 124:126–135. https://doi.org/10.1111/jam.13629
100. Park SK, Ha JS, Kim JM, et al. Antiamnesic Effect of Broccoli (*Brassica oleracea* var. Italica) Leaves on Amyloid Beta (Aβ) 1-42 – Induced Learning and Memory Impairment. Journal of Agricultural and Food Chemistry. 2016;64:3353–3361. https://doi.org/10.1021/acs.jafc.6b00559
101. Vinha AF, Alves RC, Barreira SVP, et al. Impact of Boiling on Phytochemicals and Antioxidant Activity of Green Vegetables Consumed in the Mediterranean Diet. Food & Function. 2015;6:1157–1163. https://doi.org/10.1039/c4fo01209g
102. Owis AI. Broccoli: The Green Beauty: A Review. Journal of Pharmaceutical Sciences and Research. 2015;7:696–703.

103. Mahn A, Reyes A. An Overview of Health- Promoting Compounds of Broccoli (*Brassica oleracea* var. Italica) and the Effect of Processing. Food Science and Technology International. 2012;18:503–514. https://doi.org/10.1177/1082013211433073
104. Yang Y, Zhang X. In Vitro Antitumor Activity of Broccolini Seeds Extracts. Scanning. 2011;23:402–404. https://doi.org/10.1002/sca.20256
105. Conzatti A, Fróes FC, Schweigert Perry ID, Souza CG. Clinical and Molecular Evidence of the Consumption of Broccoli, Glucoraphanin and Sulforaphane in Humans. Nutrición hospitalaria 2014;31(2):559–569. https://doi.org/10.3305/nh.2015.31.2.7685
106. Abdel-Aal el., Akhtar H, Zaheer K, Ali R. Dietary Sources of Lutein and Zeaxanthin Carotenoids and Their Role in Eye Health. Nutrients. 2013;5(4):1169–1185. https://doi.org/10.3390/nu5041169
107. Wang J, Fang X, Ge L, et al. Antitumor, Antioxidant and Anti-Inflammatory Activities of Kaempferol and Its Corresponding Glycosides and the Enzymatic Preparation of Kaempferol. PLoS One. 2018;13(5):e0197563. https://doi.org/10.1371/journal.pone.0197563
108. Liu X, Lv K. Cruciferous Vegetables Intake Is Inversely Associated with Risk of Breast Cancer: A Meta-Analysis. Breast. 2013;22(3):309–313. https://doi.org/10.1016/j.breast.2012.07.013
109. Liu B, Mao Q, Cao M, et al. Cruciferous Vegetables Intake and Risk of Prostate Cancer: A Meta-Analysis. International Journal of Urology. 2012;19(2):134–141. https://doi.org/10.1111/j.1442-2042.2011.02906.x
110. Wu QJ, Yang Y, Wang J, et al. Cruciferous Vegetable Consumption and Gastric Cancer Risk: A Meta-Analysis of Epidemiological Studies. Cancer Science. 2013;104(8):1067–1073. https://doi.org/10.1111/cas.12195
111. Wu QJ, Yang Y, Vogtmann E, et al. Cruciferous Vegetables Intake and the Risk of Colorectal Cancer: A Meta-Analysis of Observational Studies. Annals of Oncology. 2013;24(4):1079–1087. https://doi.org/10.1093/annonc/mds601
112. Liu B, Mao Q, Wang X, et al. Cruciferous Vegetables Consumption and Risk of Renal Cell Carcinoma: A Meta-Analysis. Nutrition and Cancer. 2013;65(5):668–676. https://doi.org/10.1080/01635581.2013.795980
113. Liu B, Mao Q, Lin Y, Zhou F, Xie L. The Association of Cruciferous Vegetables Intake and Risk of Bladder Cancer: A Meta-Analysis. World Journal of Urology. 2013;31(1):127–133. https://doi.org/10.1007/s00345-012-0850-0
114. Bahadoran Z, Mirmiran P, Hosseinpanah F, et al. Broccoli Sprouts Powder Could Improve Serum Triglyceride and Oxidized LDL/LDL-Cholesterol Ratio in Type 2 Diabetic Patients: A Randomized Double-Blind Placebo-Controlled Clinical Trial. Diabetes Research and Clinical Practice. 2012;96(3):348–354. https://doi.org/10.1016/j.diabres.2012.01.009

Dietary Supplements and Functional Foods in the Management of Endocrine Disorders

Elham Razmpoosh, Shima Abdollahi, and Sepideh Soltani

Contents

Endocrine disorders 138
Nutraceuticals and functional foods 138
Role of nutraceuticals in endocrine disorders 139
Metabolic syndrome 139
Omega-3 fatty acids 139
Probiotics 140
Flavonoids 141
Vitamins 142
Diabetes mellitus 143
Type 2 diabetes mellitus 143
Flavonoids 144
Vitamin D 144
Vitamin C 145
Vitamin E 145
Dietary fiber 145
Type 1 diabetes 146
Flavonoids 146
Vitamin D 147
Vitamin C 147
Vitamin E 147
Dietary fiber 148
Gestational diabetes mellitus 148
Flavonoids 148
Vitamin D 149
Vitamin C 149
Vitamin E 150
Dietary fiber 150

DOI: 10.1201/9781003220053-8

Fatty liver 151
Antioxidants 151
Herbal medicine 152
Turmeric 152
Thyroid disorders 153
Myoinositol 153
Carnitine 153
Iodine 154
Soy 154
Conclusion 154
References 155

Endocrine disorders

Endocrine and metabolic disorders are one of the most common contemporary human afflictions. Based on the epidemiology of endocrine disorders, including the disease burden (prevalence) and risk of disease development (incidence), there are both common and unusual endocrine and metabolic diseases (1). Accordingly, the prevalence and incidence of major endocrine disorders, including metabolic syndrome (2) and diabetes mellitus (3), have been widely defined in large population-based studies. Additionally, fatty liver disease (4) and thyroid disorders (5) are also on the top of the list of endocrine disorders with major complications and higher prevalence/incidence rates. Moreover, dietary intervention, as one of the important potential factors in the management of endocrine disorders for both the treatment and prevention of the aforementioned diseases, has been extensively reported compared with other endocrine disorders (6).

Nutraceuticals and functional foods

It is clear from the evidence that dietary interventions can provide useful approaches for the management of endocrine disorders (7). Meanwhile, a new diet–health paradigm is recently evolving, emphasizing the positive aspects of diet and 'nutraceuticals', a hybrid of 'nutrition' and 'pharmaceutical', defined as "a food (or a part of food, such as supplement) that provides medical or health benefits, including the prevention and or treatment of a disease" (8). It is also defined as "a product prepared from foods, but sold in the form of pills, or powder (potions) or in other medicinal forms, not usually associated with foods" (9). The food products used as nutraceuticals can be categorized as dietary fiber, prebiotics, probiotics, polyunsaturated fatty acids, antioxidants, and other different types of herbal/natural foods (10). The major effects of these nutraceuticals and dietary supplements in the aforementioned major endocrine disorders are discussed as followed.

Role of nutraceuticals in endocrine disorders

Metabolic syndrome

Metabolic syndrome is a cluster of conditions that occur together, including increased blood glucose, high blood pressure, dyslipidemia, and excess body fat storage around the waist. Three or more of these conditions result in an increase in inflammation, oxidative stress, and insulin resistance, which makes the body prone to cardiovascular diseases, diabetes, and other health problems. However, a healthy lifestyle, including a normal body weight, being physically active, and eating a healthy diet, can effectively prevent or slow the development of metabolic syndrome. Along with lifestyle modifications, drug therapies such as insulin-sensitizing or lipid- and blood pressure-lowering drugs may also be a part of the treatment of patients with metabolic syndrome. In addition, studies have shown that some functional foods and nutraceuticals can improve the condition of metabolic syndrome. In this section, we review the effects of some of these foods/nutrients.

Omega-3 fatty acids

Omega-3 fatty acids, notably EPA (eicosapentaenoic acid, 20:5n-3) and DHA (docosahexaenoic acid, 22:6n-3) – essential fatty acids in humans – have been considered by researchers and clinicians as a beneficial group of fatty acids used as an adjunct in the prevention and/or treatment of cardiovascular disease and metabolic syndrome (11). A meta-analysis of observational studies also demonstrated that higher omega-3 consumption is associated with a 26% lower risk for metabolic syndrome (12). In a healthy diet, the omega-3 to omega-6 ratio appears to be 1:1 to 1:4; while, in a typical Western diet, this ratio has risen to about 1:17 or more. It has been suggested that higher omega-6 consumption is associated with higher adipogenesis, systemic inflammation, and developing obesity (13). Omega-3 fatty acids can reduce the risk of metabolic syndrome through improvement in metabolic syndrome components (11). Several studies have shown that omega-3 fatty acids can stimulate nuclear receptors, including peroxisome proliferator-activated receptor (PPAR) α, regulating the important genes needed for lipid metabolism. This receptor is also a target for lipid-lowering drugs, and its activation by omega-3 fatty acids can reduce triglyceride levels by about 30%, in a dose-dependent manner (14). Conversely, omega-3 fatty acids seem to reduce the activity of another receptor (PPARγ) involved in lipogenesis in adipocytes. Activation of PPARγ also has been shown to stimulate consumption of high-fat foods; as a result, omega-3 can control body weight by reducing the activity of this receptor (15). The American Heart Association (16) as well as the Endocrine Society (17) recommend 2 to 4 g/day of EPA and DHA for people with hypertriglyceridemia. However, it has been shown that higher intake of EPA+DHA agents might increase blood low-density lipoprotein cholesterol (LDL-C), with a concurrent decrease in triglyceride. Notably, despite the hypotriglyceridemic effect of EPA, it does not exert the aforementioned effects alone (18). On the other hand, it has been reported that hepatic lipogenesis and very low-density lipoprotein cholesterol (VLDL-C) secretion, as well as hepatic enzymes involved in triacylglycerol synthesis, are decreased following omega-3 supplementation (19). Moreover, meta-analysis studies concluded that omega-3 fatty acids exert beneficial effects on blood pressure by blocking angiotensin receptors (11). In addition, inflammatory cytokines also decreased, leading to an improvement in endothelial function (20). The antihypertensive effect of omega-3s also contributes to reduced production of thromboxane, increased prostacyclin, and promoting the relaxation of vascular muscles (21). Miller et al. postulated that omega-3 fatty acids may be more effective than other lifestyle modifications, including a low-sodium diet or higher physical activity, in patients with hypertension (22, 23).

Regarding the antidiabetic effects of omega-3 fatty acids, controversial results have been reported. A recent comprehensive cumulative study reported no or little beneficial effects of omega-3 on glucose metabolism (24); while some others concluded significant improvement in fasting glucose and insulin resistance (25).

However, the question that arises is, does a low dose of dietary omega-3 consumption have such beneficial effects? A prospective cohort study found that fish consumption was not significantly associated with the incidence of hypertension in a large adult population, over a 16-year follow-up period (26). However, there is some other strong evidence supporting the cardioprotective or antidiabetic effects of dietary omega-3 intake. The concern about seafood consumption focuses on environmental contaminants such as methylmercury and polychlorinated biphenyl (27), which are reported to be associated with impaired insulin signaling and glucose metabolism in animal studies (28). Therefore, the recommendation on consuming high doses of dietary omega-3 fatty acids in order to benefit from the effects of metabolic syndrome should be with caution, especially during pregnancy.

Probiotics

The collection of microbes that naturally live in our digestive tracks (probiotics) has been shown to have a wide beneficial effect on metabolism, immune function, as well as digestion (29). Studies showed that probiotics or probiotics-containing foods can affect all the components of metabolic syndrome, energy balance, and body weight. Probiotics also have been found to improve lipid profiles, glycemic indices, blood pressure, and inflammation (30–35). Moreover, it has been demonstrated that gut dysbiosis, with a lower ratio of *Firmicutes* and a higher proportion of *Bacteroidetes* and *Proteobacteria*, is associated with insulin resistance and type 2 diabetes (36–38). The antihypertensive property of probiotics has been also observed, by inhibiting angiotensin converting enzyme (ACE) (32, 39).

Although the exact molecular mechanisms of health benefits of probiotics are still unknown, one of the underlying mechanisms is related to their anti-inflammatory effects. It has long been suggested that the short-chain fatty acids (SCFAs) – produced as a result of bacterial fermentation – are responsible for the immune health benefits of probiotics. Acetate, propionate, and butyrate, the main SCFAs, have been indicated to suppress pro-inflammatory cytokines, tumor necrosis factor-α (TNF-α), interleukin-1β (IL-1β), and interferon-γ (IFN-γ) signaling pathways (40). Moreover, dysbiosis can raise opportunistic pathogens, and increase intestinal permeability, leading to the penetration of bacteria into the circulation. The bacteria migration induces inflammation in insulin-sensitive organs, resulting in insulin resistance, impaired metabolism, and the development of metabolic syndrome (41). Studies also demonstrated a hypocholesterolemic effect of probiotics, through lower cholesterol absorption, enterohepatic circulation of bile salt, and inhibition of hepatic cholesterol synthesis (42). Previous meta-analyses also reported about a 5–8 mg/dL decline in total cholesterol and LDL-C (43–45), 9–15 mg/dL reduction in fasting blood glucose, 0.3–0.5% in Hemoglobin A1C (HbA1c) levels, as well as a significant improvement in Homeostatic Model Assessment for Insulin Resistance (HOMA-IR) and insulin levels following probiotic supplementation (46–48).

It is reported that different strain of probiotics may exhibit different health benefits on the host, which might be due to the different metabolites or survival rates among strains. However, there is limited evidence on the strain-specific effects of probiotics. It seems there is still a long way to go for recommending long-term use of probiotic supplementation, since some of the harmful aspects of probiotics such as gene modification and so on may not be clear.

Flavonoids

Flavonoids belong to a large family of polyphenolic compounds naturally produce by plants (49). The antioxidant capacity of flavonoids has long been described, and their health benefits are attributed to their free radical scavenging property. There is some evidence supporting the beneficial effects of dietary flavonoids on obesity (50); however, some other studies using flavonoid supplements failed to show a significant effect (51–53). A meta-analysis study that investigated the effect of green tea flavonoids (catechins) on body weight measures found no significant related improvement; however, when supplementation was combined with caffeine, a significant reduction in body mass index, body weight, and waist circumference was observed (54). Another meta-analysis revealed that green tea supplementation improved body weight, body mass index, and body fat when a low-dose (dosage ≤800 mg/day) and long-term intervention (>8 weeks) was conducted in overweight and obese patients with type 2 diabetes (55). The results of a recent umbrella review explored possible advantages of resveratrol supplementation on waist circumference in patients with metabolic syndrome; body weight and inflammatory cytokines in patients with nonalcoholic fatty liver disease (NAFLD); and blood pressure, lipid profile, glycemic indices, insulin resistance, and HbA1c levels in patients with type 2 diabetes (56). Conversely, another meta-analysis summarizing the results of 19 interventions reported no significant effect of resveratrol supplementation on body weight measures in a general population (57). Flavonoids have been reported to exert favorable effects in several ways, including inhibiting the activity of some digestive enzymes (pancreatic lipase, α-amylase, and α-glucosidase) (58–60), increased enzymes involved in lipolysis, decreased enzymes involved in lipogenesis, and activation of adiponectin-signaling pathways, resulting in lower dietary intake and weight loss, consequently (61).

Altogether, short-term intake of flavonoids from supplements does not appear to cause a clinical improvement in metabolic syndrome components. However, prolonged intake of flavonoid-rich foods makes us benefit from fiber, vitamins, and other antioxidants in addition to flavonoids. More studies are needed to confirm this conclusion.

Vitamins

Chronic inflammation and excessive oxidative stress are known as the core mechanisms of insulin resistance and metabolic syndrome. Therefore, nutrients with antioxidant properties are mostly considered in prevention or improvement of metabolic syndrome components. Most of the previous observational studies have shown that the higher intake of vitamins A, C, E, D, and β-carotene from diet is associated with lower risk of chronic diseases (62, 63). However, results were inconsistence when the antioxidants were administered in a single nutrient supplement form. A recent meta-analysis pooled data from observational studies and found that an increase of 25 nmol/L in serum vitamin D levels was associated with a 20% reduction in the risk of metabolic syndrome (63), while another meta-analysis from interventional studies reported that vitamin D supplementation had no significant effect on lipid profile in patients with metabolic syndrome (64). It seems that the natural form of nutrients taken through healthy diets is more effective, compared to a high-dose synthetic form. It may be related to the synergistic associations between micronutrients and fiber content of foods, which is less seen in supplements. Moreover, adherence to a healthy diet has been shown to be associated with a lower intake of sugary foods, high-fat foods, red meat, processed meat, and salty snacks, and a higher intake of vegetables, fruits, and whole grains, compared with increasing the dietary intakes of particular micronutrients (i.e., taking supplements along with an unhealthy diet). However, supplements may be useful in patients with vitamin deficiency, older adults, individuals with any conditions of food deprivation, or individuals with an impaired digestive system or a poor appetite.

Although taking these supplements is suggested for the treatment of any deficiencies, intakes greater than the recommended daily dose exert no future beneficial effects.

Diabetes mellitus

Evidence widely supports the important role of foods and bioactive peptides derived from foods in the prevention, treatment, and overall management of diabetes mellitus (7). According to animal studies, diabetes complications including neural and renal dysfunction may be prevented or delayed via supplying adequate dietary antioxidants, which provide protection against oxidative stress (8, 65). However, despite the many publications in human studies, there is not a unified conclusion regarding the correlation between nutraceuticals and the management of diabetes (66–69).

Major nutraceuticals that have been shown to affect the pathogenesis of diabetes include flavonoids; vitamins and minerals such as vitamins C, D, and E; chromium and magnesium, as well as dietary fibers, phytoestrogens, omega-3 fatty acids, conjugated linoleic acid, and α-lipoic acid (70). Here, we discussed the efficacies of flavonoids; vitamins D, C, and E; and dietary fibers in detail, and provide a summary of the effects of other nutraceuticals in diabetes mellitus.

Type 2 diabetes mellitus

Type 2 diabetes mellitus (T2DM) is one of the most widespread chronic diseases in the world. According to the American Diabetes Association (ADA), T2DM is a condition in which the body's cells do not respond well to the hormone insulin, that is why it was formerly called non-insulin-dependent diabetes or adult-onset diabetes (71). More than 95% of people with diabetes have T2DM. The prevalence of T2DM in adults worldwide was estimated to be 9.3% (463 million people) in 2019, and expected to rise to 10.2% (578 million) by 2030 and 10.9% (700 million) by 2045 (72) due to the population aging and a constant increase in obesity (73).

Epidemiologic studies have established an association between increased inflammatory biomarkers and the occurrence of T2DM (74). Diabetes type 2 is closely associated with obesity. Higher adipose tissue is associated with increased production of a number of hormone-like compounds that can increase insulin resistance (75). Additionally, excess visceral fat, together with central obesity, are associated with higher production of inflammatory cytokines and, hence, chronic inflammation (76). Both systemic inflammation and oxidative stress, which result from increased generation of reactive oxygen species (ROS), lead to pancreatic β-cell dysfunction and insulin resistance, and finally the pathogenesis of T2DM and metabolic syndrome (70). Dietary interventions, including nutraceuticals and functional foods, have a potential role in the management of T2DM (77), which are as followed.

Flavonoids

Flavonoids are natural substances with polyphenol structures that are prominent components of fruits, vegetables, herbs, and legumes (78). Flavonoids have shown to have potential anti-inflammatory properties through radical scavenging activities and inhibition of reactive oxygen species (ROS) production, modulation of platelet activation, and modulation of pro-inflammatory gene expression (79). Evidence suggests that a diet rich in flavonoids (mostly anthocyanins, quercetin, kaempferol, epigallocatechin gallate, and naringin (80)) may help ameliorate T2DM risk, mostly through reducing body fat percentage (81). A recent systematic review and meta-analysis of 28 randomized controlled trials (RCTs) showed that flavonoid intake had a modest but statistically significant benefit on key biomarkers of T2DM, including fasting blood glucose, HbA1c, HOMA-IR, total cholesterol, triglycerides, and LDL-C (82). They also reported that flavonoids had non-significant improvements in the levels of 2 h-postprandial glucose (2 h-PPG), insulin, and high-density lipoprotein-C (HDL-C) (82). An experimental study also revealed that nobiletin, a flavonoid found in citrus fruits, exhibited anti-diabetic properties through anti-apoptotic and insulinotropic effects on β-cells and, hence, alleviated the development of T2DM (83). Overall, due to different types of flavonoids and their bioactive compounds, a medium dose and short-term intervention would have modest positive effects in T2DM. More studies are needed along different types of flavonoids and their role in T2DM.

Vitamin D

There are variety findings on the association between vitamin D and T2DM. Longitudinal and cohort investigations have described that glycemic control in populations with T2DM is affected by the exposure to sunlight and hence, has a seasonal variation (84). Vitamin D may modulate the development of T2DM through increasing insulin secretion and decreasing insulin resistance, which is possibly caused by altering the balance between extracellular and intracellular calcium in β cells (85). Evidence through large RCT studies also found that a combination of calcium and vitamin D treatment may lower the risk of T2DM (84). Results from clinical investigations also showed that vitamin D supplementation may improve insulin secretion, HbA1C levels, and glucose tolerance among vitamin D-deficient populations with T2DM (84, 86). A recent meta-analysis of observational studies revealed that vitamin D may promote β-cell function, which increases insulin sensitivity (87). However, another meta-analysis revealed that in spite of the favorable effects of vitamin D supplementation on circulating levels of inflammatory cytokines such as tumor necrosis factor-alpha (TNF-α), high-sensitivity C-reactive protein (hs-CRP), and interleukin-6 (IL-6) in individuals with T2DM, no significant correlation was observed between the dosage of vitamin D supplements and the concentrations of these inflammatory biomarkers (88). In general, despite the positive effects of vitamin D in T2DM, more studies with different doses and duration should be performed to reach a coherent conclusion.

Vitamin C

The role of vitamin C in the management of T2DM has been observed through antioxidant and scavenging activities of vitamin C, leading to a reduction in protein glycation (89, 90). Results from animal studies also showed that vitamin C resulted in a reduction in lipid peroxidation and diabetes-induced sorbitol accumulation erythrocytes (89). However, according to a human study, 800 mg/day of vitamin C partially increased vitamin C levels in patients with T2DM who were with low vitamin C levels, although it did not improve insulin resistance (90). A recent meta-analysis on 28 studies showed that improvements in HbA1c and diastolic blood pressure had a very low evidence of certainty, and that is why vitamin C supplementation cannot be recommended as a therapy in T2DM, and high-quality clinical trial investigations with larger sample sizes and long-term duration of intervention are needed to confirm these findings (91).

Vitamin E

There has been conflicting evidence about vitamin E levels in patients with T2DM (92, 93). Short-term vitamin E supplementation (600 mg daily for two weeks) was performed to assess the reversibility of lipid peroxidation and platelet activation in T2DM (94). Results indicated that vitamin E supplementation enhanced lipid peroxidation and persistent platelet activation (94). Prospective cohort studies suggested that higher dietary intake of vitamin E in combination with vitamins A, C, folic acid, and zinc was associated with a lower risk of cardiovascular disease in patients with T2DM (94). Vitamin E and vitamin B supplementation has also shown to be associated with an increase in vascular events and a more rapid decrease in glomerular filtration rate in comparison with placebo in patients with T2DM who had nephropathy (95). Recently, vitamin E proved to have beneficial effects in the treatment of non-alcoholic steatohepatitis in adults with prediabetes mellitus (96). More studies still need to be conducted on the association between vitamin E and T2DM.

Dietary fiber

There is a body of evidence showing that patients with T2DM do not have adequate daily intake of dietary fiber (97). Epidemiologic studies suggest an inverse association between dietary fiber intake and the risk of T2DM (98). Among different types of dietary fibers, soluble dietary fiber intake has shown to be associated with increased insulin sensitivity and decreased postprandial blood glucose levels in patients with T2DM, which is mostly due to the viscous properties of soluble fiber (99). Findings from clinical studies indicated that an increased intake of soluble fiber for a medium-long duration (more than ten weeks) resulted in considerable improvements in insulin resistance and fasting and postprandial blood glucose levels, without any improvement in the secretory function of the islets of Langerhans (99). However, concerning the reduced risk of incidence of T2DM, it has been reported that higher consumption of insoluble fiber, and not the soluble fiber intake, resulted in a reduced risk of T2DM (97). A meta-analysis of six cohort studies showed that an increase of two servings/day in consumption of whole grains would reduce the risk of T2DM by 21% (97). Concerning that, more studies are needed to be performed regarding the duration, dose, and the type of dietary fiber intake and its benefits in the management and prevention of T2DM.

Type 1 diabetes

According to the definition by the World Health Organization (WHO), type 1 diabetes (T1DM) (previously known as insulin-dependent, juvenile, or childhood-onset diabetes) is characterized by deficient insulin production (77). It is an autoimmune disease that leads to the destruction of insulin-producing pancreatic β-cells. Importantly, insulin allows glucose to enter muscle and adipose cells, and stimulates the liver to store glucose as glycogen and synthesize fatty acids. Without insulin, diabetic ketoacidosis (DKA) develops, which is life threatening for these patients. That is why patients with T1DM require life-long insulin replacement therapy (100). According to WHO, T1DM has a lower incidence rate compared to T2DM. In 2017, there were nine million people with T1DM, mostly living in high-income countries (77). A recent systematic review and meta-analysis showed that the incidence of T1DM was 15 per 100,000 people and the prevalence was 9.5% (95% CI: 0.07 to 0.12) in the world, which was statistically significant (101). The exact cause or the means to prevent it are unknown, though there are currently biochemical islet autoantibodies measured in the serum directed against insulin, including islet antigen, glutamic decarboxylase, and zinc transporter. Since development of islet autoantibodies occurs before clinical diagnosis of T1DM, this disease night be a predictable disease in an individual with two or more autoantibodies (102). Recent evidence showed that not only the management of T1DM but also the related primary prevention involved dietary interventions (103).

Flavonoids

Common flavonoids, including quercetin, kaempferol, apigenin, and resveratrol have been shown to have protection activities against β-cells in T1DM (65). Flavonoids reduced β-cell damage through inhibiting oxidative pathways and protecting against lipid peroxidation (104).

An animal study reported that naringenin intervention (25 and 50 mg/kg) resulted in significant improvement in urine parameters in diabetic mice, and renal damage was attenuated significantly (105). Another animal study revealed that resveratrol is an important flavonoid modulating T1DM and even can decrease the risk of T1DM mainly via ameliorating LDL-oxidation (106). A recent clinical trial study showed that a 60-day resveratrol supplementation resulted in a significant decrease in fasting blood glucose and HbA1c in patients with T1DM. Beneficial and significant improvements in oxidative stress, total anti-oxidant capacity, and malondialdehyde were observed in this population, while the changes in the levels of insulin, HOMA-IR, as well as the parameters of liver and kidney function, were not statistically significant (107). Above all, more studies should be conducted on the types and exact mechanisms of the effect of flavonoids in the management of T1DM.

Vitamin D

Evidence showed that the prevalence of vitamin D insufficiency/deficiency in patients with T1DM, mostly in children, is very high. In addition, vitamin D deficiency in these patients significantly increases the prevalence of coronary disease (108). Meanwhile, an animal study showed that vitamin D supplementation significantly decreased fasting blood glucose concentrations and improved pancreatic β-cell function and reduced inflammatory markers in mice with T1DM (109). An RCT study reported that a daily supplementation of cholecalciferol (70 IU/kg body weight/day) for 12 months led to an improvement in the function of Tregs in patients with T1DM, and as a result could serve as an agent in the development of immunomodulatory combination therapies for T1DM (74). Evidence also showed that increased vitamin D intake by infants may reduce the risk of developing T1DM. An open-labeled RCT found that vitamin D supplementation in adult patients with new onset T1DM, temporarily reduced the required insulin dose (84, 110). However, larger clinical trial studies need to be conducted to confirm these findings.

Vitamin C

The endothelial impairment in patients with T1DM has been reported in several studies (111, 112). There is also evidence that some intervention could improve this process and consequently delay the early development of atherosclerosis and other related micro- and macro-vascular complications (113). Studies also showed the protective role of vitamin C in the prevention of T1DM, where higher plasma ascorbic acid levels can serve as a protective agent against islet autoimmunity in children at risk for T1DM (114). Findings indicated that treatment with ascorbic acid would block acute hyperglycemic impairment of endothelial function in adolescents with T1DM (112). Overall, although vitamin C is an essential antioxidant in humans, its effective role in patients with T1DM as a high-risk population either with or without appropriate glycemic range was not sufficiently powered (115). Further studies are warranted to confirm these findings.

Vitamin E

Patients with T1DM, with poorly controlled status, surprisingly had elevated plasma vitamin E levels and increased MDA levels (85). On the other hand, supplementation with vitamin E in patients with T1DM resulted in a significant decrease in MDA levels and significant increase in glutathione and vitamin E concentrations. In fact, vitamin E was shown to have improvements in antioxidant defense systems and a reducing effect on oxidative stress in T1DM patients. However, vitamin E was not shown to have any advantage for metabolic parameters (116). Meanwhile, evidence showed that vitamin E supplementation for an additional three to nine months resulted in no further beneficial changes in serum vitamin E and lipoprotein peroxidability (117). More investigations should be conducted on the exact dose and duration intervention of vitamin E supplementation in T1DM.

Dietary fiber

The beneficial role of dietary fiber has been widely reported in patients with T1DM. In fact, a significant modification in pre-prandial and post-prandial glycemic control has been shown in patients with T1DM who consumed foods with high fiber contents and low glycemic load (118). Similar with findings on the association between dietary fiber intake and T2DM, soluble fiber has been shown to have more efficacies in the management of T1DM, compared to insoluble fiber. Additionally, the role of fermentable fiber has been observed in decreasing the incidence T1DM through conferring microbiota-dependent increases in short-chain fatty acids (SCFA) and IL-22 (119). Moreover, β-glucan consumption was also reported to exert promising effects on the glycemic control of children and adolescents with T1DM. These benefits were only observed when the β-glucan supplementation dose was 6 g/day, while the 3 g/day of β-glucan supplementation was similar to that of the standard diet alone (120). Therefore, a designation of more long-term and large-sized investigations are necessary to evaluate the exact effect of different types of fibers on glycemic control, and other glycemic parameters in patients with T1DM.

Gestational diabetes mellitus

Gestational diabetes mellitus (GDM) occurs during pregnancy. It is defined as hyperglycemia with blood glucose values above normal but below the diagnostic values of diabetes. Gestational diabetes is diagnosed through prenatal screening, rather than through reported symptoms. Women with GDM are at an increased risk of complications during pregnancy and at delivery. These women and possibly their children are also at increased risk of T2DM in the future (77). According to the Centers for Disease Control (CDC), about 6% to 9% of pregnant women develop GDM, which has an increasing trend. Evidence showed that the percentage of pregnant women with GDM increased 56% from 2000 to 2010 (121).

Flavonoids

As mentioned previously, flavonoids, as dietary polyphenols, can positively modulate insulin signaling pathways by alleviating high blood glucose and insulin resistance, reducing inflammatory adipokines, as well as modifying microRNA profiles. Regarding the association between dietary flavonoids and GDM, very few data are available (122). An animal study showed that naringenin, a flavonoid, is shown to improve GDM-associated oxidative stress, insulin sensitivity, and inflammation, and can be considered as a novel preventive therapeutic in GDM (123). Concerning the preventive effect of flavonoids against GDM, a meta-analysis of observational studies reported that higher dietary intake of total polyphenols and flavonoids from fruits is associated with lower GDM risk (124). Additionally, data suggests that a natural diet rich in total flavonoids and anthocyanidin has a beneficial impact in pregnant women with GDM. In fact, pregnant women taking a higher total flavonoid intake and anthocyanidin were shown to be less obese compared to those women with lower intakes (125), and hence would have fewer complications related to obesity and GDM. More studies, mostly including the investigation on the 'omics', are needed to better understand the interaction between flavonoids and GDM.

Vitamin D

A meta-analysis provides further evidence on the beneficial role of vitamin D in GDM, showing the inverse association between serum 25(OH)D concentrations and the risk of GDM, which may be useful for the prevention of pregnancy complications (126). Besides, a significant association between vitamin D deficiency and an increased risk of GDM was observed. In fact, pregnant women with serum vitamin D levels of 40–90 nmol/L were shown to have the lowest risk of GDM (127). However, another meta-analysis reported that there was no difference in the postpartum period in pregnant women with previous GDM who had taken vitamin D supplements in the prenatal period (67). A recent RCT reported that combined supplementation of vitamin D and omega-3 fatty acids for six weeks among women with GDM significantly reduced blood sugar, improved insulin resistance, and effectively improved lipid metabolism (128). Another RCT showed that daily vitamin D intake of 5000 units in the first and second trimesters of pregnancy among women who had at least one risk factor of GDM, reduced glucose tolerance test and GDM risk (129). Despite the outstanding role of vitamin D in pregnancy, more studies are needed to examine the beneficial and safe dose of vitamin D supplementation in pregnant women with GDM.

Vitamin C

Cohort studies revealed that higher dietary intake of vitamin C during pregnancy is independently related with a lower risk of GDM. In fact, taking more than 200 mg/day of foods rich in vitamin C, mostly through vegetables and fruits, would reduce the risk of GDM (130). In addition, a recent RCT study reported that supplementation of a medium dose of ascorbic acid considerably reversed the oxidative stress status in women with GDM and improved neonatal outcome (68). It was also found that a short-term oral ascorbic acid supplementation improved endothelium-dependent vasodilatation in women with GDM (131). However, it was stated that diet alone may not be enough to provide adequate levels of vitamin C, and routine supplementation according to the RDA recommendation as well as routine monitoring should be performed (132). Overall, vitamin C supplementation is correlated with decreased lipid peroxidation and a significant increase in antioxidant capacity, which is correlated with reduced consequence of GDM. Not all the investigations support routine daily supplementation with vitamin C in women with GDM, since excess maternal intake may lead to elevated fetal plasma concentrations (133). Further studies are needed to examine the beneficial and safe dose and duration of vitamin C supplementation in women with GDM.

Vitamin E

A systematic review showed that serum concentration of vitamin E was significantly lower in the third trimester of pregnancy in women with GDM compared with the healthy pregnant women (69). Another study proved that a combination of vitamins E and A resulted in more reduction in oxidative stress markers as well as reduced independent risk factors affecting the pregnancy outcomes in women with GDM. That is why the clinical monitoring of oxidative stress levels should be performed for taking the adequate quantity of vitamin E and other supplements, as these may be of great significance in improving pregnancy outcomes in women with GDM (134). A recent network meta-analysis including 16 studies of 1173 women with GDM revealed that a combination of probiotics and omega-3 with vitamin E compared with placebo had a significant increasing effect on serum total anti-oxidant capacity levels (66). Hence, administration of an adequate combination of vitamin E with other beneficial dietary supplements in women with GDM would be helpful in limiting the oxidative stress associated with GDM (66).

Dietary fiber

According to the findings, having a diet rich in total fiber and fruit fiber may play a beneficial role in the management of GDM, though findings are controversial. Concerning the preventive role of dietary fiber in the incidence of GDM, taking total fiber and fruit fiber in both 13–16 gestational weeks (GWs) and in 21–24 GWs were showed to be significantly correlated with decreased risk of GDM. On the contrary, Xu et al. reported that consumption of cereal fiber during GWs 21–24 was positively correlated with GDM risk (135). Li et al. believed that appropriate quantity taking of fruit and vegetable during pregnancy would induce or prevent the development of GDM (136). They showed that higher amounts of grape, potatoes, melon, as well as fruit juice consumption were positively correlated with the incidence of GDM. In contrast, the similar quantity of orange, apple, and vegetable consumption, except for potatoes, were negatively associated with the incidence of GDM (136). An RCT study reported that taking higher total dietary fiber led to an increased serum glutathione and antioxidant capacity, and decreased MDA levels in women with GDM. Additionally, dietary fiber intakes decreased serum plasminogen activator inhibitor-1 and visfatin, as biomarkers of insulin resistance and adipocyte dysfunction. These findings support the need for supplementing maternal diets with fiber for reducing oxidative stress and risks of metabolic complications in women with GDM (137). Moreover, a recent systematic review showed that taking simultaneous higher amounts of dietary fiber and folic acid were associated with reduced blood glucose and overall GDM-associated complications in women with GDM (138). However, taking higher doses of fiber supplements during pregnancy must be approached with caution.

Fatty liver

Fatty liver disease is a condition characterized by excess fat accumulation in hepatocytes (more than 5% of hepatocytes), which may be due to excessive alcohol intake (alcoholic fatty liver) or metabolic disorder (including NAFLD). Nonalcoholic fatty liver disease is the most common chronic liver disease, affecting about 30% of the adult population (139). The development of NAFLD may lead to cardiometabolic complications, cirrhosis, and carcinoma (140). The current guidelines emphasize lifestyle modifications, losing weight, and using proper medications. Recently, adjuvant therapy has also gained interest as an alternative approach in the treatment of NAFLD. Here, we briefly discuss some of the most widely used supplements concerning nutraceuticals in patients with NAFLD.

Antioxidants

Oxidative damage is known as an involving mechanism related to NAFLD. Several sources of oxidative stress in NAFLD are known, including β-oxidation, oxidative phosphorylation, improper activity of cytochrome P450, and microsomal metabolism. In this condition, cellular response to oxidative stress is induced, resulting to injury- and death-induced inflammation (141–144). Antioxidants have been suggested to act as potential therapeutic agents in this situation.

Vitamin E, a fat-soluble vitamin, is a common supplement studied in patients with NAFLD. Clinical trials showed a significant improvement in hepatic steatosis and inflammation following vitamin E supplementation. A large clinical trial reported greater health-related benefits of vitamin E supplements than pioglitazone in patients with steatohepatitis (145). However, other evidence has expressed concerns about the long-term side effects. A meta-analysis study found that a high dose of vitamin E (>800 IU/day) may be associated with a higher risk for cancers and stroke, because of its angiogenic effects (146).

Vitamin C, another vitamin with antioxidant properties, has been shown to have protective effects on hepatocytes against oxidative stress by scavenging excessive free radicals. There are also epidemiological studies supporting the association between higher dietary vitamin C intake and improved insulin resistance and glucose metabolism (147–149). A recent clinical trial study also indicated that vitamin C supplementation had favorable effects on insulin resistance, glycemic parameters, and hepatic enzymes (98). However, the limited number of studies makes us unable to draw a firm conclusion.

Another group of antioxidants that has received an increasing scientific attention is flavonoids, especially resveratrol. Studies have shown that resveratrol can activate AMPKα/SIRT1 pathway, resulting in downregulation of inflammatory cytokines and, consequently, a reduction in hepatic steatosis (99). Resveratrol also has been shown to reduce the stress of endoplasmic reticulum, lipotoxicity, fatty acid beta-oxidation, and hepatocyte damages (150). However, the results from meta-analysis studies do not support the beneficial effect of resveratrol on hepatic enzymes, glucose hemostasis, lipid profile, and body weight in patients with NAFLD (150–152).

Herbal medicine

Milk thistle (*Silybum marianum*) has been extensively used to treat NAFLD. It is postulated that silymarin – the biologically active compound found in milk thistle – is responsible for its hepatoprotective effects. Silymarin has been found to act through activating antioxidant enzymes, including glutathione peroxidase, glutathione reductase, superoxide dismutase, and catalase as well as inhibiting inflammatory markers such as nuclear factor kappa (NFκB) and cyclooxygenase (153, 154). Silymarin also known as an anti-fibrotic agent by maintaining cell membrane stability and preventing toxic substances from entering the cell (155). A cumulative study pooled data from trials and showed a significant reduction in alanine transaminase (ALT) and aspartate transaminase (AST) levels following silymarin supplementation in patients with NAFLD (156). However, another meta-analysis declared that there is no clinical advantage over silymarin supplementation in patients with liver diseases, despite significant improvement in ALT and AST enzymes (157).

Green tea is another herb consumed largely in patients with NAFLD. The health benefits of green tea are mostly contributed to the contents of epigallocatechin-3-gallate (EGCG), which is the main active component of green tea. EGCG may exert its protective effects by attenuating mitochondrial lipid peroxidation, hepatic cyclooxygenase-2, and prostaglandin E2 production, as well as increasing the activation of antioxidative enzymes such as icotinamide adenine dinucleotide phosphate (NADPH) oxidase and cytochrome P450 2E1 (158–161). Pooling data from clinical trials also confirm positive effects of green tea in patients with NAFLD (162, 163), but not in healthy participants (162).

Turmeric

Turmeric is a traditional spice mostly used in India and Eastern countries. Turmeric is known as a powerful antioxidant and anti-inflammatory agent, which has been studied in various diseases. The hepatoprotective effects of turmeric have been proven in previous meta-analyses, through a significant reduction in ALT, AST, serum total cholesterol, LDL-C, fasting blood sugar, HOMA-IR, serum insulin, and waist circumference (164, 165). The potential favorable effects of turmeric are attributed to a phenolic agent named 'curcumin'. It is proposed that curcumin can modulate NAFLD risk factors, including dyslipidemia, insulin resistance, and oxidative stress. Curcumin also can downregulate PPARγ, and upregulate PPARα, resulting in lipid hemostasis improvement and decreasing hepatic steatosis (166, 167).

Thyroid disorders

Thyroid disorders vary considerably among different communities, based on dietary iodine, an essential element needed for thyroid hormone synthesis. People who live in areas with severe iodine deficiency suffer from some degree of thyroid dysfunction, ranging from hypothyroidism, mostly caused by Hashimoto's thyroiditis, to thyrotoxicosis, which is caused by Graves' disease. Decreased T3 and T4 levels in the bloodstream stimulate the secretion of thyroid stimulating hormone (TSH) from the pituitary gland, which, indeed, leads to an overproduction of thyroid hormones. A number of nutraceuticals have been shown to affect TSH and thyroid hormone production.

Myoinositol

It is proposed that myoinositol may act as a TSH stimulator. Red rice, citrus foods, and beans are known as sources for myoinositol. It is also produced by the liver and kidney from glucose (168, 169). Myoinositol is an important factor in phosphatidyl inositol 3 kinases (PI3K) signaling pathway, a necessary cascade for immunity response of cells. The impairment in this network can lead to immune system dysfunction and, hence, autoimmune diseases (170). Myoinositol is also involved in H_2O_2-related iodination of thyroid hormones in response to TSH, and disruption in this pathway can lead to TSH resistance and hypothyroidism (171). Some clinical trials examined the effect of myoinositol on autoimmune hypothyroidism and showed a significant reduction in TSH and AbTPO, and increased thyroid hormones (172–174).

Carnitine

Carnitine is an amino acid derivative that has both endogenous and exogenous sources. The primary role of carnitine is to transport fatty acids across the inner mitochondrial membrane during the oxidation of long-chain fatty acids. Studies suggest that there is an association between thyroid hormones and carnitine. Thyroid hormones stimulate carnitine-dependent beta-oxidation, and increase the urinary excretion of carnitine. Carnitine deficiency may be observed in both patients with hypothyroidism and hyperthyroidism, which are mostly due to the decreased carnitine synthesis and the increased beta-oxidation, which each use carnitine (175). Carnitine deficiency in these patients may cause fatigue symptoms, specifically in patients taking medications. There are limited clinical trials supporting this hypothesize (176, 177). However, carnitine has been proposed to act as a peripheral antagonist of thyroid hormone (178), which raises concerns about the carnitine-drug interactions in hypothyroidism.

Iodine

The association between iodine deficiency and thyroid disorders has been long known. Although the salt iodization program has reduced iodine deficiency in many countries, there are still regions with endemic goiter. Scientific data proposed that taking iodine-rich foods, including sea foods, kelp, and iodized salt, improves thyroid function. On the other hand, vegetables rich in thiocyanate, such as those in the Brassicaceae family, can suppress the iodine uptake by thyroid cells (179). However, the exact quantity of consuming Brassicaceae vegetables that results in thyroid dysfunction is not clear.

Soy

Studies demonstrated isoflavone-rich foods, such as soy products, may inhibit thyroid peroxidase enzyme, which is needed for hormone production. Although the impairment in thyroid function in patients with iodine deficiency has been shown to be more affected by soy products, a recent meta-analysis on the general population reported no significant harmful effect of soy supplementation on thyroid hormones, and only a modest increase in TSH levels was observed (180).

Conclusion

There are a number of functional foods and dietary supplements, including nutraceuticals, that are used as adjunct therapy in endocrine disorders. Some of them are essential nutrients that the body needs to stay healthy and to perform optimally. It seems that taking supplements beyond the daily requirement has no greater benefit in patients with endocrine disorders who do not have nutritional deficiencies. However, patients with an increased need for extra nutrients, or those who cannot meet the daily requirements of nutrients through routine dietary intakes, may benefit more from taking supplements. Altogether, nutraceuticals may be effective in mediating endocrine disorders, though their full benefits are still not clear. That is why the primary choice for the management of endocrine disorders is to follow a healthy diet with medications prescribed by expert physicians.

References

1. Golden SH, Robinson KA, Saldanha I, Anton B, Ladenson PW. Clinical review: prevalence and incidence of endocrine and metabolic disorders in the United States: a comprehensive review. The journal of clinical endocrinology and metabolism. 2009; 94(6):1853–78.
2. Ford ES, Li C, Zhao G, Pearson WS, Mokdad AH. Prevalence of the metabolic syndrome among U.S. adolescents using the definition from the International Diabetes Federation. Diabetes care. 2008;31(3):587–9.
3. Cowie CC, Rust KF, Byrd-Holt DD, Eberhardt MS, Flegal KM, Engelgau MM, et al. Prevalence of diabetes and impaired fasting glucose in adults in the U.S. population: National Health and Nutrition Examination Survey 1999–2002. Diabetes care. 2006; 29(6):1263–8.
4. Bellentani S, Scaglioni F, Marino M, Bedogni G. Epidemiology of non-alcoholic fatty liver disease. Digestive diseases (Basel, Switzerland). 2010;28(1):155–61.
5. Bjoro T, Holmen J, Krüger O, Midthjell K, Hunstad K, Schreiner T, et al. Prevalence of thyroid disease, thyroid dysfunction and thyroid peroxidase antibodies in a large, unselected population. The Health Study of Nord-Trondelag (HUNT). European journal of endocrinology. 2000;143(5):639–47.
6. Kreisberg RA, Owen WC, Siegal AM. Nutrition and endocrine disease. The medical clinics of North America. 1970;54(6):1473–94.
7. Antony P, Vijayan R. Bioactive peptides as potential nutraceuticals for diabetes therapy: a comprehensive review. International journal of molecular sciences. 2021; 22(16):9059.
8. Maddi V, Aragade P, Digge V, Nitalikar M. Phcog Rev.: short review importance of nutraceuticals in health management. Pharmacognosy reviews. 2007;1(2):377–9.
9. Wildman RE, Wildman R, Wallace TC. Handbook of nutraceuticals and functional foods: Boca Raton, Florida: CRC Press; 2016.
10. Das L, Bhaumik E, Raychaudhuri U, Chakraborty R. Role of nutraceuticals in human health. Journal of food science and technology. 2012;49(2):173–83.
11. Poudyal H, Panchal SK, Diwan V, Brown L. Omega-3 fatty acids and metabolic syndrome: effects and emerging mechanisms of action. Progress in lipid research. 2011; 50(4):372–87.
12. Jang H, Park K. Omega-3 and omega-6 polyunsaturated fatty acids and metabolic syndrome: a systematic review and meta-analysis. Clinical nutrition. 2020;39(3):765–73.
13. Simopoulos AP. An increase in the omega-6/omega-3 fatty acid ratio increases the risk for obesity. Nutrients. 2016;8(3):128.
14. Harris WS, Miller M, Tighe AP, Davidson MH, Schaefer EJ. Omega-3 fatty acids and coronary heart disease risk: clinical and mechanistic perspectives. Atherosclerosis. 2008;197(1):12–24.
15. Ryan KK, Li B, Grayson BE, Matter EK, Woods SC, Seeley RJ. A role for central nervous system PPAR-γ in the regulation of energy balance. Nature medicine. 2011; 17(5):623–6.
16. Miller M, Stone NJ, Ballantyne C, Bittner V, Criqui MH, Ginsberg HN, et al. Triglycerides and cardiovascular disease: a scientific statement from the American Heart Association. Circulation. 2011;123(20):2292–333.
17. Berglund L, Brunzell JD, Goldberg AC, Goldberg IJ, Sacks F, Murad MH, et al. Evaluation and treatment of hypertriglyceridemia: an Endocrine Society clinical practice guideline. The journal of clinical endocrinology & metabolism. 2012;97(9):2969–89.
18. Skulas-Ray AC, Wilson PW, Harris WS, Brinton EA, Kris-Etherton PM, Richter CK, et al. Omega-3 fatty acids for the management of hypertriglyceridemia: a science advisory from the American Heart Association. Circulation. 2019;140(12):e673–e91.
19. Jacobson TA. Role of n-3 fatty acids in the treatment of hypertriglyceridemia and cardiovascular disease. The American journal of clinical nutrition. 2008;87(6): 1981S–90S.

20. Skulas-Ray AC, Kris-Etherton PM, Harris WS, Vanden Heuvel JP, Wagner PR, West SG. Dose-response effects of omega-3 fatty acids on triglycerides, inflammation, and endothelial function in healthy persons with moderate hypertriglyceridemia. The American journal of clinical nutrition. 2011;93(2):243–52.
21. Limbu R, Cottrell GS, McNeish AJ. Characterisation of the vasodilation effects of DHA and EPA, n-3 PUFAs (fish oils), in rat aorta and mesenteric resistance arteries. PloS one. 2018;13(2):e0192484.
22. Dickinson HO, Mason JM, Nicolson DJ, Campbell F, Beyer FR, Cook JV, et al. Lifestyle interventions to reduce raised blood pressure: a systematic review of randomized controlled trials. Journal of hypertension. 2006;24(2):215–33.
23. Miller PE, Van Elswyk M, Alexander DD. Long-chain omega-3 fatty acids eicosapentaenoic acid and docosahexaenoic acid and blood pressure: a meta-analysis of randomized controlled trials. American journal of hypertension. 2014;27(7):885–96.
24. Brown TJ, Brainard J, Song F, Wang X, Abdelhamid A, Hooper L. Omega-3, omega-6, and total dietary polyunsaturated fat for prevention and treatment of type 2 diabetes mellitus: systematic review and meta-analysis of randomised controlled trials. BMJ. 2019;366.
25. Delpino FM, Figueiredo LM, da Silva BGC, da Silva TG, Mintem GC, Bielemann RM, et al. Omega-3 supplementation and diabetes: a systematic review and meta-analysis. Critical reviews in food science and nutrition. 2021:1–14.
26. Matsumoto C, Yoruk A, Wang L, Gaziano JM, Sesso HD. Fish and omega-3 fatty acid consumption and risk of hypertension. Journal of hypertension. 2019;37(6):1223–9.
27. Levine KE, Levine MA, Weber FX, Hu Y, Perlmutter J, Grohse PM. Determination of mercury in an assortment of dietary supplements using an inexpensive combustion atomic absorption spectrometry technique. Journal of automated methods and management in chemistry. 2005;2005(4):211–6.
28. Chen YW, Huang CF, Tsai KS, Yang RS, Yen CC, Yang CY, et al. The role of phosphoinositide 3-kinase/Akt signaling in low-dose mercury–induced mouse pancreatic β-cell dysfunction in vitro and in vivo. Diabetes. 2006;55(6):1614–24.
29. Aggarwal J, Swami G, Kumar M. Probiotics and their effects on metabolic diseases: an update. Journal of clinical and diagnostic research: JCDR. 2013;7(1):173.
30. Ardeshirlarijani E, Tabatabaei-Malazy O, Mohseni S, Qorbani M, Larijani B, Jalili RB. Effect of probiotics supplementation on glucose and oxidative stress in type 2 diabetes mellitus: a meta-analysis of randomized trials. DARU Journal of pharmaceutical sciences. 2019;27(2):827–37.
31. Kocsis T, Molnár B, Németh D, Hegyi P, Szakács Z, Bálint A, et al. Probiotics have beneficial metabolic effects in patients with type 2 diabetes mellitus: a meta-analysis of randomized clinical trials. Scientific reports. 2020;10(1):1–14.
32. Qi D, Nie X-L, Zhang J-J. The effect of probiotics supplementation on blood pressure: a systemic review and meta-analysis. Lipids in health and disease. 2020;19(1):1–11.
33. Yao K, Zeng L, He Q, Wang W, Lei J, Zou X. Effect of probiotics on glucose and lipid metabolism in type 2 diabetes mellitus: a meta-analysis of 12 randomized controlled trials. Medical science monitor: international medical journal of experimental and clinical research. 2017;23:3044.
34. Zheng HJ, Guo J, Jia Q, Huang YS, Huang W-J, Zhang W, et al. The effect of probiotic and synbiotic supplementation on biomarkers of inflammation and oxidative stress in diabetic patients: a systematic review and meta-analysis of randomized controlled trials. Pharmacological research. 2019;142:303–13.
35. Zheng HJ, Guo J, Wang Q, Wang L, Wang Y, Zhang F, et al. Probiotics, prebiotics, and synbiotics for the improvement of metabolic profiles in patients with chronic kidney disease: a systematic review and meta-analysis of randomized controlled trials. Critical reviews in food science and nutrition. 2021;61(4):577–98.
36. Han J-L, Lin H-L. Intestinal microbiota and type 2 diabetes: from mechanism insights to therapeutic perspective. World journal of gastroenterology. 2014;20(47):17737.

37. Hulston CJ, Churnside AA, Venables MC. Probiotic supplementation prevents high-fat, overfeeding-induced insulin resistance in human subjects. British journal of nutrition. 2015;113(4):596–602.
38. Larsen N, Vogensen FK, Van Den Berg FW, Nielsen DS, Andreasen AS, Pedersen BK, et al. Gut microbiota in human adults with type 2 diabetes differs from non-diabetic adults. PloS one. 2010;5(2):e9085.
39. Seppo L, Jauhiainen T, Poussa T, Korpela R. A fermented milk high in bioactive peptides has a blood pressure–lowering effect in hypertensive subjects. The American journal of clinical nutrition. 2003;77(2):326–30.
40. Cox MA, Jackson J, Stanton M, Rojas-Triana A, Bober L, Laverty M, et al. Short-chain fatty acids act as antiinflammatory mediators by regulating prostaglandin E2 and cytokines. World journal of gastroenterology. 2009;15(44):5549.
41. Kekkonen RA, Lummela N, Karjalainen H, Latvala S, Tynkkynen S, Järvenpää S, et al. Probiotic intervention has strain-specific anti-inflammatory effects in healthy adults. World journal of gastroenterology. 2008;14(13):2029.
42. Pereira DI, Gibson GR. Effects of consumption of probiotics and prebiotics on serum lipid levels in humans. Critical reviews in biochemistry and molecular biology. 2002; 37(4):259–81.
43. Agerholm-Larsen L, Bell ML, Grunwald G, Astrup A. The effect of a probiotic milk product on plasma cholesterol: a meta-analysis of short-term intervention studies. European journal of clinical nutrition. 2000;54(11):856–60.
44. Cho YA, Kim J. Effect of probiotics on blood lipid concentrations: a meta-analysis of randomized controlled trials. Medicine. 2015;94(43):e1714–24.
45. Guo Z, Liu X, Zhang Q, Shen Z, Tian F, Zhang H, et al. Influence of consumption of probiotics on the plasma lipid profile: a meta-analysis of randomised controlled trials. Nutrition, metabolism and cardiovascular diseases. 2011;21(11):844–50.
46. Hu Y-m, Zhou F, Yuan Y, Xu Y-c. Effects of probiotics supplement in patients with type 2 diabetes mellitus: a meta-analysis of randomized trials. Medicina Clínica (English Edition). 2017;148(8):362–70.
47. Sun J, Buys NJ. Glucose-and glycaemic factor-lowering effects of probiotics on diabetes: a meta-analysis of randomised placebo-controlled trials. British journal of nutrition. 2016;115(7):1167–77.
48. Zhang Q, Wu Y, Fei X. Effect of probiotics on glucose metabolism in patients with type 2 diabetes mellitus: a meta-analysis of randomized controlled trials. Medicina. 2016;52(1):28–34.
49. Manach C, Scalbert A, Morand C, Rémésy C, Jiménez L. Polyphenols: food sources and bioavailability. The American journal of clinical nutrition. 2004;79(5):727–47.
50. Barth SW, Koch TC, Watzl B, Dietrich H, Will F, Bub A. Moderate effects of apple juice consumption on obesity-related markers in obese men: impact of diet–gene interaction on body fat content. European journal of nutrition. 2012;51(7):841–50.
51. Almoosawi S, Fyfe L, Ho C, Al-Dujaili E. The effect of polyphenol-rich dark chocolate on fasting capillary whole blood glucose, total cholesterol, blood pressure and glucocorticoids in healthy overweight and obese subjects. British journal of nutrition. 2010;103(6):842–50.
52. Bell ZW, Canale RE, Bloomer RJ. A dual investigation of the effect of dietary supplementation with licorice flavonoid oil on anthropometric and biochemical markers of health and adiposity. Lipids in health and disease. 2011;10(1):1–10.
53. Knab AM, Shanely RA, Jin F, Austin MD, Sha W, Nieman DC. Quercetin with vitamin C and niacin does not affect body mass or composition. Applied physiology, nutrition, and metabolism. 2011;36(3):331–8.
54. Phung OJ, Baker WL, Matthews LJ, Lanosa M, Thorne A, Coleman CI. Effect of green tea catechins with or without caffeine on anthropometric measures: a systematic review and meta-analysis. The American journal of clinical nutrition. 2010;91(1): 73–81.

55. Asbaghi O, Fouladvand F, Gonzalez MJ, Aghamohammadi V, Choghakhori R, Abbasnezhad A. Effect of green tea on anthropometric indices and body composition in patients with type 2 diabetes mellitus: a systematic review and meta-analysis. Complementary medicine research. 2021;28(3):244–51.
56. Zeraattalab-Motlagh S, Jayedi A, Shab-Bidar S. The effects of resveratrol supplementation in patients with type 2 diabetes, metabolic syndrome, and nonalcoholic fatty liver disease: an umbrella review of meta-analyses of randomized controlled trials. The American journal of clinical nutrition. 2021;114(5):1675–85.
57. Delpino FM, Figueiredo LM, Caputo EL, Mintem GC, Gigante DP. What is the effect of resveratrol on obesity? A systematic review and meta-analysis. Clinical nutrition ESPEN. 2021;41:59–67.
58. Kawaguchi K, Mizuno T, Aida K, Uchino K. Hesperidin as an inhibitor of lipases from porcine pancreas and Pseudomonas. Bioscience, biotechnology, and biochemistry. 1997;61(1):102–4.
59. Rains TM, Agarwal S, Maki KC. Antiobesity effects of green tea catechins: a mechanistic review. The journal of nutritional biochemistry. 2011;22(1):1–7.
60. Sbarra V, Ristorcelli E, Le Petit-Thévenin J, Teissedre P-L, Lombardo D, Vérine A. In vitro polyphenol effects on activity, expression and secretion of pancreatic bile salt-dependent lipase. Biochimica et Biophysica Acta (BBA)-molecular and cell biology of lipids. 2005;1736(1):67–76.
61. Galleano M, Calabro V, Prince PD, Litterio MC, Piotrkowski B, Vazquez-Prieto MA, et al. Flavonoids and metabolic syndrome. Annals of the New York Academy of Sciences. 2012;1259(1):87–94.
62. Guo H, Ding J, Liu Q, Li Y, Liang J, Zhang Y. Vitamin C and metabolic syndrome. A meta-analysis of observational studies. Frontiers in nutrition. 2021:709.
63. Lee K, Kim J. Serum vitamin D status and metabolic syndrome: a systematic review and dose-response meta-analysis. Nutrition research and practice. 2021;15(3):329.
64. AlAnouti F, Abboud M, Papandreou D, Mahboub N, Haidar S, Rizk R. Effects of vitamin d supplementation on lipid profile in adults with the metabolic syndrome: a systematic review and meta-analysis of randomized controlled trials. Nutrients. 2020;12(11):3352.
65. Apaya MK, Kuo TF, Yang MT, Yang G, Hsiao CL, Chang SB, et al. Phytochemicals as modulators of β-cells and immunity for the therapy of type 1 diabetes: recent discoveries in pharmacological mechanisms and clinical potential. Pharmacological research. 2020;156:104754.
66. Chatzakis C, Sotiriadis A, Tsakmaki E, Papagianni M, Paltoglou G, Dinas K, et al. The effect of dietary supplements on oxidative stress in pregnant women with gestational diabetes mellitus: a network meta-analysis. Nutrients. 2021;13(7):2284–2311.
67. Kron-Rodrigues MR, Rudge MVC, Lima SAM. Supplementation of vitamin D in the postdelivery period of women with previous gestational diabetes mellitus: systematic review and meta-analysis of randomized trials. Revista brasileira de ginecologia e obstetricia: revista da Federacao Brasileira das Sociedades de Ginecologia e Obstetricia. 2021;43(9):699–709.
68. Maged AM, Torky H, Fouad MA, GadAllah SH, Waked NM, Gayed AS, et al. Role of antioxidants in gestational diabetes mellitus and relation to fetal outcome: a randomized controlled trial. The journal of maternal-fetal & neonatal medicine: the official journal of the European Association of Perinatal Medicine, the Federation of Asia and Oceania Perinatal Societies, the International Society of Perinatal Obstetricians. 2016;29(24):4049–54.
69. Sharifipour F, Abedi P, Ciahkal SF, Jahanfar S, Mohaghegh Z, Zahedian M. Serum vitamin E level and gestational diabetes mellitus: a systematic review and meta-analysis. Journal of diabetes and metabolic disorders. 2020;19(2):1787–95.
70. Davì G, Santilli F, Patrono C. Nutraceuticals in diabetes and metabolic syndrome. Cardiovascular therapeutics. 2010;28(4):216–26.

71. ADA. American Diabetes Association. Classification and diagnosis of diabetes: Standards of medical care in diabetes-2020. Diabetes Care. 2020 Jan;43(Suppl 1):S14–S31. doi: 10.2337/dc20-S002. PMID: 31862745.
72. Saeedi P, Petersohn I, Salpea P, Malanda B, Karuranga S, Unwin N, et al. Global and regional diabetes prevalence estimates for 2019 and projections for 2030 and 2045: results from the International Diabetes Federation Diabetes Atlas, 9th edition. Diabetes research and clinical practice. 2019;157:107843.
73. King H, Aubert RE, Herman WH. Global burden of diabetes, 1995-2025: prevalence, numerical estimates, and projections. Diabetes care. 1998;21(9):1414–31.
74. Treiber G, Prietl B, Fröhlich-Reiterer E, Lechner E, Ribitsch A, Fritsch M, et al. Cholecalciferol supplementation improves suppressive capacity of regulatory T-cells in young patients with new-onset type 1 diabetes mellitus - A randomized clinical trial. Clinical immunology (Orlando, Fla). 2015;161(2):217–24.
75. Derosa G, Limas CP, Macías PC, Estrella A, Maffioli P. Dietary and nutraceutical approach to type 2 diabetes. Archives of medical science. 2014;10(2):336–44.
76. Ellulu MS, Patimah I, Khaza'ai H, Rahmat A, Abed Y. Obesity and inflammation: the linking mechanism and the complications. Archives of medical science. 2017; 13(4):851–63.
77. WHO. World Health Organization: Diabetes 2021 [Available from: https://www.who.int/news-room/fact-sheets/detail/diabetes].
78. Panche AN, Diwan AD, Chandra SR. Flavonoids: an overview. Journal of nutritional science. 2016;5:e47–65.
79. Kumar S, Pandey AK. Chemistry and biological activities of flavonoids: an overview. ScientificWorldJournal. 2013;2013:162750.
80. Blahova J, Martiniakova M, Babikova M, Kovacova V, Mondockova V, Omelka R. Pharmaceutical drugs and natural therapeutic products for the treatment of type 2 diabetes mellitus. Pharmaceuticals (Basel, Switzerland). 2021;14(8):806–38.
81. Bondonno NP, Dalgaard F, Murray K, Davey RJ, Bondonno CP, Cassidy A, et al. Higher habitual flavonoid intakes are associated with a lower incidence of diabetes. The journal of nutrition. 2021;151(11):3533–42.
82. Liu F, Sirisena S, Ng K. Efficacy of flavonoids on biomarkers of type 2 diabetes mellitus: a systematic review and meta-analysis of randomized controlled trials. Critical reviews in food science and nutrition. 2021:1–27.
83. Y KK, Kan T, Ishikawa T. [Citrus flavonoids as a target for the prevention of pancreatic β-cells dysfunction in diabetes.]. Nihon yakurigaku zasshi Folia pharmacologica Japonica. 2020;155(4):209–13.
84. Danescu LG, Levy S, Levy J. Vitamin D and diabetes mellitus. Endocrine. 2009; 35(1):11–7.
85. Ahmed LHM, Butler AE, Dargham SR, Latif A, Robay A, Chidiac OM, et al. Association of vitamin D2 and D3 with type 2 diabetes complications. BMC endocrine disorders. 2020;20(1):65.
86. Mirzavandi F, Talenezhad N, Razmpoosh E, Nadjarzadeh A, Mozaffari-Khosravi H. The effect of intramuscular megadose of vitamin D injections on E-selectin, CRP and biochemical parameters in vitamin D-deficient patients with type-2 diabetes mellitus: a randomized controlled trial. Complementary therapies in medicine. 2020;49:102346.
87. Pittas AG, Lau J, Hu FB, Dawson-Hughes B. The role of vitamin D and calcium in type 2 diabetes. A systematic review and meta-analysis. The journal of clinical endocrinology and metabolism. 2007;92(6):2017–29.
88. Dashti F, Mousavi SM, Larijani B, Esmaillzadeh A. The effects of vitamin D supplementation on inflammatory biomarkers in patients with abnormal glucose homeostasis: a systematic review and meta-analysis of randomized controlled trials. Pharmacological research. 2021;170:105727.
89. Mangge H, Becker K, Fuchs D, Gostner JM. Antioxidants, inflammation and cardiovascular disease. World journal of cardiology. 2014;6(6):462–77.

90. Chen H, Karne RJ, Hall G, Campia U, Panza JA, Cannon RO, 3rd, et al. High-dose oral vitamin C partially replenishes vitamin C levels in patients with type 2 diabetes and low vitamin C levels but does not improve endothelial dysfunction or insulin resistance. American journal of physiology heart and circulatory physiology. 2006;290(1):H137–45.
91. Mason SA, Keske MA, Wadley GD. Effects of vitamin C supplementation on glycemic control and cardiovascular risk factors in people with type 2 diabetes: a GRADE-assessed systematic review and meta-analysis of randomized controlled trials. Diabetes care. 2021;44(2):618–30.
92. Sesso HD, Buring JE, Christen WG, Kurth T, Belanger C, MacFadyen J, et al. Vitamins E and C in the prevention of cardiovascular disease in men: the Physicians' Health Study II randomized controlled trial. JAMA. 2008;300(18):2123–33.
93. Davì G, Falco A, Patrono C. Lipid peroxidation in diabetes mellitus. Antioxidants & redox signaling. 2005;7(1–2):256–68.
94. Davì G, Ciabattoni G, Consoli A, Mezzetti A, Falco A, Santarone S, et al. In vivo formation of 8-iso-prostaglandin f2alpha and platelet activation in diabetes mellitus: effects of improved metabolic control and vitamin E supplementation. Circulation. 1999;99(2): 224–9.
95. House AA, Eliasziw M, Cattran DC, Churchill DN, Oliver MJ, Fine A, et al. Effect of B-vitamin therapy on progression of diabetic nephropathy: a randomized controlled trial. JAMA. 2010;303(16):1603–9.
96. Sanyal AJ, Chalasani N, Kowdley KV, McCullough A, Diehl AM, Bass NM, et al. Pioglitazone, vitamin E, or placebo for nonalcoholic steatohepatitis. The New England journal of medicine. 2010;362(18):1675–85.
97. Papathanasopoulos A, Camilleri M. Dietary fiber supplements: effects in obesity and metabolic syndrome and relationship to gastrointestinal functions. Gastroenterology. 2010;138(1):65–72.e1-2.
98. Xie Y, Gou L, Peng M, Zheng J, Chen L. Effects of soluble fiber supplementation on glycemic control in adults with type 2 diabetes mellitus: a systematic review and meta-analysis of randomized controlled trials. Clinical nutrition (Edinburgh, Scotland). 2021;40(4):1800–10.
99. Chen C, Zeng Y, Xu J, Zheng H, Liu J, Fan R, et al. Therapeutic effects of soluble dietary fiber consumption on type 2 diabetes mellitus. Experimental and therapeutic medicine. 2016;12(2):1232–42.
100. Bullard KM, Cowie CC, Lessem SE, Saydah SH, Menke A, Geiss LS, et al. Prevalence of diagnosed diabetes in adults by diabetes type - United States, 2016. MMWR Morbidity and mortality weekly report. 2018;67(12):359–61.
101. Mobasseri M, Shirmohammadi M, Amiri T, Vahed N, Hosseini Fard H, Ghojazadeh M. Prevalence and incidence of type 1 diabetes in the world: a systematic review and meta-analysis. Health promotion perspective. 2020;10(2):98–115.
102. Simmons KM, Michels AW. Type 1 diabetes: a predictable disease. World journal of diabetes. 2015;6(3):380–90.
103. Jacobsen LM, Haller MJ, Schatz DA. Understanding pre-type 1 diabetes: the key to prevention. Frontiers in endocrinology (Lausanne). 2018;9:70.
104. Lee YJ, Suh KS, Choi MC, Chon S, Oh S, Woo JT, et al. Kaempferol protects HIT-T15 pancreatic beta cells from 2-deoxy-D-ribose- induced oxidative damage. Phytotherapy research. 2010;24(3):419–23.
105. Kulkarni YA, Suryavanshi SV. Combination of naringenin and lisinopril ameliorates nephropathy in type-1 diabetic rats. Endocrine, metabolic & immune disorders drug targets. 2021;21(1):173–82.
106. Darwish MA, Abdel-Bakky MS, Messiha BAS, Abo-Saif AA, Abo-Youssef AM. Resveratrol mitigates pancreatic TF activation and autophagy-mediated beta cell death via inhibition of CXCL16/ox-LDL pathway: a novel protective mechanism against type 1 diabetes mellitus in mice. European journal of pharmacology. 2021;901:174059.

107. Movahed A, Raj P, Nabipour I, Mahmoodi M, Ostovar A, Kalantarhormozi M, et al. Efficacy and safety of resveratrol in type 1 diabetes patients: a two-month preliminary exploratory trial. Nutrients. 2020;12(1):161–83.
108. Buksińska-Lisik M, Kwasiborski PJ, Ryczek R, Lisik W, Mamcarz A. Vitamin D deficiency as a predictor of a high prevalence of coronary artery disease in pancreas transplant candidates with type 1 diabetes. Frontiers in endocrinology (Lausanne). 2021;12:714728.
109. Lai X, Liu X, Cai X, Zou F. Vitamin D supplementation induces CatG-mediated CD4(+) T cell inactivation and restores pancreatic beta-cell function in mice with type 1 diabetes. American journal of physiology endocrinology and metabolism. 2021;322(1):e74–84.
110. Palomer X, González-Clemente JM, Blanco-Vaca F, Mauricio D. Role of vitamin D in the pathogenesis of type 2 diabetes mellitus. Diabetes, obesity & metabolism. 2008;10(3):185–97.
111. Järvisalo MJ, Raitakari M, Toikka JO, Putto-Laurila A, Rontu R, Laine S, et al. Endothelial dysfunction and increased arterial intima-media thickness in children with type 1 diabetes. Circulation. 2004;109(14):1750–5.
112. Hoffman RP, Dye AS, Bauer JA. Ascorbic acid blocks hyperglycemic impairment of endothelial function in adolescents with type 1 diabetes. Pediatric diabetes. 2012;13(8):607–10.
113. Beckman JA, Goldfine AB, Gordon MB, Garrett LA, Keaney JF, Jr., Creager MA. Oral antioxidant therapy improves endothelial function in type 1 but not type 2 diabetes mellitus. American journal of physiology heart and circulatory physiology. 2003;285(6):H2392–8.
114. Mattila M, Erlund I, Lee H-S, Niinistö S, Uusitalo U, Andrén Aronsson C, et al. Plasma ascorbic acid and the risk of islet autoimmunity and type 1 diabetes: the TEDDY study. Diabetologia. 2020;63(2):278–86.
115. Sabri MR, Tavana EN, Ahmadi A, Hashemipour M. The effect of vitamin C on endothelial function of children with type 1 diabetes: an experimental study. International journal of preventive medicine. 2014;5(8):999–1004.
116. Gupta S, Sharma TK, Kaushik GG, Shekhawat VP. Vitamin E supplementation may ameliorate oxidative stress in type 1 diabetes mellitus patients. Clinical laboratory. 2011;57(5–6):379–86.
117. Engelen W, Keenoy BMy, Vertommen J, De Leeuw I. Effects of long-term supplementation with moderate pharmacologic doses of vitamin E are saturable and reversible in patients with type 1 diabetes. The American journal of clinical nutrition. 2000; 72(5): 1142–9.
118. Fuller S, Beck E, Salman H, Tapsell L. New horizons for the study of dietary fiber and health: a review. Plant foods for human nutrition. 2016;71(1):1–12.
119. Zou J, Reddivari L, Shi Z, Li S, Wang Y, Bretin A, et al. Inulin fermentable fiber ameliorates type I diabetes via IL22 and short-chain fatty acids in experimental models. Cellular and molecular gastroenterology and hepatology. 2021;12(3):983–1000.
120. Bozbulut R, Şanlıer N, Döğer E, Bideci A, Çamurdan O, Cinaz P. The effect of beta-glucan supplementation on glycemic control and variability in adolescents with type 1 diabetes mellitus. Diabetes research and clinical practice. 2020;169:108464.
121. Kim SY, Deputy NP, Robbins CL. Diabetes during pregnancy: surveillance, preconception care, and postpartum care. Journal of Women's Health (Larchmt). 2018 May;27(5):536–541. doi: 10.1089/jwh.2018.7052. Epub 2018 May 1. PMID: 29715050.
122. Santangelo C, Zicari A, Mandosi E, Scazzocchio B, Mari E, Morano S, et al. Could gestational diabetes mellitus be managed through dietary bioactive compounds? Current knowledge and future perspectives. The British journal of nutrition. 2016; 115(7):1129–44.
123. Nguyen-Ngo C, Willcox JC, Lappas M. Anti-diabetic, anti-inflammatory, and anti-oxidant effects of naringenin in an in vitro human model and an in vivo Murine model of gestational diabetes mellitus. Molecular nutrition & food research. 2019;63(19):e1900224.

124. Gao Q, Zhong C, Zhou X, Chen R, Xiong T, Hong M, et al. Inverse association of total polyphenols and flavonoids intake and the intake from fruits with the risk of gestational diabetes mellitus: a prospective cohort study. Clinical nutrition (Edinburgh, Scotland). 2021;40(2):550–9.
125. Balbi MA, Crivellenti LC, Zuccolotto DCC, Franco LJ, Sartorelli DS. The relationship of flavonoid intake during pregnancy with excess body weight and gestational diabetes mellitus. Archives of endocrinology and metabolism. 2019;63(3):241–9.
126. Zhao R, Zhou L, Wang S, Xiong G, Hao L. Association between maternal vitamin D levels and risk of adverse pregnancy outcomes: a systematic review and dose-response meta-analysis. Food & function. 2021;1(13):14–37.
127. Milajerdi A, Abbasi F, Mousavi SM, Esmaillzadeh A. Maternal vitamin D status and risk of gestational diabetes mellitus: a systematic review and meta-analysis of prospective cohort studies. Clinical nutrition (Edinburgh, Scotland). 2021;40(5):2576–86.
128. Huang S, Fu J, Zhao R, Wang B, Zhang M, Li L, et al. The effect of combined supplementation with vitamin D and omega-3 fatty acids on blood glucose and blood lipid levels in patients with gestational diabetes. Annals of palliative medicine. 2021;10(5):5652–8.
129. Shahgheibi S, Farhadifar F, Pouya B. The effect of vitamin D supplementation on gestational diabetes in high-risk women: results from a randomized placebo-controlled trial. Journal of research in medical sciences. 2016;21(2):1–6.
130. Liu C, Zhong C, Chen R, Zhou X, Wu J, Han J, et al. Higher dietary vitamin C intake is associated with a lower risk of gestational diabetes mellitus: a longitudinal cohort study. Clinical nutrition (Edinburgh, Scotland). 2020;39(1):198–203.
131. Lekakis JP, Anastasiou EA, Papamichael CM, Stamatelopoulos KS, Dagre AG, Alevizaki MC, et al. Short-term oral ascorbic acid improves endothelium-dependent vasodilatation in women with a history of gestational diabetes mellitus. Diabetes care. 2000;23(9):1432–4.
132. Kozlowska A, Jagielska AM, Okreglicka KM, Dabrowski F, Kanecki K, Nitsch-Osuch A, et al. Dietary vitamin and mineral intakes in a sample of pregnant women with either gestational diabetes or type 1 diabetes mellitus, assessed in comparison with Polish nutritional guidelines. Ginekologia polska. 2018;89(11):581–6.
133. Brown B, Wright C. Safety and efficacy of supplements in pregnancy. Nutrition review. 2020;78(10):813–26.
134. Ma H, Qiao Z, Li N, Zhao Y, Zhang S. The relationship between changes in vitamin A, vitamin E, and oxidative stress levels, and pregnancy outcomes in patients with gestational diabetes mellitus. Annals of palliative medicine. 2021;10(6):6630–6.
135. Xu Q, Tao Y, Zhang Y, Zhang X, Xue C, Liu Y. Dietary fiber intake, dietary glycemic load, and the risk of gestational diabetes mellitus during the second trimester: a nested case-control study. Asia Pacific journal of clinical nutrition. 2021;30(3):477–86.
136. Li H, Xie S, Zhang X, Xia Y, Zhang Y, Wang L. Mid-pregnancy consumption of fruit, vegetable and fruit juice and the risk of gestational diabetes mellitus: a correlation study. Clinical nutrition ESPEN. 2021;46:505–9.
137. Basu A, Crew J, Ebersole JL, Kinney JW, Salazar AM, Planinic P, et al. Dietary blueberry and soluble fiber improve serum antioxidant and adipokine biomarkers and lipid peroxidation in pregnant women with obesity and at risk for gestational diabetes. Antioxidants (Basel, Switzerland). 2021;10(8):1318–19.
138. Cui Y, Liao M, Xu A, Chen G, Liu J, Yu X, et al. Association of maternal pre-pregnancy dietary intake with adverse maternal and neonatal outcomes: a systematic review and meta-analysis of prospective studies. Critical reviews in food science and nutrition. 2021;October 20:1–22.
139. Marjot T, Moolla A, Cobbold JF, Hodson L, Tomlinson JW. Nonalcoholic fatty liver disease in adults: current concepts in etiology, outcomes, and management. Endocrine reviews. 2020;41(1):66–117.
140. Chalasani N, Younossi Z, Lavine JE, Charlton M, Cusi K, Rinella M, et al. The diagnosis and management of nonalcoholic fatty liver disease: practice guidance from the American Association for the Study of Liver Diseases. Hepatology. 2018;67(1):328–57.

141. Abe Y, Hines IN, Zibari G, Pavlick K, Gray L, Kitagawa Y, et al. Mouse model of liver ischemia and reperfusion injury: method for studying reactive oxygen and nitrogen metabolites in vivo. Free radical biology and medicine. 2009;46(1):1–7.
142. Al-Asmari A, Khan A, Al-Masri N. Mitigation of 5-fluorouracil–induced liver damage in rats by vitamin C via targeting redox–sensitive transcription factors. Human & experimental toxicology. 2016;35(11):1203–13.
143. Cichoż-Lach H, Michalak A. Oxidative stress as a crucial factor in liver diseases. World journal of gastroenterology. 2014;20(25):8082.
144. Ore A, Akinloye OA. Oxidative stress and antioxidant biomarkers in clinical and experimental models of non-alcoholic fatty liver disease. Medicina. 2019;55(2):26.
145. Sanyal AJ, Chalasani N, Kowdley KV, McCullough A, Diehl AM, Bass NM, et al. Pioglitazone, vitamin E, or placebo for nonalcoholic steatohepatitis. The New England journal of medicine. 2010;362(18):1675–85.
146. Miller III ER, Pastor-Barriuso R, Dalal D, Riemersma RA, Appel LJ, Guallar E. Meta-analysis: high-dosage vitamin E supplementation may increase all-cause mortality. Annals of internal medicine. 2005;142(1):37–46.
147. Hirashima O, Kawano H, Motoyama T, Hirai N, Ohgushi M, Kugiyama K, et al. Improvement of endothelial function and insulin sensitivity with vitamin C in patients with coronary spastic angina: possible role of reactive oxygen species. Journal of the American College of Cardiology. 2000;35(7):1860–6.
148. Park S, Ham J-O, Lee B-K. Effects of total vitamin A, vitamin C, and fruit intake on risk for metabolic syndrome in Korean women and men. Nutrition. 2015;31(1):111–8.
149. Zhou C, Na L, Shan R, Cheng Y, Li Y, Wu X, et al. Dietary vitamin C intake reduces the risk of type 2 diabetes in Chinese adults: HOMA-IR and T-AOC as potential mediators. Plos one. 2016;11(9):e0163571.
150. Ding S, Jiang J, Zhang G, Bu Y, Zhang G, Zhao X. Resveratrol and caloric restriction prevent hepatic steatosis by regulating SIRT1-autophagy pathway and alleviating endoplasmic reticulum stress in high-fat diet-fed rats. PloS one. 2017;12(8):e0183541.
151. Elgebaly A, Radwan IA, AboElnas MM, Ibrahim HH, Eltoomy MF, Atta AA, et al. Resveratrol supplementation in patients with non-alcoholic fatty liver disease: systematic review and meta-analysis. Journal of gastrointestinal & liver diseases. 2017;26(1):59–67.
152. Wei S, Yu X. Efficacy of resveratrol supplementation on liver enzymes in patients with non-alcoholic fatty liver disease: a systematic review and meta-analysis. Complementary therapies in medicine. 2021;57(November 30):102635.
153. KÖksal E, Gülçin I, Beyza S, Sarikaya O, Bursal E. In vitro antioxidant activity of silymarin. Journal of enzyme inhibition and medicinal chemistry. 2009;24(2):395–405.
154. Pradhan S, Girish C. Hepatoprotective herbal drug, silymarin from experimental pharmacology to clinical medicine. Indian journal of medical research. 2013;137(2):491–504.
155. Salamone F, Galvano F, Cappello F, Mangiameli A, Barbagallo I, Volti GL. Silibinin modulates lipid homeostasis and inhibits nuclear factor kappa B activation in experimental nonalcoholic steatohepatitis. Translational research. 2012;159(6):477–86.
156. Zhong S, Fan Y, Yan Q, Fan X, Wu B, Han Y, et al. The therapeutic effect of silymarin in the treatment of nonalcoholic fatty disease: a meta-analysis (PRISMA) of randomized control trials. Medicine. 2017;96(49):e9061–8.
157. de Avelar CR, Pereira EM, de Farias Costa PR, de Jesus RP, de Oliveira LPM. Effect of silymarin on biochemical indicators in patients with liver disease: systematic review with meta-analysis. World journal of gastroenterology. 2017;23(27):5004.
158. Li J, Sapper TN, Mah E, Moller MV, Kim JB, Chitchumroonchokchai C, et al. Green tea extract treatment reduces NFκB activation in mice with diet-induced nonalcoholic steatohepatitis by lowering TNFR1 and TLR4 expression and ligand availability. The journal of nutritional biochemistry. 2017;41:34–41.
159. Pan MH, Yang G, Li S, Li MY, Tsai ML, Wu JC, et al. Combination of citrus polymethoxyflavones, green tea polyphenols, and Lychee extracts suppresses obesity and hepatic steatosis in high-fat diet induced obese mice. Molecular nutrition & food research. 2017;61(11):1601104.

160. Tan Y, Kim J, Cheng J, Ong M, Lao W-G, Jin X-L, et al. Green tea polyphenols ameliorate non-alcoholic fatty liver disease through upregulating AMPK activation in high fat fed Zucker fatty rats. World journal of gastroenterology. 2017;23(21):3805.
161. Zhu W, Chen S, Chen R, Peng Z, Wan J, Wu B. Taurine and tea polyphenols combination ameliorate nonalcoholic steatohepatitis in rats. BMC complementary and alternative medicine. 2017;17(1):1–12.
162. Mahmoodi M, Hosseini R, Kazemi A, Ofori-Asenso R, Mazidi M, Mazloomi SM. Effects of green tea or green tea catechin on liver enzymes in healthy individuals and people with nonalcoholic fatty liver disease: a systematic review and meta-analysis of randomized clinical trials. Phytotherapy research. 2020;34(7):1587–98.
163. Mansour-Ghanaei F, Hadi A, Pourmasoumi M, Joukar F, Golpour S, Najafgholizadeh A. Green tea as a safe alternative approach for nonalcoholic fatty liver treatment: a systematic review and meta-analysis of clinical trials. Phytotherapy research. 2018;32(10):1876–84.
164. Jalali M, Mahmoodi M, Mosallanezhad Z, Jalali R, Imanieh MH, Moosavian SP. The effects of curcumin supplementation on liver function, metabolic profile and body composition in patients with non-alcoholic fatty liver disease: a systematic review and meta-analysis of randomized controlled trials. Complementary therapies in medicine. 2020;48:102283.
165. Wei Z, Liu N, Tantai X, Xing X, Xiao C, Chen L, et al. The effects of curcumin on the metabolic parameters of non-alcoholic fatty liver disease: a meta-analysis of randomized controlled trials. Hepatology international. 2019;13(3):302–13.
166. Li YY, Tang D, Du YL, Cao CY, Nie YQ, Cao J, et al. Fatty liver mediated by peroxisome proliferator-activated receptor-α DNA methylation can be reversed by a methylation inhibitor and curcumin. Journal of digestive diseases. 2018;19(7):421–30.
167. Yan C, Zhang Y, Zhang X, Aa J, Wang G, Xie Y. Curcumin regulates endogenous and exogenous metabolism via Nrf2-FXR-LXR pathway in NAFLD mice. Biomedicine & pharmacotherapy. 2018;105:274–81.
168. Clements Jr RS, Darnell B. Myo-inositol content of common foods: development of a high-myo-inositol diet. The American journal of clinical nutrition. 1980;33(9):1954–67.
169. Hooper NM. Glycosyl-phosphatidylinositol anchored membrane enzymes. Clinica chimica acta. 1997;266(1):3–12.
170. Kashiwada M, Lu P, Rothman PB. PIP3 pathway in regulatory T cells and autoimmunity. Immunologic research. 2007;39(1–3):194.
171. Grasberger H, Van Sande J, Hag-Dahood Mahameed A, Tenenbaum-Rakover Y, Refetoff S. A familial thyrotropin (TSH) receptor mutation provides in vivo evidence that the inositol phosphates/Ca2+ cascade mediates TSH action on thyroid hormone synthesis. The journal of clinical endocrinology & metabolism. 2007;92(7):2816–20.
172. Ferrari S, Fallahi P, Di Bari F, Vita R, Benvenga S, Antonelli A. Myo-inositol and selenium reduce the risk of developing overt hypothyroidism in patients with autoimmune thyroiditis. European review for medical and pharmacological sciences. 2017; 21(2 Suppl):36–42.
173. Nordio M, Basciani S. Myo-inositol plus selenium supplementation restores euthyroid state in Hashimoto's patients with subclinical hypothyroidism. European review for medical and pharmacological sciences. 2017;21(Suppl 2):51–9.
174. Nordio M, Pajalich R. Combined treatment with Myo-inositol and selenium ensures euthyroidism in subclinical hypothyroidism patients with autoimmune thyroiditis. Journal of thyroid research. 2013;2013:1–5.
175. Galland S, Georges B, Le Borgne F, Conductier G, Dias JV, Demarquoy J. Thyroid hormone controls carnitine status through modifications of γ-butyrobetaine hydroxylase activity and gene expression. Cellular and molecular life sciences. 2002;59(3):540–5.
176. An JH, Kim YJ, Kim KJ, Kim SH, Kim NH, Kim HY, et al. L-carnitine supplementation for the management of fatigue in patients with hypothyroidism on levothyroxine treatment: a randomized, double-blind, placebo-controlled trial. Endocrine journal. 2016:EJ16–0109.

177. Benvenga S, Ruggeri RM, Russo A, Lapa D, Campenni A, Trimarchi F. Usefulness of L-carnitine, a naturally occurring peripheral antagonist of thyroid hormone action, in iatrogenic hyperthyroidism: a randomized, double-blind, placebo-controlled clinical trial. The journal of clinical endocrinology & metabolism. 2001;86(8):3579–94.
178. Benvenga S, Lakshmanan M, Trimarchi F. Carnitine is a naturally occurring inhibitor of thyroid hormone nuclear uptake. Thyroid. 2000;10(12):1043–50.
179. Felker P, Bunch R, Leung AM. Concentrations of thiocyanate and goitrin in human plasma, their precursor concentrations in brassica vegetables, and associated potential risk for hypothyroidism. Nutrition Reviews. 2016;74(4):248–58.
180. Otun J, Sahebkar A, Östlundh L, Atkin SL, Sathyapalan T. Systematic review and meta-analysis on the effect of soy on thyroid function. Scientific reports. 2019;9(1):1–9.

Changes in the Regulation of Energy Metabolism in Chronic Diseases Using Functional Foods and Nutraceuticals

Aparoop Das, Manash Pratim Pathak, Kalyani Pathak, Urvashee Gogoi, and Riya Saikia

Contents

Energy metabolism and its regulation 168
Mechanism of energy metabolism in different physiological setups 170
Brain regulation 170
Cardiovascular regulation 170
Gut microbiota regulation 171
Skeletal muscle regulation 171
Factors that alter the regulation of energy metabolism 172
Basal and resting metabolism 172
Dietary-induced thermogenesis 173
Energy intake 173
Chronic diseases that alter the regulation of energy metabolism 173
Chronic liver disease 173
Chronic renal failure 174
Cancer 174
Cardiovascular diseases 175
Respiratory diseases 176
Diabetes mellitus (DM) 176
Obesity 177
Thyroid disease 177
Alzheimer's disease 178
Role of functional foods and nutraceuticals in ameliorating the changes made in energy metabolism by chronic diseases 179
Benefits to health 182
Chronic liver disease 182
Chronic renal failure 183

DOI: 10.1201/9781003220053-9

Cancer..183
Cardiovascular diseases 184
Diabetes mellitus...185
Obesity 186
Thyroid disease 187
Alzheimer's disease 187
References...188

Energy metabolism and its regulation

Energy can be defined as a quantitative property that is utilized and is required in most of the activities carried out at the cellular level. Free energy is required for the organization and the synthesis of cell walls and other cellular components, which in turn contributes to the overall growth of a cell. Energy is also involved in various maintenance processes like disposal of protein and maintaining ion concentration and integrity of cellular membranes. The energy obtained from the protein phosphorylation that includes the transfer of phosphate groups from the energy-rich molecules is utilized by different signaling pathways. A huge amount of the energy (around 40%) is also required to carry out the most elemental functions of living organisms, that is, the transcription of genome followed by the synthesis of protein. In this context, the energy produced after the breakdown of glucose substrate is transferred into the endergonic reactions. The main energy currency in living organism is adenosine triphosphate (ATP), which is continuously utilized by the cell to carry out its functional and fundamental activities like signal processing and generating appropriate responses. Thus, energy is regarded as the chief player in the complex system of the living cell. The science that studies the different metabolic reactions that make use of the available substrates to release the energy in the form of ATP is regarded as energy metabolism (EM). However, living organisms preserve these metabolic pathways, indicating its importance and significance in maintaining and running various cellular functions (1, 2). Glycolysis is considered to be the most common pathway in all cell species; it can easily convert sugar molecules (mainly GLC) into pyruvate (PYR), producing two ATP molecules. The PYR units can be further processed through oxidative phosphorylation (OP), which in return yields 30 units of ATP against each molecule of glucose (GLC). The process of OP takes place in the mitochondria of higher organisms and eukaryotes and is regarded as the energy generator of the cell; mitochondria are regarded as the powerhouse of the cell. However, OP occurs shortly in the cytosolic fluid and cell membrane of the prokaryotic organisms. Even though OP and glycolysis are considered to be the main and primary pathways for the generation of energy in tissues and cells, they also exhibit some other mechanisms through which energy is produced and stored. For example, during the high-demand periods, phosphagen molecules like phosphocreatine (PCr) can effectively work as a temporary buffer. Subsequently, this mechanistic procedure was displayed as an important mechanism taken up by the tissues of muscle under different conditions of metabolic regime and energy demand (3). Further studies into the field indicated that the process of glycolysis not only involves GLC as the energy substrate, but some other forms of GLC, like starch found in plants and glycogen found in mammalian tissues, are also used during glycolysis that are stored by some of the specific cells under different conditions. These local reservoirs can be employed during a nutrient limitation period or during the high-demand periods. Also, there are some other substrates that can make an entry into the cells at the time of EM. For example, lactate (LAC) derived in the cerebral tissue plays an important role during the metabolism of energy that was previously considered as an unwanted by-product produced during anaerobic glycolysis (4, 5). The reverse action produced by the enzyme lactate dehydrogenase (LDH) can easily convert the LAC residues into PYR that can further feed OP. This mechanistic reaction can take place inside the mitochondria in neurons, indicating the fact that instead of PYR, the LAC residues can be used as the for OP. Further research suggested that in order to produce PYR residues, GLC units have to pass through nine sequential reactions of glycolysis. However, considering the utility of LAC in the regulation of EM, it was discovered that an increase in the extracellular concentration of LAC leads to an enhanced production of ATP in the cells (6).

In case of chronic disorders like chronic liver disease, chronic renal failure (CRF), cancer, cardiovascular disorders, respiratory diseases, diabetes mellitus (DM), obesity, thyroid disease and Alzheimer's disease (AD), the energy homeostasis is completely disturbed, affecting the energy balance equation. Studies have indicated that the presence of pathologies induces certain metabolic responses that not only contribute toward disease-associated undernutrition but also results in co-morbidity due to poor immune responses. However, these anomalies associated with EM can be overcome by the means of nutritional support that could prevent any sort of energy drain (7, 8). Nutraceuticals and functional foods such as fruits, legumes, spices, vegetables and cereals have been reported to modulate various metabolic processes that can protect and prevent the effects of several chronic disorders like chronic liver disease, CRF, cancer, cardiovascular disorders, respiratory diseases, DM, obesity, thyroid disease and AD. Thus, the consumption of such dietary products is found to be related with lower cases of these chronic disorders. Biochemical, epidemiological and clinical studies of these bioactive compounds have indicated that they exhibit different mechanisms through which they produce several potential activities like antiproliferative, antihypertensive, antioxidant, antidiabetic, antimicrobial and anti-Alzheimer's activities in human beings (9–24). The chemical composition of the bioactives present in these functional foods might have affinities toward specific proteins that can modulate and inhibit the action of certain enzymes, which can improve health and prevent the onset of diseases (25–28). This chapter deals with the role of nutraceuticals and functional foods in ameliorating the changes in the regulation of EM produced by various chronic disorders.

Mechanism of energy metabolism in different physiological setups

Brain regulation

The brain controls a variety of energy-consuming processes, including locomotion, skeletal muscle fatty acid oxidation and thermogenesis. Energy intake and energy expenditure is balanced in a healthy individual; however, the same becomes disbalanced in an obese condition, which results in hyperphagia and finally obesity (29). Anorexigenic neurons present in the hypothalamus express proopiomelanocortin (POMC) and the neurons producing POMC also produce α-melanocyte stimulating hormone (α-MSH) that binds to melanocortin receptors 3 and 4 (MC3R and MC4R). Binding of α-MSH to MC3R and MC4R results in increased energy expenditure and reduction in the intake of food (30). Thermogenesis occurs in the brown adipose tissue (BAT) that induces energy expenditure. Exposure to cold as well as intracerebroventricular (ICV) co-injection of insulin and leptin leads to browning of white adipose tissue (WAT) followed by thermogenesis (31). However, inhibition of WAT browning leads to decreased thermogenesis followed by decreased energy expenditure, and results in obesity (32).

Cardiovascular regulation

The heart needs a continuous energy supply in the form of ATPs to execute its functions, and the high energy demand of the heart is fulfilled by the oxidation of fatty acids and carbohydrates like glucose and lactate executed by mitochondria (33). The adult mammalian heart depends on the long-chain fatty acids (LCFA), whereas the fetal heart depends upon catabolism of glucose and lactate as substrates for ATP production (34). The heart utilizes the required energy in the form of ATP by the domestication of mitochondria through a cytosolic control network where the cytosolic adenosine diphosphate (ADP) is converted to form ATP in the mitochondria after the oxidation of metabolic substrates. Cytosolic Ca^{2}+ dependent parallel activation scheme for metabolism is one of the proposed models that aligns with many reports. According to this model, cytosolic Ca^{2} regulates both the use of ATP by work-producing ATPases and the generation of ATP by mitochondria (35). The capacity of the mammalian postnatal heart to generate energy by using fats is influenced in part by the expression of nuclear genes encoding enzymes involved in mitochondrial fatty acid β-oxidation (FAO) (34). Patients with heart failure have a deficiency of cardiac energy as well as anatomical and physiological abnormalities. Obesity and type 2 DM are chronic pathophysiological diseases that promote heart failure by altering metabolic pathways, modifying cardiac energetics and affecting cardiac contractility (36). Obesity causes the failing heart to become overly reliant on fatty acid oxidation as a source of ATP generation, at least in part by blocking AMPKTyr172/ACCSer79 signaling and increasing acetylation of β-oxidation enzymes; however, it has no effect on the cardiac energy deficit (37). Weight loss is an effective way for improving the failing heart in terms of improving cardiac function and EM. A recent study demonstrated that weight loss following administration of a low-fat diet or caloric restriction improves the condition of a failing heart in obese animals by ameliorating FAO, acetylation of ß-oxidation enzymes and AMPKTyr172/ACCSer79 signaling (37).

Gut microbiota regulation

Gut microbiota play an important role in the regulation of energy in both the lean and obese conditions. Composition of gut microbiota differs in lean and in obese humans. In healthy humans over 1000 phylotypes are reported, which are classified into six bacterial divisions viz., Firmicutes, Bacteroidetes, Proteobacteria, Fusobacteria, Actinobacteria and Verrucomicrobia (38). Firmicutes and Bacteroidetes make up the majority of microbes inhabiting the human intestine as compared to actinobacteria, proteobacteria, verrucomicrobia, fusobacteria and euryarchaeota (39). Firmicutes (Clostridium), Prevotella and Methanobrevibacter predominate in the gut microbiota of hosts with a high-fat/carbohydrate diet as compared to Bacteroides, Bifidobacterium, Lactobacillus and Akkermansia. It boosts the host's ability to extract energy from digested food as well as promotes production of metabolites such as short-chain fatty acids (SCFAs), secondary bile acids and microbial products like lipopolysaccharides. These products of gut microbiota aid in the signaling pathway during modulation of appetite and energy expenditure, uptake and storage (38). SCFA can be absorbed in the gut and used as an energy source, resulting in a higher energy harvest from the food consumed and is thought to account for 10% of all energy consumed by people in the Western world (40). Moreover, SCFAs are reported to regulate the lipogenesis by activating the carbohydrate responsive element-binding protein (ChREBP) and the sterol regulatory element-binding transcription factor 1 (SREBP1) (41).

Skeletal muscle regulation

People's energy expenditure differs depending on their body size and composition, and those with a "low" metabolic rate appear to be at a higher risk of gaining weight. Development of obesity is due to the differences in the balance of energy intake and energy expenditure. In humans, although the short-term success rate is high, the long-term success rate of reducing energy intake is notably low, probably due to lower resting metabolic rate (RMR) and exercise-related energy expenditure favoring an unhealthy energy consumption pattern (42, 43). If body size and composition are considered fully, rather than thermogenesis, which appeared to be a critical determinant, RMR is connected with maximal aerobic capacity, according to a study (44). The availability of ATP is critical for skeletal muscle contraction during exercise. To maintain a constant supply of ATP, skeletal muscle has a range of metabolic pathways, some aerobic and others anaerobic. According to a report, both aerobic and anaerobic energy provision from PCr breakdown and carbohydrate utilization in the glycolytic pathway allow athletes to satisfy the high energy demands of specific events or sports (45).

Factors that alter the regulation of energy metabolism (46, 47)

Physical activity can act as one of the vital factors that can prominently affect the metabolic rate of individuals when compared to other factors. The fact remains the same in case of two individuals having identical weight, body composition, age, height, sex and body composition. However, there are also individuals who spend a greater portion of the day in sedentary activities; hence, in such circumstances, other factors become vital and superior in calculating the extent of energy expenditure associated with an individual.

Basal and resting metabolism

Basal metabolic rate (BMR) can be defined as the amount of energy that has been spent by an individual in a post-absorptive state, that is, after a minimum time span of 12 hours from the last meal and at rest both physically and mentally in a warm comfortable room. The basic aim is to calculate the amount of energy expenditure related to the obligatory minimum requirement for energy in a thermally stable and neutral environment. Without an increase in the production of heat, the temperature of the body should be regulated and monitored constantly by a vasomotor tone, and this can occur at a temperature below the critical temperature known as ambient temperature. However, it was thought that metabolic rate is not affected by the energy obtained from the food consumed 12 hours before the measurement, but studies have indicated that there is an impact from the energy obtained from previous day. Also, it was thought that the subject is at rest both physically and mentally, but later on it was difficult to establish whether the requirements has been actually met. Thus, to overcome these hindrances, the term "RMR" was introduced, whereby measurements were carried out under circumstances that are at very close proximity to the standard BMR definition. However, it is important to take necessary precautions, as RMR is used to determine the resting metabolism at any time cycle, including the effects of circadian rhythm and intake of meals. Thus, it is very important to indicate and state the conditions under which RMR is to be measured. RMR is also affected by other factors like sex, body size, stress factor, age and health. Further problems arise when the RMR of obese people is to be compared with the average size people, as one needs to accurately measure the body composition using indirect techniques. The stage of the menstrual cycle, time of the day and time of the year also affect the RMR, because these parameters must be controlled during the experimental studies. In experimental studies including the animals that variation can be minimized by keeping the animals under controlled activity and diet plans. However, this becomes a challenging task in human beings because the effect of the previous day or previous weeks' activity and food intake is not known.

Dietary-induced thermogenesis

The effect on the metabolic rate upon taking a meal was studied and reported. Studies reported that after taking a large meal an individual can feel hotter. Thereafter, different terms like thermic effect of a meal, specific dynamic effect, heat increment of feeding and specific dynamic action were introduced to define the increase in production of heat followed by a meal. Also, in many species it was observed that there is an increase in the 24-hour metabolism upon intake of high energy for several days or weeks. In such experimental studies, the rate of metabolism has been immediately calculated after 24 hours or after feeding. Hence, each of the components is regarded as dietary-induced thermogenesis.

Energy intake

Different experimental studies, including overfeeding for varying time length, were carried out by a group of researchers. The main focus of the study was to determine whether some group of people can easily adapt to overeating without any weight gain or a minimal gain in weight that could be easily predicted from the additional intake of energy. One of the major drawbacks of such a study is that it is practically impossible to monitor the expenditure of energy continuously for 24 hours, making it difficult to obtain precise and accurate energy balances. The problems of overfeeding will continue until newer techniques, including long-term studies of energy balance, are designed and carried out.

Chronic diseases that alter the regulation of energy metabolism

All creatures require energy to function properly because it orchestrates essential biological activities. As a result, it's no wonder that disturbances in EM are at the root of a variety of disorders. EM gives a precise measurement of our health, both temporally and functionally. Changes in EM are identified in a wide range of hereditary and environmental disorders, both rare and common (48).

Chronic liver disease

Insulin and other metabolic hormones regulate the metabolic activity of the liver, which is a key metabolic organ. In chronic liver illness, abnormalities in protein and EM are common, resulting in malnutrition. In individuals with liver cirrhosis, protein-energy malnutrition is a typical observation (49). Nonalcoholic fatty liver disease (NAFLD) refers to a group of chronic liver illnesses that are not caused by alcohol. Increased lipid input into the liver, enhanced de novo hepatic lipogenesis and poor lipid utilization all contribute to hepatic lipid buildup, resulting in NAFLD. Chronic dietary excesses of fructose and saturated fatty acids increase lipid metabolite buildup, oxidative and endoplasmic reticulum stress, and cytokine production, all of which promote NAFLD pathogenesis (50). Hepatic steatosis is linked to oxidative changes in mitochondria. According to one study, fatty liver mitochondrial oxidative changes are linked to a significant decrease in F_0F_1-ATP synthase. These alterations, which are amplified by starvation, could explain the lower hepatic ATP generation seen in the presence of fatty invasion (51). Mitochondria are involved in lipid metabolism and oxidative stress in the liver. In liver tissues from patients with alcohol-related and non-related liver disorders, microscopic mitochondrial lesions, aberrant mitochondrial dynamics, reduced respiratory chain complex activity and reduced ability to generate ATP have been identified. Enhanced lipogenesis combined with reduced fatty acid b-oxidation gives rise to triglyceride accumulation in hepatic cells, which, when coupled with rising levels of reactive oxygen species (ROS), directly contribute to insulin resistance in steatohepatitis patients (52). In patients with chronic liver disease, the ability to metabolize proteins is closely related to the hepatic functional reserve, and hypoalbuminemia and hyperammonemia develop as the disease progresses. Zinc deficiency, which is common in patients with chronic liver disease, has a significant impact on protein metabolism (53).

Chronic renal failure

The leading causes of death in patients with chronic kidney disease (CKD) are increased atherosclerosis and cardiovascular disease (CVD). During the progression of CKD, the plasma lipid profile frequently changes (54). CRF is linked with hypertriglyceridemia and dysfunctional clearance of very-low-density lipoprotein (VLDL) and chylomicrons, which is caused primarily by lipoprotein lipase (LPL) nutritional deficiencies (55). LPL is an enzyme that regulates plasma lipoprotein and EM. Renal dysfunction is also linked to numerous changes in lipoprotein metabolism that result in dyslipidemia and atherogenic particle accumulation (56, 57). End-Stage Renal Disease (ESRD) has a significant impact on EM. This is because impaired LPL activity and VLDL receptor deficiency can limit the availability of lipid fuel for power generation in skeletal muscle and thus physical function. Similarly, these deficiencies can reduce energy storage in adipose tissue and contribute to the wasting syndrome (58).

Cancer

Cancer cells have distiıct metabolic demands. An increased rate of glucose uptake is observed in most tumor types, and this serves as a realistic guideline for understanding selective metabolism in tumors (59). Otto Warburg described the first and most well-known cancer metabolic anomaly in the 1920s. Dr. Warburg discovered that tumors had unusually high rates of glucose uptake and lactate production when compared to normal body tissue, even when oxygen was present (Warburg effect) (60). The roles of transcriptional regulators (TRs) in tumor formation and metastasis have been extensively studied (61). In cancer cells, energy metabolic pathways have a dual purpose. They control ATP levels and serve as precursors to energy-intensive (proliferation, differentiation, migration, invasion, colonization, ion homeostasis) and anabolic (biogenesis of nucleic acids, proteins, phospholipids and cholesterol) processes. Multiple TRs control the energy metabolic pathways in response to oxygen and substrate (glucose, glutamine, fatty acids) availability (62). Signal transducer and activator of transcription (STAT) proteins have been shown to play a key role in metabolic control in recent years (63). STAT molecules (especially STAT3 and STAT5) are inherently activated in a wide range of cancers. STAT3 and STAT5 are present in mitochondria, and their effects on metabolic enzyme regulation are mostly mediated by an increase in Hypoxia-inducible factor-1alpha (HIF-1) (64). The most malignant of cancers, which are frequently hypoxic, rely on intensified glycolysis to fulfill the increased need for ATP and biosynthetic precursors, as well as robust pH-regulating systems to resist excessive lactic and carbonic acid production (65). Eagle discovered the special role of glutamine in proliferating cells in 1955 (66). Eagle discovered that numerous cell lines absorbed glutamine at tenfold higher rates than any other amino acid and that many cell lines could not multiply or maintain viability without it. Experiments later revealed that glutamine has a role in oxidative mitochondrial metabolism (67).

Cardiovascular diseases

A sufficient supply of oxygen and oxidizable substrates is required for normal cardiac function to generate enough ATP to meet the organ's energy need. This energy is predominantly obtained in the adult heart by fatty acid oxidation via OP (68, 69). However, several factors may influence this energy source including availability of substrate, energy requirements, oxygen supply, distinct metabolic condition, alterations in glucose oxidation and transport developed in the diabetic heart. Owing to this, compromised myocardial performance may occur in situations where ATP provided by glycolysis is important, such as ischemia and reperfusion. Derangement of fatty acid supply to mitochondria and change of some critical enzymes of EM reduces ATP synthesis in several cardiac conditions such as ischemic cardiomyopathy, heart failure, hypertrophy and dilated cardiomyopathy. Energy depletion can also be caused by a lack of certain cofactors like L-carnitine and creatine. Other mitochondrial enzymes including the creatine kinase system are also impacted (70). Coenzyme Q10 (CoQ10) is an endogenous antioxidant synthesized by all cells and is important for EM and antioxidant protection. The distribution of CoQ10 in different organs is not uniform, and the heart has the highest concentration. CoQ10 supplementation has been shown to lower oxidative stress, cardiovascular mortality and enhance the clinical outcome in patients following coronary artery bypass surgery (71). In a range of disease models, including pressure overload, ischemia/reperfusion and metabolic disruption, accumulating data shows that dysfunction in mitochondrial dynamics leads to myocardial injury and cardiac disease development. These results show that modifying mitochondrial dynamics could be a viable treatment strategy for cardiovascular disorders (72).

Respiratory diseases

COPD, or chronic obstructive pulmonary disease, is a spectrum of disorders that cause airflow obstruction and breathing difficulty (73). COPD is one of the world's most prevalent chronic diseases. Emphysema, chronic bronchitis and small airway blockage are all part of this condition. Inflammation, apoptosis and senescence are all influenced by abnormal mitochondrial ROS (mtROS) generation, decreased matrix metalloproteinase (MMP), mitochondrial calcium overload, mitochondrial DNA (mtDNA) mutation and dysregulated mitophagy. COPD, pulmonary fibrosis and asthma are all chronic respiratory illnesses that are linked to these biological processes (74). COPD patients have a large disparity in aerobic and anaerobic EM in a resting state, as well as a high likelihood of anaerobic energy supply mechanism, which correlates favorably with disease development (75). Nutritional status is generally poor in COPD patients, and it is an independent predictor of morbidity and mortality (76). Malnutrition is linked to higher energy expenditure in COPD patients due to hypermetabolism, which is caused by increased respiratory muscle exertion, oxygen demand, inflammation and poorer dietary patterns (77). The metabolism of glucose, fatty acids and glutamine has been shown to play a key role in the etiology of chronic lung diseases. Recent research shows that metabolic reprogramming occurs in chronic lung disease patients and animal models, implying that metabolic dysregulation may play a role in the development and progression of this disease (78).

Diabetes mellitus (DM)

DM is a family of metabolic illnesses marked by hyperglycemia caused by faults in insulin production, insulin action, or both. The most important hormones that affect glucose, protein and lipid metabolism are insulin, glucagon and GH (somatotropin). During insulin deprivation, persons with type 1 DM have significant metabolic alterations. Increased basal energy expenditure (BEE) and decreased mitochondrial function are two of them. Furthermore, insulin deprivation has a considerable impact on protein metabolism. There is a higher increase in protein breakdown than protein synthesis, resulting in a net protein loss (79). Despite a gain in whole-body oxygen consumption, insulin insufficiency and accompanying metabolic alterations lower muscle mitochondrial ATP production rate (MAPR) and expression of OP genes in type 1 diabetes. Increased transcript levels of genes implicated in the vascular endothelial growth factor (VEGF), inflammation, cytoskeleton and integrin signaling pathways show that vascular factors and cell proliferation may probably happen in conjunction with mitochondrial alterations (80). Insulin deprivation causes significant metabolic alterations, such as whole-body protein catabolism and muscle wastage. Insulin treatment can easily reverse the abnormalities in metabolism and body composition that occur in insulin-dependent DM (IDDM) (81). A study observed that the BEE of diabetic patients was higher than that of healthy people. This disparity appears to be attributable to a rise in fasting blood glucose, which leads to increased glycosuria or gluconeogenesis. Furthermore, those with diabetes appear to have a lower activity energy expenditure (AEE) than healthy people (82).

Obesity

Obesity and related metabolic illnesses are becoming a major public health concern around the world. The condition causes substantial comorbidities like diabetes, fatty liver disease and cardiovascular disease, as well as a considerable reduction in the length and quality of a person's life (83). An overabundance of adipose tissue characterizes obesity. Adipose tissue is an important endocrine organ that produces hormones that control body metabolism. Increased fat cell mass causes hormonal imbalances, which can have a variety of metabolic consequences (84). Leptin, visfatin, apelin, resistin and adiponectin are the hormones that play a big part in body weight management. Energy homeostasis, glucose and lipid metabolism, vascular homeostasis, immunological response and reproductive processes are all influenced by the hormones released by adipose tissue. The synthesis of the cytokines IL-6, TNF- and leptin, all of which play important roles in the development of obesity and insulin resistance, has recently been revealed to also be influenced by adipose tissue. Obese people have a different profile of circulating adipokines like adiponectin due to their changed adipose tissue composition. Insulin-mimetic and insulin-sensitizing properties of adiponectin include stimulation of glucose uptake in skeletal muscle and reduction of glucose synthesis in the liver (85). Many genes are activated or deactivated in obesity to modulate EM, and epigenetic factors are the key pathways for gene expression modulation. The epigenome, which comprises DNA methylation, histone changes and RNA-mediated activities, does indeed have a big impact on energy metabolic control in obese individuals (86).

Thyroid disease

Thyroid hormone (TH) regulates metabolic processes that are necessary for normal growth and development along with metabolism (87, 88). TH levels are linked to body weight and energy expenditure (89) and this has been well documented (90). Increased hepatic glucose output, increased futile cycling of glucose degradation products between the liver and skeletal muscle, decreased glycogen stores in the liver and skeletal muscle, altered oxidative and nonoxidative glucose metabolism, decreased active insulin output from the pancreas and increased renal insulin clearance are all effects of TH on glucose homeostasis. TH influences adipokines and adipose tissue, making the patient more susceptible to ketosis. If left untreated, thyrotoxicosis can change glucose metabolism in a type 2 diabetes patient to the point that diabetic ketoacidosis develops (91). Subclinical hypothyroidism (SH) is linked to lipid abnormalities such as normal or slightly higher total cholesterol, increased LDL and reduced HDL values. Thyroid illnesses cause significant changes in the content and distribution of lipoproteins. Hypercholesterolemia and a significant rise in low-density lipoproteins (LDL) and apolipoprotein B (apo A) are signs of overt hypothyroidism. Furthermore, hypothyroidism enhances the oxidation of plasma cholesterol, owing to a changed binding pattern and higher cholesterol levels, which provide a substrate for oxidative stress (92). One of the most common endocrine diseases is autoimmune thyroid disease, which includes Graves' disease (GD) and Hashimoto's thyroiditis (HT). Thyroid dysfunction caused by autoimmune thyroid illness has a significant impact on serum metabolomic patterns. Hypothyroidism had a considerable impact on alanine, aspartate and glutamate metabolism, while hyperthyroidism had a strong impact on arginine and proline metabolism and aminoacyl-transfer ribonucleic acid (tRNA) biosynthesis (93).

Alzheimer's disease

AD is a degenerative disease that puts a strain on memory and other cognitive abilities. Overall hypometabolism may have a role in cognitive impairment in AD. Indeed, early-stage AD is marked by a decrease in glucose consumption and mitochondrial dysfunction, both of which have negative effects on neurons due to increased formation of ROS, ATP depletion and activation of cell death mechanisms. Recent data suggests that mitochondrial defects in this neurodegenerative condition are linked to abnormalities in mitochondrial dynamics, which can be triggered by amyloid-beta (Aβ), which builds up in this organelle over time and acts as a direct toxin. Furthermore, Aβ causes glutamate N-methyl-D-aspartate receptors (NMDARs) activation and/or excessive calcium release from the endoplasmic reticulum (ER), both of which may contribute to mitochondrial calcium dyshomeostasis, disrupting organelle function and, ultimately, harming neurons (94). In atypical AD, researchers discovered an early and severe impairment in brain glucose metabolism, which worsens as dementia symptoms progress (95). Alterations in EM have been linked to aging and the development of late-onset AD (LOAD). According to a study published in Nature, neural progenitor cells and astrocytes segmented from LOAD patient-derived induced pluripotent stem cells to show various interrelated bioenergetic alterations, including impact on energy production by mitochondrial respiration versus glycolysis, as a result of changes in bioenergetic substrate processing and transfer of reducing agents, decreased levels of NAD/NADH, lowered uptake of glucose, and response rates to insulin

(INS)/IGF-1 signaling, decreased INS receptor and glucose transporter 1 density, and changes in the metabolic transcriptome (96). Deficiencies in various enzyme complexes involved in the mitochondrial oxidation of substrates to produce energy are among the hallmark abnormalities in the AD brain. The pyruvate dehydrogenase complex (PDHC), the α-ketoglutarate dehydrogenase complex (KGDHC) and the electron transport chain's (ETC's) Complex IV are among them (COX). The AD brain's inherent impairment of glucose oxidation may be assumed to interact constructively with an inadequate supply of oxygen and glucose to cause brain damage (97). There is a definitive link between existing AD biomarkers (Aβ, tau protein and phosphorylated tau protein) and mitochondrial abnormalities (98, 99), implying that mitochondrial functions play a role in both the pathogenesis and diagnosis of AD. Many mitochondrial proteins are degraded, including citric acid cycle enzymes and ETC complexes (100–103).

Role of functional foods and nutraceuticals in ameliorating the changes made in energy metabolism by chronic diseases

Phytochemicals, or plant-based bioactive dietary components, have recently gained popularity due to their association with chronic illnesses such as heart disease, cancer and degenerative diseases. Many diseases have been linked to oxidative stress, which can damage DNA, proteins and lipids. The availability of bioactive phytochemicals, widely used in medicinal applications, appears to be linked to the health benefits of specific plant meals, mainly fruits and vegetables. According to studies, ROS-mediated pathways appear to minimize the risk of chronic disease. Consumption of these foods has become increasingly popular to boost man's health. Plants' nutraceutical value and phytotherapy efficacy can be increased by biotechnological means. Dietary changes have a significant impact on one's health, the environment and society.

Obesity, liver disease, osteoporosis, cancer, diabetes, renal impairment, allergies, thyroid disease, AD, respiratory disorders, cardiovascular diseases and dental issues are caused due to malnutrition and insufficient diet. Other issues include aging populations, high-energy foods and imbalanced diets. Chronic diseases are noncommunicable diseases (NCDs) that last a long time and have a gradual onset or progression. These illnesses can be avoided and managed, but they cannot be cured. They are called degenerative diseases because the structure or function of the affected tissues or organs deteriorates over time, whether as a result of regular wear and tear or lifestyle choices like exercise or dietary habits. They are frequently associated with aging, a process synonymous with mitochondrial degradation, resulting from poor nutrition, stress and toxic overload. Degenerative diseases have a major negative impact on one's health, quality of life and lifespan (104). These diseases are quickly spreading throughout the world, accounting for nearly 60% of the world's 56.5 million total recorded fatalities and around 46% of the global burden of disease. Around half of all deaths from chronic diseases are caused by cardiovascular diseases, and a sizable proportion can also be related to obesity and diabetes, which are now prevalent in childhood (105). Epidemiological and experimental studies have established a strong link between a high intake of fruits, vegetables, spices, beverages, legumes, whole grains and fish, as well as a high-fiber diet and other food-related products and a decreased risk of chronic diseases such as cardiovascular disease, cancer, diabetes, AD, sexual dysfunction, cataracts and age-related functional decline. Foods that contain medicinal components necessary for human health promotion and disease prevention might be considered functional foods in addition to their nutritional value. Additionally, plant extracts containing bioactive chemicals can be employed as functional food additives and in the pharmaceutical and food sectors for the manufacturing of pharmaceuticals (106). In recent years, more than 80% of bioactive chemicals found in functional foods and more than 30% of pharmaceuticals have been synthesized from bioactive natural ingredients. Secondary metabolites are substances that have bioactivity in cells and other organs of the body and are referred to as phytochemicals. Extracts of plants containing bioactive chemicals may be employed as functional food ingredients or nutraceuticals for the treatment and/or management of a variety of degenerative disorders (107, 108).

Functional foods and nutraceuticals play an essential role in reducing the effects of chronic diseases on EM, as well as in the prevention and treatment of these disorders (109). Nutraceuticals, functional foods, value-added food products and whole plant foods are examples of foods or food items with health claims depicting this feature. Consumption of these foods lowers healthcare expenses and promotes economic development, particularly in rural areas (110). Additionally, an attempt has been made to summarize the biotechnological approaches being used to increase the level of bioactive compounds in food crops in order to improve their nutraceutical and functional values, which are important factors in nutritional therapy for cancer, cataracts, immune system decline, cardiovascular disease, brain dysfunction and atherosclerosis. CVDs, particularly atherosclerotic Congestive Heart Disease (CHD) and stroke, are the major causes of disability and mortality in industrialized countries and are expected to overtake them in underdeveloped countries by 2020. Arterial atherosclerosis, which predominantly affects the heart, kidneys and brain, develops gradually until a catastrophic cardiovascular event occurs, resulting in morbidity and mortality. Oxidative stress permanently alters the genetic material, causing degenerative or chronic diseases like atherosclerosis and cancer (111). Many scientific studies have shown that certain bioactive chemicals in foods have disease-fighting properties, which has piqued the attention of consumers all over the world in food's health benefits. Healthy eating standards have advised the general people to consume more fresh fruits, vegetables, low fat and high fiber diets throughout the world (112).

Nutraceuticals have gained worldwide popularity, have been defined as purified or concentrated food and food products with health-promoting and disease-preventing properties beyond their basic nutritional function. When consumed, they enhance health and quality of life for people of all ages. Nutraceuticals range from isolated nutrients, dietary supplements and diets to naturally occurring substances such as herbals, vitamins, amino acids or their formulations and processed products such as cereals, soups and beverages (111, 112). Nutraceuticals, functional foods and value-added food products are therefore rich repositories of health-promoting bioactive compounds. Scientific research shows that a large proportion of these compounds often act as antioxidants and have a role in vivo in modulating disease development by inhibiting ROS-mediated reactions, which have been associated with the initiation and progression of several pathological processes (107, 108).

The major active ingredients found in plants are flavonoids. Flavonoids are a family of low molecular weight polyphenolic compounds present in cereals, vegetables, fruits and drinks of plant origin, such as red wine, tea, cocoa and coffee. The compounds are derived from parent compounds known as flavones. About 4000 flavonoids are known, classified into five major groups, namely flavones, isoflavones, flavanones and flavanols [flavan-3-ols), and anthocyanins flavanonols. They are potent antioxidants or free radical scavengers that offer protection against cardiovascular disease by reducing the oxidation of LDL. It has been reported that flavonoids may exert local anticarcinogenic effects by acting as intraluminal antioxidants in the intestine (113). They may also exert cardioprotective action by preventing or retarding oxidative reactions in cells, which is a predisposing factor for developing CVDs. In particular, studies have shown that Mediterranean diets rich in resveratrol can lower the risk for the development and progression of obesity disease in humans. The major bioactive compounds in food crops have been classified based on their activities. Examples include carotenoids (α- and β-carotene, β-cryptoxanthin, lutein, lycopene and zeaxanthin), phenols (flavonoids), cyclic phenolics (chlorogenic acid, ellagic acid and coumarins), glucosinolates (sulforaphane, indole-3 carbinol), saponins, phytosterols (campesterol, β-sitosterol and stigmasterol), sulfides and thiols, and phytoestrogens (isoflavones, daidzein, genistein and lignans). The carotenoids and flavonoids have been primarily implicated in the prevention, control and management of chronic diseases. Carotenoids comprise an extended group of natural pigments, numbering about 600, of which carotene is found to be most prominent. Many studies have shown that individuals with higher dietary intakes of carotenoids have a reduced risk of several chronic diseases. The carotenes are generally tissue-specific in their biological activity, and the xanthophylls serve to protect other antioxidants. The mechanism by which carotenoids protect cells against ROS-mediated damage depends mainly on physical quenching, a process in which the energy of the excited oxygen is transferred to the carotenoid molecule (114, 115).

Benefits to health

According to research, regular consumption of functional foods and nutraceuticals has been linked to a lower risk of coronary heart disease and other age-related degenerative diseases like Parkinson's and Alzheimer's. Bioactive chemicals accumulate in the plasma and tissues of consumers in proportion to dietary intakes and play a vital function in suppressing processes mediated by ROS. These chemicals are anticarcinogenic, anti-inflammatory, antidiabetic, antioxidant, antifungal, antipyretic, anti-apoptotic, chemopreventive, hepatoprotective, hypolipidemic and analog a CNS stimulant (109, 115). The antioxidant tocopherol slows the course of heart attack and heart failure. Vitamins reduce platelet adhesion, raise HDL levels and inhibit smooth muscle cell growth, which is a risk factor for atherosclerosis. Nutraceuticals, functional foods and value-added food items are increasingly accepted as alternatives to conventional medications and pharmaceuticals (114–116). Here are some brief biological activities of functional food and nutraceuticals playing an important role in ameliorating the changes made in EM by chronic diseases.

Chronic liver disease

Current food tendencies, suboptimal dietary habits and a sedentary lifestyle are spreading metabolic disorders worldwide. Consequently, the prevalence of liver pathologies is increasing, as it is the main metabolic organ in the body. With NAFLD as the main cause, chronic liver diseases have an alarming prevalence of around 25% worldwide. Otherwise, the consumption of certain drugs leads to an acute liver failure (ALF), with drug-induced liver injury (DILI) as its main cause, or alcoholic liver disease (ALD). Although programs carried out by authorities and scientific bodies are focused on improving dietary habits and lifestyle, the long-term compliance of the patient makes them difficult to follow. Thus, supplementation with certain substances may represent a more easy-to-follow approach for patients. In this context, the consumption of polyphenol-rich food represents an attractive alternative as these compounds have been characterized to be effective in ameliorating liver pathologies (117). Numerous prior researches have examined the apparent health advantages of natural product metabolites such as phenolic chemicals, lipids, proteins, carbohydrate-dietary fibers and vitamins using experimental model systems. In general, current research suggests that lifestyle interventions such as increased exercise should be combined with so-called functional foods such as quercetin, polyunsaturated fatty acids, mung bean protein, Kefir peptide, vitamin E, astaxanthin, dietary fructo-oligosaccharides, pectin and Ginkgo polysaccharide (118, 119). Various physiological and metabolic processes have been implicated as contributing factors to NAFLD and thus potential therapeutic targets. Among the recognized mechanisms relevant to NAFLD intervention are those managing cholesterol homeostasis by absorption, biosynthesis and excretion and those governing triglyceride and fatty acid metabolism via absorption, lipogenesis and oxidation (120).

Chronic renal failure

CKD is a global health problem with an ever-increasing prevalence. The biology of CKD is complex and not entirely understood. However, it is believed that increased oxidative stress plays a critical role in the development of this disease. Additionally, CKD is believed to be an inflammatory condition in which uremic toxins contribute to the establishment of an inflammatory milieu. A healthy, balanced diet promotes long-term health by lowering the chance of developing chronic diseases such as chronic renal disease, DM and hypertension (121). Numerous studies have established those functional molecules and nutrient such as fatty acids and fiber and nutraceuticals such as curcumin, steviol glycosides and resveratrol have favorable effects on both pro-inflammatory and anti-inflammatory pathways as on the gut mucosa. Numerous animal experiments have been conducted in the field of nephrology in an attempt to discover the "golden substance." Al-Okbi et al. (122) investigated the preventive effects of avocado, walnut, flaxseed and Eruca sativa seed extracts in a rat model of renal failure produced by intraperitoneal cisplatin. Pre-treatment with various nutraceuticals was related with significant protection against kidney dysfunction, as measured by a significant drop in plasma creatinine and urea levels, as well as an increase in plasma albumin, plasma total protein and creatinine clearance. The administration of the examined nutraceuticals resulted in an increase in total antioxidant capacity and catalase activity, a decrease in free radicals and an improvement in kidney function as a result of the prevention of damage and an inflammatory process (121, 122). The histopathological changes revealed corroborated the effects observed in that study. A decrease in free radicals generated by nutraceuticals also resulted in a decrease in chromosomal abnormalities. Consumption of nutraceuticals may positively affect the development and progression of renal diseases (123, 124).

Cancer

Cancer is the second leading cause of death in most wealthy countries, after cardiovascular disease. More than half of all cancers and deaths worldwide are considered preventable. Since its inception, the Disease Control Priority Suite has focused on delivering cost-effective health interventions that can significantly reduce death and disability. Functional foods are foods and dietary ingredients that provide health benefits and those associated with essential nutrition. There is some evidence indicating a link between functional foods and cancer. Plant extracts containing daidzein, biochanin, isoflavones and genistein were found to inhibit the growth of prostate cancer cells (125). Lycopene is considered a powerful antioxidant and a singlet oxygen quencher due to its unsaturated nature. Lycopene is found in high concentrations in the prostate, testes, skin and adrenal glands, where it protects against cancer. The link between carotenoids and cancer and cardiovascular disease prevention has increased the importance of vegetables and fruits in the human diet. Lycopene-rich vegetables and fruits protect against cancer by lowering oxidative stress and DNA damage. Lycopene is a major carotenoid found only in tomatoes, guava, pink grapefruit, watermelon and papaya. β-carotene has antioxidant properties and helps to prevent cancer and other diseases. β-carotene has the highest antioxidant activity of all carotenes (126, 127). Chronic inflammation is linked to an increased risk of cancer. Chronic inflammation is also linked to immune suppression, which is linked to cancer risk. Ginseng is an anti-inflammatory molecule that targets many of the key players in the inflammation-to-cancer pathway (128). Clinical investigations on a large scale indicate that certain medicines, such as green tea, Vitamins D and E, selenium, lycopene, soy, anti-inflammatory medications and 5a-reductase inhibitors, are beneficial for preventing prostate cancer. Ongoing trials may aid in the development of novel chemoprevention strategies (129, 130).

Cardiovascular diseases

Functional foods, containing physiologically active components either from plant or animal sources, are marketed with the claim of their ability to reduce heart disease risk, focusing primarily on established risk factors: blood cholesterol, diabetes and hypertension. Functional foods exert their cardioprotective effects mainly through lipid-lowering effects, antioxidant actions and decreased homocysteine levels. Vegetable and fruit fibers (with pectin), garlic and oily seeds (walnut, almonds, etc.) and fish oils have lipid-lowering effects in humans through inhibition of fat absorption and suppression hepatic cholesterol synthesis (131). Homocysteine increases the risk of both cardiovascular and cerebrovascular disorders by enhancing arteriolar constriction and decreasing endothelial vasodilation. A higher intake of folate, antioxidant vitamins, whole grains and phytochemicals has been reported to abrogate the deleterious vascular effects of homocysteine in the heart. A significant cardiovascular benefit of phytochemicals (polyphenols in wine, grapes and teas), vitamins (ascorbate, tocopherol) and minerals (selenium, magnesium) in foods is thought to be the capability of scavenging free radicals produced during atherogenesis. Several functional foods are thought to be of benefit in treating and preventing CVD. The most common functional foods that have been studied in cardiovascular patients are long-chain n-3 fatty acids, dietary fiber and phytochemicals, as well as nutrients based on or enriched with vegetable proteins, mainly soy. Plant foods contain many bioactive compounds known as "phytochemicals." Some groups of phytochemicals that have or appear to have significant health potentials are carotenoids, phenolic compounds (flavonoids, phytoestrogens, phenolic acids), phytosterols, tocotrienols, organosulfur compounds and nondigestible carbohydrates (dietary fiber and prebiotics). Isoflavones are found in high concentration in soybean, soybean products (e.g., tofu) and red clover. Lignans are mainly found in flaxseed (132–133). Numerous nutraceuticals being studied for the prevention and treatment of cardiovascular disease are well tolerated by patients. However, data on long-term safety and effectiveness against clinical outcomes such as myocardial infarction and mortality are frequently poor. Additional clinical research should be performed to determine which nutraceuticals are the most clinically and economically successful in preventing and treating CVD (134).

Diabetes mellitus

DM is a chronic metabolic disorder characterized by high blood glucose levels and impaired carbohydrate, protein and lipid metabolism. Type 2 diabetes is the most common non-insulin-dependent form of this chronic disease. Type 1 diabetes is insulin-dependent and occurs in only 10% of persons with DM (135). Insulin secretion or action deficiencies or changes cause hyperglycemia. Increased oxidative stress has been linked to diabetes initiation and progression. Peroxidation of cellular organelles, activation of apoptotic pathways and oxidative damage to pancreatic cells have all been linked to oxidative stress. Long-term exposure to excessive glucose, free fatty acids, or both can cause cell dysfunction. Extracellular hyperglycemia caused by low glucose levels in muscle and adipose tissue can induce tissue damage and diabetes consequences include heart disease, atherosclerosis, cataract formation, neurological disorders and diabetic retinopathy (135, 136). The various pathways leading to DM incidence and pathophysiology may be therapeutic targets for functional food bioactive substances. Preventing carbohydrate absorption after a meal reduces postprandial hyperglycemia. Hydrolyzing enzymes like amylase and glucosidase break down complex polysaccharides and oligosaccharides to glucose, which is then absorbed into the intestinal epithelium and circulated (137, 138). Multiple pharmacological and non-pharmacological interventions have been developed to improve glycemic control and prevent diabetes complications based on current understanding of the pathophysiology of insulin resistance and type 2 DM; in this area, the use of functional foods and their bioactive components has recently been considered a novel approach. Nutraceutical delivers medical-health benefits, such as illness prevention and therapy for diabetes. Nutraceuticals and functional foods or bioactive phytochemicals with health-promoting, disease-preventive, or therapeutic characteristics, such as vitamin C, vitamin E, calcium, alpha-lipoic acid, chromium, fiber and so on, obtained from herbs or plants can be used effectively for the management of DM (139). They contain biologically active elements associated with physiological health advantages for avoiding and controlling chronic diseases, such as type 2 diabetic mellitus (T2DM). A regular consumption of functional foods may be associated with higher antioxidant, anti-inflammatory, insulin sensitivity and anti-cholesterol capabilities, which are considered important to prevention and control of T2DM. Components of the Mediterranean diet (MD) – such as fruits, vegetables, oily fish, olive oil and tree nuts – serve as a model for functional foods based on their natural amounts of nutraceuticals, including polyphenols, terpenoids, flavonoids, alkaloids, sterols, pigments and unsaturated fatty acids (140). Polyphenols inside MD and polyphenol-rich herbs such as coffee, green tea and black tea have shown clinically meaningful advantages on metabolic and microvascular activity, cholesterol and fasting glucose reduction, and anti-inflammation and antioxidation in high-risk and T2DM patients. However, combining exercise with functional food consumption can trigger and increase various metabolic and cardiovascular protective advantages, although it is under-investigated in people with T2DM and bariatric surgery patients (141). Detecting functional food benefits can now rely on an "omics" biological profiling of individuals' molecular components, genetics, transcriptomics, proteomics and metabolomics, but is under-investigated in multicomponent interventions. A tailored approach for preventing and controlling T2DM should integrate biological and behavioral models, and embed nutrition instruction as part of lifestyle diabetes preventive studies. Functional foods may bring additional benefits to such an approach (139, 141).

Obesity

Obesity is a chronic disease that threatens havoc on the world's public health system. The condition is caused by excessive nutrition, poor physical activity, environmental factors, the Western diet and genetics. It has been identified as a critical mediator in the progression of various metabolic syndromes, including cancer, congestive heart failure, stroke, hypertension, rheumatoid arthritis, hyperlipidemia, DM and insulin resistance. Increased calorie consumption results in fat accumulation in adipose tissue (142). Adipose tissue is an endocrine organ surrounded by blood capillaries that controls the production of hormones, cytokines and growth factors such as tumor necrosis factor-alpha (TNF-α), interleukin-6 (IL-6) and adipokines that promote inflammation progression. Adipocyte enlargement is characterized by increased plasma, hypoxia, hyperplasia and inflammation. Nutritional elements aimed at reducing energy balance by lowering energy intake and increasing energy expenditures appear to be an alternate treatment for the disease. Nutraceutical is a term that refers to a substance extracted from food that is typically used as a medicine and traded on the market. Nutraceuticals exert physiological activity by defending against a variety of illnesses (143). The administration of nutraceuticals containing saturated fats aids in the treatment of obesity and also slows the growth of oxidative stress and inflammation, hence preventing the onset of obesity. Food provides the human body with energy, lipids, proteins, carbs, vitamins and antioxidants. Although newer substances such as carotenoids, flavonoids, nutraceuticals and glucosinolates have been studied in the treatment of a variety of illnesses. Adding galactomannan (Sylvester), a carbohydrate supplement, to the diet of overweight people promotes fat loss. Researchers are investigating if natural remedies for obesity could be effective. Future effective and safe anti-obesity drugs could be developed using this method instead of traditional approaches. Caloric restriction and increased physical exercise have been demonstrated to be highly effective in controlling obesity. So, researchers and obese people are turning to nutraceuticals and medications to assist them lose weight. Weight loss requires a nutraceutical that can enhance energy expenditure and/or decrease caloric consumption (142, 144). Caffeine, ephedrine, chitosan, ma huang-guarana and green tea are efficient herbal stimulants for weight loss. Green tea extract and 5-hydroxytryptophan both boost energy expenditure while decreasing hunger. It's possible that dietary phytochemicals could be used to treat obesity by limiting the growth of adipocytes, lowering preadipocyte differentiation, increasing lipolysis and inducing the death of existing adipocytes. Nutritional supplements and functional foods containing polyphenols (such as resveratrol and curcumin), flavonoids (such as curcumin), alkaloid compounds (such as capsaicin) and fiber (such as ephedrine) can be used to treat obesity (144, 145).

Thyroid disease

Nutraceuticals may be used to prevent a variety of pathological ailments, including thyroid disease and related problems. Along with iodine, a crucial ingredient for thyroid function, various dietary components like carnitine, flavonoids, melatonin, omega-3, resveratrol, selenium, vitamins, zinc and inositol were discovered to contribute to thyroid homeostasis, implying that they may be involved in clinical thyroidology. In some cases of autoimmune thyroid disorders, such as Graves' orbitopathy and perhaps postpartum thyroiditis, selenium status has a significant impact on clinical markers and quality of life (146, 147). It is still a possible option for supplementation to enhance these markers and quality of life. Prior to making definite recommendations for widespread routine clinical use, more solid evidence is required. The lack of clinical data confirming the efficacy of nutraceuticals in prevention and therapy is the primary concern when it comes to their appropriate and effective use. Another drawback is the mismatch between the label's advertised concentration and the actual concentration of the ingredient (148).

Alzheimer's disease

Age is frequently associated with a decline in a broad range of cognitive capacities, including reasoning, memory, perceptual quickness and language. When more than one of these activities is impaired over an extended period and is coupled with functional loss, this is referred to as dementia. AD is the most prevalent and severe kind of dementia, accounting for around 70% of all dementia cases and experiencing a dramatic epidemic as a result of the world's rapidly growing elderly population. In recent decades, there has been a surge of interest in interventions that may assist in improving cognitive function in older adults or, at the very least, delay the onset of dementia (149). Due to the lack of a cure for dementia and AD, the public health priority has shifted more recently to cognitive decline prevention. Although the molecular cascade of neurodegeneration in AD is unknown, vascular pathology and risk factors have recently played a critical role in AD pathogenesis, alongside genetic and environmental variables. As a result, lifestyle interventions that benefit neurodegeneration and vascularity, such as natural nutrition and nutritional supplementation, cognitive and social interaction, and physical activity, have been identified as potential target alternatives for AD prevention. There is compelling evidence that a diet high in certain nutritious food groups (fruit, seafood and vegetables) can help reduce the incidence and prevalence of several major clinical outcomes, including neurodegenerative illnesses (150). These specialized nutritional food groups are high in micronutrients and vitamins due to their nutritional and health benefits (similar to medications). Among the various types of diets, the Mediterranean pattern has garnered substantial interest in recent decades as a result of massive epidemiological and laboratory research demonstrating its high nutraceutical content. The Mediterranean diet is defined by a high intake of plant foods, fish and olive oil as key sources of monounsaturated fat and a moderate consumption of alcohol. Due to the synergistic effects of its components, this type of food intake pattern may be particularly healthful. Synergistic interactions between food components are thought to be responsible for the neuroprotective properties of some nutrients and nutraceuticals (151). Additionally, cardiovascular diseases such as DM, hypertension and lipid disorders, as well as lesions of the white substance, are extremely susceptible to micronutrient alterations. The role of vascular disease and risk factors in both AD and its prodromal phase, mild cognitive impairment, has recently been proved to be important (152). Regardless of how the Mediterranean diet helps, a few of the studies have recently shown that it can help prevent MCI and AD. Flavonoids and their metabolites regulate different neurological processes, according to research showing a connection with neuronal-glial signaling pathways critical for neuronal survival and function. They also change cerebral blood flow, stimulate the expression of synaptic plasticity and neuronal-repair-related antioxidant enzymes and proteins, as well as hinder AD-related neuropathological processes in affected brain regions (153, 154).

References

1. Lotharius J, Brundin P. Pathogenesis of Parkinson's disease: dopamine, vesicles and alpha-synuclein. Nature Review. Neuroscience. 2002,3:932–942.
2. Parihar MS, Brewer GJ. Mitoenergetic failure in Alzheimer disease. American Journal of Physiology. Cell Physiology. 2007,292:8–23.
3. Hochachka P, McClelland G. Cellular metabolic homeostasis during large-scale change in ATP turnover rates in muscles. The Journal of Experimental Biology. 1997, 200:381–386.
4. Pellerin L, Magistretti PJ. Glutamate uptake into astrocytes stimulates aerobic glycolysis: a mechanism coupling neuronal activity to glucose utilization. Proceedings of the National Academy of Sciences of the United States of America. 1994,91:10625–10629.
5. Schurr A. Lactate: the ultimate cerebral oxidative energy substrate? Journal of Cerebral Blood Flow and Metabolism. 2006,26:142–152.
6. Ainscow EK, Mirshamsi S, Tang T, et al. Dynamic imaging of free cytosolic ATP concentration during fuel sensing by rat hypothalamic neurones: evidence for ATP-independent control of ATPsensitive Kþ channels. Journal of Physiology. 2002,544:429–445.
7. Sofi F, Abbate R, Gensini G, Casini A. Importance of diet on disease prevention. International Journal of Medicine and Medical Sciences. 2013,5(2):55–59.
8. Somrongthong R, Hongthong D, Wongchalee S, Wongtongkam N. The influence of chronic illness and lifestyle behaviors on quality of life among older Thais. Biomedical Research International. 2016,2016:7.

9. Adedayo BC, Oboh G, Akindahunsi AA. Changes in the total phenol content and antioxidant properties of pepperfruit (Dennettia tripetala) with ripening. African Journal of Food Science. 2010,4(6):403–409.
10. Adefegha SA, Oboh G. Enhancement of total phenolics and antioxidant properties of some tropical green leafy vegetables by steam cooking. Journal of Food Processing and Preservation. 2011,35(5):615–622.
11. Adefegha SA, Oboh G. Acetylcholinesterase (AChE) inhibitory activity, antioxidant properties and phenolic composition of two Aframomum species. Journal of Basic and Clinical Physiology and Pharmacology. 2012a,23(4):153–161.
12. Adefegha SA, Oboh G. Inhibition of key enzymes linked to type-2 diabetes and sodium nitroprusside-induced lipid peroxidation in rat pancreas by water extractable phytochemicals from some tropical spices. Pharmaceutical Biology. 2012b,50(7): 857–865.
13. Adefegha SA, Oboh G. Phytochemistry and mode of action of some tropical spices in the management of type-2 diabetes and hypertension. African Journal of Pharmacy and Pharmacology. 2013,7(7):332–346.
14. Adefegha SA, Oboh G, Adefegha OM, Boligon AA, Athayde ML. Antihyperglycemic, hypolipidemic, hepatoprotective and antioxidative effects of dietary clove (Szyzgium aromaticum) bud powder in a high-fat diet/streptozotocin-induced diabetes rat model. Journal of the Science of Food and Agriculture. 2014,94:2726–2737.
15. Adefegha SA, Oboh G, Ejakpovi II, Oyeleye SI. Antioxidant and antidiabetic effects of gallic and protocatechuic acids: a structure–function perspective. Comparative Clinical Pathology. 2015,24:1579–1585.
16. Adefegha SA, Oboh G, Odubanjo T, Ogunsuyi OB. A Comparative study on the antioxidative activities, anticholinesterase properties and essential oil composition of Clove (Syzygium aromaticum) bud and Ethiopian pepper (Xylopia aethiopica). La Rivista Italiana Delle Sostanze Grasse. 2015,92(1):257–268.
17. Adefegha SA, Oboh G, Omojokun OS, Jimoh TO, Oyeleye SI. In vitro antioxidant activities of African birch (Anogeissus leiocarpus) leaf and its effect on α-amylase and α-glucosidase inhibitory properties of Acarbose. Journal of Taibah University Medical Sciences. 2016,11(3):236–242.
18. Adefegha SA, Oboh G, Molehin OR, Saliu JA, Athayde ML, Boligon AA. Chromatographic fingerprint analysis, acetylcholinesterase inhibitory properties and antioxidant activities of Redflower Ragleaf (Crassocephalum crepidioides) extract. Journal of Food Biochemistry. 2016,40:109–119.
19. Adefegha SA, Oboh G, Oyeleye SI, Dada FA, Ejakpovi II, Boligon AA. Cognitive enhancing and antioxidative potentials of velvet beans(Mucuna pruriens) and horseradish (Moringa oleifera) seeds extracts: a comparative study. Journal of Food Biochemistry. 2016. doi:10.1111/jfbc.12292.
20. Adefegha SA, Oboh G, Omojokun OS, Adefegha OM. Alterations of Na+/K+-ATPase, cholinergic and antioxidant enzymes activity by protocatechuic acid in cadmium-induced neurotoxicity and oxidative stress in Wistar rats. Biomedicine and Pharmacotherapy. 2016,83:559–568.
21. Adefegha SA, Oboh G, Olasehinde TA. Alkaloid extracts from shea butter and breadfruit as potential inhibitors of monoamine oxidase, cholinesterases, and lipid peroxidation in rats' brain homogenates: a comparative study. Comparative Clinical Pathology. 2016,25(6):1213–1219. doi:10.1007/s00580-016-2331-0.
22. Adefegha SA, Oboh G, Adefegha OM. Ashanti pepper (Piper guineense Schumach et Thonn) attenuates carbohydrate hydrolyzing, blood pressure regulating and cholinergic enzymes in experimental type 2 diabetes rat model. Journal of Basic Clinical Pharmacology and Physiology. 2017,28(1):19–30.
23. Adefegha SA, Oboh G, Oyeleye SI, Ejakpovi II. Erectogenic, antihypertensive, antidiabetic, antioxidative properties and phenolic compositions of almond fruit (Terminalia catappa L.) parts (hull and drupe) – in vitro. Journal of Food Biochemistry. 2017. doi:10.1111/jfbc.12309.

24. Adefegha SA, Olasehinde TA, Oboh G. Essential oil composition, antioxidant, antidiabetic and antihypertensive properties of two Afromomum species. Journal of Oleo Sciences. 2017,66(1):51–63.
25. Liu RH. Health benefits of fruit and vegetables are from additive and synergistic combinations of phytochemicals. American Journal of Clinical Nutrition. 2003,78: 517S–520S.
26. Hasler CM. Functional foods: their role in disease prevention and health promotion. Food Technology. 1998,52(11):63–70.
27. Pandey KB, Rizvi SI. Plant polyphenols as dietary antioxidants in human health and disease. Oxidative Medicine and Cellular Longevity. 2009,2(5):270–278.
28. Pang G, Xie J, Chen Q, Hu Z. How functional foods play critical roles in human health. Food Science and Human Wellness. 2012,1:26–60.
29. Roh E, Kim, MS, et al. Brain regulation of energy metabolism. Endocrinology and metabolism. 2016,31(4):519–524.
30. Cowley MA, Pronchuk N, Fan W, et al. Integration of NPY, AGRP, and melanocortin signals in the hypothalamic paraventricular nucleus: evidence of a cellular basis for the adipostat. Neuron. 1999,24(1):155–163.
31. Spiegelman BM, Flier JS, et al. Obesity and the regulation of energy balance cell. Cell. 2001,104(4):531–543.
32. Seale P, Conroe HM, Kajimura S, et al. Prdm16 determines the thermogenic program of subcutaneous white adipose tissue in mice. The Journal of Clinical Investigation. 2011;121(1):96–105.
33. Lopaschuk GD, Belke DD, Gamble J, et al. Regulation of fatty acid oxidation in the mammalian heart in health and disease. Biochimica et Biophysica Acta (BBA)-Lipids and Lipid Metabolism. 1994;1213(3):263–276.
34. Barger PM, Kelly DP, et al. PPAR signaling in the control of cardiac energy metabolism. Trends in Cardiovascular Medicine. 2000,10(6):238–245.
35. Balaban RS, et al. Cardiac energy metabolism homeostasis: role of cytosolic calcium. Journal of Molecular and Cellular Cardiology. 2002,34(10):1259–1271.
36. Guo CA, Guo S, et al. Insulin receptor substrate signaling controls cardiac energy metabolism and heart failure. The Journal of Endocrinology. 2017,233(3):131–143.
37. Karwi QG, Zhang L, Altamimi TR, et al. Weight loss enhances cardiac energy metabolism and function in heart failure associated with obesity. Diabetes, Obesity and Metabolism. 2019,21(8):1944–1955.
38. Backhed F, Roswall J, Peng Y, et al. Dynamics and stabilization of the human gut microbiome during the first year of life. Cell Host & Microbe. 2015,15(5):690–703.
39. Heiss CN, Olofsson LE, et al. Gut microbiota-dependent modulation of energy metabolism. Journal of Innate Immunity. 2018,10(3):163–171.
40. McNeil NI, et al. The contribution of the large intestine to energy supplies in man. The American Journal of Clinical Nutrition. 1984,39(2):338–342.
41. Al-Assal K, Martinez AC, Torrinhas RS, et al. Gut microbiota and obesity. Clinical Nutrition Experimental. 2018,20:60–64.
42. Van den Berg SA, Van Marken Lichtenbelt W, Van Dijk KW, et al. Skeletal muscle mitochondrial uncoupling, adaptive thermogenesis and energy expenditure. Current Opinion in Clinical Nutrition & Metabolic Care. 2011,14(3):243–249.
43. Rosenbaum M, Kissileff HR, Mayer LE, et al. Energy intake in weight-reduced humans. Brain Research. 2010,1350:95–102.
44. Gavini CK, Mukherjee S, Shukla C, et al. Leanness and heightened nonresting energy expenditure: role of skeletal muscle activity thermogenesis. American Journal of Physiology-Endocrinology and Metabolism. 2014,306(6):635–647.
45. Hargreaves M, Spriet LL, et al. Skeletal muscle energy metabolism during exercise. Nature Metabolism. 2020,2(9):817–828.
46. Dauncey M. J. Factors affecting energy metabolism. Nutrition and Food Science. 1979, 79(5):2–4.

47. Dauncey M. Energy metabolism in man and the influence of diet and temperature. Journal of Human Nutrition. 1979,33:259–269.
48. Appanna VD, Auger C, et al. Energy, the driving force behind good and ill health. Frontiers in Cell and Developmental Biology. 2014,2:28.
49. Lautz HU, Selberg O, et al. Protein-calorie malnutrition in liver cirrhosis. The Clinical Investigator. 1992,70(6):478–486.
50. Koliaki C, Roden M. Hepatic energy metabolism in human diabetes mellitus, obesity and non-alcoholic fatty liver disease. Molecular and Cellular Endocrinology. 2013,379(1–2): 35–42.
51. Vendemiale G, Grattagliano I, et al. Mitochondrial oxidative injury and energy metabolism alteration in rat fatty liver: effect of the nutritional status. Hepatology. 2001,33(4):808–815.
52. Mansouri A, Gattolliat CH, et al. Mitochondrial dysfunction and signaling in chronic liver diseases. Gastroenterology. 2018,155(3):629–647.
53. Katayama K. Zinc and protein metabolism in chronic liver diseases. Nutrition Research. 2020,74:1–9.
54. Vaziri ND, Norris K. Lipid disorders and their relevance to outcomes in chronic kidney disease. Blood Purification. 2011,31(1–3):189–196.
55. Conjard A, Ferrier B, et al. Effects of chronic renal failure on enzymes of energy metabolism in individual human muscle fibers. Journal of the American Society of Nephrology. 1995,6(1):68–74.
56. Vaziri ND, Liang K. Down-regulation of tissue lipoprotein lipase expression in experimental chronic renal failure. Kidney International. 1996,50(6):1928–1935.
57. Bulbul MC, Dagel T, et al. Disorders of lipid metabolism in chronic kidney disease. Blood Purification. 2018,46(2):144–152.
58. Vaziri ND. Innovation in the treatment of uremia: Proceedings from the Cleveland Clinic Workshop: causes of dysregulation of lipid metabolism in chronic renal failure. Seminar in Dialysis. 2009,22(6):644–651.
59. Locasale JW, Cantley LC. Altered metabolism in cancer. BMC Biology. 2010,8(1): 1–3.
60. Warburg O. On the origin of cancer cells. Science. 1956,123(3191):309–314.
61. Ell B, Kang Y. Transcriptional control of cancer metastasis. Trends in Cell Biology. 2013,23(12):603–611.
62. Rodríguez-Enríquez S, Marín-Hernández Á, Gallardo-Pérez JC, Pacheco-Velázquez SC, Belmont-Díaz JA, Robledo-Cadena DX, Vargas-Navarro JL, Peña NAC de la, Saavedra E, Moreno-Sánchez R. Transcriptional regulation of energy metabolism in cancer cells. Cells. 2019,8(10):1225.
63. Camporeale A, Demaria M, Monteleone E, Giorgi C, Wieckowski MR, Pinton P, Poli V. STAT3 activities and energy metabolism: dangerous liaisons. Cancers. 2014, 6(3): 1579–1596.
64. Gao S, Chen M, et al. Crosstalk of mTOR/PKM2 and STAT3/c-Myc signaling pathways regulate the energy metabolism and acidic microenvironment of gastric cancer. Journal of Cellular Biochemistry. 2019,120(2):1193–1202.
65. Valle-Mendiola A, Soto-Cruz I. Energy metabolism in cancer: the roles of STAT3 and STAT5 in the regulation of metabolism-related genes. Cancers. 2020,12(1):124.
66. Parks SK, Chiche J, Pouysségur J. Disrupting proton dynamics and energy metabolism for cancer therapy. Nature Reviews. Cancer 2013,13(9):611–623.
67. Eagle H. Nutrition needs of mammalian cells in tissue culture. Science. 1955,122(3168): 501–504.
68. Kovačević Z, Morris HP. The role of glutamine in the oxidative metabolism of malignant cells. Cancer Research. 1972,32(2).
69. Neely JR, Morgan HE. Relationship between carbohydrate and lipid metabolism and the energy balance of heart muscle. Annual Review of Physiology. 1974,36: 413–459.

70. Jeffrey FM, Diczku V, et al. Substrate selection in the isolated working rat heart: effects of reperfusion, afterload, and concentration. Basic Research in Cardiology. 1995,90(5):388–396.
71. Carvajal K, Moreno-Sánchez R. Heart metabolic disturbances in cardiovascular diseases. Archives of Medical Research. 2003,34(2):89–99.
72. Rabanal-Ruiz Y, Llanos-González E, Alcain FJ. The use of coenzyme Q10 in cardiovascular diseases. Antioxidants. 2021,10(5):755.
73. Forte M, Schirone L, et al. The role of mitochondrial dynamics in cardiovascular diseases. British Journal of Pharmacology. 2021,178(10):2060–2076.
74. Chronic Obstructive Pulmonary Disease (COPD) | CDC https://www.cdc.gov/copd/index.html (accessed Oct 19, 2021).
75. Zhou WC, Qu J, et al. Mitochondrial dysfunction in chronic respiratory diseases: implications for the pathogenesis and potential therapeutics. Oxidative Medicine and Cellular. Longevity. 2021,2021:5188306.
76. Xue M, Zeng Y, et al. Metabolomic profiling of anaerobic and aerobic energy metabolic pathways in chronic obstructive pulmonary disease. Experimental Biology and Medicine. 2021,246(14):1586–1596.
77. Congleton J. The pulmonary cachexia syndrome: aspects of energy balance. Proceeding of the Nutrition Society. 1999,58(2):321–328.
78. Rawal G, Yadav S. Nutrition in chronic obstructive pulmonary disease: A review. Journal of Translational Internal Medicine. 2015 Oct,3(4):151.
79. Zhao H, Dennery PA, et al. Metabolic reprogramming in the pathogenesis of chronic lung diseases, including BPD, COPD, and pulmonary fibrosis. American Journal of Physiology. Lung Cellular and Molecular Physiology. 2018 Apr 1,314(4):L544–54. https://doi.org/10.1152/ajplung.00521.2017
80. Hebert SL, Nair KS. Protein and energy metabolism in type 1 diabetes. Clinical Nutrition. 2010,29(1):13.
81. Karakelides H, Asmann YW, et al. Effect of insulin deprivation on muscle mitochondrial atp production and gene transcript levels in type 1 diabetic subjects. Diabetes. 2007,56(11):2683–2689.
82. Charlton M, Nair KS. Protein metabolism in insulin-dependent diabetes mellitus. Journal of Nutrition. 1998,128(2):323S–327S.
83. Caron N, Peyrot N, et al. Energy expenditure in people with diabetes mellitus: a review. Frontiers in Nutrition. 2016,3:56.
84. Engin A. The definition and prevalence of obesity and metabolic syndrome. Advances in Experimental Medicine and Biology. 2017,960:1–17.
85. Singla P, Bardoloi A, et al. Metabolic effects of obesity: a review. World Journal of Diabetes. 2010,1(3):76.
86. Fang X, Sweeney G. Mechanisms regulating energy metabolism by adiponectin in obesity and diabetes. Biochemical Society Transactions. 2006,34(5):798–801.
87. Gao W, Liu J-L, et al. Epigenetic regulation of energy metabolism in obesity. Journal of Molecular Cell Biology. 2021,13(7):480–499.
88. Cheng SY, Leonard JL, et al. Molecular aspects of thyroid hormone actions. Endocrine Reviews. 2010,31(2):139–170.
89. Brent GA. Mechanisms of thyroid hormone action. The Journal of Clinical Investigation. 2012,122(9):3035–3043.
90. Iwen KA, Schröder E, et al. Thyroid hormones and the metabolic syndrome. European Thyroid Journal. 2013,2(2):83–92.
91. Fox CS, Pencina MJ, et al. Relations of thyroid function to body weight: cross-sectional and longitudinal observations in a community-based sample. Archives of Internal Medicine. 2008,168(6):587–592.
92. Potenza M, Via MA, et al. Excess thyroid hormone and carbohydrate metabolism. Endocrine Practice. 2009,15(3):254–262.
93. Duntas LH. Thyroid disease and lipids. Thyroid. 2002,12(4):287–293.

94. Liu J, Fu J, Jia Y, Yang N, Li J, Wang G. Serum metabolomic patterns in patients with autoimmune thyroid disease. Endocrine Practice. 2020,26(1):82–96.
95. L Ferreira I, Resende R, et al. Multiple defects in energy metabolism in Alzheimer's disease. Current Drug Targets. 2010,11(10):1193–1206.
96. Hoyer S. Brain glucose and energy metabolism abnormalities in sporadic Alzheimer disease. causes and consequences: an update. Experimental Gerontology. 2000,35(9–10): 1363–1372.
97. Ryu W-I, Bormann MK, et al. Brain cells derived from Alzheimer's disease patients have multiple specific innate abnormalities in energy metabolism. Molecular Psychiatry. 2021:1–13.
98. Blass JP, Sheu RK, et al. Inherent abnormalities in energy metabolism in Alzheimer disease: interaction with cerebrovascular compromise. Annals of the New York Academy of Sciences. 2000,903(1):204–221.
99. Chaturvedi RK, Beal MF. Mitochondrial diseases of the brain. Free Radical Biology & Medicine. 2013,63:1–29.
100. Cummings JL. Biomarkers in Alzheimer's disease drug development. Alzheimer's & Dementia. 2011,7(3):e13–e44.
101. Karbowski M, Neutzner A. Neurodegeneration as a consequence of failed mitochondrial maintenance. Acta Neuropathologica. 2012,123(2):157–171.
102. Gibson GE, Sheu KF et al. Abnormalities of mitochondrial enzymes in Alzheimer disease. Journal of Neural Transmission. 1998,105(8–9):855–870.
103. Cardoso SM, Proença MT et al. Cytochrome c oxidase is decreased in Alzheimer's disease platelets. Neurobiology of Aging. 2004,25(1):105–110.
104. Kalra EK. Nutraceutical – definition and introduction. AAPS PharmSci. 2003,5:E25.
105. Zhao J. Nutraceuticals, nutritional therapy, phytonutrients, and phytotherapy for improvement of human health: a perspective on plant biotechnology application. Recent Patents on Biotechnology. 2007,1(1), 75–97.
106. Chauhan B, Kumar G, Kalam N, et al. Current concepts and prospects of herbal nutraceutical: a review. Journal of Advanced Pharmaceutical Technology & Research. 2013,4(1):4–8.
107. Zeisel SH. Regulation of nutraceuticals. Science. 1999,285(5435):1853–1855.
108. Hardy G. Nutraceuticals and functional foods: introduction and meaning. Nutrition. 2000,16(7–8):688–689.
109. Baradaran A, Madihi Y, Merrikhi A, et al. Serum lipoprotein (a) in diabetic patients with various renal function not yet on dialysis. Pakistan Journal of Medical Sciences. 2013,29(1):354–357.
110. Nasri H. Impact of diabetes mellitus on parathyroid hormone in hemodialysis patients. Journal of Parathyroid Disease. 2013,1:9–11.
111. Madihi Y, Merrikhi A, Baradaran A, et al. Impact of sumac on postprandial high-fat oxidative stress. Pakistan Journal of Medical Sciences. 2013,29:340–345.
112. Setorki M, Rafieian-Kopaei M, Merikhi A, et al. Suppressive impact of Anethum graveolens consumption on biochemical risk factors of atherosclerosis in hypercholesterolemic rabbits. International Journal of Preventive Medicine. 2013,4(8): 889–895.
113. Khosravi-Boroujeni H, Mohammadifard N, Sarrafzadegan N, et al. Potato consumption and cardiovascular disease risk factors among Iranian population. International Journal of Food Sciences and Nutrition. 2012,63(8):913–920.
114. Khosravi-Boroujeni H, Sarrafzadegan N, Mohammadifard N, et al. White rice consumption and CVD risk factors among Iranian population. Journal of Health, Population, and Nutrition. 2013,31(2):252–261.
115. Brouns F. Soya isoflavones: a new and promising ingredient for the health foods sector. Food Research International. 2002,35(2–3):187–193.
116. Losso JN. Targeting excessive angiogenesis with functional foods and nutraceuticals. Trends in Food Science & Technology. 2003,14(11):455–468.

117. Cicero AFG, Colletti A, Bellentani S. Nutraceutical approach to non-alcoholic fatty liver disease (NAFLD): the available clinical evidence. Nutrients. 2018,10(9):1153.
118. Leung PS. The gastrointestinal system: gastrointestinal, nutritional and hepatobiliary physiology. New York, NY: Springer; 2016. pp. 24–36.
119. Araújo AR, Rosso N, Bedogni G, et al. Global epidemiology of non-alcoholic fatty liver disease/non-alcoholic steatohepatitis: what we need in the future. Liver International. 2018,38(Suppl 1):47–51.
120. Nasri H, Rafieian-Kopaei M. Oxidative stress and aging prevention. International Journal of Preventive Medicine. 2013,4(9):1101–1102.
121. Rafieian-Kopaei M, Baradaran A, Rafieian M. Plants antioxidants: from laboratory to clinic. Journal of Nephropathology. 2013,2(2):152–153.
122. Al-Okbi SY, Mohamed DA, Hamed TE, et al. Prevention of renal dysfunction by nutraceuticals prepared from oil rich plant foods. Asian Pac J Trop Biomed. 2014,4(8): 618–627.
123. Aggarwal BB. Targeting inflammation-induced obesity and metabolic diseases by curcumin and other nutraceuticals. Annual Review of Nutrition. 2010,30:173–199.
124. Prasad S, Gupta SC, Tyagi AK, et al. Curcumin, a component of golden spice: from bedside to bench and back. Biotechnology Advances. 2014,32(6):1053–1064.
125. Willis MS, Wians FH. The role of nutrition in preventing prostate cancer: a review of the proposed mechanism of action of various dietary substances. Clinica Chimica Acta. 2003,330(1-2):57–83.
126. Shirzad H, Kiani M, Shirzad M. Impacts of tomato extract on the mice fibrosarcoma cells. Journal of Herbmed Pharmacology. 2013,2:13–16.
127. Stahl W, Sies H. Bioactivity and protective effects of natural carotenoids. Biochimica et Biophysica Acta. 2005,1740(2):101–107.
128. Shirzad H, Taji F, Rafieian-Kopaei M. Correlation between antioxidant activity of garlic extracts and WEHI-164 fibrosarcoma tumor growth in BALB/c mice. Journal of Medicinal Food. 2011,14(9):969–974.
129. Shirzad H, Shahrani M, Rafieian-Kopaei M. Comparison of morphine and tramadol effects on phagocytic activity of mice peritoneal phagocytes in vivo. International Immunopharmacology. 2009,9(7–8):968–970.
130. Limer JL, Speirs V. Phyto-estrogens and breast cancer chemoprevention. Breast Cancer Research. 2004,6(3):119–127.
131. Asgary S, Keshvari M, Sahebkar A, et al. Clinical investigation of the acute effects of pomegranate juice on blood pressure and endothelial function in hypertensive individuals. ARYA Atherosclerosis. 2013,9(6):326–331.
132. Nasri H, Sahinfard N, Rafieian M, et al. Effects of *Allium sativum* on liver enzymes and atherosclerotic risk factors. Journal of Herbmed Pharmacology. 2013,2:23–28.
133. Rafieian-Kopaei M. Medicinal plants and the human needs. Journal of Herbmed Pharmacology. 2012,1:1–2.
134. Hu FB, Willett WC. Optimal diets for prevention of coronary heart disease. JAMA. 2002,288(20):2569–2578.
135. Roshan B, Stanton RC. A story of microalbuminuria and diabetic nephropathy. Journal of Nephropathology. 2013,2(4):234–240.
136. Rahimi-Madiseh M, Heidarian E, Rafieian-kopaei M. Biochemical components of *Berberis lycium* fruit and its effects on lipid profile in diabetic rats. Journal of Herbmed Pharmacology. 2014,3:15–19.
137. Rafieian-Kopaei M, Nasri H. Ginger and diabetic nephropathy. Journal of Renal Injury Prevention. 2013,2(1):9–10.
138. Sirtori CR, Galli C. N-3 fatty acids and diabetes. Biomedicine & Pharmacotherapy. 2002,56(8):397–406.
139. Coleman MD, Eason RC, Bailey CJ. The therapeutic use of lipoic acid in diabetes: a current perspective. Environmental Toxicology and Pharmacology. 2001,10(4): 167–172.

140. Nasri H, Rafieian-Kopaei M, Mardani S, et al. Herbal medicine and diabetic kidney disease. Journal of Nephropharmacology. 2013,2(1):1–2.
141. Kazemi S, Asgary S, Moshtaghian J, et al. Liver-protective effects of hydroalcoholic extract of Allium hirtifolium Boiss. In rats with alloxan-induced diabetes mellitus. ARYA Atherosclerosis. 2010,6(1):11–15.
142. WHO: World Health Organization. Fact Sheet for World Wide Prevalence of Obesity.
143. Calle EE, Kaaks R. Overweight, obesity and cancer: epidemiological evidence and proposed mechanisms. Nature Reviews Cancer. 2004,4(8):579–591.
144. Shimizu M, Weinstein IB. Modulation of signal transduction by tea catechins and related phytochemicals. Mutation Research. 2005,591(1–2):147–160.
145. Kao YH, Chang HH, Lee MJ, et al. Tea, obesity, and diabetes. Molecular Nutrition and Food Research. 2006,50(2):188–210.
146. Alexander EK, Pearce EN, Brent GA, et al. 2017 Guidelines of the American Thyroid Association for the Diagnosis and Management of Thyroid Disease During Pregnancy and the Postpartum. Thyroid. 2017,27(3):315–389.
147. Drutel A, Archambeaud F, Caron P. Selenium and the thyroid gland: more good news for clinicians. Clinical Endocrinology (Oxf). 2013,78(2):155–164.
148. Benvenga S, Feldt-Rasmussen U, Bonofiglio D, et al. Nutraceutical supplements in the thyroid setting: health benefits beyond basic nutrition. Nutrients. 2019 Sep 13, 11(9):2214.
149. Sadhukhan P, Saha S, Dutta S, et al. Nutraceuticals: an emerging therapeutic approach against the pathogenesis of Alzheimer's disease. Pharmacology Research. 2018,129: 100–114.
150. Chauhan NB, Mehla J. Ameliorative effects of nutraceuticals in neurological disorders. In: Bioactive nutraceuticals and dietary supplements in neurological and brain disease. Amsterdam: Elsevier; 2015, pp. 245–260.
151. Bhaskarachary K. Traditional foods, functional foods and nutraceuticals. Proceedings of the Indian National Science Academy. 2016,82:1565–1577.
152. Engel RR, Satzger W, Guünther W, et al. Double-blind cross over study of phosphatidylserine vs. placebo in patients with early dementia of Alzheimer type. European Neuropsychopharmacology. 1992,2(2):149–155.
153. Falinska AM, Colmbo CB, Irina AG, et al. The role of omega 3-fatty acid in brain function and ameliorating Alzheimer's disease: opportunities for biotechnology in the development of nutraceuticals. Biocatalysis and Agricultural Biotechnology. 2011,1:159–166.
154. Hager K, Marahrens A, Kenklies M, Riederer P, Muünch G. Alpha-lipoic acid as a new treatment option for Alzheimer type dementia. Archives of Gerontology and Geriatrics. 2001,32(3):275–282.

Functional Foods for the Prevention and Treatment of Cardiovascular Diseases Including Hypertension

Nahid Ramezani-Jolfaie and Mohammad Mohammadi

Contents

Cardiovascular diseases (CVDs) 197
Functional foods 198
Plant sterols 198
Garlic 199
Chocolate 201
Whole grains 202
Fruits and vegetables 203
Nuts 205
Tea 206
Dietary fiber 208
Polyunsaturated fatty acids (fish oil) 210
Olive oil 211
Soy 212
Probiotics 214
Curcumin 216
References 217

Cardiovascular diseases (CVDs)

Cerebrovascular disease, coronary heart disease (CHD), congenital heart disease, and many other conditions constitute a group of disorders named cardiovascular diseases (CVDs) that involve the heart and blood vessels (1). Representing the main cause of mortality around the world, CVDs were responsible for 31% of global deaths in 2012. Of these, 14.1 million were caused by stroke and CHD. As reported by the World Health Organization (WHO), it is projected that these two conditions will remain the principal causes of disease in the future,

DOI: 10.1201/9781003220053-10

with about 23.6 million people dying due to CVDs in 2030 (1). Some of the risk factors for CVDs are non-modifiable, including ethnicity, family history, and age; however, most cases occur due to modifiable factors like unhealthy diets, low physical activity, smoking, and excess use of alcohol (2). During the last 20 years, effective control of the fast-growing burden of CVDs through policies and key actions has been on the agenda all around the world. The main pre-emptive approach has been to promote a healthy lifestyle in populations (3). In this context, evidence shows that consumption of a high-potassium low-sodium diet, with lots of fruits, vegetables, and healthy fats strongly diminishes the risk of CVDs (4).

Functional foods

The term "functional foods," which was introduced in the 1970s in Japan and the United States, refers to foods fortified with probiotics or various microorganisms, natural or processed food that has biologically active components and should have special and evidence-based benefits for health promotion (5). Using food labels that say a specific food is functional is becoming increasingly prevalent in the food industry. Although it may be thought that some functional foods have short-term health benefits, the benefits are generally associated with reduced risk of chronic diseases in the long term. There are animal- or plant-based components found in functional foods that are biologically active and reduce the risk of heart diseases, with the primary effects being on the established risk factors, including diabetes, dyslipidemia, and high blood pressure. Functional foods are believed to exert protection against heart problems through antioxidative effects and lipid profile-modifying actions (6).

According to epidemiologic findings, cardiovascular health can be promoted by certain dietary patterns (7–10). Research on the potential cardioprotective effects of dietary components can be helpful in the process of functional food development (11). This chapter aims to examine the role of individual bioactive dietary items in cardiovascular protection.

Plant sterols

In the non-saponifiable fraction of plant oils, some naturally occurring plant sterols exist; they are called phytosterols. The structure of phytosterols is comparable to cholesterol, with the only difference being the C24 position substitutions on the sterol side chain. These compounds are not abundant in human tissues, because humans cannot synthesize them, they have a poor absorption rate, and the excretion rate through the liver is faster than that of cholesterol. Sitosterol, stigmasterol, and campesterol are the best-known dietary phytosterols. Humans usually consume about 200–400 mg of plant sterols daily. Beta-sitosterol is the most common phytosterol in Western diets; however, the tissue and plasma concentrations of this compound in humans are up to 1000 times lower as compared to endogenous cholesterol (12). In recent years, phytosterols have gained attention in regard to chronic disease risk reduction. Inhibition of cholesterol uptake, both dietary and biliary, from intestinal cells is hypothesized to be the mechanism of action. As a result, decreased serum total and LDL-cholesterol (LDL-C) levels occur (13). However, there is no evidence of HDL-cholesterol and triglycerides being affected by dietary phytosterol intake (13).

In one study, administration of 1.7 g oil phytosterols/day (primarily sitosterol and campesterol) (14), led to 15.5% more reduction of LDL-C in hypercholesterolemic men compared with the control diet. In a double-blind, crossover study, two table spreads fortified with non-esterified vegetable oil sterols (mainly from soybean oil) or shea nut oil sterols were assessed for their effects on cholesterol levels in a period of nine weeks (15). A total of 9.8% reduction in total and LDL-C levels were observed by soybean oil sterols compared to the controls. Another dataset reported that commercial corn oil contains natural phytosterols that significantly reduce the absorption of cholesterol in human subjects (16). Also, when it is especially important that LDL-C levels are targeted, studies have shown that adding other dietary factors such as fish oil (17), psyllium (18), beta-glucan (19), or statin drugs (20, 21) to phytosterols could be beneficial in secondary prevention of heart disease. However, a systematic review and meta-analysis of 17 datasets failed to provide any evidence about the link between serum concentrations of plant sterols and the risk of CVDs among more than 1000 subjects (22).

There are several theories about how phytosterols exert their cholesterol-lowering effect (23). According to one of them, cholesterol (already marginally soluble) in the intestine and in the presence of added phytosterols and stanols, is precipitated into a non-absorbable state. Another theory says that cholesterol must enter mixed micelles containing bile salt and phospholipids through intestinal cells to be absorbed into the bloodstream. Finally, the function of the main transporters involved in cholesterol absorption may be modulated by phytosterols.

Overall, it appears that dietary phytosterols provocatively exhibit protective effects against CVDs through playing an important role in the regulation of serum cholesterol. However, more research is to be carried out for researchers to reach a firm conclusion in this regard.

Garlic

Central Asia is believed to be the origin of a group of vegetables of the family Alliaceae, of which garlic (*Allium sativum*) is the focus of this chapter (24, 25). There is well-documented evidence showing that garlic has always had medicinal applications. Garlic was used by the Indians, Chinese, Romans, and Egyptians for curing various types of ailments. It was thought to be an effective solution for increasing performance in laborers, for respiratory and gastrointestinal illnesses, and for certain infectious diseases (26). Garlic cloves, essential garlic oil, garlic powder, garlic oil macerates, garlic extract, and aged garlic extract are different forms of garlic preparations (26).

RCTs are indicative of the beneficial role of garlic in the treatment of hypertension. A meta-analysis of nine double-blind trials with 482 individuals disclosed a more effective reduction in systolic blood pressure (SBP) and diastolic blood pressure (DBP) in individuals treated with garlic preparations as compared to placebo. However, high heterogeneity was present, which may have influenced the results (27). In another meta-analysis including seven randomized, placebo-controlled trials, significant SBP-lowering (−6.71 mmHg) and DBP-lowering (−4.79 mmHg) effects of garlic were confirmed (28). The effect of garlic on blood pressure was also investigated through an updated meta-analysis including 20 trials with 970 participants. It also indicated a mean decrease in systolic blood pressure (SBP) and diastolic blood pressure (DBP) of 5.1 and 2.5 mm Hg compared with placebo (29), respectively.

The bulb portion of garlic produces sulfur-containing compounds that are thought to be involved in the antihypertensive effects observed after consumption of the plant. The highest sulfur content is found in one of garlic's main bioactive compounds, allicin (allyl 2-propenethiosulfinate), which is produced from the breakdown of S-allyl cysteine-S-oxide (alliin) (30). As a vasodilating agent, allicin has been able in vitro to inhibit angiotensin-converting enzyme (involved in the production of angiotensin, a potent vasoconstrictor), which could be another mechanism for the blood pressure-lowering effect of garlic in individuals with hypertension (31). It is believed that garlic products exert their effect on blood pressure through their ability to produce allicin, depending on how they were prepared. The method of preparation can affect this ability since heat inactivates the enzyme responsible for the production of allicin from alliin (32). Accordingly, all garlic products do not contain the same amount of allicin (32). According to ongoing research, other compounds in addition to allicin can be involved in the medicinal properties of garlic. Odor-free phenolic and steroidal compounds in garlic also have pharmacological effects and, contrary to allicin, are not sensitive to heat processes (30, 33, 34).

The exact estimate of blood pressure reduction by garlic, specifically in hypertensive individuals, is yet to be defined. Furthermore, the possible impact of garlic on cardiovascular morbidity and mortality remains unknown as well. In future trials, hypertensive patients should be randomized into groups receiving different doses of garlic versus placebo. These trials should be large enough to determine the differences in cardiovascular morbidity and serious adverse events and mortality.

Chocolate

Cocoa is a rich source of flavanols (epicatechin, catechin, and procyanidins) and has been proposed to reduce CVD risk by lowering blood pressure (35, 36). Other plant-derived products such as beans, apricots, blackberries, apples, and tea leaves are also sources of flavanols, but the concentration in these products is much lower than in cocoa products (37). Other factors responsible for the amount of flavanol intake are serving size and the processing method of the products. Different flavors and smoothness of chocolate are the results of the processing being fine-tuned by different chocolate manufacturers through the years, but this has also altered flavanol content in various cocoa products. Dark chocolate contains 50–85% cocoa, while only 20–30% of milk chocolate is made of cocoa. The variety and ripeness of cocoa beans and the manufacturing steps are among other factors influencing the content and composition of flavanols. Fresh and fermented cocoa beans contain 100 mg flavonols per gram (about 10%), the Kuna Indians consume a cocoa powder containing 3.6% flavanols, and dark chocolate rich in cocoa has about 0.5% flavanols (38, 39). Moreover, the flavanol content can be reduced to less than 10 mg per 100 grams by alkalizing of chocolate to pH 7–8 (heavy ditching).

According to evidence, the monomeric flavanols, the epicatechin and catechin, and to a lesser extent the polymeric flavanols, the procyanidins, have blood pressure-lowering and vasoactive effects (40). The monomeric flavanol content of cocoa and the epicatechin/catechin ratio are changed by the modern methods of processing (41). Depending on the growing region, variety, and harvesting practices, 2.5 and 16.5 mg epicatechin per gram are found in fresh and fermented cocoa beans (42), whereas roasting and dutching of processed cocoa reduces its epicatechin content to only 2–18% (41). Since the flavanol content of cocoa and chocolate is highly variable, it is critical that clinical trials examining the effect of cocoa on blood pressure compare the dosages of flavanols rather than simply the administered amounts of cocoa or chocolate products.

In contrast to previous findings, moderate-quality evidence by a Cochrane review claims that chocolate and cocoa products rich in flavanols only cause a short-term small reduction in blood pressure, mainly in healthy adults (43). However, 16 prospective studies were included in a meta-analysis and the results confirmed that the highest versus the lowest category of chocolate intake had a protective effect against CVD, with an overall risk ratio of 0.77 (44).

The cocoa mechanism of action has been linked to the formation of endothelial nitric oxide (NO), leading to promoting vasodilation and, as a result, lower blood pressure levels. Through the insulin-mediated signaling pathway, upregulation of NO synthase causes higher NO production (45). A number of datasets have shown that insulin sensitivity is improved after cocoa intake (46–49). Also, angiotensin-converting enzyme (ACE) activity has been indicated to be diminished by cocoa flavanols, which leads to a lower blood pressure level (50, 51).

Finally, evidence suggests an indirect antioxidant effect of cocoa flavanols within the cardiovascular system, which causes the upregulation of NO synthase activity and hence blood pressure reduction (52–54).

Overall, available studies suffer from significant statistical heterogeneity, and further investigation regarding the most appropriate dose and the long-term side effect profile is warranted in case we seek to recommend cocoa products as an option in hypertension and CVD control.

Whole grains

When the edible parts of a natural grain kernel are not removed, it is called whole grain. All whole grains are similar in structure and include the endosperm, germ, and bran. Whole grains are a rich source of dietary fiber, resistant starch, antioxidants, phytoestrogens, and other important nutrients such as vitamins and folic acid (55). Most of the bran and some of the germ is removed in the grain-refining process, and the content of dietary fiber, minerals, vitamins, phytoestrogens, lignans, phenolic compounds, and phytic acid is highly diminished. Finally, refined white flours are produced by grounding the remaining starchy endosperm.

The Western diet contains important grains such as rice, wheat, maize, oats, barley, and rye. In wholemeal foods, whole grains have a finer texture and do not remain whole in the ultimate product. According to the definition by the EU Healthgrain consortium, whole grains "consist of the intact, ground, cracked or flaked kernel after the inedible parts such as the hull and husk have been removed. Whole grains and the intact kernel have the same main anatomical components - the starchy endosperm, germ, and bran-in the same relative proportions" (56). Small losses of components during processing have also been allowed in this definition. Some specific grains have also been listed as whole grain in this definition. Based on the evidence, biologically important compounds stay intact after such processing of whole grains (57). Whole grains and wholemeal foods are similar nutritionally.

A minimum of 27 g whole grains/100 g is necessary for foods (breakfast cereals, biscuits, bread, pasta, and grain-based snack foods) made from whole grain if they are to be called whole grain food (58). This recommendation was made since previous definitions provided by authorities about whole-grain foods were inconsistent (58).

According to a recent systematic review and meta-analysis, subjects who eat three servings of whole grains have a lower risk of developing stroke, cardiovascular disease, and CHD (59). Also, a number of cohort studies were included in two different meta-analyses and the results indicated a 21% reduction in CVD risk when an average of four servings/day of whole grains compared to lower intakes (60), or 2.5 servings/day compared to 0.2 servings/day were consumed (61). Moreover, when subjects recruited in the Nurses' Health Study consumed more whole grains, a decrease in the risk of CHD was documented (62). The contribution of the diet to intakes of dietary fiber, vitamin B6, folic acid, and vitamin E did not fully explain the reduced risk associated with higher whole grain intake. Pursuant to the Atherosclerosis Risk in Communities (ARIC) study, whole grain consumption was found to be beneficial in reducing the total incidence and mortality of coronary artery disease but not the ischemic stroke risk (63). In this study, 15,792 people aged 45 to 64 were followed for more than ten years. The consumption of whole grains in relation to CVD risk was reviewed and a strong inverse association was revealed (64). There are also reports about the associations between whole grain consumption and risk factors for CHD. In the Framingham Offspring study, consuming diets rich in whole grains led to lower levels of body mass index, total cholesterol, and low-density lipoprotein (LDL) cholesterol (65).

Despite evidence showing an inverse association between cereal fiber intake and risk of CVD (66, 67), researchers still need to figure out how fiber or other components found in whole grains (phytochemicals and micronutrients) (68, 69) exert beneficial influences on CVD and its risk factors (58). Small changes in body fat, but no alteration in body weight parameters were documented in a recent systematic review of randomized controlled trials (RCTs) (70). In another systematic review, whole grains were investigated in relation to type 2 diabetes and its risk factors (71). They found only one relevant randomized trial that indicated a minor improvement in insulin sensitivity (72).

Currently, there is no proof from RCTs showing that whole-grain diets are beneficial in cardiovascular events and mortality. Due to this clear gap in the evidence, carefully designed studies with adequate statistical power and sufficient duration are necessary to see if whole-grain diets could reduce the risk of cardiovascular events and associated mortality. According to the Global Burden of Disease Study, 1.7 million deaths in 2015 were due to ischemic heart disease globally, while 3.1 million all-cause deaths were attributed to consumption of diets without enough whole grains (73). Longitudinal cohort studies were used to derive these figures, so RCTs are needed in this area. A Cochrane review reported that the lipid profile or blood pressure of subjects was beneficially influenced by whole-grain diets, but most studies were highly biased with follow-up durations less than 16 weeks (74). Also, well-designed RCTs with adequate power and ≥ 12 months follow-up periods are required to determine the position of whole grains in the prevention of CVDs, both primary and secondary.

Fruits and vegetables

Many complex factors determine how much fruits and vegetables are consumed by a person. Consequently, various conceptual frameworks have been used to design interventions aimed at encouraging people to consume more fruits and vegetables (75). One such conceptual framework may suggest the higher effectiveness of interventions that target cultural and personal parameters in promoting fruit and vegetable consumption as compared to interventions that are aimed at personal determinants only. A framework based on behavior change theories at different levels of influence involving both social and ecological aspects is suggested to be the best approach for addressing dietary intake changes, including fruit and vegetable consumption (76).

According to observational and experimental studies, CVD prevention and treatment may be feasible through consuming ≥ five portions of fruits and vegetables a day (more than 400 g) (77). However, it is not precisely known what the mechanisms are. Protective elements found in fruit and vegetables such as vitamins, minerals, micronutrients, antioxidants, and phytochemicals may be involved (78). These elements may cause protection by reducing blood pressure, lowering serum LDL cholesterol, fighting antioxidant stress, and improving the regulation of hemostasis (79, 80).

There are theories, primarily based on the Health Belief Model (81), the Theory of Planned Behavior (82), social cognitive theory (83), or the Stages of Change Model (84), to explain by which mechanisms fruit and vegetable intakes are influenced by lifestyle modifications such as advice provision and fruit and vegetable interventions. These theories all emphasize that beliefs have a dynamic nature and suggest that behavioral change is dependent on the change in a person's perceived attitudes, norms, knowledge, expectancies, and skills change (85). There are also social-ecological theories about the mechanisms by which interventions targeting fruit and vegetable intake may be effective. According to these theories, a multitude of factors including inter- and intrapersonal factors, community and organizational parameters, and those relating to public policy influence a person's health behavior (86).

Availability, demographics, lifestyle factors, and sensory appeal determine how much fruits and vegetables are consumed in adulthood (87, 88). Based on observational studies, a considerable body of information exists regarding the determinants of fruits and vegetables to aid in the development of interventions; however, the effectiveness of interventions to increase fruit and vegetable consumption has not been investigated. There have been some attempts to systematically review the findings in this regard (89–91). Pomerleau et al. (91), for example, systematically reviewed the existing literature and reported the largest increase in fruit and vegetable consumption as a result of interventions that selected their sample from high-risk populations or those who were already diagnosed with a certain disease. Compared to this, healthy adults showed a much smaller increase in fruit and vegetable intake following interventions. Similarly, Brunner et al. (92) reported 1.25 servings per day increase in the consumption of fruits and vegetables due to dietary advice as compared to no advice in healthy adults. The findings of recent systematic review and meta-analysis of 15 RCTs to determine the effectiveness of orange juice intake on major cardiometabolic markers also showed a significant reduction in total cholesterol and HOMA-IR, but reported no positive effect of orange juice in improving other cardiovascular risk factors such as blood pressure, anthropometric indices, other blood lipids and blood glucose control indicators, and inflammatory markers (93).

RCTs exclusively on the effects of providing families with fruit and vegetables and giving them dietary guidance to eat more fruit and vegetable for primary prevention of CVD are lacking. This is unexpected considering that people should eat at least five servings of vegetables and fruits every day, as constantly recommended by national and international guidelines. In particular, and above all, RCTs looking at the sustainability of behavioral change and the effects on CVD events through interventions that solely target the long-term increased consumption of fruit and vegetable are limited. Increased fruit and vegetable consumption in the format of multifactorial dietary interventions has shown no benefits on CVD events (94), which may imply that no benefit may come from promoting higher fruit and vegetable consumption as a standalone intervention, albeit if we assume that other components of the dietary intervention do no cause any harm.

Nuts

Humans have consumed nuts for thousands of years. According to records, pistachio nuts have been used as far back as 7000 BC (95). Today, we can see the same pattern, and people eat nuts all around the world, both as snacks and as ingredients in recipes (95). However, various amounts of nuts are eaten globally. For instance, nut consumption in countries with an American diet is half that of those with a Mediterranean diet (96, 97).

Nuts are botanically considered a dry one-seeded fruit with a hard pericarp or shell (96). Nuts contain around an average of 60% total fat, which makes them an energy-dense food (98). An average of 10% saturated fatty acid is found in nuts, with unsaturated fatty acids making almost 50% of their total fat content (99). Also, they contain an average of 20% protein (100) and provide some vitamins and minerals, such as selenium, folic acid, niacin, and zinc (101). Antioxidants are also found in many nuts (102). For example, walnuts are a good source of polyphenols and tocopherols, while flavonoids such as flavonols, catechins, and flavonones are abundant in almonds (100). In particular, a high antioxidant efficacy has been revealed for raw walnuts and toasted almonds (102). This is why most trials use almonds and walnuts as the main types of nuts studied.

According to epidemiological studies, CVD risk factors could be beneficially modified by nut consumption (103, 104). The Iowa Women's Health Study (IWHS), for example, indicated lower mortality due to CVD in people who consumed five or more times nuts/peanut butter per week (100). Also, there is evidence from systematic reviews of observational studies confirming the beneficial effects of nuts. Based on a systematic review by Mukuddem and

Petersen regarding the effect of nuts on the lipid profile of normal and hyperlipidemic participants (105), total cholesterol and bad cholesterol (LDL-C) were significantly reduced in those eating a heart-healthy diet supplemented with an average of 2.5 portions of nuts at least five times a week. In particular, in three studies, consuming an average of 50 g almonds daily lowered total cholesterol and LDL-C by 4% to 17% and 7% to 19%, respectively, in both hypercholesterolemia and non-cholesterolemic subjects (106). Moreover, RCTs provide some evidence that people with CVD risk factors may benefit from consuming nuts (107). In one RCT a recommended cholesterol-lowering diet supplemented with 84 g walnuts per day for four weeks led to a 12% reduction in serum TC levels (108). According to a meta-analysis of 21 RCTs, reduction in SBP as a result of nut consumption was not observed in the total population, but participants without type 2 diabetes showed a significant SBP improvement. It was also suggested that, among different nut types, only pistachios significantly diminish SBP. Pursuant to this study, a significant DBP-reducing effect was documented for pistachios and mixed nuts (109).

It is not exactly known how nuts reduce the risk of CVD. However, due to a number of nutritional attributes, nuts have been linked to cardio protection (98). For instance, walnuts are a valuable source of n-3 fatty acids with established cardioprotective properties (98). The benefits of nuts for combating CVD and its risk factors could be attributed to individual nutrients found in them, the composite of these nutrients, or both (110). Moreover, another major consideration when discussing the health benefits associated with frequent nut intake is that nuts have a high amount of unsaturated fatty acids and a low amount of saturated fatty acids (98).

Currently, very limited evidence exists about the impact of nut consumption on CVD risk factors and no evidence for effects on the CVD clinical events in primary prevention. It is not feasible to draw conclusions, and more trials with better designs are needed to answer the review question.

Tea

According to records from Deka et al. (111), medicinal use of tea goes as far back as the 10th century and now it is consumed all around the world. The plant *Camellia sinesis* is the origin of tea leaves and there are three main types of tea including black, green, and oolong. Depending on how the leaves are processed, the type of tea produced from the leaves is different. For instance, oolong tea is made from partially fermented leaves, while black tea and green tea are produced from fermented and nonfermented leaves, respectively (111).

All types of *Camellia sinesis*-based teas are rich in water-soluble vegetative pigments named flavonoids, which belong to a family of polyphenolic compounds (112, 113). Flavanols are the main class of flavonoids found in tea, including epigallocatechin gallate (EGCG), epigallocatechin (EGC), epicatechin gallate

(ECG), and epicatechin (EC) (114). While the chemical structures of green and black tea are different, both have similar total flavonoid content. The reason could be that in the oxidation process, complex compounds including thearubigins and theaflavins are produced from the flavonoids found in black tea (111, 115). About 80% to 90% of total flavonoids in green tea are catechin, whereas they constitute around 20% to 30% of total flavonoids in the black variety (115).

The vitamin and mineral content of green tea is also high, and a person can get 5% to 10% of the daily requirement for riboflavin, niacin, folic acid, and pantothenic acid by consuming five cups of green tea per day. Also, 45%, 25%, and 5% of the daily requirements of manganese, potassium, and magnesium are provided with daily consumption of this much green tea a day, respectively (116).

According to epidemiological, observational, and experimental evidence, cardiovascular function may benefit from the consumption of green and black varieties of tea (111, 117–120). In particular, CVD risk has been reduced in observational studies using a high intake of both black and green tea (117, 120). For example, a prospective cohort study by de Koning Gans et al. (121) in the Netherlands, involving 37,514 subjects, reported a reduction in the risk of CHD mortality as a result of consuming three to six cups of mainly black tea on a daily basis. Mineharu et al. (118) also reported that consuming more than six cups of green tea daily and CVD mortality were inversely related among 76,979 Japanese adults. When interpreting the results of these observational findings caution is warranted because there are potential confounding factors associated with tea intake, such as healthier lifestyles, that might have played a role in the observed inverse associations. Indeed, the intake of tea and CVD risk has not been related in some studies (122).

The inverse relationship between tea and CVD risk observed in individual studies has been confirmed by meta-analysis studies (123, 124). Seven case-control and ten cohort studies regarding the association between tea and CVD were analyzed by Peters and colleagues (124), and results indicated that consuming ≥3 cups of tea/day reduced the risk of myocardial infarction by 11%.

According to intervention studies, tea consumption is beneficial in reducing CVD risk factors (119, 125). In a randomized double-blind placebo-controlled study, Fujita et al. (126) investigated the benefits of black tea extract in borderline hypercholesterolemia among 47 Japanese subjects. Low-density lipoprotein (LDL)-cholesterol and serum total cholesterol levels were significantly diminished by black tea extract consumption. According to a pooled analysis of 11 studies, regular consumption of four to five cups of black tea per day significantly improved SBP and DBP in 378 subjects (127).

Despite ambiguities about the exact mechanisms of CVD risk reduction by tea consumption, the high levels of polyphenols found in tea, in particular flavonoids, may be the response. These constituents have been linked to improved insulin sensitivity, dyslipidemia, and endothelial function by reduced oxidative stress, platelet inhibition, and anti-inflammatory effects. They also have been proved to induce weight loss (111). Finally, the antioxidant properties of tea and its flavonoid content may be helpful in lowering CVD risk (111, 128).

RCTs regarding the effects of black and green tea consumption on the primary prevention of CVD are scarce. In particular, more RCTs are needed to examine the effects of long-term consumption of black and green tea and to elucidate the effects of such interventions on CVD events.

Dietary fiber

Due to discrepancies about which plant-derived substances should be considered as dietary fiber and how values for dietary fiber are determined, no globally accepted single definition could be presented for dietary fiber at the moment (129). In general, a variety of plant substances resistant to the action of digestive enzymes are referred to as dietary fiber (130). There are two main categories of dietary fiber: one is soluble fiber, found in bran, flaxseeds, oat cereal, and pears, which dissolves in water, and by forming a gel, delays the emptying of the stomach and slows digestion (129). The other one is insoluble fiber that does not dissolve in water and accelerates food and waste passage through the stomach. Barley, brown rice, celery, cabbage, and whole grains are good sources of insoluble fiber (129). Different types of fiber have different beneficial roles in the human body, so consuming a healthy diet which has both soluble and insoluble fiber resources is important (131).

The global consumption of fiber is generally low. For example, the average fiber intake in the United Kingdom between 2008 and 2011 was 12.8 and 14.8 g/day for women and men, respectively (132). An average of 15.9 g fiber intake per day was reported in the United States in 2007–2008 (133). Similar patterns have been indicated in Malaysia and Japan as well (134, 135). This is despite the fact that, according to current guidelines, fiber intake should be between 18 and 40 g/day (132). The adverse effects of fiber intake over time are not fully recognized (136); however, minor adverse events have been reported when various doses of Arabic gum or psyllium were administered in different populations (137, 138). For instance, fiber supplementation in individuals with fecal incontinence resulted in belching, bloating, flatus, and fullness, as reported by Bliss et al. (136).

It is not known how exactly and through what mechanisms dietary fiber reduces CVD risk. However, there are some speculations. When soluble fiber is exposed to water, a gel is formed in the stomach and as a result a number of effects occur, including reduced gastric emptying, increased small intestine movement, and controlled nutrient absorption. In doing so, soluble fiber prevents sudden increases in postprandial blood glucose and lipid levels, both of which are known risk factors for CVD (139). Furthermore, it is thought that both soluble and insoluble fiber increase satiety through an effect on gut hormones and also enhance gastric distension. As a result, food intake is diminished, which promotes weight loss and improved glucose metabolism in long term (140, 141). It has also been proved that the rate of bile acid excretion is increased by dietary fiber, leading to reduced levels of total and LDL cholesterol. In addition, dietary fiber inhibits the synthesis of cholesterol by short-chain fatty acid production as a by-product of fiber fermentation in the colon (140, 141). Finally, plaque stability may be influenced by dietary fiber. The proposed mechanism is that fiber intake is associated with decreased levels of pro-inflammatory cytokines; it is known that plaque stability is affected by these cytokines (141).

According to a recent meta-analysis of prospective cohort studies, the risk of total mortality was reduced by increased fiber intake (142). A recent observational cohort analysis of the PREDIMED trial also confirmed this finding (143). Moreover, a beneficial association between dietary fiber and CVD risk factors has been shown in a number of observational studies (144–146). In one study involving 39,876 female health professionals, dietary fiber intake was inversely linked to CVD risk (147). Another study showed a lower risk of incident ischemic CVD as a result of high fiber intake in both men and women (148). Systematic reviews of observational studies further confirm the beneficial association between dietary fiber and blood pressure and lipid levels. For instance, higher dietary fiber intake was associated with a lower risk of both CHD and CVD in a systematic review by Threapleton et al. (139).

There are also promising results from experimental studies about the beneficial effect of dietary fiber on CVD risk factors (149, 150). For instance, oat-derived beta-glucan was found to improve mild to moderate hypercholesterolemia in male participants, as a part of the American Heart Association Step 2 diet (151). In another study, six weeks of oat cereal consumption in hypertensive and hyperinsulinemic participants significantly reduced systolic and diastolic blood pressure as compared to a low-fiber cereal (152). Experimental studies have been also systematically reviewed. In a systematic review and meta-analysis of 24 RCTs, Streppel et al. showed a significant reduction in diastolic blood pressure but a nonsignificant reduction in systolic blood pressure following fiber supplementation (153). Brown et al. also conducted a systematic review of 67 controlled trials including 2990 participants and found significantly reduced LDL and total cholesterol levels in persons with diets rich in soluble fiber (154). A reduced risk of mortality from CVD and all cancers was also reported in a recent meta-analysis as a result of high dietary fiber intake. Another meta-analysis also showed an 18% lower risk of CVD mortality in relation to consumption of cereal fiber (155). Moreover, Charlotte et al. concluded that lower SBP and DBP are possible through higher consumption of beta-glucan fiber (156).

Evidence from RCTs proving the benefits of dietary fiber in CVD clinical events is lacking. Despite some evidence suggesting reduced total and LDL cholesterol and diastolic blood pressure due to dietary fiber intake, there are still questions to be answered about the best type (soluble or insoluble) or the most beneficial source of fiber (fiber supplements as opposed to natural foods high in fiber). So, recommendations to change the current practice are not possible. Overall, future RCTs to ascertain the effects of fiber type and source on CVD and its risk factors need to be carefully conducted and have longer durations.

Polyunsaturated fatty acids (fish oil)

Polyunsaturated fatty acids (PUFAs) are fats whose long hydrocarbon chains contain unsaturated carbon bonds. Therefore, unlike saturated fats, these types are liquid at room temperature because they do not pack well. There are three possible types of PUFAs, including omega-3 (the first double bond and the methyl-carbon end of the molecule are three carbons apart), omega-6, and omega-9. Omega-3 and omega-6 PUFAs are usually found in fish and plant oils, respectively. Essential omega-3 and omega-6 PUFAs for human subjects are alpha-linolenic acid and linoleic acid, respectively. Fish-derived long-chain omega-3 fatty acids are docosahexaenoic acid (DHA), eicosapentaenoic acid (EPA), and docosapentaenoic acid (DPA, 22:5). In contrast, omega-3s derived from grass-fed meat and plants are short-chain fatty acids such as alpha-linolenic acid (ALA), and the human body converts this type of fat to long-chain omega-3 fatty acids. The effectiveness of this conversion is controversial, since assessment over the short or long term or on other dietary factors may exert differences (157, 158). Due to this, the effectiveness of long-chain omega-3 fats may be different from that of ALA.

Ever since low mortality rate due to ischemic heart disease among the Greenland Inuit people was linked to the high level of omega-3 fatty acids in their diet, there has been a considerably increased interest in the protective role and possible mechanism of action of marine unsaturated fats (17), ALA-rich plant seeds and oils (chia seed, flaxseed, and canola oils) (159), margarines (one of their derivatives), purslane leaves, and walnuts (more than other nuts) (160).

Omega-3 fats are proposed to protect against CVDs through beneficial effects on the lipid profile, especially serum triglyceride concentration, blood pressure, thrombotic tendency, arterial lipoprotein lipase levels, inflammation and arrhythmia, vascular endothelial function and insulin sensitivity, and paraoxonase levels and plaque stability (161).

Since oily fish or fish oil capsules (often fish liver) are the most common form of consumed omega-3 fats, there have been concerns about high levels of toxic compounds such as dioxins, mercury, and polychlorinated biphenyls (PCBs) in these sources (162). Due to their nature as fat-soluble fats and their ability to increase over time in the body, long-term fish oil consumption or supplementation may be harmful. According to animal and human studies with subjects suffering from exposure to PCBs and dioxins, sub-fertility problems may result from prenatal exposure, and the total number of cancer cases may increase due to exposures in adults (163). Neurological problems occurred in individuals exposed to high levels of mercury (164). Exploring the possibly detrimental impact of fish-derived omega-3 is crucial, since consuming oily fish or fish oil supplements once or twice a week is common among so many people. There is also the possibility of harm by omega-3 fats per se, through inducing suppressed immune responses or prolonged bleeding times (165).

Overall, cardiovascular effects of taking a fish oil supplement are different from eating an oily fish because a considerable amount of nutrients other than omega-3 fats are found in fish as compared to fish oil, including protein, iodine, selenium, zinc, and calcium. A variety of other foods including sources of saturated or trans fats may be removed from the diet by consuming fish instead of other types of meat, which could reduce CVD risk with other mechanisms (166).

Olive oil

Olive oil, long considered an extraordinary precious and sacred gift, is produced from the fruits of the olive tree (*Olea europaea*) with the Middle East and Eastern Mediterranean regions considered as the origins of this tree. In the last decades, however, the awareness about the favorable effect of olive oil (extra-virgin olive oil in particular) on well-being and against CVDs has increased, which is why nontraditional markets outside the Mediterranean regions have also become increasingly interested in olive oil consumption (167). Up to 83% of the total fatty acids found in olive oil are oleic acid, a monounsaturated fatty acid (MUFA) (168). Only 2% of the total weight of olive oil is composed of various phytochemicals, such as sterols, triterpenes, tocopherols, phenolic compounds, and carotenoids (169). Tyrosol, hydroxytyrosol, and their derivatives are the most common phenolic compounds found in extra virgin olive oil (168). Polyphenols found in olive oil exert their beneficial effects through anti-inflammatory, antiatherosclerotic, and antioxidant characteristics (25). Since many population-based datasets have confirmed that olive oil polyphenols are advantageous, the European Food Safety Authority announced the protective effects of olive oil and extra-virgin olive oil polyphenols on blood lipids, if consumed on a daily basis (170).

Reactive oxygen species (ROS) are constantly produced by the cells in the cardiovascular system, which impair the vascular function and structure and damage or alter the DNA, lipids, and proteins (171, 172). The postprandial hemostatic profile is altered by the virgin olive oil's bioactive compounds and has anti-oxidative effects (173). As confirmed by a number of studies, tangible amounts of biologically active compounds found in all different types of olive such as olive oil (174), extra-virgin olive oil (175, 176), and virgin olive oil (174, 177) cause them to strongly protect against oxidative stress. High concentrations of biologically active compounds such as MUFAs and polyphenols are the reason for the prophylactic properties of olive oil. Recently, it has been reported by numerous researchers that consumption of olive oil has unique benefits toward controlling blood pressure (BP) and favorably modulates endothelial function in the initial levels of hypertension (178–182). Up to 80% oleic acid found in olive oil and virgin olive oil is thought to play a significant role in the management of high BP (183). Upon olive oil and virgin olive oil consumption, the membrane concentration of oleic acid increases, leading to alternations in membrane lipids (H_{II} phase propensity) and consequently initiating a G protein-mediated signaling. As a result, a decrease in BP results from the regulation of phospholipase C and adenylyl cyclase (184).

The use of olive oil and its products for nutritional and medicinal purposes in everyday life has a long history. The benefits of olive oil and its metabolites on CVD prevention have been revealed in a large number of studies. Different mechanisms have been proposed, mostly focusing on the important bioactive compounds found in the olive oil varieties besides energy and fat-soluble vitamins. Olive oil polyphenols have received particular attention in recent years, with studies showing beneficial effects on HDL levels, oxidative stress, thrombogenic activity, endothelial function, BP, and inflammation. It has been also demonstrated that these compounds alter the atherosclerosis process through effects on gene expression (185, 186). According to these, the recommendation to consume olive oil and its products every day is both based on its beneficial fatty acid profile and also important bioactive components found in it with advantageous effects on human health.

Soy

The main reason for the cultivation of soybeans (Glycine max) is their lipid content, which makes them the number one oilseed produced globally (187). As a rich source of nutrients, soybeans also contain about 40% high-quality protein, 18% polyunsaturated fatty acids, and dietary fibers. The primary types of carbohydrates contained in soybeans are sucrose, stachyose, and raffinose (188). For a thousand years, soy products have had a part in the human diet, but in recent years, the associated health benefits of soy have gained considerable attention, particularly its potential for lowering the LDL-C in order to reduce CVD risk. The hypocholesterolemic potential of bioactive peptides in soy protein has been the focus of research. These peptides exert their effects mainly via pathways involving bile acid regulation and the LDL-C receptor (189, 190). Several meta-analyses support these findings (187), claiming that consumption of 25 g of soy protein did not reduce the risk of CHD in Europe, but it was beneficial in the United States and Canada (187). However, soy has other constituents with health benefits, which makes it even more interesting for further investigation. Among these are phytochemicals such as isoflavones, phytosterols, and lecithins, along with soluble fibers, polysaccharides, and saponins, which may exert unique health benefits collectively or via independent mechanisms (191, 192). Further, reduction of LDL-C seems to be one of many health benefits of soy proteins; it also protects against renal dysfunction (193) and oxidative stress (194) and improves the markers of endothelial function (195). Finally, soybeans are the single greatest dietary source of isoflavones, and their health benefits have also been extensively studied.

It has been shown that isoflavones target vasodilation and relieve hypertension as a result. In particular, endogenous NO production by endothelial nitric oxide (NO) synthase is enhanced, and better brachial artery flow happens (196). Isoflavone supplementation in postmenopausal women for six months resulted in improved endothelial vasodilation and cellular adhesion molecules such as vascular cell adhesion protein 1, intercellular adhesion molecule 1, and E selectin (197). Animal studies support these findings and say that renal blood flow and sodium excretion are increased by soy isoflavones. The mechanism is through inhibiting angiotensin-converting enzyme activity in the renin-angiotensin-aldosterone system when they interact with estrogen receptors (198). However, there is controversy in the clinical evidence regarding the role of isoflavones in hypertension management. Fourteen RCTs were investigated in a meta-analysis and the authors concluded that 25–375 mg isoflavone extract intake/day during 2–24 weeks in adults with normal blood pressure resulted in significant SBP reduction (–1.92 mmHg, 95% CI: –3.45 to –0.39 mmHg), but not DBP (199). In another meta-analysis, consumption of 65–153 mg of soy isoflavones/day in hypertensive subjects for 1–12 months significantly reduced blood pressure; however, this result was not observed in adults with normal blood pressure (200). According to this, persons with established hypertension benefit the most from isoflavone supplementation. Lately, 71 trials about the effect of phytoestrogen therapy on arterial hypertension were included in a meta-analysis. No significant reductions in SBP and DBP were reported (201). Hypotensive properties have been also documented for other constituents of soy, besides isoflavones. Soy pulp, rich in fiber and oligopeptides, can suppress angiotensin-converting enzyme activity, as exhibited in in vitro studies. This provides mechanistic evidence for a hypotensive effect (202). According to previous systematic reviews, it is difficult to determine the effect of the protein content of soy on blood pressure due to the dearth of studies focusing exclusively on the soy protein as the active agent (195, 203). In pre-diabetic, postmenopausal, hypertensive women, the combination of soy protein and isoflavones apparently lowers SBP as compared to milk protein (204, 205). DBP is also reduced by soy protein in patients with type 2 diabetes and metabolic syndrome, according to a new meta-analysis (206). Further, blood pressure was not improved in type 2 diabetes women receiving a bread fortified with soybean flour (207), and brachial artery flow-mediated dilation was not ameliorated by soy lecithin alone or in combination with isoflavone-rich soy protein isolate, despite observed improvements in the plasma lipid profile (208). The hypotensive effects of dietary soy could also be attributed to its amino acid composition. Legumes, and soy products in particular, are valuable sources of arginine, which converts into NO in the L-arginine-nitric oxide pathway (209). Arginine is believed to increase the production of NO and improve its bioavailability in the vascular endothelium, and, in this way, it regulates blood pressure (210). According to two meta-analyses, blood pressure and endothelial function were significantly enhanced in adults receiving L-arginine supplements (210, 211). The hypotensive activity of soy-containing foods may be influenced by the fact that the arginine content of soy foods is more than their lysine content. Competition between these amino acids for the same luminal transporter in the lumen is the reason that when lysine is increased relative to arginine, the uptake of the latter is limited. As a result, the bio-conversion of arginine and its downstream hypotensive effects of arginine change (212).

Overall, it appears that obesity, blood pressure, glycemic control, and inflammation are improved with isoflavones and their metabolites. Limited data using RCTs exist on the effects of other components found in soy, such as fiber, protein, lecithin, and saponins, on hypertension, which according to current observations are not effective as expected. However, a synergy between these constituents and soy protein may result in the modulation of plasma lipids as compared to just soy protein consumption. Although convincing mechanistic evidence shows that minor soy constituents exert an effect on glycemic indices, there are few human studies in this regard. Promising results regarding the effect of isoflavones on adiposity have been reported mainly in animal studies; however, soy and milk proteins do not seem to differ in terms of their effects on body composition. According to a limited body of literature, satiety may be improved by soy protein and its other components. Collectively, these findings demonstrate that, besides soy proteins, several other bioactive constituents of soy are involved in reducing CVD risk as well. These components need to be assessed individually and the role of the microbiome in alleviating these effects should be clarified if we want to define exactly how health-promoting these novel bioactive compounds are.

Probiotics

The WHO defines probiotics as "microbes with a beneficial impact on the health of the individual, should sufficient amounts are consumed" (213). It was in the early 1900s when the Nobel Prize winner Elie Metchnikoff, in her study about the longevity of Bulgarian peasants, first theorized that certain live microorganisms might have a beneficial effect on the human body (214). Probiotic products available to consumers are versatile, including fermented milk and yogurt products and different types of supplements. *Lactobacillus* sp., *Bifidobacterium* sp., *Enterococcus* sp., and *Streptococcus* sp. are four general species of lactic acid bacteria usually used in these products. Each product is different in terms of the composition and probiotic bacteria type (215).

It was in the 1970s that the effects of fermented milk products containing a wild Lactobacillus strain on lipid profile were discovered by Mann. As he suggested, inhibition of acetate conversion into cholesterol was the mechanism of action (216). Following this study, numerous datasets reported the cholesterol-lowering effects of lactic acid bacteria (LAB), particularly Lactobacillus and Bifidobacterium strains (217). Many other studies successfully repeated the same results and showed considerable reductions in LDL-cholesterol and total cholesterol levels after consumption of probiotics (218–220). Some mechanisms appear to justify this hypocholesterolemic effect, including bile-salt hydrolase (BSH)-mediated deconjugation of bile salts, assimilation of the bacterial cell membrane by cholesterol, and production of short-chain fatty acids (221, 222).

According to recent evidence, probiotics and their metabolites can be used for high blood pressure management through mechanisms associated with LDL cholesterol and total cholesterol improvement, blood glucose and insulin resistance control, and regulation of the renin-angiotensin system (222). There is evidence that SBP is significantly reduced by some probiotic strains such as *L. helveticus*, *L. plantarum*, *Streptococcus thermophilus*, and *Lactobacillus casei* (223). Khalesi et al. conducted a systematic review of nine trials in order to determine the effects of probiotics on BP. Probiotic consumption significantly improved SBP and DBP as compared to the controls. When probiotic strains were administered in combination, the effect was more prominent as compared to single strains. Also, in studies where the baseline blood pressure was ≥130/85 mmHg, DBP improvement was greater than those with the baseline values of <130/85. Moreover, when probiotics were administered for less than eight weeks with daily dose of $<10^{11}$ CFU, reductions in blood pressure levels did not happen. Therefore, a modest improvement in BP levels seems to occur after probiotic consumption, and significant reductions happen when the baseline BP is high, with a combination of strains, daily doses $\geq 10^{11}$ CFU, and for periods equal to or longer than eight weeks. It is also suggested that future studies focus on different strains with different doses in order to confirm these conclusions (222).

Overall, probiotics seem to be among dietary supplements with many health benefits. In a report by the WHO (2001), it was mentioned that probiotic consumption has no negative acute effects (213). Based on these findings, probiotic products are turning into one of the over-the-counter diet supplements and functional foods with very high sale rates (224). Probiotics are mainly known for their effects on the gastrointestinal tract, but there is novel evidence that probiotics have beneficial effects on other parts of the human body, including the cardiovascular system (225, 226). Based on animal and human studies, probiotics can reduce cholesterol if the right type of microbes with the right dose at the right time is administered. Probiotic effects against hypertension, diabetes, and obesity have also yielded promising results. The benefits of probiotic consumption for cardiovascular functions need to be addressed in more studies. Microencapsulation and cell immobilization are new techniques used for incorporating probiotics into food, and future studies need to focus on these aspects as well (227).

Curcumin

Asian countries widely use turmeric as a preservative, natural food dye, and spice (228). The yellow color of turmeric comes from a polyphenol called curcumin (diferuloylmethane). The fraction of turmeric with yellow pigmentation has a component chemically associated with curcumin, called curcuminoids (229). Curcumin possesses antifungal, antiviral, anti-inflammatory, and antioxidant properties. According to studies, no toxicity associated with curcumin use has been reported in humans (230).

Studies have shown that serum levels of cholesterol and lipid peroxide are influenced by curcumin (231). Following administration of 500 mg curcumin per day for a week in healthy subjects, serum lipid peroxides and cholesterol levels decreased while serum HDL-C showed improvement (232). Similarly, 28 days of curcumin use (20 mg/day) in atherosclerotic patients with atherosclerosis significantly improved the serum LDL-C (decrease) and HDL (increase) levels (231). Since abnormal lipid metabolism is involved in atherosclerosis, these observations suggest that curcumin can indirectly protect against atherosclerotic diseases (231). The therapeutic effects of curcumin have been widely investigated, especially in anti-inflammatory conditions (233–235). Besides, high doses of curcumin (12 g/day) are tolerated well and are inexpensive and not toxic (236). Based on evidence, various CVDs can be potentially prevented by curcumin. Adriamycin-induced cardiotoxicity (237) and diabetic cardiovascular complications (238) can be significantly reduced by the antioxidant effects of curcumin. The pathological changes seen in atherosclerosis can be prevented by the cholesterol-lowering, anti-inflammatory, anti-proliferative, anti-thrombotic effects of curcumin (6). Furthermore, calcium homeostasis can be corrected by curcumin, which could result in a trial and ventricular arrhythmias prevention (239, 240).

There are limited rodent-based studies about the anti-inflammatory properties of curcumin in relation to vascular dysfunction. However, the effects of curcumin on endothelial inflammation have been addressed in in vitro studies. Nuclear Factor-κB (NFκB) signaling was inhibited by curcumin in human umbilical vein endothelial cells, with the resultant suppression of pro-inflammatory cytokine expression, reactive oxygen species production, and signal transducer and activator of transcription (STAT)-3 and mitogen-activated protein kinases (MAPK) (241) pathways. In another dataset using human umbilical vein endothelial cells, it was revealed that curcumin down-regulates toll-like receptors (TLR)-2 and -4, which leads to anti-inflammatory effects on high mobility group box 1 (HMGB1) protein (242). These in vitro studies along with subsequent clinical trials indicate that curcumin attenuates CVD risk factors through anti-inflammatory properties.

References

1. Organization WH. Cardiovascular disease. http://wwwwhoint/cardiovascular_diseases/en/. 2017.
2. Siti HN, Kamisah Y, Kamsiah J. The role of oxidative stress, antioxidants and vascular inflammation in cardiovascular disease (a review). Vascular pharmacology. 2015; 71:40–56.
3. Pascual-Teresa D, Moreno DA, García-Viguera C. Flavanols and anthocyanins in cardiovascular health: a review of current evidence. International journal of molecular sciences. 2010;11:1679–703.
4. Organization WH. Diet, nutrition, and the prevention of chronic diseases: report of a joint WHO/FAO expert consultation: World Health Organization; Geneva, Switzerland. 2003.
5. Bjelakovic G, Gluud C. Surviving antioxidant supplements. Journal of the National Cancer Institute. 2007;99:742–3.
6. Asgary S, Rastqar A, Keshvari M. Functional food and cardiovascular disease prevention and treatment: a review. Journal of the American College of Nutrition. 2018; 37:429–55.
7. Abuajah CI, Ogbonna AC, Osuji CM. Functional components and medicinal properties of food: a review. Journal of food science and technology. 2015;52:2522–9.
8. Davidson MH, Maki KC, Dicklin MR, Feinstein SB, Witchger M, Bell M, et al. Effects of consumption of pomegranate juice on carotid intima-media thickness in men and women at moderate risk for coronary heart disease. The American journal of cardiology. 2009;104:936–42.
9. Prentice RL, Aragaki AK, Van Horn L, Thomson CA, Beresford SA, Robinson J, et al. Low-fat dietary pattern and cardiovascular disease: results from the Women's Health Initiative randomized controlled trial. The American journal of clinical nutrition. 2017;106:35–43.
10. Caprara G. Mediterranean-type dietary pattern and physical activity: The winning combination to counteract the rising burden of non-communicable diseases (NCDS). Nutrients. 2021;13:429.
11. Goldberg I. Functional foods: designer foods, pharmafoods, nutraceuticals: Springer Science & Business Media; New York. 2012.
12. Jones PJ, AbuMweis SS. Phytosterols as functional food ingredients: linkages to cardiovascular disease and cancer. Current opinion in clinical nutrition & metabolic care. 2009;12:147–51.
13. Abumweis S, Barake R, Jones P. Plant sterols/stanols as cholesterol lowering agents: a meta-analysis of randomized controlled trials. Food & nutrition research. 2008;52:1811.
14. Jones PJ, Ntanios FY, Raeini-Sarjaz M, Vanstone CA. Cholesterol-lowering efficacy of a sitostanol-containing phytosterol mixture with a prudent diet in hyperlipidemic men. The American journal of clinical nutrition. 1999;69:1144–50.
15. Sierksma A, Weststrate JA, Meijer GW. Spreads enriched with plant sterols, either esterified 4, 4-dimethylsterols or free 4-desmethylsterols, and plasma total-and LDL-cholesterol concentrations. British journal of nutrition. 1999;82:273–82.
16. Ostlund Jr RE, Racette SB, Okeke A, Stenson WF. Phytosterols that are naturally present in commercial corn oil significantly reduce cholesterol absorption in humans. The American journal of clinical nutrition. 2002;75:1000–4.
17. Krittanawong C, Isath A, Hahn J, Wang Z, Narasimhan B, Kaplin SL, et al. Fish consumption and cardiovascular health: a systematic review. The American journal of medicine. 2021;134:713–20.
18. Franco EAN, Sanches-Silva A, Ribeiro-Santos R, de Melo NR. Psyllium (Plantago ovata Forsk): from evidence of health benefits to its food application. Trends in food science & technology. 2020;96:166–75.
19. Theuwissen E, Mensink RP. Simultaneous intake of β-glucan and plant stanol esters affects lipid metabolism in slightly hypercholesterolemic subjects. The journal of nutrition. 2007;137:583–8.

20. Blair SN, Capuzzi DM, Gottlieb SO, Nguyen T, Morgan JM, Cater NB. Incremental reduction of serum total cholesterol and low-density lipoprotein cholesterol with the addition of plant stanol ester-containing spread to statin therapy. The American journal of cardiology. 2000;86:46–52.
21. De Jong A, Plat J, Bast A, Godschalk R, Basu S, Mensink R. Effects of plant sterol and stanol ester consumption on lipid metabolism, antioxidant status and markers of oxidative stress, endothelial function and low-grade inflammation in patients on current statin treatment. European journal of clinical nutrition. 2008;62:263–73.
22. Genser B, Silbernagel G, De Backer G, Bruckert E, Carmena R, Chapman MJ, et al. Plant sterols and cardiovascular disease: a systematic review and meta-analysis. European heart journal. 2012;33:444–51.
23. Rozner S, Garti N. The activity and absorption relationship of cholesterol and phytosterols. Colloids and surfaces A: physicochemical and engineering aspects. 2006; 282:435–56.
24. Rahman K, Lowe GM. Garlic and cardiovascular disease: a critical review. The journal of nutrition. 2006;136:736S–40S.
25. Ansary J, Forbes-Hernández TY, Gil E, Cianciosi D, Zhang J, Elexpuru-Zabaleta M, et al. Potential health benefit of garlic based on human intervention studies: a brief overview. Antioxidants. 2020;9:619.
26. Ashfaq F, Ali Q, Haider M, Hafeez M, Malik A. Therapeutic activities of garlic constituent phytochemicals. Biological and clinical sciences research journal. 2021; 2021:e007.
27. Rohner A, Ried K, Sobenin IA, Bucher HC, Nordmann AJ. A systematic review and metaanalysis on the effects of garlic preparations on blood pressure in individuals with hypertension. American journal of hypertension. 2015;28:414–23.
28. Xiong X, Wang P, Li S, Li X, Zhang Y, Wang J. Garlic for hypertension: a systematic review and meta-analysis of randomized controlled trials. Phytomedicine. 2015;22: 352–61.
29. Ried K. Garlic lowers blood pressure in hypertensive individuals, regulates serum cholesterol, and stimulates immunity: an updated meta-analysis and review. The journal of nutrition. 2016;146:389S–96S.
30. Amagase H, Petesch BL, Matsuura H, Kasuga S, Itakura Y. Intake of garlic and its bioactive components. The journal of nutrition. 2001;131:955S–62S.
31. Banerjee SK, Maulik SK. Effect of garlic on cardiovascular disorders: a review. Nutrition journal. 2002;1:1–14.
32. Stabler SN, Tejani AM, Huynh F, Fowkes C. Garlic for the prevention of cardiovascular morbidity and mortality in hypertensive patients. The Cochrane database of systematic reviews. 2012;2012:CD007653.
33. Lanzotti V. The analysis of onion and garlic. Journal of chromatography A. 2006; 1112:3–22.
34. Matsuura H. Saponins in garlic as modifiers of the risk of cardiovascular disease. The journal of nutrition. 2001;131:1000S–5S.
35. Corti R, Flammer AJ, Hollenberg NK, Lüscher TF. Cocoa and cardiovascular health. Circulation. 2009;119:1433–41.
36. Heiss C, Kelm M. Chocolate consumption, blood pressure, and cardiovascular risk. European heart journal. 2010;31:1554–56.
37. Fernández-Murga L, Tarín J, García-Perez M, Cano A. The impact of chocolate on cardiovascular health. Maturitas. 2011;69:312–21.
38. Keen CL, Holt RR, Polagruto JA, Wang JF, Schmitz HH. Cocoa flavanols and cardiovascular health. Phytochemistry reviews. 2002;1:231–40.
39. Chevaux KA, Jackson L, Villar ME, Mundt JA, Commisso JF, Adamson GE, et al. Proximate, mineral and procyanidin content of certain foods and beverages consumed by the Kuna Amerinds of Panama. Journal of food composition and analysis. 2001;14:553–63.

40. Schroeter H, Heiss C, Balzer J, Kleinbongard P, Keen CL, Hollenberg NK, et al. (–)-Epicatechin mediates beneficial effects of flavanol-rich cocoa on vascular function in humans. Proceedings of the National Academy of Sciences. 2006;103:1024–9.
41. Payne MJ, Hurst WJ, Miller KB, Rank C, Stuart DA. Impact of fermentation, drying, roasting, and Dutch processing on epicatechin and catechin content of cacao beans and cocoa ingredients. Journal of agricultural and food chemistry. 2010;58:10518–27.
42. Melo TS, Pires TC, Engelmann JVP, Monteiro ALO, Maciel LF, da Silva Bispo E. Evaluation of the content of bioactive compounds in cocoa beans during the fermentation process. Journal of food science and technology. 2021;58:1947–57.
43. Ried K, Sullivan TR, Fakler P, Frank OR, Stocks NP. Effect of cocoa on blood pressure. The Cochrane database of systematic reviews. 2012;8:CD008893.
44. Gianfredi V, Salvatori T, Nucci D, Villarini M, Moretti M. Can chocolate consumption reduce cardio-cerebrovascular risk? A systematic review and meta-analysis. Nutrition. 2018;46:103–14.
45. Addison S, Stas S, Hayden MR, Sowers JR. Insulin resistance and blood pressure. Current hypertension reports. 2008;10:319.
46. Davison K, Coates A, Buckley J, Howe P. Effect of cocoa flavanols and exercise on cardiometabolic risk factors in overweight and obese subjects. International journal of obesity. 2008;32:1289–96.
47. Grassi D, Lippi C, Necozione S, Desideri G, Ferri C. Short-term administration of dark chocolate is followed by a significant increase in insulin sensitivity and a decrease in blood pressure in healthy persons. The American journal of clinical nutrition. 2005;81:611–4.
48. Grassi D, Necozione S, Lippi C, Croce G, Valeri L, Pasqualetti P, et al. Cocoa reduces blood pressure and insulin resistance and improves endothelium-dependent vasodilation in hypertensives. Hypertension. 2005;46:398–405.
49. Grassi D, Desideri G, Necozione S, Lippi C, Casale R, Properzi G, et al. Blood pressure is reduced and insulin sensitivity increased in glucose-intolerant, hypertensive subjects after 15 days of consuming high-polyphenol dark chocolate. The journal of nutrition. 2008;138:1671–6.
50. Actis-Goretta L, Ottaviani JI, Fraga CG. Inhibition of angiotensin converting enzyme activity by flavanol-rich foods. Journal of agricultural and food chemistry. 2006;54: 229–34.
51. Persson IA-L, Persson K, Hägg S, Andersson RG. Effects of cocoa extract and dark chocolate on angiotensin-converting enzyme and nitric oxide in human endothelial cells and healthy volunteers—a nutrigenomics perspective. Journal of cardiovascular pharmacology. 2011;57:44–50.
52. Fraga CG, Oteiza PI. Dietary flavonoids: role of (–)-epicatechin and related procyanidins in cell signaling. Free radical biology and medicine. 2011;51:813–23.
53. Keen CL, Holt RR, Oteiza PI, Fraga CG, Schmitz HH. Cocoa antioxidants and cardiovascular health. The American journal of clinical nutrition. 2005;81:298S–303S.
54. Mehrabani S, Arab A, Mohammadi H, Amani R. The effect of cocoa consumption on markers of oxidative stress: a systematic review and meta-analysis of interventional studies. Complementary therapies in medicine. 2020;48:102240.
55. Joye IJ. Dietary fibre from whole grains and their benefits on metabolic health. Nutrients. 2020;12:3045.
56. Van der Kamp JW, Poutanen K, Seal CJ, Richardson DP. The HEALTHGRAIN definition of 'whole grain'. Food & nutrition research. 2014;58:22100.
57. Zhang D, Wang L, Tan B, Zhang W. Dietary fibre extracted from different types of whole grains and beans: a comparative study. International journal of food science & technology. 2020;55:2188–96.
58. Ferruzzi MG, Jonnalagadda SS, Liu S, Marquart L, McKeown N, Reicks M, et al. Developing a standard definition of whole-grain foods for dietary recommendations: summary report of a multidisciplinary expert roundtable discussion. Advances in nutrition. 2014;5:164–76.

59. Aune D, Keum N, Giovannucci E, Fadnes LT, Boffetta P, Greenwood DC, et al. Whole grain consumption and risk of cardiovascular disease, cancer, and all cause and cause specific mortality: systematic review and dose-response meta-analysis of prospective studies. BMJ. 2016;353:i2716.
60. Ye EQ, Chacko SA, Chou EL, Kugizaki M, Liu S. Greater whole-grain intake is associated with lower risk of type 2 diabetes, cardiovascular disease, and weight gain. The journal of nutrition. 2012;142:1304–13.
61. Mellen PB, Walsh TF, Herrington DM. Whole grain intake and cardiovascular disease: a meta-analysis. Nutrition, metabolism and cardiovascular diseases. 2008;18:283–90.
62. Liu S, Stampfer MJ, Hu FB, Giovannucci E, Rimm E, Manson JE, et al. Whole-grain consumption and risk of coronary heart disease: results from the Nurses' Health Study. The American journal of clinical nutrition. 1999;70:412–9.
63. Steffen LM, Jacobs Jr DR, Stevens J, Shahar E, Carithers T, Folsom AR. Associations of whole-grain, refined-grain, and fruit and vegetable consumption with risks of all-cause mortality and incident coronary artery disease and ischemic stroke: the Atherosclerosis Risk in Communities (ARIC) Study. The American journal of clinical nutrition. 2003;78:383–90.
64. Seal CJ. Whole grains and CVD risk. Proceedings of the Nutrition Society. 2006; 65:24–34.
65. McKeown NM, Meigs JB, Liu S, Wilson PW, Jacques PF. Whole-grain intake is favorably associated with metabolic risk factors for type 2 diabetes and cardiovascular disease in the Framingham Offspring Study. The American journal of clinical nutrition. 2002;76:390–8.
66. Barrett EM, Batterham MJ, Beck EJ. Whole grain and cereal fibre intake in the Australian Health Survey: associations to CVD risk factors. Public health nutrition. 2020;23:1404–13.
67. Swaminathan S, Dehghan M, Raj JM, Thomas T, Rangarajan S, Jenkins D, et al. Associations of cereal grains intake with cardiovascular disease and mortality across 21 countries in Prospective Urban and Rural Epidemiology study: prospective cohort study. BMJ. 2021;372:m4948.
68. Fardet A. New hypotheses for the health-protective mechanisms of whole-grain cereals: what is beyond fibre? Nutrition research reviews. 2010;23:65–134.
69. Okarter N, Liu RH. Health benefits of whole grain phytochemicals. Critical reviews in food science and nutrition. 2010;50:193–208.
70. Pol K, Christensen R, Bartels EM, Raben A, Tetens I, Kristensen M. Whole grain and body weight changes in apparently healthy adults: a systematic review and meta-analysis of randomized controlled studies. The American journal of clinical nutrition. 2013;98:872–84.
71. Priebe M, van Binsbergen J, de Vos R, Vonk RJ. Whole grain foods for the prevention of type 2 diabetes mellitus. The Cochrane database of systematic reviews. 2008;1:CD006061.
72. Pereira MA, Jacobs Jr DR, Pins JJ, Raatz SK, Gross MD, Slavin JL, et al. Effect of whole grains on insulin sensitivity in overweight hyperinsulinemic adults. The American journal of clinical nutrition. 2002;75:848–55.
73. Collaborators GRF. Global, regional, and national comparative risk assessment of 79 behavioural, environmental and occupational, and metabolic risks or clusters of risks, 1990–2015: a systematic analysis for the Global Burden of Disease Study 2015. Lancet (London, England). 2016;388:1659.
74. Kelly SA, Hartley L, Loveman E, Colquitt JL, Jones HM, Al-Khudairy L, et al. Whole grain cereals for the primary or secondary prevention of cardiovascular disease. The Cochrane database of systematic reviews. 2017;8:CD005051.
75. Hodder RK, O'Brien KM, Tzelepis F, Wyse RJ, Wolfenden L. Interventions for increasing fruit and vegetable consumption in children aged five years and under. The Cochrane database of systematic reviews. 2020;5:CD008552.

76. Peterson K, Sorensen G, Pearson M, Hébert JR, Gottlieb B, McCormick M. Design of an intervention addressing multiple levels of influence on dietary and activity patterns of low-income, postpartum women. Health education research. 2002;17:531–40.
77. Ness AR, Powles JW. Fruit and vegetables, and cardiovascular disease: a review. International journal of epidemiology. 1997;26:1–13.
78. Zurbau A, Au-Yeung F, Blanco Mejia S, Khan TA, Vuksan V, Jovanovski E, et al. Relation of different fruit and vegetable sources with incident cardiovascular outcomes: a systematic review and meta-analysis of prospective cohort studies. Journal of the American Heart Association. 2020;9:e017728.
79. Åsgård R, Rytter E, Basu S, Abramsson-Zetterberg L, Möller L, Vessby B. High intake of fruit and vegetables is related to low oxidative stress and inflammation in a group of patients with type 2 diabetes. Scandinavian journal of food and nutrition. 2007;51:149–58.
80. Dauchet L, Amouyel P, Hercberg S, Dallongeville J. Fruit and vegetable consumption and risk of coronary heart disease: a meta-analysis of cohort studies. The journal of nutrition. 2006;136:2588–93.
81. Rosenstock IM. Why people use health services. The Milbank Quarterly. 2005;83:1–32.
82. Ajzen I. The theory of planned behavior. Organizational behavior and human decision processes. 1991;50:179–211.
83. Bandura A. Social cognitive theory: an agentic perspective. Annual review of psychology. 2001;52:1–26.
84. Prochaska JO, DiClemente CC. The transtheoretical approach: crossing traditional boundaries of therapy: Krieger Publishing Company; Florida. 1994.
85. Ogden J. EBOOK: Health Psychology, 6e: McGraw Hill; New York. 2019.
86. Robinson T. Applying the socio-ecological model to improving fruit and vegetable intake among low-income African Americans. Journal of community health. 2008;33: 395–406.
87. Pollard J, Kirk SL, Cade JE. Factors affecting food choice in relation to fruit and vegetable intake: a review. Nutrition research reviews. 2002;15:373–87.
88. Hartley L, Igbinedion E, Holmes J, Flowers N, Thorogood M, Clarke A, et al. Increased consumption of fruit and vegetables for the primary prevention of cardiovascular diseases. The Cochrane database of systematic reviews. 2013;2013:CD009874.
89. Ammerman AS, Lindquist CH, Lohr KN, Hersey J. The efficacy of behavioral interventions to modify dietary fat and fruit and vegetable intake: a review of the evidence. Preventive medicine. 2002;35:25–41.
90. Rees K, Dyakova M, Ward K, Thorogood M, Brunner E. Dietary advice for reducing cardiovascular risk. The Cochrane database of systematic reviews. 2013;12:CD002128.
91. Pomerleau J, Lock K, Knai C, McKee M. Interventions designed to increase adult fruit and vegetable intake can be effective: a systematic review of the literature. The journal of nutrition. 2005;135:2486–95.
92. Brunner E, Rees K, Ward K, Burke M, Thorogood M. Dietary advice for reducing cardiovascular risk. The Cochrane database of systematic reviews. 2007;4:CD002128.
93. Motallaei M, Ramezani-Jolfaie N, Mohammadi M, Shams-Rad S, Jahanlou AS, Salehi-Abargouei A. Effects of orange juice intake on cardiovascular risk factors: a systematic review and meta-analysis of randomized controlled clinical trials. Phytother Res. 2021;35:5427–39.
94. Howard BV, Van Horn L, Hsia J, Manson JE, Stefanick ML, Wassertheil-Smoller S, et al. Low-fat dietary pattern and risk of cardiovascular disease: the Women's Health Initiative Randomized Controlled Dietary Modification Trial. JAMA. 2006;295: 655–66.
95. King JC, Blumberg J, Ingwersen L, Jenab M, Tucker KL. Tree nuts and peanuts as components of a healthy diet. The journal of nutrition. 2008;138:1736S–40S.
96. Sabaté J, Ros E, Salas-Salvadó J. Nuts: nutrition and health outcomes. British journal of nutrition. 2006;96:S1–S2.

97. Mieziene B, Emeljanovas A, Fatkulina N, Stukas R. Dietary pattern and its correlates among Lithuanian young adults: Mediterranean diet approach. Nutrients. 2020; 12:2025.
98. Sabate J, Wien M. Nuts, blood lipids and cardiovascular disease. Asia Pacific journal of clinical nutrition. 2010;19:131–6.
99. Ros E, Mataix J. Fatty acid composition of nuts–implications for cardiovascular health. British journal of nutrition. 2006;96:S29–S35.
100. Blomhoff R, Carlsen MH, Andersen LF, Jacobs DR. Health benefits of nuts: potential role of antioxidants. British journal of nutrition. 2006;96:S52–S60.
101. Brufau G, Boatella J, Rafecas M. Nuts: source of energy and macronutrients. British journal of nutrition. 2006;96:S24–S8.
102. Vinson JA, Cai Y. Nuts, especially walnuts, have both antioxidant quantity and efficacy and exhibit significant potential health benefits. Food & function. 2012;3:134–40.
103. Fraser GE, Sabate J, Beeson WL, Strahan TM. A possible protective effect of nut consumption on risk of coronary heart disease: the Adventist Health Study. Archives of internal medicine. 1992;152:1416–24.
104. Albert CM, Gaziano JM, Willett WC, Manson JE. Nut consumption and decreased risk of sudden cardiac death in the Physicians' Health Study. Archives of internal medicine. 2002;162:1382–7.
105. Mukuddem-Petersen J, Oosthuizen W, Jerling JC. A systematic review of the effects of nuts on blood lipid profiles in humans. The journal of nutrition. 2005;135:2082–9.
106. Martin N, Germanò R, Hartley L, Adler AJ, Rees K. Nut consumption for the primary prevention of cardiovascular disease. The Cochrane database of systematic reviews. 2015;9:CD011583.
107. Nishi SK, Viguiliouk E, Blanco Mejia S, Kendall CW, Bazinet RP, Hanley AJ, et al. Are fatty nuts a weighty concern? A systematic review and meta-analysis and dose–response meta-regression of prospective cohorts and randomized controlled trials. Obesity reviews. 2021;22:e13330.
108. Sabate J, Fraser GE, Burke K, Knutsen SF, Bennett H, Lindsted KD. Effects of walnuts on serum lipid levels and blood pressure in normal men. New England journal of medicine. 1993;328:603–7.
109. Mohammadifard N, Salehi-Abargouei A, Salas-Salvadó J, Guasch-Ferré M, Humphries K, Sarrafzadegan N. The effect of tree nut, peanut, and soy nut consumption on blood pressure: a systematic review and meta-analysis of randomized controlled clinical trials. The American journal of clinical nutrition. 2015;101:966–82.
110. Kris-Etherton PM, Hu FB, Ros E, Sabaté J. The role of tree nuts and peanuts in the prevention of coronary heart disease: multiple potential mechanisms. The journal of nutrition. 2008;138:1746S–51S.
111. Deka A, Vita JA. Tea and cardiovascular disease. Pharmacological research. 2011; 64:136–45.
112. Corradini E, Foglia P, Giansanti P, Gubbiotti R, Samperi R, Laganà A. Flavonoids: chemical properties and analytical methodologies of identification and quantitation in foods and plants. Natural product research. 2011;25:469–95.
113. Scalbert A, Johnson IT, Saltmarsh M. Polyphenols: antioxidants and beyond. The American journal of clinical nutrition. 2005;81:215S–7S.
114. Kris-Etherton PM, Hecker KD, Bonanome A, Coval SM, Binkoski AE, Hilpert KF, et al. Bioactive compounds in foods: their role in the prevention of cardiovascular disease and cancer. The American journal of medicine. 2002;113:71–88.
115. Stangl V, Lorenz M, Stangl K. The role of tea and tea flavonoids in cardiovascular health. Molecular nutrition & food research. 2006;50:218–28.
116. Shukla Y. Tea and cancer chemoprevention: a comprehensive review. Asian Pacific journal of cancer prevention. 2007;8:155–66.
117. Kuriyama S. The relation between green tea consumption and cardiovascular disease as evidenced by epidemiological studies. The journal of nutrition. 2008;138:1548S–53S.

118. Mineharu Y, Koizumi A, Wada Y, Iso H, Watanabe Y, Date C, et al. Coffee, green tea, black tea and oolong tea consumption and risk of mortality from cardiovascular disease in Japanese men and women. Journal of epidemiology and community health. 2011; 65:230.
119. Nagao T, Hase T, Tokimitsu I. A green tea extract high in catechins reduces body fat and cardiovascular risks in humans. Obesity. 2007;15:1473–83.
120. Chieng D, Kistler PM. Coffee and tea on cardiovascular disease (CVD) prevention. Trends in cardiovascular medicine. 2021;32:399–405.
121. Gans JMdK, Uiterwaal CSPM, Schouw YTvd, Boer JMA, Grobbee DE, Verschuren WMM, et al. Tea and coffee consumption and cardiovascular morbidity and mortality. Arteriosclerosis, thrombosis, and vascular biology. 2010;30:1665–71.
122. Abe SK, Inoue M. Green tea and cancer and cardiometabolic diseases: a review of the current epidemiological evidence. European journal of clinical nutrition. 2021;75: 865–76.
123. Arab L, Liu W, Elashoff D. Green and black tea consumption and risk of stroke. Stroke. 2009;40:1786–92.
124. Peters U, Poole C, Arab L. Does tea affect cardiovascular disease? A meta-analysis. American journal of epidemiology. 2001;154:495–503.
125. Brown AL, Lane J, Coverly J, Stocks J, Jackson S, Stephen A, et al. Effects of dietary supplementation with the green tea polyphenol epigallocatechin-3-gallate on insulin resistance and associated metabolic risk factors: randomized controlled trial. British journal of nutrition. 2008;101:886–94.
126. Fujita H, Yamagami T. Antihypercholesterolemic effect of Chinese black tea extract in human subjects with borderline hypercholesterolemia. Nutrition research. 2008;28: 450–6.
127. Greyling A, Ras RT, Zock PL, Lorenz M, Hopman MT, Thijssen DHJ, et al. The effect of black tea on blood pressure: a systematic review with meta-analysis of randomized controlled trials. PLOS one. 2014;9:e103247.
128. Gardner EJ, Ruxton CHS, Leeds AR. Black tea – helpful or harmful? A review of the evidence. European journal of clinical nutrition. 2007;61:3–18.
129. Buttriss J, Stokes C. Dietary fibre and health: an overview. Nutrition bulletin. 2008; 33:186–200.
130. Eastwood M, Kritchevsky D. Dietary fiber: how did we get where we are? Annual review of nutrition. 2005;25:1–8.
131. NHS N. Choices: non-alcoholic fatty liver disease. http://www.nhs.uk/Conditions/fatty-liver-disease/2022.
132. Hartley L, May MD, Loveman E, Colquitt JL, Rees K. Dietary fibre for the primary prevention of cardiovascular disease. The Cochrane database of systematic reviews. 2016;2016:CD011472.
133. King DE, Mainous III AG, Lambourne CA. Trends in dietary fiber intake in the United States, 1999-2008. Journal of the Academy of Nutrition and Dietetics. 2012;112:642–8.
134. Nakaji S, Sugawara K, Saito D, Yoshioka Y, MacAuley D, Bradley T, et al. Trends in dietary fiber intake in Japan over the last century. European journal of nutrition. 2002;41:222–7.
135. Shahadan SZ, Daud A, Ibrahim M, Draman S. Association between dietary macronutrient intake and high-sensitivity C- reactive protein levels among obese women in Kuantan, Malaysia. Makara journal of science. 2020;24:5.
136. Bliss DZ, Savik K, Jung H-JG, Whitebird R, Lowry A. Symptoms associated with dietary fiber supplementation over time in individuals with fecal incontinence. Nursing research. 2011;60:S58.
137. Jenkins DJ, Kendall CW, Vuksan V, Vidgen E, Parker T, Faulkner D, et al. Soluble fiber intake at a dose approved by the US Food and Drug Administration for a claim of health benefits: serum lipid risk factors for cardiovascular disease assessed in a randomized controlled crossover trial. The American journal of clinical nutrition. 2002;75:834–9.

138. Vuksan V, Jenkins AL, Jenkins DJ, Rogovik AL, Sievenpiper JL, Jovanovski E. Using cereal to increase dietary fiber intake to the recommended level and the effect of fiber on bowel function in healthy persons consuming North American diets. The American journal of clinical nutrition. 2008;88:1256–62.
139. Threapleton DE, Greenwood DC, Evans CE, Cleghorn CL, Nykjaer C, Woodhead C, et al. Dietary fibre intake and risk of cardiovascular disease: systematic review and meta-analysis. BMJ. 2013;347:f6879.
140. Satija A, Hu FB. Cardiovascular benefits of dietary fiber. Current atherosclerosis reports. 2012;14:505–14.
141. Lattimer JM, Haub MD. Effects of dietary fiber and its components on metabolic health. Nutrients. 2010;2:1266–89.
142. Kim Y, Je Y. Dietary fiber intake and total mortality: a meta-analysis of prospective cohort studies. American journal of epidemiology. 2014;180:565–73.
143. Buil-Cosiales P, Zazpe I, Toledo E, Corella D, Salas-Salvado J, Diez-Espino J, et al. Fiber intake and all-cause mortality in the Prevención con Dieta Mediterránea (PREDIMED) study. The American journal of clinical nutrition. 2014;100:1498–507.
144. Eshak ES, Iso H, Date C, Kikuchi S, Tamakoshi A, Watanabe Y, et al. Dietary fiber intake is associated with reduced risk of mortality from cardiovascular disease among Japanese men and women. The journal of nutrition. 2010;140:1445–53.
145. Kokubo Y, Iso H, Saito I, Yamagishi K, Ishihara J, Inoue M, et al. Dietary fiber intake and risk of cardiovascular disease in the Japanese population: the Japan Public Health Center-based study cohort. European journal of clinical nutrition. 2011;65:1233–41.
146. Janzi S, Dias JA, Martinsson A, Sonestedt E. Association between dietary fiber intake and risk of incident aortic stenosis. Nutrition, metabolism and cardiovascular diseases. 2020;30:2180–5.
147. Liu S, Buring JE, Sesso HD, Rimm EB, Willett WC, Manson JE. A prospective study of dietary fiber intake and risk of cardiovascular disease among women. Journal of the American College of Cardiology. 2002;39:49–56.
148. Wallström P, Sonestedt E, Hlebowicz J, Ericson U, Drake I, Persson M, et al. Dietary fiber and saturated fat intake associations with cardiovascular disease differ by sex in the Malmö Diet and Cancer Cohort: a prospective study. PloS one. 2012;7:e31637.
149. Berg A, König D, Deibert P, Grathwohl D, Berg A, Baumstark MW, et al. Effect of an oat bran enriched diet on the atherogenic lipid profile in patients with an increased coronary heart disease risk. Annals of nutrition and metabolism. 2003;47:306–11.
150. Saltzman E, Das SK, Lichtenstein AH, Dallal GE, Corrales A, Schaefer EJ, et al. An oat-containing hypocaloric diet reduces systolic blood pressure and improves lipid profile beyond effects of weight loss in men and women. The journal of nutrition. 2001;131:1465–70.
151. Reyna-Villasmil N, Bermúdez-Pirela V, Mengual-Moreno E, Arias N, Cano-Ponce C, Leal-Gonzalez E, et al. Oat-derived β-glucan significantly improves HDLC and diminishes LDLC and non-HDL cholesterol in overweight individuals with mild hypercholesterolemia. American journal of therapeutics. 2007;14:203–12.
152. Keenan JM, Pins JJ, Frazel C, Moran A, Turnquist L. Oat ingestion reduces systolic and diastolic blood pressure in patients with mild or borderline hypertension: a pilot trial. The journal of family practice. 2002;51:369.
153. Streppel MT, Arends LR, van't Veer P, Grobbee DE, Geleijnse JM. Dietary fiber and blood pressure: a meta-analysis of randomized placebo-controlled trials. Archives of internal medicine. 2005;165:150–6.
154. Brown L, Rosner B, Willett WW, Sacks FM. Cholesterol-lowering effects of dietary fiber: a meta-analysis. The American journal of clinical nutrition. 1999;69:30–42.
155. Hajishafiee M, Saneei P, Benisi-Kohansal S, Esmaillzadeh A. Cereal fibre intake and risk of mortality from all causes, CVD, cancer and inflammatory diseases: a systematic review and meta-analysis of prospective cohort studies. British journal of nutrition. 2016;116:343–52.

156. Evans CE, Greenwood DC, Threapleton DE, Cleghorn CL, Nykjaer C, Woodhead CE, et al. Effects of dietary fibre type on blood pressure: a systematic review and meta-analysis of randomized controlled trials of healthy individuals. Journal of hypertension. 2015;33:897–911.
157. Pawlosky RJ, Hibbeln JR, Novotny JA, Salem N. Physiological compartmental analysis of α-linolenic acid metabolism in adult humans. Journal of lipid research. 2001; 42:1257–65.
158. Wang Z-G, Zhu Z-Q, He Z-Y, Cheng P, Liang S, Chen A-M, et al. Endogenous conversion of n-6 to n-3 polyunsaturated fatty acids facilitates the repair of cardiotoxin-induced skeletal muscle injury in fat-1 mice. Aging (Albany NY). 2021;13:8454.
159. Nettleton JA. ω-3 fatty acids: comparison of plant and seafood sources in human nutrition. Journal of the American Dietetic Association. 1991;91:331–7.
160. Simopoulos AP, Norman HA, Gillaspy JE, Duke JA. Common purslane: a source of omega-3 fatty acids and antioxidants. Journal of the American College of Nutrition. 1992;11:374–82.
161. Jaca A, Durão S, Harbron J. Omega-3 fatty acids for the primary and secondary prevention of cardiovascular disease. South African medical journal. 2020;110:1158–9.
162. Bourdon J, Bazinet T, Arnason T, Kimpe L, Blais J, White P. Polychlorinated biphenyls (PCBs) contamination and aryl hydrocarbon receptor (AhR) agonist activity of Omega-3 polyunsaturated fatty acid supplements: implications for daily intake of dioxins and PCBs. Food and chemical toxicology. 2010;48:3093–7.
163. Joint F, Additives WECoF. Summary of the Fifty-seventh Meeting of the Joint FAO. WHO Expert Committee on Food Additives (JECFA), Rome. 2001;514.
164. U.S. Department of Health and Human Services, U.S. Environmental Protection Agency. What you need to know about mercury in fish and shellfish. Available at: https://nepis.epa.gov/Exe/ZyNET.exe/2004.
165. US Food and Drug Administration. Letter regarding dietary supplement health claim for omega-3 fatty acids and coronary heart disease (docket no. 91n-0103). Available at: Accessed December. 2000;7.
166. Abdelhamid AS, Brown TJ, Brainard JS, Biswas P, Thorpe GC, Moore HJ, et al. Omega-3 fatty acids for the primary and secondary prevention of cardiovascular disease. The Cochrane database of systematic reviews. 2018;11:CD003177.
167. Turkekul B, Gunden C, Abay C, Miran B. A market share analysis of virgin olive oil producer countries with special respect to competitiveness. European Association of Agricultural Economists (EAAE); Spain. 2007.
168. Boskou D, Blekas G, Tsimidou M. Olive oil composition. Olive Oil: Elsevier; New York. 2006. pp. 41–72.
169. Ortega R. Importance of functional foods in the Mediterranean diet. Public health nutrition. 2006;9:1136–40.
170. Bellumori M, Cecchi L, Innocenti M, Clodoveo ML, Corbo F, Mulinacci N. The EFSA health claim on olive oil polyphenols: acid hydrolysis validation and total hydroxytyrosol and tyrosol determination in Italian virgin olive oils. Molecules. 2019;24:2179.
171. Leopold JA, Loscalzo J. Oxidative risk for atherothrombotic cardiovascular disease. Free radical biology and medicine. 2009;47:1673–706.
172. Lubos E, Loscalzo J, Handy DE. Glutathione peroxidase-1 in health and disease: from molecular mechanisms to therapeutic opportunities. Antioxidants & redox signaling. 2011;15: 1957–97.
173. Ruano J, López-Miranda J, de la Torre R, Delgado-Lista J, Fernández J, Caballero J, et al. Intake of phenol-rich virgin olive oil improves the postprandial prothrombotic profile in hypercholesterolemic patients. The American journal of clinical nutrition. 2007; 86:341–6.
174. Covas M-I, Nyyssönen K, Poulsen HE, Kaikkonen J, Zunft H-JF, Kiesewetter H, et al. The effect of polyphenols in olive oil on heart disease risk factors: a randomized trial. Annals of internal medicine. 2006;145:333–41.

175. Borges TH, Cabrera-Vique C, Seiquer I. Antioxidant properties of chemical extracts and bioaccessible fractions obtained from six Spanish monovarietal extra virgin olive oils: assays in Caco-2 cells. Food & function. 2015;6:2375–83.
176. Kouka P, Priftis A, Stagos D, Angelis A, Stathopoulos P, Xinos N, et al. Assessment of the antioxidant activity of an olive oil total polyphenolic fraction and hydroxytyrosol from a Greek Olea europea variety in endothelial cells and myoblasts. International journal of molecular medicine. 2017;40:703–12.
177. Quintero-Flórez A, Pereira-Caro G, Sánchez-Quezada C, Moreno-Rojas JM, Gaforio JJ, Jimenez A, et al. Effect of olive cultivar on bioaccessibility and antioxidant activity of phenolic fraction of virgin olive oil. European journal of nutrition. 2018;57:1925–46.
178. Ghibu S, Morgovan C, Vostinaru O, Olah N, Mogosan C, Muresan A. 0347: Diuretic, antihypertensive and antioxidant effect of olea europaea leaves extract, in rats. Archives of cardiovascular diseases supplements. 2015;7:184.
179. Lockyer S, Rowland I, Spencer JPE, Yaqoob P, Stonehouse W. Impact of phenolic-rich olive leaf extract on blood pressure, plasma lipids and inflammatory markers: a randomised controlled trial. European journal of nutrition. 2017;56:1421–32.
180. Romero M, Toral M, Gómez-Guzmán M, Jiménez R, Galindo P, Sánchez M, et al. Antihypertensive effects of oleuropein-enriched olive leaf extract in spontaneously hypertensive rats. Food & function. 2016;7:584–93.
181. Schwingshackl L, Christoph M, Hoffmann G. Effects of olive oil on markers of inflammation and endothelial function—a systematic review and meta-analysis. Nutrients. 2015;7:7651–75.
182. Valero-Muñoz M, Martín-Fernández B, Ballesteros S, de la Fuente E, Quintela JC, Lahera V, et al. Protective effect of a pomace olive oil concentrated in triterpenic acids in alterations related to hypertension in rats: mechanisms involved. Molecular nutrition & food research. 2014;58:376–83.
183. Rodriguez-Rodriguez R, Perona JS, Herrera MD, Ruiz-Gutierrez V. Triterpenic compounds from "orujo" olive oil elicit vasorelaxation in aorta from spontaneously hypertensive rats. Journal of agricultural and food chemistry. 2006;54:2096–102.
184. Teres S, Barceló-Coblijn G, Benet M, Alvarez R, Bressani R, Halver JE, et al. Oleic acid content is responsible for the reduction in blood pressure induced by olive oil. Proceedings of the National Academy of Sciences. 2008;105:13811–6.
185. Mehmood A, Usman M, Patil P, Zhao L, Wang C. A review on management of cardiovascular diseases by olive polyphenols. Food science & nutrition. 2020;8:4639–55.
186. Massaro M, Scoditti E, Carluccio MA, Calabriso N, Santarpino G, Verri T, et al. Effects of olive oil on blood pressure: epidemiological, clinical, and mechanistic evidence. Nutrients. 2020;12:1548.
187. Ramdath DD, Padhi EMT, Sarfaraz S, Renwick S, Duncan AM. Beyond the cholesterol-lowering effect of soy protein: a review of the effects of dietary soy and its constituents on risk factors for cardiovascular disease. Nutrients. 2017;9:324.
188. Singh P, Kumar R, Sabapathy S, Bawa A. Functional and edible uses of soy protein products. Comprehensive reviews in food science and food safety. 2008;7:14–28.
189. Torres N, Torre-Villalvazo I, Tovar AR. Regulation of lipid metabolism by soy protein and its implication in diseases mediated by lipid disorders. The journal of nutritional biochemistry. 2006;17:365–73.
190. Maki KC, Butteiger DN, Rains TM, Lawless A, Reeves MS, Schasteen C, et al. Effects of soy protein on lipoprotein lipids and fecal bile acid excretion in men and women with moderate hypercholesterolemia. Journal of clinical lipidology. 2010;4:531–42.
191. McCue P, Shetty K. Health benefits of soy isoflavonoids and strategies for enhancement: a review. Critical reviews in food science and nutrition. 2004;44:361–7.
192. Manach C, Scalbert A, Morand C, Rémésy C, Jiménez L. Polyphenols: food sources and bioavailability. The American journal of clinical nutrition. 2004;79:727–47.
193. McGraw NJ, Krul ES, Grunz-Borgmann E, Parrish AR. Soy-based renoprotection. World journal of nephrology. 2016;5:233.

194. Omoni AO, Aluko RE. Soybean foods and their benefits: potential mechanisms of action. Nutrition reviews. 2005;63:272–83.
195. Rebholz CM, Friedman EE, Powers LJ, Arroyave WD, He J, Kelly TN. Dietary protein intake and blood pressure: a meta-analysis of randomized controlled trials. American journal of epidemiology. 2012;176:S27–S43.
196. Jackson RL, Greiwe JS, Schwen RJ. Emerging evidence of the health benefits of S-equol, an estrogen receptor β agonist. Nutrition reviews. 2011;69:432–48.
197. Colacurci N, Chiàntera A, Fornaro F, de Novellis V, Manzella D, Arciello A, et al. Effects of soy isoflavones on endothelial function in healthy postmenopausal women. Menopause. 2005;12:299–307.
198. Patten GS, Abeywardena MY, Bennett LE. Inhibition of angiotensin converting enzyme, angiotensin II receptor blocking, and blood pressure lowering bioactivity across plant families. Critical reviews in food science and nutrition. 2016;56:181–214.
199. Taku K, Lin N, Cai D, Hu J, Zhao X, Zhang Y, et al. Effects of soy isoflavone extract supplements on blood pressure in adult humans: systematic review and meta-analysis of randomized placebo-controlled trials. Journal of hypertension. 2010;28:1971–82.
200. Liu X, Li S, Chen J, Sun K, Wang X, Wang X, et al. Effect of soy isoflavones on blood pressure: a meta-analysis of randomized controlled trials. Nutrition, metabolism and cardiovascular diseases. 2012;22:463–70.
201. Garcia Garcia M, Arenas R, Martinez C, Perez L. Usefulness of phytoestrogens in treatment of arterial hypertension: systematic review and meta-analysis. Archives of clinical hypertension. 2016;1:29–34.
202. Hügel HM, Jackson N, May B, Zhang AL, Xue CC. Polyphenol protection and treatment of hypertension. Phytomedicine. 2016;23:220–31.
203. Engberink MF, Brink EJ, Van Baak MA, Bakker SJ, Navis G, Van't Veer P, et al. Dietary protein and blood pressure: a systematic review. PLoS one. 2010;5:e12102.
204. Liu Z-M, Ho SC, Chen Y-M, Woo J. Effect of soy protein and isoflavones on blood pressure and endothelial cytokines: a 6-month randomized controlled trial among postmenopausal women. Journal of hypertension. 2013;31:384–92.
205. Welty FK, Lee KS, Lew NS, Zhou J-R. Effect of soy nuts on blood pressure and lipid levels in hypertensive, prehypertensive, and normotensive postmenopausal women. Archives of internal medicine. 2007;167:1060–7.
206. Zang Y, Zhang L, Igarashi K, Yu C. The anti-obesity and anti-diabetic effects of kaempferol glycosides from unripe soybean leaves in high-fat-diet mice. Food & function. 2015;6:834–41.
207. Salari Moghaddam A, Entezari MH, Iraj B, Askari G, Sharifi Zahabi E, Maracy MR. The effects of soy bean flour enriched bread intake on anthropometric indices and blood pressure in type 2 diabetic women: a crossover randomized controlled clinical trial. International journal of endocrinology. 2014;2014:240760.
208. Evans M, Njike VY, Hoxley M, Pearson M, Katz DL. Effect of soy isoflavone protein and soy lecithin on endothelial function in healthy postmenopausal women. Menopause. 2007;14:141–9.
209. Rajapakse NW, Giam B, Kuruppu S, Head GA, Kaye DM. Impaired l-arginine-nitric oxide pathway contributes to the pathogenesis of resistant hypertension. Clinical Science. 2019;133:2061–7.
210. Dong J-Y, Qin L-Q, Zhang Z, Zhao Y, Wang J, Arigoni F, et al. Effect of oral L-arginine supplementation on blood pressure: a meta-analysis of randomized, double-blind, placebo-controlled trials. American heart journal. 2011;162:959–65.
211. Bai Y, Sun L, Yang T, Sun K, Chen J, Hui R. Increase in fasting vascular endothelial function after short-term oral L-arginine is effective when baseline flow-mediated dilation is low: a meta-analysis of randomized controlled trials. The American journal of clinical nutrition. 2009;89:77–84.
212. Vasdev S, Gill V. The antihypertensive effect of arginine. International journal of angiology. 2008;17:07–22.

213. Hotel ACP, Cordoba A. Health and nutritional properties of probiotics in food including powder milk with live lactic acid bacteria. Prevention. 2001;5:1–10.
214. Kopp-Hoolihan L. Prophylactic and therapeutic uses of probiotics: a review. Journal of the American Dietetic Association. 2001;101:229–41.
215. Parvez S, Malik KA, Ah Kang S, Kim HY. Probiotics and their fermented food products are beneficial for health. Journal of applied microbiology. 2006;100:1171–85.
216. Mann GV. A factor in yogurt which lowers cholesteremia in man. Atherosclerosis. 1977;26:335–40.
217. Tsai C-C, Lin P-P, Hsieh Y-M, Zhang Z-y, Wu H-C, Huang C-C. Cholesterol-lowering potentials of lactic acid bacteria based on bile-salt hydrolase activity and effect of potent strains on cholesterol metabolism in vitro and in vivo. The Scientific World Journal. 2014;2014:690752.
218. Fava F, Lovegrove J, Gitau R, Jackson K, Tuohy K. The gut microbiota and lipid metabolism: implications for human health and coronary heart disease. Current medicinal chemistry. 2006;13:3005–21.
219. Kiessling G, Schneider J, Jahreis G. Long-term consumption of fermented dairy products over 6 months increases HDL cholesterol. European journal of clinical nutrition. 2002;56:843–9.
220. Ejtahed HS, Mohtadi-Nia J, Homayouni-Rad A, Niafar M, Asghari-Jafarabadi M, Mofid V. Probiotic yogurt improves antioxidant status in type 2 diabetic patients. Nutrition. 2012; 28:539–43.
221. Thushara RM, Gangadaran S, Solati Z, Moghadasian MH. Cardiovascular benefits of probiotics: a review of experimental and clinical studies. Food & function. 2016; 7:632–42.
222. Khalesi S, Sun J, Buys N, Jayasinghe R. Effect of probiotics on blood pressure: a systematic review and meta-analysis of randomized, controlled trials. Hypertension. 2014;64:897–903.
223. Ebel B, Lemetais G, Beney L, Cachon R, Sokol H, Langella P, et al. Impact of probiotics on risk factors for cardiovascular diseases. A review. Critical reviews in food science and nutrition. 2014;54:175–89.
224. Sanders ME. Probiotics: definition, sources, selection, and uses. Clinical infectious diseases. 2008;46:S58–S61.
225. Shahriari A, Karimi E, Shahriari M, Aslani N, khooshideh M, Arab A. The effect of probiotic supplementation on the risk of gestational diabetes mellitus among high-risk pregnant women: A parallel double-blind, randomized, placebo-controlled clinical trial. Biomedicine & pharmacotherapy. 2021;141:111915.
226. Hadi A, Arab A, Khalesi S, Rafie N, Kafeshani M, Kazemi M. Effects of probiotic supplementation on anthropometric and metabolic characteristics in adults with metabolic syndrome: a systematic review and meta-analysis of randomized clinical trials. Clinical nutrition. 2021;40:4662–73.
227. Saini R, Saini S. Probiotics: the health boosters. Journal of cutaneous and aesthetic surgery. 2009;2:112.
228. Sasikumar B. Genetic resources of Curcuma: diversity, characterization and utilization. Plant genetic resources. 2005;3:230–51.
229. Agarwal S, Mishra R, Gupta AK, Gupta A. Turmeric: isolation and synthesis of important biological molecules. Synthesis of Medicinal Agents from Plants: Elsevier; Amsterdam, The Netherlands. 2018. pp. 105–25.
230. Akram M, Shahab-Uddin AA, Usmanghani K, Hannan A, Mohiuddin E, Asif M. Curcuma longa and curcumin: a review article. Romanian journal of plant biology. 2010;55:65–70.
231. Soler A, Carrión-Gutiérrez M, Diaz-Alperi J, Bernd A, Miquel J. An hydroalcoholic extract of Curcuma longa lowers the abnormally high values of human-plasma fibrinogen. Mechanisms of ageing and development. 2000;114:207–10.
232. Soni K, Kutian R. Effecf of oral curcumin administranon on serum peroxides and cholesterol levels in human volunteers. Indian journal of physiology and pharmacology. 1992;36:273–5.

233. Duvoix A, Blasius R, Delhalle S, Schnekenburger M, Morceau F, Henry E, et al. Chemopreventive and therapeutic effects of curcumin. Cancer letters. 2005;223:181–90.
234. Rezvanirad A, Mardani M, Ahmadzadeh SM, Asgary S, Naimi A, Mahmoudi G. Curcuma longa: A review of therapeutic effects in traditional and modern medical references. Journal of chemical and pharmaceutical sciences. 2016;9:3438–48.
235. Rahmani S, Asgary S, Askari G, Keshvari M, Hatamipour M, Feizi A, et al. Treatment of non-alcoholic fatty liver disease with curcumin: a randomized placebo-controlled trial. Phytotherapy research. 2016;30:1540–8.
236. Cheng A, Hsu C, Lin J, Hsu M, Ho Y, Shen T, et al. Phase 1 clinical trial of curcumin, a chemopreventive agent, in patients with high-risk of pre-malignant lesions. Anticancer research. 2001;21:2895–2900.
237. Al Fatease A, Shah V, Nguyen DX, Cote B, LeBlanc N, Rao DA, et al. Chemosensitization and mitigation of adriamycin-induced cardiotoxicity using combinational polymeric micelles for co-delivery of quercetin/resveratrol and resveratrol/curcumin in ovarian cancer. Nanomedicine: nanotechnology, biology and medicine. 2019;19:39–48.
238. Daosukho C, Chen Y, Noel T, Sompol P, Nithipongvanitch R, Velez JM, et al. Phenylbutyrate, a histone deacetylase inhibitor, protects against adriamycin-induced cardiac injury. Free radical biology and medicine. 2007;42:1818–25.
239. Schoonderwoerd BA, Smit MD, Pen L, Van Gelder IC. New risk factors for atrial fibrillation: causes of 'not-so-lone atrial fibrillation'. Europace. 2008;10:668–73.
240. Phrommintikul A, Chattipakorn N. Roles of cardiac ryanodine receptor in heart failure and sudden cardiac death. International journal of cardiology. 2006;112:142–52.
241. Kim YS, Ahn Y, Hong MH, Joo SY, Kim KH, Sohn IS, et al. Curcumin attenuates inflammatory responses of TNF-α-stimulated human endothelial cells. Journal of cardiovascular pharmacology. 2007;50:41–9.
242. Kim D-C, Lee W, Bae J-S. Vascular anti-inflammatory effects of curcumin on HMGB1-mediated responses in vitro. Inflammation research. 2011;60:1161–8.

Functional Foods and Natural Products for Obesity Management

Idris Adewale Ahmed, Najihah Mohd Hashim, and Rozana Othman

Contents

Background 231
Etiology of obesity 233
Clinical presentation and assessment of obesity 234
Pathogenesis of obesity 235
Complications of obesity 236
Current medical management of obesity 236
Natural products and functional foods for obesity management 238
Studies supporting the anti-obesity potential of functional foods 241
Common myths about obesity 243
Conclusion 244
References 245

Background

Chronic and noncommunicable diseases remain the greatest threat to humanity. Thus, there has always been a steady increase in the number of patients with cardiovascular diseases and diabetes associated with obesity (1). Currently, billions of people are affected by metabolic and related diseases (2). And one of the largest global health challenges currently facing the world all over is obesity. Being a metabolic syndrome indicator, obesity is closely associated with cancer, cardiovascular disease, hyperlipidemia, hypertension, and type 2 diabetes (3).

The major contributors to the global burden of chronic diseases and their complications are obesity and overweight (4). According to the World Health Organization (WHO) (5), both overweight and obesity involve abnormal or excessive fat accumulation that may impair health. Overweight is when a body mass index (BMI) is greater than or equal to 25 kg/m^2, while obesity is when a BMI is greater than or equal to 30 kg/m^2. Obesity ranks fifth among

DOI: 10.1201/9781003220053-11

the greatest causes of noncommunicable diseases (6). There are more than 1.1 billion overweight adults, which includes over 312 million obese adults. Obesity is now considered an "epidemic" of noncommunicable pathology, according to the estimates from the International Obesity Task Force (IOTF). The rate of increase in the prevalence of obesity in some industrialized countries is alarmingly doubling or almost tripling among the general population, no thanks to the rapid globalization and proliferated adoption of the so-called Western lifestyle, through the consumption of high-refined carbohydrate and high-fat diets as well as increasing sedentary and inactive lifestyle daily routines. For instance, in the United States alone, about 68% of the adult population are overweight with over half of them having health risks (7, 8).

The rates of obesity, however, are increasing faster in developing countries than in developed countries, with Asia having the highest prevalence rates of obesity among adolescents. The prevalence of obesity plateaued during the first decade of the 21st century in the United States. The overall prevalence of overweight, including obesity, in school children in European countries, was estimated at 20.5%, while the proportions of overweight and obesity were 24.5% and 11.9% in Eastern Asia countries and the Western Asia regions, respectively (9).

From an energy homeostasis perspective, more energy intake than expenditure (exercise) leads to obesity, which, in turn, reduces the quality of life and is also linked closely to a high incidence of lipidemia, diabetes, metabolic diseases, and other several life-threatening disorders (2, 3, 8). Nevertheless, there is a wide variation in the main influencing and determining factors affecting an individual's energy intake and output (10). From a clinical perspective, an excessive accumulation of body fat, which impairs health status, typifies obesity (11).

The rapidly growing body mass indices and waistlines in modern society are also associated with a precarious rise in the prevalence of obesity-associated metabolic imbalances across the globe. Both obesity and metabolic syndrome are continuously putting a strain and a great burden on the global economic and social setting, thus necessitating fundamental but sustainable changes in both lifestyle and nutritional standards (7) such as the adoption of naturally derived ingredients and anti-obesity foods (8).

The human gut microbiome, for instance, comprises trillions of microbial cells colonizing the gut through a close symbiotic relationship to the host while promoting notable immune function and physiological homeostasis. Such a huge prokaryotic population outnumbers the total human body cells by an order of magnitude. The human microbiome also mediates a variety of critical communications between the gut, the brain, and the enteric nervous system. Thus, dysbiosis, which is a perturbation to the gut microbiome, leads to different cognitive, gastrointestinal, and metabolic pathologies (7). A deviation from the normal body weight values due to lack of a balanced diet can also result in a health decline due to weight-related imbalances (12). Strict adherence to healthy eating patterns can improve metabolic health, reduce the risk of cardiovascular diseases and mortality from all causes, and, thus, has been regarded as an effective preventive approach to the transition to metabolic unhealthy obese phenotypes from metabolically healthy obesity (13). It is, therefore, very relevant to appraise the role of functional foods.

Functional foods are wholesome, enriched, fortified, or enhanced foods with potential health benefits beyond the provision of essential nutrients upon regular consumption at efficacious levels and as part of a varied diet (14). Functional foods or nutraceuticals also include processed foods with nutritive value as well as disease-preventing and/or health-promoting benefits (15). This review, thus, discusses the role of functional foods in obesity management vis-à-vis the etiology, clinical presentation and assessment, pathogenesis, complications, current medical management, commonly used natural products and functional foods, studies supporting the anti-obesity potential of functional foods, and common myths about obesity. The data and information on obesity and functional foods were collated from various resources and literature databases such as Google, Google Scholar, Inflibnet, PubMed, Science Direct, Wiley, Scopus, Springer, and Taylor & Francis.

Etiology of obesity

The role of obesity and its contributions to various chronic diseases such as cardiovascular diseases, cerebrovascular incidents, hypertension, type 2 diabetes, hyperlipidemia, and obstructive sleep apnea has been reported in the literature (16). Though excess nutrients in combination with lack of physical activity are the primary causes of obesity, other factors such as cravings, elevated BMI, endocrine disorders, hereditary, mental illness, medications, hormonal disruptors, inadequate sleep, smoking habits, pregnancy at a later age, inherited risk factors, and the variability of ambient temperature also play some contributing roles (17).

The systemic energy metabolism of the body is, however, regulated by the adipose tissues, which are of two primary types, namely, brown adipose tissue (BAT) and white adipose tissue (WAT), as illustrated in Table 9.1. WAT is

Table 9.1 Regulation of the Body Systemic Energy Metabolism

Properties	White Adipose Tissue	Brown Adipose Tissue
Morphology	Spherical	Elliptical
Localization	Subcutaneous, intra-abdominal, epicardial, and gonadal	Interscapular, paravertebral, perirenal, cervical, and supraclavicular
Cell composition	Single lipid droplet, few mitochondria, little endoplasmic reticulum, and flattened peripheral nucleus	Multiple small lipid droplets, oval central nucleus, and a large number of mitochondria
Function	Energy storing and maintenance of energy homeostasis	Heat production and energy dissipation

found in the subcutaneous, intra-abdominal, epicardial, and gonadal parts of the body, while BAT is found in the interscapular, paravertebral, perirenal, cervical, and supraclavicular body parts (18). BAT acts primarily through energy dissipation to produce heat, while WAT, being an energy storage site, is important for the maintenance of energy homeostasis, through endocrine communication. The presence of characteristically multilocular lipid droplets and large amounts of mitochondria is responsible for a high lipid oxidation rate in BAT, while WAT has a single large lipid droplet. Brite or beige adipocyte, an inducible thermogenic adipocyte, has also been found in WAT depots and shares many similar metabolic and morphologic characteristics with BAT (3).

Fat cells, besides storing energy, also secrete a variety of cytokines for the regulation of signal transduction in both adipose tissues and muscles. Adipocytes' differentiation is mainly related to changes in cell morphology, gene expression, and hormone sensitivity. The early adipogenesis stage is enhanced by the expression of certain transcription factors such as peroxisome proliferator-activated receptor γ (PPARγ) and CCAAT-enhancer-binding proteins α (C/EBPα). Sterol regulatory element-binding protein 1c (SREBP-1c) is also involved in the metabolism of fatty acid and lipid biosynthesis (8). Adiponectin suppresses obesity by phosphorylating and activating AMP-dependent protein kinase (AMPK), acting as a regulator for the maintenance of homeostasis of various cellular energetics (1).

The hypothalamus mainly regulates the energy balance by integrating nutritional signals and circulating hormones. The arcuate nucleus (ARC) of the hypothalamus particularly plays a major role in the energy balance control using its primary order neurons: orexigenic neuropeptide Y (NPY)/agouti-related peptide (AgRP) and the anorexigenic proopiomelanocortin (POMC) neurons, which sense glucose and adiposity signals. The NPY/AgRP neurons upregulate feeding behaviors and increase energy expenditures, while the POMC neurons downregulate them. The dysfunction of the melanocortin system is linked with obesity and its associated disorders such as type 2 diabetes mellitus, cancer, cardiovascular diseases, and neurodegenerative diseases (2).

Clinical presentation and assessment of obesity

Obesity is a multifactorial condition with several underlying causes, such as the type and number of calories consumed; energy expenditure; genetic predisposition; epigenetic, metabolic processes; and physiological, sociocultural, and psychosocial influences (11). Some of the clinical manifestations of obesity in males include double chin, gynecomastia, a round face, pendulous abdomen, polydactyly in the foot and hand, while females tend to show features like enlarged breast, round face, early menarche, and pendulous abdomen (19–21).

Obesity is also thought to be responsible for idiopathic genu valgum, which progresses with skeletal maturation (22).

The diagnoses of overweight and obesity are currently accomplished using either BMI or waist circumference calculation. The BMI is usually expressed as a weight (kg) divided by the square of height (m^2) (23). Admittedly, there is a concern over the accuracy of BMI for predicting health at an individual level. Nevertheless, it is useful for population measure with other measures such as waist-to-hip ratio and waist circumference (24). The categories of BMI, according to the National Heart, Lung, and Blood Institute, are shown in Table 9.2.

Table 9.2 The Categories of BMI

Body Mass Index (kg/m^2)	Body Habitus Description
<18.5	Underweight
18.5–24.9	Normal
25–29.9	Overweight
30–34.9	Obese (class I)
35–39.9	Obese (class II)
≥40	Extreme obesity (class III)

Source: National Heart, Lung, and Blood Institute (Salazar, 2006).

Pathogenesis of obesity

High caloric diets and inactive lifestyles, according to most epidemiological researchers, are responsible for the increasing prevalence of obesity (25). Obesity is caused by several factors (Figure 9.1), involving a complex interaction among environment, drugs, genetics, epigenetics, hormones, inflammation, metabolism, microbiome dysbiosis, physiological, sociocultural, and several other reasons (11, 26). Multiple candidate genes reportedly implicated in the pathogenesis of obesity include beta-3-adrenergic receptor gene,

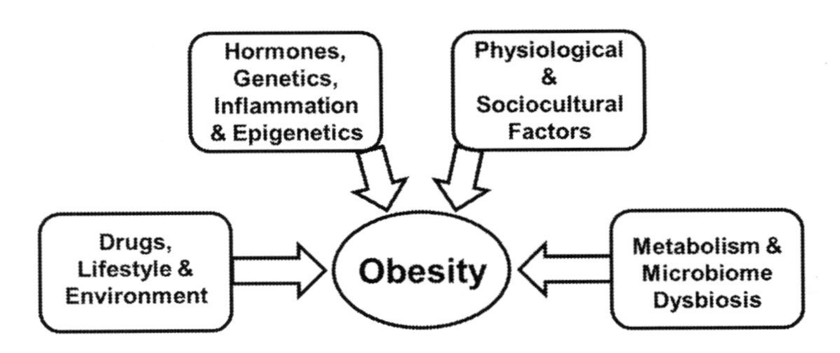

Figure 9.1 *Factors responsible for obesity.*

chromosome 10p, melanocortin-4 receptor gene, peroxisome-proliferator-activated receptor gamma 2 gene, and other genetic polymorphisms, while hormones such as adipokines gut-related hormones and others have also been implicated. Of utmost importance is ghrelin, while some drugs and neuroendocrine diseases can also lead to obesity (27). Both ghrelin and leptin internally mediate feeding and hunger in humans. Ghrelin, a peptide hormone synthesized by the epsilon cells of the pancreas and fundus lining of the stomach, regulates temporary appetite control, while leptin, another peptide hormone synthesized by adipocytes, regulates long-standing appetite and plays a key role in the storage of fat in the body (17).

Complications of obesity

The economic costs of obesity and its related health consequences are enormous, necessitating several hospitalizations, physician visits, and other related expenses (4). There is a strong link between obesity and most chronic diseases such as cancer, cardiovascular disease, hyperlipidemia, hypertension, inflammation, and type 2 diabetes (3, 24). Obesity-induced inflammatory stress stimulates cytokines release, which, in turn, leads to tissue damage (28). Other complications that have been linked to obesity include obesity-related osteoarthritis (OA), which contributes to increased morbidity and mortality. Two-thirds of obese individuals are reported to have OA, while the incidence of OA tends to increase with an increased BMI. OA is a complex biopsychosocial condition that increases the family financial burden and the healthcare economy. The most significant factor in the pathogenesis of OA is obesity-induced inflammation, as evidenced by the occurrence of OA and impaired metabolism. There are other complications, such as obesity-related systemic factors on the knee and hand. Obesity also contributes to the initiation of the degradation process of the osteoarthritic joint and cartilage (29). Obese children are not only at least twice as likely to be obese adults but also to have an increased risk of cancer, premature death, disability in adulthood, fractures, hypertension, insulin resistance, cardiovascular disease markers, and other psychological issues (24). Impotence and infertility have been linked with abdominal obesity in men, while some reproductive complications of obesity such as dystocia, gestational diabetes, macrosomia, and increased rates of cesarean sections reportedly occur in pregnancy and labor (30). Other common complications of obesity reported in the literature include chronic kidney disease, diastolic heart failure, and nonalcoholic fatty liver disease (31).

Current medical management of obesity

BMI is the basis for most guidelines for obesity treatment (23). Therefore, weight loss remains the cornerstone approach to obesity management. It occurs through the generation of a negative energy balance by consuming fewer calories while expending more energy. The major nonpharmacological recommendations for the management of obesity are behavioral therapy, lifestyle management, diet therapy, and physical activity. The use of a tiered system composed of integrated lifestyle interventions, strengthened obesity education and training, medical treatments, use of advanced electronic health technologies, as well as the involvement of health management centers could potentially improve obesity management (32). Nevertheless, pharmacotherapy, as well as endoscopic and bariatric surgery, are also commonly employed (27, 33). Previously approved drugs for the clinical management of obesity and its related metabolic syndromes, in the long term, include sibutramine, rimonabant, and orlistat. Though only orlistat is being used and approved for long-term obesity treatment, both orlistat and phentermine are the only medications approved by the US Food and Drug Administration for obesity treatment in adolescents (34). According to the literature, however, body weight and drug treatment interact in four distinct ways: unintentional influence of drugs on body weight (as a side effect), intentional and direct effect of drugs for weight reduction, alteration and adjustment of drug's dose for massively obese patients for pharmacodynamic purpose, and the alteration in drug's pharmacokinetic characteristics due to body weight's increase (Figure 9.2). Furthermore, despite obesity being a chronic disease and requiring long-term treatment, weight loss through drugs should be limited to six to eight months (35). The severe cardiovascular adverse effects of some of the drugs, such as fenfluramine, rimonabant, and sibutramine, have culminated in their withdrawal (4, 16). The role of policy makers in tackling the menace of obesity cannot be overemphasized. Governments can enact and enforce laws and regulations to modify the food environment by putting in place some restrictions, regulations, and taxation (31).

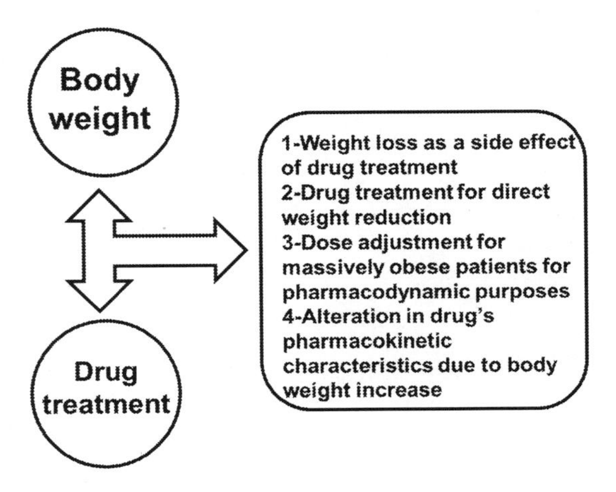

Figure 9.2 *Drug treatment-body weight interactions.*

Natural products and functional foods for obesity management

Despite the controversy regarding the extent to which changes in diet composition drive the obesity epidemic, the consumption of highly processed foods with relatively low fiber but higher levels of sugars and fats, in particular, theoretically affect energy balance. Dietary carbohydrates, especially refined sugars, tend to increase insulin secretion, which, in turn, suppresses lipolysis (36). Excessive food consumption and lack of physical activity could lead to an accumulation of the extra energy as fat (11, 37). On the other hand, several medicinal plants, fruits, and vegetables, which are natural products, are being studied both for preventive and therapeutic management of obesity and as food supplements for weight-loss promotion. Curcumin (from *Curcuma longa* rhizomes, Figure 9.3), capsaicin (from *Capsicum annuum*, Figure 9.4), and celastrol (from *Tripterygium wilfordii* roots) are common examples of plant-derived anti-obesity natural products (4).

Figure 9.3 *Curcuma longa* L. *rhizomes.*

Figure 9.4 *Capsicum annuum* L.

Common natural products such as chromium, cinnamon extract, dark chocolate, dietary fiber, flaxseed oil, ω-3 polyunsaturated fatty acids found in fish, resveratrol in red wine, soy protein, and traditional Chinese herbs have been widely used as dietary interventions against metabolic syndrome (38). On the other hand, the term "functional food" could be traced to the early 1980s when it was coined in Japan. Though there is no universal definition of functional food, it refers to "processed foods with nutritive value as well as disease-preventing and/or health-promoting benefits". In other words, functional foods overlap with other terminologies such as nutraceuticals, medical foods, pharmafoods, probiotics, and designer foods (15). They are enriched, fortified, enhanced, or wholesome foods of either animal or plant origin with several potential health benefits beyond the basic provision of essential nutrients when consumed regularly at efficacious levels and as part of a varied diet (14). Functional foods are potent dietary supplementations that not only support weight management but also protect against the metabolic consequences of obesity.

Honey (Figure 9.5), for instance, is a nutrient-rich, great natural sweetener and medicinal food with several health benefits. The daily supplementation of honey reportedly reversed the formation of hepatic steatosis due to their antioxidative and lipid-lowering effects as well as weight-reducing ability in obese induced rats (39). The anti-obesity potential of honey has been reported in several in vivo and randomized clinical trials (38). The most commercialized honey throughout the world is produced by the Apis bees (*Apis mellifera*). Other honey-producing bees, albeit less known, are the stingless bees within the family Apidae, subfamily Meliponinae, and tribe Meliponini. The Apis bees

Figure 9.5 *Honey.*

are found almost everywhere except in Antarctica and desert parts of the world, while stingless bees, with about 500 species across 32 genera, are mainly found in subtropical and tropical regions. The most common species of stingless bees with high economic importance include *Meliponula togoensis, M. bocandei, M. lendliana, M. ferruginea, Liotrigona* sp., and *Plebeina armata* (40), as well as *Tetragonula laeviceps* and *T. biroi* (41). In Malaysia, the stingless bee *Trigona* spp. produces Kelulut honey, while the rock bee (*Apis dorsata*) produces a tropical multifloral rain forest Tualang honey (42).

The increasing consumption of nuts, on a daily basis, is associated with a lower risk of obesity and less long-term weight gain in adults (43). The consumption of nuts (Figure 9.6), in the long term, is also associated with lower weight gain due to their richness in proteins, dietary fiber, and unsaturated fats, all of which increase thermogenesis and resting energy expenditure. Nuts contain a variety of antioxidants, minerals, phytosterols, and vitamins (44). However, comprehensive information about the safety and potential risks of nuts and seeds is important because of potential and unexpected severe IgE sensitization and allergic reactions among children (45).

Figure 9.6 *Common commercial nuts.*

The anti-obesity of both dietary and medicinal mushrooms has also been recognized. *Auricularia, Flammulina, Grifola, Hericium, Lentinus, Pleurotus,* and *Tremella* species are common dietary mushrooms with beneficial effects on the gut microbiota, while *Ganoderma* and *Trametes* are some of the commonly used medicinal mushrooms. Mushrooms are rich in fiber, high-quality polyunsaturated fatty acids, protein, minerals, and vitamins (17).

Polyphenols, which are found in various fruits, herbs, and vegetables, are one of the most prominent bioactive compounds. The bioavailability of phenolic compounds is the main determinant for their preventative potential (28). Furthermore, methylxanthines and chlorogenic acids (phenolic acid derivatives) from animals and plants (such as caffeine, theobromine, and theophylline) are also getting increasing attention. Their anti-obesity activity is reported to be through nonspecific antagonism of adenosine receptors (A_1R) lipolysis stimulation (11). The common mechanisms of most active ingredients include decreasing adipogenesis and enhancing energy expenditure, interfering with nutrient absorption, modifying the intestinal microbiota composition, suppressing the appetite, and increasing fat excretion (16). Some common natural products and functional foods for the management of obesity are highlighted in Tables 9.3 and 9.4, respectively.

Studies supporting the anti-obesity potential of functional foods

The treatment of differentiated Murine 3T3-L1 cells with some flavonoid compounds (quercitrin, isoquercitrin, kaempferol, and afzelin) reportedly decreased triglyceride levels and significantly downregulated several adipogenic transcription factors, inhibited adipogenesis in pre-adipocytes, reduced intracellular lipid accumulation in mature adipocytes, and resulted in lower expression of lipogenesis-related proteins, while resveratrol reversed free fatty acid-induced insulin resistance (28). *Moringa oleifera* leaves also showed anti-obesity activities through bodyweight loss, improving the lipid profiles, regulating significant genes associated with adipogenesis, insulin resistance, glucose uptake, and hormones in different studies, including clinical trials where it impacted on BMI, low-density lipoprotein, total cholesterol, and postprandial blood glucose (50).

Several microalgae such as *Euglena gracilis, Phaeodactylum tricornutum, Spirulina platensis, Spirulina maxima,* and *Nitzschia laevis* have also been reported to have anti-obesity effects in vitro and in vivo through inhibition of pre-adipocyte differentiation, reduction of de novo lipogenesis, and triglyceride assembly (49).

The extracts of *Coptis chinensis, Mahonia aquifolium, Berberis vulgaris,* and *Chelidonium majus* containing berberine and other alkaloids resulted in reduced adipocyte differentiation, neutral lipid content, and lipolysis rate (51). Ginseng seed oil attenuated intracellular triglyceride levels and accumulation

Table 9.3 Common Natural Products for Obesity Management

Natural Products	Sources	Functions	References
Methylxanthines	Tea leaves (*Camellia sinensis* L.), cocoa (*Theobroma cacao* L.), and coffee beans (*Coffea* sp.)	Multiple physiological effects in the human body	(11)
Chlorogenic acids	Coffee and tea	Anti-inflammatory activity, protective role against diabetes mellitus and obesity through modulation of lipid and glucose metabolism	(11)
Prebiotics & Probiotics	Prebiotics are nondigestible food ingredients. Probiotics such as *Lactobacillus* spp., *Bifidobacterium* spp., and *Akkermansia muciniphila*	For reducing fat deposition and food intake, improving energy metabolism, treating and enhancing insulin sensitivity, and treating obesity	(4, 46)
Calcium Supplements	Dairy products	Inhibition of adipogenic differentiation, lipogenesis, and acceleration of lipolysis	(47)
Cinnamic Acids	Cinnamon bark or benzoin	Improvement of blood lipid profile, hepatic steatosis, and adipose hypertrophy	(48)
Microbial Products	Lipstatin (from *Streptomyces toxytricini*), gut microbiota, and probiotic bacteria	For prolonging satiation, reducing food intake and fat deposition, improving energy metabolism, treating and enhancing insulin sensitivity, and treating obesity	(4)
Resveratrol	Grapes and red wine	Anti-obesity, antidiabetic, and feeding behavior modulations	(25)
Marine Products	Palinurin (from sponge *Ircinia dendroides*), callyspongynic acid (from sponge *Callyspongia truncate*), dysidine (from sponge *Dysidea* sp.), questinol and citreorosein (from sponge *Stylissa flabelliformis*), brown algae, green algae, and microalgae	Increase in energy expenditure, appetite suppressant effect, inhibition of digestive enzyme activity, regulation of adipocyte differentiation, and lipid metabolism	(4)
Epigallocatechin gallate	Green tea	Suppression of pre-adipocyte proliferation, inhibition of adipocyte differentiation and adipogenesis, stimulation of lipolysis and fatty acid β-oxidation	(25)

of lipid droplets in both HepG2 cells and rat hepatocytes (52). Similarly, the extracts of *Centella asiatica* (Asian pennywort), *Morinda citrifolia* fruit (noni), and *Momordica charantia* (bitter gourd) have been reported to inhibit the activity of lipoprotein lipase in vitro (53). The anti-obesity and antioxidant activities of *Cosmos caudatus* ethanolic extract have also been reported (54). Functional snacks with the addition of nanoencapsulated resveratrol prepared

Table 9.4 Common Functional Foods for Obesity Management

Functional Foods	Sources	Functions	References
Honey	A naturally sweet substance produced by *Apis mellifera* bees from different materials	Antioxidative, lipid-lowering effects and weight-reducing ability	(38)
Nuts	Nuts	Increases oxidation, decreases body fat accumulation, increases thermogenesis and resting energy expenditure, delays gastric emptying and subsequent absorption, and suppresses hunger	(44)
Mushrooms	Dietary and medicinal mushrooms	Augments anti-obesity effects, regulates dysbiosis and composition of the microbiota	(17)
Microalgae	*Euglena gracilis, Phaeodactylum tricornutum, Spirulina platensis, Spirulina maxima,* or *Nitzschia laevis*	Inhibits pre-adipocyte differentiation, reduces de novo lipogenesis and triglyceride assembly	(49)

from horse-chestnut, water-chestnut, and lotus-stem starch particles demonstrated significantly higher antioxidant, anti-obesity, and antidiabetic properties than snacks containing no or free resveratrol (55). Some probiotics such as *Bifidobacterium*, most *Lactobacillus*, and some *Bacteroidetes* reportedly show anti-obesity activities, while some dietary plants, such as apple, berries, grapes, chili, soy, sorghum, turmeric, and barley, show anti-obesity efficacy by downregulating obesogenic gut microbiota, upregulating anti-obesity gut microbiota, and increasing the diversity of gut microbiota (56). The justification for concluding causation and treatment effects from nonexperimental data are rarely met in both clinical and public proposals with regards to obesity. Furthermore, observational associations and relationships germane to the causes, prevention, and treatment of obesity are subject to substantial confounding and inconsistency, and are fraught with measurement problems (57). Therefore, it would be relevant to review some of the common myths about obesity and its management.

Common myths about obesity

The sheer number of existing anti-obesity diet regimens and the increasing obesity pandemic would suggest that no single diet has been universally successful at maintaining or inducing weight loss. Most of these dietary programs do follow contemporary principles of weight loss and are grounded on sound scientific pieces of evidence, while others simply recommend the consumption of a particular food type at the expense of other foods or disregard one or more essential food groups (58). Thus, there are a few but common myths and presumptions about obesity, some of which (such as the effects of eating breakfast daily, snacking, and eating more fruits and vegetables) can be tested with standard study designs. Most of the findings seem to be indefinite, while some of the trials have either been completed or are in progress. One of the most common myths about obesity management is that it is very important to set realistic goals for weight loss to prevent patients from becoming frustrated and thus lose less weight. Empirical data, however, indicate no consistent or reliable negative association between ambitious goals and weight loss or program completion. Altering unrealistic goals led to more realistic weight-loss expectations but did not improve outcomes in some studies. Another myth is that breastfeeding is protective against obesity. On the contrary, existing data indicate that though breastfeeding should be encouraged because it has important potential benefits both for the infant and mother, it, however, does not have important anti-obesity effects in children (57). It is a myth that losing weight quickly will predispose to greater weight regain relative to losing weight more slowly. Though rapid weight loss may lead to other serious health problems, only permanent lifestyle changes and adoption of a healthier way of life, such as increasing physical activity and making healthful food choices, promote long-term weight loss. One other common myth is that genes do not contribute to the obesity epidemic. Obesity like other complex diseases, however, is caused by a complex interaction between environmental, genetic, and behavioral factors. Though there is certainly an important genetic component to obesity, changes in environmental influences also have a major impact on the recent epidemic of obesity (59).

Conclusion

Obesity and its complications have several economic, psychological, and sociocultural effects. Diet, lifestyle modification, and exercise are the best approaches for both prevention and treatment. The side effects and high costs of common pharmacological drugs for treating obesity have led to the intensified efforts on the development of alternative drugs from natural products such as active ingredients from plants, microbial sources, and marine sponges owing to their huge health benefits, and remarkable anti-obesity potential. Several functional foods, medicinal plants, fruits, and vegetables have also been studied, both for weight loss promotion as well as preventive and therapeutic management of obesity. The common mechanisms of most active ingredients are interference with nutrient absorption, decrease in adipogenesis and enhance energy expenditure, suppression of the appetite, modification of the intestinal microbiota composition, and increase in fat excretion. Nevertheless, further studies are required to evaluate their efficacy, bioavailability, and safety, in both animal and human subjects. There is also a dire need for systematic targeted clinical studies before the incorporation of these functional foods and natural compounds into the mainstream therapy for obesity management.

References

1. Moriyasu, Y, et al., Validation of antiobesity effects of black soybean seed coat powder suitable as a food material: comparisons with conventional yellow soybean seed coat powder. Foods, 2021. 10(4): 841.
2. Kim, EA, et al., Anti-obesity effect of pine needle extract on high-fat diet-induced obese mice. Plants (Basel), 2021. 10(5): 837.
3. Zhang, K, et al., Functional ingredients present in whole-grain foods as therapeutic tools to counteract obesity: effects on brown and white adipose tissues. Trends in Food Science & Technology, 2021. 109: 513–526.
4. Arya, A, et al., Chapter thirteen - anti-obesity natural products, in Annual Reports in Medicinal Chemistry, S.D. Sarker and L. Nahar, Editors. 2020. London: Academic Press. 411–433.
5. WHO, The Double Burden of Malnutrition: Priority Actions on Ending Childhood Obesity. 2020. New Delhi: World Health Organization, Regional Office for South-East Asia.
6. Oussaada, SM, et al., The pathogenesis of obesity. Metabolism, 2019. 92: 26–36.
7. Green, M, K Arora, and S Prakash, Microbial medicine: prebiotic and probiotic functional foods to target obesity and metabolic syndrome. International Journal of Molecular Sciences, 2020. 21(8): 1–28.
8. Song, JH, et al., *In vivo* evaluation of *Dendropanax morbifera* leaf extract for anti-obesity and cholesterol-lowering activity in mice. Nutrients, 2021. 13(5): 1424.
9. Mazidi M, et al., Prevalence of childhood and adolescent overweight and obesity in Asian countries: a systematic review and meta-analysis. Archives of Medical Science, 2018. 14(6): 1185–1203.
10. Salazar SS, Assessment and management of the obese adult female: a clinical update for providers. Journal of Midwifery & Women's Health, 2006. 51(3): 202–207.
11. Carrageta DF, et al., Anti-obesity potential of natural methylxanthines. Journal of Functional Foods, 2018. 43: 84–94.
12. Dalili D, et al., The role of body composition assessment in obesity and eating disorders. European Journal of Radiology, 2020. 131: 109227.
13. Vilela DLS, et al., Influence of dietary patterns on the metabolically healthy obesity phenotype: a systematic review. Nutrition, Metabolism and Cardiovascular Diseases, 2021. 31(10): 2779–2791.
14. Hasler CM, Functional foods: benefits, concerns and challenges-a position paper from the American council on science and health. The Journal of Nutrition, 2002. 132(12): 3772–3781.
15. Arihara K, Functional foods, in Encyclopedia of Meat Sciences (Second Edition), M. Dikeman and C. Devine, Editors. 2014. Academic Press: Oxford. 32–36.
16. Fu C et al., Natural products with anti-obesity effects and different mechanisms of action. Journal of Agricultural and Food Chemistry, 2016. 64(51): 9571–9585.
17. Ganesan K and Xu B, Anti-obesity effects of medicinal and edible mushrooms. Molecules, 2018. 23(11): 2880.
18. El Hadi H, et al., Food ingredients involved in white-to-brown adipose tissue conversion and in calorie burning. Frontiers in Physiology, 2019. 9(1954): 1–8.
19. Thadchanamoorthy V, Jayasekara N and Dayasiri KJC, Primary hypertension as the presenting feature of Laurence-Moon-Bardet-Biedl syndrome: a report of two children. Cureus, 2021. 13(1): e12617.
20. Kulshreshtha B, et al., Adolescent gynecomastia is associated with a high incidence of obesity, dysglycemia, and family background of diabetes mellitus. Indian Journal of Endocrinology and Metabolism, 2017. 21(1): 160–164.
21. Li W, et al., Association between obesity and puberty timing: a systematic review and meta-analysis. International Journal of Environmental Research and Public Health, 2017. 14(10): 1266.

22. Walker JL, et al., Idiopathic genu valgum and its association with obesity in children and adolescents. Journal of Pediatric Orthopedics, 2019. 39(7): 347–352.
23. Hu X, et al., Marine-derived bioactive compounds with anti-obesity effect: a review. Journal of Functional Foods, 2016. 21: 372–387.
24. Kinlen D, Cody D, and O'Shea D, Complications of obesity. QJM: An International Journal of Medicine, 2018. 111(7): 437–443.
25. Lai CS, Wu JC and Pan MH, Molecular mechanism on functional food bioactives for anti-obesity. Current Opinion in Food Science, 2015. 2: 9–13.
26. So I and Yadav H, Obesity and its complications pathogenesis, in Pathophysiology of Obesity-Induced Health Complications. 2020. Springer Nature Switzerland AG (Cham, Switzerland). 43–56.
27. Kaila B and Raman M, Obesity: a review of pathogenesis and management strategies. Canadian Journal of Gastroenterology = Journal canadien de gastroenterologie, 2008. 22(1): 61–68.
28. Sandner G, et al., Functional foods - dietary or herbal products on obesity: application of selected bioactive compounds to target lipid metabolism. Current Opinion in Food Science, 2020. 34: 9–20.
29. Chen L, et al., Pathogenesis and clinical management of obesity-related knee osteoarthritis: impact of mechanical loading. Journal of Orthopaedic Translation, 2020. 24: 66–75.
30. Segula D, Complications of obesity in adults: a short review of the literature. Malawi Medical Journal: the Journal of Medical Association of Malawi, 2014. 26(1): 20–24.
31. Ansari S, Haboubi H, and Haboubi N, Adult obesity complications: challenges and clinical impact. Therapeutic Advances in Endocrinology and Metabolism, 2020. 11:1–14.
32. Zeng Q, et al., Clinical management and treatment of obesity in China. The Lancet Diabetes & Endocrinology, 2021. 9(6): 393–405.
33. Conforti F and Pan MH, Natural products in anti-obesity therapy. Molecules (Basel, Switzerland), 2016. 21(12): 1750.
34. Woodard K, Louque L, and Hsia DS, Medications for the treatment of obesity in adolescents. Therapeutic Advances in Endocrinology and Metabolism, 2020. 11: 1–12.
35. May M, Schindler C, and Engeli S, Modern pharmacological treatment of obese patients. Therapeutic Advances in Endocrinology and Metabolism, 2020. 11: 1–19.
36. Schwartz MW, et al., Obesity pathogenesis: an endocrine society scientific statement. Endocrine Reviews, 2017. 38(4): 267–296.
37. Ahmed IA, et al., Lifestyle interventions for non-alcoholic fatty liver disease. Saudi Journal of Biological Sciences, 2019. 26(7): 1519–1524.
38. Ramli NZ, et al., A review on the protective effects of honey against metabolic syndrome. Nutrients, 2018. 10(8): 1009.
39. Samat S, et al., Four-week consumption of Malaysian honey reduces excess weight gain and improves obesity-related parameters in high fat diet-induced obese rats. Evidence-based Complementary and Alternative Medicine, 2017. 2017: 1–9.
40. Mokaya HO, et al., Characterization of honeys produced by sympatric species of Afrotropical stingless bees (Hymenoptera, Meliponini). Food Chemistry, 2022. 366: 130597.
41. Sahlan M, et al., The effects of stingless bee (Tetragonula biroi) honey on streptozotocin-induced diabetes mellitus in rats. Saudi Journal of Biological Sciences, 2020. 27(8): 2025–2030.
42. Ranneh Y, et al., Malaysian stingless bee and Tualang honeys: a comparative characterization of total antioxidant capacity and phenolic profile using liquid chromatography-mass spectrometry. LWT, 2018. 89: 1–9.
43. Liu X, et al., Changes in nut consumption influence long-term weight change in US men and women. BMJ Nutrition, Prevention, & Health, 2019. 2019: 000034.
44. Jackson CL and Hu FB, Long-term associations of nut consumption with body weight and obesity. The American Journal of Clinical Nutrition, 2014. 100(Suppl 1): 408S–411S.

45. Santos AF, et al., Basophil activation test reduces oral food challenges to nuts and sesame. The Journal of Allergy and Clinical Immunology: In Practice, 2021. 9(5): 2016–2027.e6.
46. Liu Y, et al., Nondigestible oligosaccharides with anti-obesity effects. Journal of Agricultural and Food Chemistry, 2020. 68(1): 4–16.
47. Zhang F, et al., Anti-obesity effects of dietary calcium: the evidence and possible mechanisms. International Journal of Molecular Sciences, 2019. 20(12): 1–14.
48. Wang Z, et al., Anti-obesity effect of trans-cinnamic acid on HepG2 cells and HFD-fed mice. Food and Chemical Toxicology, 2020. 137: 111148.
49. Gómez-Zorita S, et al., Anti-obesity effects of microalgae. International Journal of Molecular Sciences, 2019. 21(1): 41.
50. Ali Redha A, et al., Novel insights on anti-obesity potential of the miracle tree, *Moringa oleifera*: a systematic review. Journal of Functional Foods, 2021. 84: 104600.
51. Haselgrübler R, et al., Hypolipidemic effects of herbal extracts by reduction of adipocyte differentiation, intracellular neutral lipid content, lipolysis, fatty acid exchange and lipid droplet motility. Scientific Reports, 2019. 9(1): 10492.
52. Kim GW, Jo HK and Chung SH, Ginseng seed oil ameliorates hepatic lipid accumulation *in vitro* and *in vivo*. Journal of Ginseng Research, 2018. 42(4): 419–428.
53. Gooda Sahib N, et al., Plants' metabolites as potential antiobesity agents. The Scientific World Journal, 2012. 2012: 436039.
54. Abdul Rahman H, et al., Anti-obesity and antioxidant activities of selected medicinal plants and phytochemical profiling of bioactive compounds. International Journal of Food Properties, 2017. 20(11): 2616–2629.
55. Ahmad M and Gani A, Development of novel functional snacks containing nano-encapsulated resveratrol with anti-diabetic, anti-obesity and antioxidant properties. Food Chemistry, 2021. 352: 129323.
56. Cao SY, et al., Dietary plants, gut microbiota, and obesity: effects and mechanisms. Trends in Food Science & Technology, 2019. 92: 194–204.
57. Casazza K, et al., Myths, presumptions, and facts about obesity. The New England Journal of Medicine, 2013. 368(5): 446–454.
58. Matarese LE and Pories WJ, Adult weight-loss diets: metabolic effects and outcomes. Nutrition in Clinical Practice, 2014. 29(6): 759–767.
59. Casazza K, et al., Weighing the evidence of common beliefs in obesity research. Critical Reviews in Food Science and Nutrition, 2015. 55(14): 2014–2053.

Noncommercial Plant-Based Edible Oil for Prevention and Treatment of Chronic Diseases

Der Jiun Ooi, Yun Ping Neo, Yin Sim Tor, and Jhi Biau Foo

Contents

Introduction 249
Chemical composition 251
Preclinical and clinical studies 251
  Pumpkin seed oil (PSO) 251
  Watermelon seed oil (WSO) 254
  Citrus fixed seed oil (CFSO) 258
  Nigella sativa seed oil (NSO) 260
  Moringa seed oil (MSO) 262
  Pistachio oil (PO) 263
  Evening primrose oil (EPO) 265
  Melon seed oil (MeSO) 267
  Cucumber seed oil (CSO) 267
  Kenaf seed oil (KSO) 268
Conclusions 269
Declaration of competing interest 283
Acknowledgments 283
References 283

Introduction

Noncommunicable diseases (NCDs), or chronic diseases, account for millions of disabilities and deaths globally. According to a report from the World Health Organization, there are approximately 41 million deaths due to NCDs annually, equivalent to 71% of all deaths globally. Cardiovascular diseases are the leading cause of death, reported to be 18 million deaths yearly. This is followed by cancers (9.3 million), respiratory diseases (4.1 million) and diabetes (1.5 million). Sadly, these chronic diseases account for approximately 80% of all the premature NCD deaths that are highly reported in both low- and middle-income countries (1). This growing threat requires great attention, as it is the major cause of

DOI: 10.1201/9781003220053-12

poverty, hindering the economic development of many countries. It is proposed that the focus of prevention and treatment of NCDs should be based on lifestyle changes, particularly modifying the diet and heightening physical activities.

Edible oil has long been documented for its usage to prevent and treat chronic diseases. For instance, the efficacy of olive oil to prevent cardiovascular disease, diabetes and cancer has been proven at the clinical level, attributable to the bioactive components present in olive oil (2, 3). On the other hand, despite controversial clinical findings reported on the use of palm oil and canola oil for the prevention of cardiovascular diseases, the health claims received much attention by the consumers and had become the basis for them to decide on the type of oil to be consumed for optimum health (4, 5). With reference to the successful commercialization of olive oil, palm oil and canola oil being endorsed by clinical research findings for their efficacies, the other noncommercial plant-based edible oils could further be explored for use in the prevention and treatment of chronic diseases. The present book chapter intends to provide an update on the preclinical and clinical research findings of some selected noncommercial plant-based edible oils. The noncommercial edible oil, being defined as plant-based, noncooking and nonrefined edible oils, include pumpkin seed oil, watermelon seed oil, citrus fixed seed oil, *Nigella sativa* seed oil, moringa seed oil, pistachio oil, evening primrose oil, melon seed oil, cucumber seed oil and kenaf seed oil (Figure 10.1).

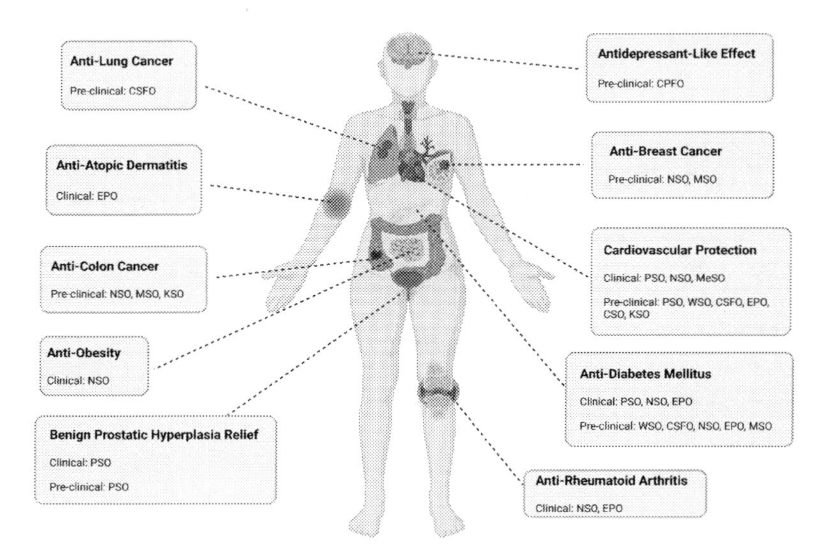

Figure 10.1 *Efficacy of selected noncommercial plant-based edible oils for prevention and treatment of chronic diseases The data are presented based on the evidence from either preclinical or clinical data. Abbreviation: Pumpkin seed oil (PSO), watermelon seed oil (WSO), citrus seed fixed seed oil (CSFO), citrus peel fixed seed oil (CPFO), Nigella sativa seed oil (NSO), moringa seed oil (MSO), evening primrose oil (EPO), melon seed oil (MeSO), cucumber seed oil (CSO), kenaf seed oil (KSO). (Created with Biorender.com.)*

Chemical composition

Edible oils are largely comprised of triacylglycerols (TAG), diacylglycerols (DAG) and monoacylglycerols (MAG), along with the presence of other lipid-soluble bioactive components that include phospholipids, isoprenoids, phenolics and vitamins. Fatty acids, being the components that often receive great attention, can further be categorized into monounsaturated (MUFA), polyunsaturated fatty acids (PUFA) and saturated (SFA), based on the chemical classifications. Fatty acids differ from each other in the number of carbon atoms. In general, the fatty acid composition varies with different plant cultivars and hence will exert different health-promoting traits in humans. A large number of epidemiological studies have shown that correct proportion of essential omega-3 PUFA (n-3 PUFA) and omega-6 PUFA (n-6 PUFA) is recommended in the regulation of inflammation for nutritional equilibrium (6). These studies have recognized that eicosanoids derived from n-3 PUFA supresses the pro-inflammatory cytokines, whereas n-6 PUFA generates inflammatory metabolites (7). Additionally, consumption of lipid soluble bioactive compounds is also reported to exhibit numerous health benefits. Among these lipid soluble bioactive compounds, isoprenoids including sterols, carotenoids and tocols (tocopherol and tocotrienols) are being actively researched and demonstrate antioxidant and anti-inflammatory properties (8–10). These compounds are especially important in the noncommercial oils that are generally unrefined.

Preclinical and clinical studies

Pumpkin seed oil (PSO)

Interest in specially formulated edible oils as functional foods and nutraceuticals is rapidly growing. PSO is regarded as one of the promising star candidates. Extracted via cold pressing, steam distillation, solvent or supercritical fluid extraction, the dark greenish-red PSO is often renowned for having a rich nutty flavor. It is also interesting to note that PSO appears from green to red depending on thickness or viscosity, mainly due to the dichromatism optical phenomenon.

Compositional analysis showed that PSO is rich in both MUFA and PUFA. While differences are observed across different pumpkin varieties, the total content of unsaturated fatty acid (73.1–80.5%) is contributed by the dominance of oleic acid and linoleic acid. Stearic acid and palmitic acid are the total SFA and were found to be in the range of 12.8–18.7%. The very long chain fatty acids with carbon chain length >18 are least abundant, with an amount in the range of 0.44–1.37%. Analysis revealed considerable α-tocopherol (27–75 mg/g) and γ-tocopherol (75–493 mg/g) contents in the seed oil. Further high-performance liquid chromatography (HPLC) analysis revealed that β-sitosterol and squalene are the other main minor compounds present in the alkaline saponified oil. Phenolic acids including tyrosol, vanillin, vanillic acid, luteolin and synapic acid are also reported.

The antioxidant capacity of PSO was reported in multiple studies. The animal study conducted by El-Boghdady demonstrated that PSO possesses anti-inflammatory and antioxidant properties. Treatment of PSO for five days prior to intraperitoneal administration of methotrexate prevented intestinal inflammation and bowel damage by improving antioxidant defence, as well as inhibiting lipid peroxidation, nitric oxide (NO) generation and neutrophils infiltration (11). On the other hand, the protective function of PSO against cytotoxicity and genotoxicity induced by both azathioprine and bisphenol-A were studied using mice model. PSO was effective in inhibiting DNA fragmentation, reducing micronucleated polychromatic erythrocytes and improving semen parameters following administration of genotoxic azathioprine (12). The adverse effects of bisphenol-A on DNA damage and histological alterations were also alleviated in PSO-treated animal groups (13). Interestingly, the radical scavenging properties of PSO may also afford a protective effect against nanoalumina-mediated oxidative stress in pregnant rats and in their progeny (14).

The potential of PSO against both acute and chronic inflammation was also studied. The resulting data revealed that topical application of PSO alleviated inflammatory responses in chemical-induced acute and chronic inflammation in mice. It was proposed that PSO modulated the cellular and molecular mediators necessitated for inflammatory response (15). Another group performed the anti-inflammatory evaluation of the seed oil using an adjuvant arthritis rat model. The PSO derived from two different varieties improved the inflammatory activity and genotoxicity changes (16).

The anti-hypertensive effect of PSO was demonstrated by the El-Mosallamy team using an N^G-Nitro-L-arginine-methyl ester (L-NAME)–induced hypertensive rat model. Intake of seed oil (40–100 mg/kg) for six weeks consecutively afforded a protective effect against the chemical-induced pathological changes in the heart and aorta. Normalized blood pressure and electrocardiogram signal, as well as reduced malondialdehyde (MDA) levels, might relate to the potential of PSO in reversing NO formation. The anti-hypertensive effect may be attributed by the specific fatty acid composition and phytoestrogen content. For instance, secoisolariciresinol has been reported to mediate the blood pressure–lowering effects and cardioprotective actions via the vascular endothelial growth factor/NO pathway (17).

Gossell-Williams and colleagues studied the potential health benefits of PSO toward blood pressure and lipid metabolism using the ovariectomized Sprague–Dawley rat models. During the 12-week intervention period, both ovariectomized and nonovariectomized rats were administered with either 40 mg/kg bodyweight of corn oil or PSO via intragastric route for five days/week. Upon completion of the experiment, the research group discovered that PSO ameliorated systolic and diastolic blood pressure. Indeed, healthy plasma lipid profile was noticed in the PSO-supplemented group (18). Significantly, the positive findings indicated potential use of PSO as a supplement to improve oestrogen deficiency-associated lipid metabolism disorder in postmenopausal women.

Another study revealed that PSO ameliorated lipid, choline, glucose, nucleotide and amino acid metabolism disorder following high-fat diet administration. The PSO-treated animals exhibited lowered unfolded protein response and effectively reduced endoplasmic reticulum stress (19).

Following the positive research findings, 35 postmenopausal women volunteers were recruited into a double-blind, randomized, placebo-controlled clinical study to further evaluate the antihypertensive and cardioprotective effects of PSO. Following a 12-week intervention period (daily 2 g PSO), the subjects who received PSO showed significant increase in high density lipoproteins and a reduction in diastolic blood pressure. The menopausal women also experienced ameliorated menopausal symptoms with less frequent headaches, joint pains and hot flashes (20). While the pilot study was presented with a few limitations, including small sample size, proper placebo choice, inability to assay selected biochemical and hormonal parameters, it provided some basis on the beneficial effect of PSO clinically and warrants further investigations.

The effect of PSO on a rat model of prostatic hyperplasia was examined by two separate preclinical studies. Daily subcutaneous injections of testosterone formulation mixture were used to induce benign prostatic hyperplasia in Sprague–Dawley rats. The animals received concurrent daily oral gavage administration of PSO (2–4 mg/kg body weight) or vehicle during the 20-day hyperplasia induction. The result showed that PSO significantly reduced the prostate size, without having significant influence on body weight (21). In a separate study conducted by Tsai et al., in a testosterone/prazosin-induced rat model, treatment with 2.5 mL/kg body weight/day of PSO significantly reduced the prostatic weight ratio and protein synthesis (22). These preclinical studies warrant future clinical studies.

Involving 2245 subjects suffering from benign prostate hypertrophy (BPH), the first clinical study reported in year 2000 showed that the PSO treatment significantly reduced the international prostate symptom score and improved quality of life over a 12-week period (23). In the study by Hong et al., a randomized, double-blind, placebo-controlled 12-month trial was conducted on 47 subjects with symptomatic BPH. While no significant changes were observed for prostate volume and serum prostate specific antigen, the oil significantly ameliorated urinary dysfunction by improving the international prostate symptom score, quality of life and maximal urinary flow rate (24). Another prospective randomized clinical trial involving 100 BPH subjects showed similar improvement following PSO intervention. While the alpha-1 blocker prazosin appeared to show better efficacy, PSO could be used as the functional food or nutraceutical in the relief of prostatitis-related symptoms (25). Interestingly, the clinical study by Nishimura et al. also demonstrated potential use to relief overactive bladder symptoms (26). All these clinical interventions further substantiated the potential use of PSO for the management of benign prostatic hyperplasia and urinary disorders.

Watermelon seed oil (WSO)

The heritable seed-size trait plays an important role in determining the uses of watermelon. Specifically, a wide variation in seed sizes present among watermelon collections, where the fruits with small seeds are preferred for consumption while the large-seed varieties are favored for production of edible seeds and planting purposes (27). Also known as Ootanga oil or Kalahari oil, WSO is considered to be one of the underexplored sources of edible oil. Analyses showed that watermelon seeds contain a considerable amount of protein and lipid contents. However, WSO demonstrated great variation in physicochemical characteristics depending on the varieties and origins. As shown in Table 10.1, the observed differences may have been attributed by the different variety among the same species, as well as specific soil and climatic condition of varying geographical regions (28). In the present analysis, the bioactivities of both *Citrullus lanatus* and *C. vulgaris* are being discussed.

Table 10.1 Composition of Noncommercial Plant-based Edible Oils

Type of Oil	Compositions		Species	References
Pumpkin seed oil (PSO)	TAG (%):	N/A	*Cucurbita maxima L.* (*var. Berrettina*), *Cucurbita maxima D.*, *Cucurbita moschata*, *Cucurbita pepo*	(29–31)
	SFA (%):	19.6–21.1		
	MUFA (%):	8.6–41.7		
	PUFA (%):	37.2–79.1		
	Phospholipids (g/kg):	N/A		
	Phytosterols (g/kg):	2.95		
	Tocopherols (mg/kg):	6.5–410.6		
	Polyphenols (mg/kg):	N/A		
Watermelon seed oil (WSO)	TAG (%):	N/A	*Citrulluslanatus* and *Citrullus vulgaris*	(32–34)
	SFA (%):	16.5–23.0		
	MUFA (%):	10.8–16.3		
	PUFA (%):	62.1–72.6		
	Phospholipids (g/kg):	N/A		
	Phytosterols (g/kg):	N/A		
	Tocopherols (mg/kg):	65.2–748.1		
	Polyphenols (mg/kg):	1428.9		
Citrus seed fixed oil (CFO)	TAG (%):	N/A	*Citrus limon*, *Citrus reticulata*, *Citrus paradisi*, *Citrus sinensis*, *Citrus limetta*	(35–38)
	SFA (%):	25.3–36.1		
	MUFA (%):	22.6–29.0		
	PUFA (%):	44.6–46.2		
	Phospholipids (g/kg):	N/A		
	Phytosterols (g/kg):	2.0–4.0		
	Tocopherols (mg/kg):	101.7–661.9		
	Polyphenols (mg/kg):	7.0–93.4		

(Continued)

Table 10.1 Composition of Noncommercial Plant-based Edible Oils *(Continued)*

Type of Oil	Compositions		Species	References
Nigella sativa seed oil (NSO)	TAG (%):	65.5	*Nigella sativa* L.	(39–41)
	SFA (%):	13.7–20.5		
	UFA (%):	68.4–83.7		
	Phospholipids (g/kg):	N/A		
	Phytosterols (g/kg):	2.9		
	Tocopherols (mg/kg):	23.3–65.4		
	Polyphenols (mg/kg):	101.0–447.2		
Moringa seed oil (MSO)	TAG (%):	N/A	*Moringa oleifera* L.	(42, 43)
	SFA (%):	17.24–23.79		
	MUFA (%):	71.70–80.70		
	PUFA (%):	0.41–2.20		
	Phospholipids (%):	6.07		
	Phytosterols (g/kg):	8.35		
	Tocopherols (mg/kg):	174.3–515.0		
	Polyphenols (mg/kg):	160		
Pistachio oil (PO)	TAG (%):	92–98	*Pistaciavera* L.	(44, 45)
	SFA (%):	10–16		
	MUFA (%):	56–77		
	PUFA (%):	10–31		
	Phospholipids (g/kg):	18		
	Phytosterols (g/kg):	2.1–7.6		
	Tocopherols (mg/kg):	300–900		
	Polyphenols (mg/kg):	16–60		
Evening primrose oil (EPO)	TAG (%):	98	*Oenothera biennis* L.	(46–49)
	SFA (%):	8.7–13.7		
	MUFA (%):	1.0–11		
	PUFA (%):	74.8–84.8		
	Phospholipids (%):	0.05		
	Phytosterols (g/kg):	9.1		
	Tocopherols (mg/kg):	183.6–445.6		
	Polyphenols (mg/kg):	N/A		
Melon seed oil (MeSO)	TAG (%):	90.7	*Cucumis melo* L.	(50–53)
	SFA (%):	14.76–19.82		
	MUFA (%):	16.23–18.92		
	PUFA (%):	61.26–69.18		
	Phospholipids (%):	0.7–2.1		
	Phytosterols (g/kg):	3.02		
	Tocopherols (mg/kg):	432–828		
	Polyphenols (mg/kg):	0.2–28.2		

(Continued)

Table 10.1 Composition of Noncommercial Plant-based Edible Oils *(Continued)*

Type of Oil	Compositions		Species	References
Cucumber seed oil (CSO)	TAG (%):	82.1–82.7	*Cucumis sativus* L.	(54–57)
	SFA (%):	20.2–24.3		
	MUFA (%):	13.1–17.5		
	PUFA (%):	55.3–70.1		
	Phospholipids (%):	2.6–3.1		
	Phytosterols (g/kg):	9.8		
	Tocopherols (mg/kg):	1086		
	Polyphenols (mg/kg):	N/A		
Kenaf seed oil (KSO)	TAG (%):	N/A	*Hibiscus cannabinus*	(58–60)
	SFA (%):	22–27		
	MUFA (%):	27–37		
	PUFA (%):	43–52		
	Phospholipids (%):	3.9–11		
	Phytosterols (g/kg):	~ 3.7		
	Tocopherols (mg/kg):	~ 847		
	Polyphenols (mg/kg):	17.5		

Notes: Triacylglycerols (TAG); saturated fatty acids (SFA); monounsaturated fatty acids (MUFA); polyunsaturated fatty acids (PUFA), not available (N/A)

The antioxidant activity of edible oil is a critical indicator of its quality. Antioxidants reduce fat oxidation and ensure oxidative stability of the oil. Being part of the dietary component, the antioxidant activity may also afford a protective effect by reducing oxidative stress *in vivo*. Briefly, the antioxidant capacity of WSO was investigated in a few studies using the same 2,2-diphenyl-1-picryl-hydrazyl-hydrate (DPPH) free radical scavenging method. The IC_{50} values obtained from these studies ranged between 7.07 and 36.04 mg/mL for *C. lanatus* and 0.31 mg/mL for *C. vulgaris* (32, 33, 61). It is suggested that the good antiradical activity efficacy values obtained for WSO may relate to the amount of phenolic compounds and tocol compositions (33).

While *C. vulgaris* WSO demonstrated potent antioxidant ability, the oil elicited concern over its moderate cytotoxicity. As evidenced via the brine shrimp lethality bioassay, a relatively high median lethal dose of 0.14 ± 0.38 μg/mL was shown upon 20 hours of incubation (32). Nevertheless, a separate study reported the protection of *C. vulgaris* WSO against cadmium-induced neurotoxicity in mice. Administration of 0.25–0.5 g/kg body weight of the seed oil for four consecutive weeks improved the acetylcholinesterase activity, alleviated oxidative stress and effectively ameliorated the cerebral edema (62).

Contrarily, ED_{50} cytotoxicity value of 241.30 μg/mL was reported for the chloroform extract of *C. lanatus* using the same bioassay following 24 hours' incubation (63). Both acute and 28-day sub-chronic administration of maximal dose of *C. lanatus* WSO showed no sign of toxicity in the rat models. Additionally, *C. lanatus* WSO demonstrated potential hepatoprotective and neuroprotective attributes by attenuating carbon tetrachloride-induced liver injuries and hippocampal dysfunction following mercury chloride intoxication (64–66).

The cardio protective effect of *C. lanatus* WSO was also investigated. It was reported that the seed oil improved serum lipid profile and inhibited the development of atherosclerosis in cholesterol-fed rats. Histopathological analysis revealed reduced foam cell formation and suppressed migration of vascular smooth muscle cells in the treatment group (67). Another study by Eke et al. showed that the seed oil exerted anti-hypolipidemic and anti-oxidative effects without compromising the liver integrity (64). Similar findings, reported twice by the same group of authors, demonstrated improved serum lipid profile and hepatic lipid composition among the *C. lanatus* WSO-treated animals (68, 69). Interestingly, oil extracted from raw and cooked *C. lanatus* seed may have varying effects on lipid metabolism and is worth further research investigation (70).

In addition, *C. lanatus* WSO also exhibited potential anti-diabetic efficacy by modulating glucose uptake. By using the alloxan-induced diabetic albino mice model, the petroleum ether-extracted seed oil exerted dose-dependent reduction on blood glucose levels. A similar observation was reported by Eke et al. in rats after 8 and 15 days of seed oil oral administration. The 2-hour oral glucose tolerance test was performed after loading the animals with 300 mg/kg of glucose solution. Improved glucose tolerance was observed (64, 71).

The *C. lanatus* WSO also depicted potent antibacterial properties against a range of gram positive and negative bacteria (72). In particular, Adewuiyi et al. had further transformed the seed oil into both anionic and nonionic biosurfactants to elicit better antibacterial activity (73). On top of that, the WSO also exerted protection against *Candida albicans* infection. Despite the fact that the WSO did not show any immunomodulatory function, the anti-inflammatory activities of the oil were present (74). The anti-inflammatory activity of *C. lanatus* WSO was earlier studied using the *in vitro* hypotonicity-induced human red blood cell membrane lysis and *in vivo* carrageenan-induced rat paw edema model. Potent anti-inflammatory potential of WSO was evidenced in these two studies (75).

Further short-term clinical assessment reported the topical safety and efficacy of applying the *C. lanatus* WSO. Skin irritation was unreported, however, the oil seems able to increase moisture retention and reduce transepidermal water loss (76). As depicted from the various reported studies, it may be recommended that both *C. lanatus* and *C. vulgaris* seed oil hold a huge potential as nutraceuticals and functional food commodities that are worth further comprehensive investigations into toxicological, nutritional and physiological functions.

Citrus fixed seed oil (CFSO)

Being one of the major fruit crops, citrus fruits, including lime (*Citrus aurantifolia*), lemon (*Citrus limon*), grapefruit (*Citrus paradise*), sweet orange (*Citrus sinensis*), tangerine (*Citrus tangerina*), mandarin orange (*Citrus reticulata*), sour orange (*Citrus aurantium*) and bergamot orange (*Citrus bergamia*), are widely cultivated all around the world. Commonly relished and consumed for its fresh and juicy taste, fruits of the citrus genus are also being used for processing in the food and beverage, pharmaceutical, cosmetic and allied industries. The latter, however, produces a large amount of unused peel, pulp, pith and seed wastage. Ongoing efforts have been undertaken to retrieve high-value utilization of these agro-industrial wastes where the potential value of citrus waste for oil extraction is highlighted (35, 77). In the present section, the biological functions of edible fixed oil pressed from the fatty portions of citrus peel and seed are discussed. The volatile citrus essential oil is not included in the current discussion.

The citrus peel- and seed-derived fixed oils are common ingredients used in food and cosmetic products, particularly for fragrance and skin-conditioning applications. Studies have reported the diverse oil content and chemical compositions of the derived fixed oil, primarily due to the complex botanicals made up of varying citrus species, cultivars, apomixes, interspecific and intergeneric hybridization as well as frequent bud mutations. Other than that, the chemical constituents of the citrus fixed oils may also be affected by geography, maturity, storage conditions and method of extraction (37, 38, 78, 79).

According to US FDA, citrus fruit-derived oils are generally recognized as safe (GRAS) under the intended use as foods for human and animal consumption. The study by Guneser & Yilmaz, however, reported that both hexane-extracted and cold-pressed lemon seed oils are not suitable for direct consumption owing to high bitterness levels. Additional refining procedures to remove the observed bitterness should be performed. In contrast, its unique aroma and being rich in bioactives suggest potential utilization in functional foods, nutraceutical, pharmaceutical and cosmeceutical applications (36). In particular, the US Pharmacopeia (USP) Food Chemicals Codex had also categorized the oil of oranges, grapefruits, bitter oranges, tangerines, limes and lemons as flavoring agents (80).

In vitro studies of lemon and sweet orange oils revealed no genotoxic effects in the Chinese hamster chromosomal aberrations, mouse lymphoma cell mutation and bacterial reverse mutation assays. Acute toxicity of sweet orange seed fixed oil reported a minimum toxic dose and maximum tolerated dose of 1732 mg/kg bodyweight via intraperitoneal injection (81).

While there are some carcinogenicity concerns, low-dose exposures to the citrus oils are deemed safe. Since the level of 5-methoxypsoralen was less than 15 ppm (0.0015%) in citrus oils, the Cosmetic Ingredient Review Expert Panel suggested that citrus oils are safe for use as cosmetic ingredients for the nonsensitizing and nonirritating formulations (78, 82). Additionally, Rosa et al. had reported the potential use of grapefruit, lemon and mandarin orange seed–derived fixed seed oils against the viability in B16F10 melanoma cells (35). Another study demonstrated the dose-dependent growth inhibition of cold-pressed mandarin orange peel oil on the proliferation of non-small cell lung cancer cells. The efficacy of mandarin orange peel oil to suppress tumor growth was further confirmed *in vivo* using a xenograft model implanting the human A549 adenocarcinomic alveolar basal epithelial cells in nude mice (83).

The antimicrobial activity of CFOs has also been investigated by several research groups. The hand-pressed oil derived from the peels of sweet orange, bitter orange, mandarin orange, bergamot orange, grapefruit and lemon demonstrated strong antimicrobial activity against a wide range of gram-positive and gram-negative bacteria, as well as fungal species (84–87). Contrarily, the cold-pressed oils derived from the seeds of sweet orange, lemon and grapefruits only demonstrated moderate antimicrobial activity against selected common foodborne pathogens (88).

Apart from antimicrobial and anticancer properties, the antioxidant activities of various citrus oils were being studied. Benelli et al. reported the antioxidant activities of sweet orange peel fixed oil extracted using carbon dioxide-aided supercritical fluid extraction, chemical-aided Soxhlet and ultrasound extraction, as well as hydrodistillation-derived essential oil. Both DPPH free radical scavenging activity and beta-carotene bleaching assays were performed in the evaluation. While no antioxidant activity was reported for the hydrodistillation-derived essential oil, substantial antioxidant activities were reported for the citrus peel fixed oils obtained by either Soxhlet, ultrasound, or supercritical fluid extraction methods. Interestingly, butylated hydroxytoluene-equivalent antioxidant potential was discovered in the supercritical fluid extracts at 250 bar/50°C and with the use of co-solvent, suggesting great antioxidant potency of oil prepared under supercritical fluid extraction conditions with relatively low critical temperature and pressure (86). The other studies reported similar observations using the hydroxyl radical and 2,2′-azino-bis(3-ethylbenzothiazoline-6-sulfonic acid) (ABTS) radical scavenging assays (35, 85).

The lemon, grapefruit and mandarin orange seed fixed oil had also demonstrated potential as an anti-melanogenic agent. *In vitro* IC_{50} tyrosinase inhibitory activity, in the range of 0.49–1.60 mg/mL, was reported for the three oils. Although the inhibitory effects of the oils were not comparable to that of kojic acid, the fixed oils may pose as promising alternative natural sources for anti-melanogenic agents, with great potential interest from the pharmaceutical, cosmetic and food-processing industries (35).

Interestingly, researchers have also attempted to explore the antidepressant-like effect of cold pressed sweet orange oil in unpredictable mild stress mice model. The oil was exposed to the mice via inhalation. During the four-week experimental period, the mice were subjected to an assortment of random mild stressors including food and water deprivation, overhang, swimming, shaking cage, whole-day light, 45° cage and housing in wet sawdust. The experimental results showed amelioration of depression-like behaviors and dyslipidemia following inhalation treatment with cold-pressed sweet orange oil. Being the most abundant compound under the oil sniffing environment, it is proposed that limonene may act by attenuating the neuroendocrine, neurotrophic and monoaminergic systems (89).

The hexane-extracted sweet orange seed fixed oil was studied for efficacy in ameliorating glucose, lipid profile and liver enzyme using the alloxan-induced rat models. The rats were treated with daily intraperitoneal injection of oil-water emulsion throughout the 28-day study period. The resulting data depicted improved fasting blood glucose level, lipid profile comprising of total cholesterol, triglycerides, HDL-C, LDL-C and VLDL-C, as well as liver enzymes activities (81). While the therapeutic potential of the sweet orange seed fixed oil was presented in the present study, further studies are still warranted to further substantiate its nutraceutical values. In particular, subchronic systemic toxicity after repeated administration of prepared oil sample via oral route for up to 90 days are necessary to reaffirm the safety for consumption.

Nigella sativa seed oil (NSO)

Nigella sativa L. is also known as black cumin or black seed. Its seed oil has been extensively studied since the 1960s among all parts of the plant. In a chemical investigation on oil extracted from *N. sativa* seed showed the presence of both fixed oil and volatile oil. Fixed oil is the major component, while the volatile component ranges between 0.5% and 1.5% of the seeds weight (90). The fatty acid composition (32–40%) of *N. sativa* seed oil (NSO) includes a predominant level of linoleic acid (44.7–60%), oleic acid (20.7–24.6%), palmitic acid (12–14.3%), arachidic acid (2–3%) and stearic acid (2.7–3%) (91). Thymoquinone, a major active constituent of NSO isolated from the volatile fraction, was reported to orchestra the pharmacological activities in various reported studies (92).

NSO has been validated by myriad experimental evidences as a potent agent to prevent and treat chronic diseases. The therapeutic potential of NSO in type 2 diabetes (T2DM) has been evaluated in both preclinical and clinical studies. Abdelmeguid et al. reported that NSO prominently elevated serum insulin and tissue superoxide dismutase, while significantly lowered the diabetes-induced increases in tissue MDA and serum glucose levels (93). The protective effect of the volatile fraction of *N. sativa* was reported by Kanter et al. employing the streptozotocin-induced diabetic rat model. The findings suggested NSO ameliorated the destruction of pancreatic β-cells. This consequently brings about a rise in insulin immunoreactivity and ultrastructural changes in comparison to the control group. It was also evident that secretory vesicles with granules increased moderately (94).

Clinically, Heshmati and colleagues examined the effect of daily supplementation of 3 g NSO or sunflower soft gel capsules. The study was carried out for 12 weeks on 72 patients with type 2 diabetes mellitus. The results revealed insignificant decline in weight and BMI accompanied by significant decrease of fasting blood sugar, HbA1c, triglyceride and LDL cholesterol in the NSO group in comparison to the placebo group (95). Another study performed centrally on 41 type 2 diabetes mellitus patients showed that daily consumption of NSO for 40 days consecutively decreased fasting blood glucose and elevated insulin levels in patients in comparison to the control. Platelet count and total leukocyte count were not affected, while reasonable hepatic and renal safety were reported in the intervention group (96).

Cancer chemopreventive activity of NSO was evaluated using Wistar rat multi-organ carcinogenesis model featuring initial treatment of five different types of carcinogens. Post-initiation treatment of NSO at 1000 or 4000 mg/L in male rats' diet for a period of 30 weeks effectively decreased sizes, incidences and multiplicities of malignant and benign tumors. NSO displayed potent suppression of tumor formation and inhibited cellular proliferation in organs such as lung, colon, forestomach and oesophageal in the post-initiation phase (97). Another study demonstrated NSO significantly reduced the occurrence of the 1,2-dimethylhydrazine-induced colonic aberrant crypt foci at both initiation and post-initiation stages in Fischer 344 rats (98). In a separate study, El-Aziz et al. demonstrated that the administration of NSO at 4 g/kg/day for three months not only reduced the frequency of mammary carcinoma in 7,12-dimethylbenz(a)anthracene-induced female Sprague–Dawley rats but also lowered the markers of tumourigenicity (total and lipid bound sialic acid), endocrine derangement (progesterone, prolactin and oestradiol) and elevated apoptotic markers (DNA fragmentation, caspase-3, tumor necrosis factor α) when compared to the control (99).

Anti-obesity characteristic of NSO has been indicated in numerous studies. A randomized controlled trial (RCT) showed that NSO (3 g/day) in combination with a low-calorie diet was able to decrease weight in obese women compared to the placebo group. NSO also leads to elevated levels of superoxide dismutase. However, glutathione peroxidase level, lipid peroxidation and total antioxidant capacity remains unchanged (100). A separate double-blinded randomized controlled trial with similar study design demonstrated that the intake of NSO (3 g/day) with low-calorie diet significantly lowered the waist circumference and weight in NSO group in comparison to placebo after eight weeks intervention. Serum levels of cholesterol, triglycerides, LDL, VLDL and atherogenic indices indicated significant reduction in the intervention group compared to the baseline. Nevertheless, only significant decreases in triglyceride and VLDL levels were observed following treatment with NSO (101).

The efficacy of NSO in female patients diagnosed with rheumatoid arthritis was evaluated by Gheita and Kenway in a placebo-controlled study. Forty patients with confirmed diagnosis consumed two placebo capsules daily for a month, followed by 1 g/day of NSO for another 30 days. The data exhibited improvement in disease activity score, alleviation of swollen joints and duration of morning sickness at the end of intervention. The effect may be attributable to the immune modulatory effect of NSO (102). Hadi et al. further demonstrated that NSO ameliorated inflammation and oxidative stress. The serum IL-6 level was heightened significantly whereas reduced serum MDA and NO level were recorded after eight weeks of NSO intervention (103).

Furthermore, NSO was reported to be an effective adjunct therapy in metabolic syndrome patients. Najmi et al. showed the adjuvant effect of supplementing 2.5 mL of NSO twice daily, along with six weeks' daily intervention of metformin (1 g/day) and atorvastatin (10 mg/day). Significant decreases of total cholesterol, LDL cholesterol and fasting blood glucose levels were recorded in metabolic syndrome patients including diabetes and dyslipidemia (104). Meanwhile, the antihypertensive effect of NSO was investigated by providing NSO (2.5 mL, twice daily) to 70 healthy volunteers in a double-blinded randomized placebo-controlled trial. The observations revealed that systolic and diastolic blood pressure significantly declined in comparison to the placebo group and baseline after eight weeks, without inducing any adverse effect (105). These findings depicted beneficial effects of NSO to be used in preventive and therapeutic therapies for various chronic diseases. Nonetheless, future studies using larger sample sizes and groups, diverse doses of NSO and longer periods of investigations are still necessitated.

Moringa seed oil (MSO)

Moringa oleifera has gained ample attention owing to the nutritional value and potential health benefits of the seed oil. *M. oleifera* mature seeds contain up to 35–45% of golden yellow liquid oil. Electronic nose analysis indicated that it possesses a mild nutty flavor resembling that of peanuts. The yield of the oil from the seeds varies due to differences in the ripening stage, seed harvesting time, extraction methods used, cultivation climate, and variety of plants (42, 43).

In terms of composition, moringa seed oil (MSO) contains a high level of fatty acids with oleic acid being the dominant fatty acid (65–75%), followed by palmitic acid (12.31%), linoleic acid (16.00%), palmitoleic acid (2.10%) and stearic acid (5.10%). The high oleic acid content of MSO is comparable to that of olive oil, high oleic canola seed oil, high oleic sunflower seed oil and high oleic safflower seed oil. The sterol fraction of the MSO is made up mainly by the presence of β-sitosterol (46.65%), stigmasterol (19.0%), campesterol (16.0%), and Δ^5-avenasterol (10.7%). These sterols account for approximately 92% of total sterols. Selected sterols, specifically cholesterol, were however not detected. Other than fatty acids and sterols, MSO is also characterized by high tocopherol content. The level of α-, γ- and δ-tocopherols were reported up to 132.3, 63.9 and 81.2 mg/kg, respectively (43). Triolein (36.7%) is the main triacylglycerol, followed by other oleic acid-containing triacylglycerols, including stearo-diolein and palmito-diolein (106).

Application of MSO in disease prevention and health promotion are ascribed to its rich bioactive compositions. A high proportion of oleic acid, a type of monounsaturated fatty acid, has been associated with reducing blood cholesterol levels in individuals without hypertriglyceridemia (107). A study demonstrated oil or diet with high oleic acid content could significantly decrease the level of LDL cholesterol and triglyceride (108). It is also suggested that the substantial amount of linoleic acid found in MSO may provide high nutritional value and has the potential to lower blood pressure, blood lipid and serum cholesterol levels. High monounsaturated to saturated fatty acids of the seed oil was also reported by Schwingshackl and Hoffmann (109). The ratio has then been linked with minimized risk of cardiovascular events, stroke, cardiovascular mortality and all-cause mortality (109). In particular, the sterol fraction of MSO is of particular research interest owing to its probable implication in the metabolism of cholesterol (110). A preclinical *in vivo* study performed by Gupta et al. suggested that the antidiabetic potential of β-sitosterol is exerted via the reduction in glycated hemoglobin and serum glucose level, along with synchronal increment of serum insulin levels in β-sitosterol-treated diabetic rats (111). Nevertheless, result data concerning the implication of sterols on cardiovascular risk remain conflicting and inconclusive (112).

Applicability of MSO in averting chronic diseases have been discussed in several preclinical investigations. MSO was found to improve the biochemical abnormalities of diabetes by significantly reducing the low density lipoprotein, triglyceride, cholesterol, blood glucose, total protein urea, serum total bilirubin and creatinine level in comparison to the diabetic control (113). Additionally, MSO displayed a protective effect against diabetic-induced nephropathy. Aside from lowering the level of serum urea nitrogen, creatinine and uric acid, which have been indicated as parameters of diabetic nephropathy development, MSO also significantly increased the glutathione level. It was suggested that MSO could scavenge streptozotocin-induced oxidative stress (114). Interestingly, nano-micelle MSO was shown to exert more selective anti-proliferative activities and apoptosis against colorectal (Caco-2 and HCT 116) and breast (MCF-7) cancer cell lines, with less cytotoxicity toward noncancerous BHK-21 kidney cells (115). While these preclinical result data implied the potential therapeutic effect of MSO in the fight against diabetes and cancer, proper translation of these experimental data for clinical studies remain warranted.

Pistachio oil (PO)

Pistachio (*Pistacia* L., *Anacardiaceae* family) consists of 11 or more tree and shrub species. It is also one of the oldest flowering nut trees. Convincing evidences reported by PREDIMED studies (Prevention with Mediterranean Diet) suggest that nuts and nut oils can lower LDL cholesterol levels, as well as reduce incidence of hypertension, metabolic syndrome, diabetes and other inflammatory conditions (116). It is worthy to highlight that the US Food and

Drug Administration sanctioned the first qualified health claim for pistachios (as part of the tree nuts) to the reduced risk of heart disease, stating that "scientific evidence suggests but does not prove that eating 1.5 oz (42.5 g) per day of most nuts, such as pistachios, as part of a diet low in SFA and cholesterol may reduce the risk of heart disease" (117). Pistachios contain a relatively low fat content in comparison to other nuts (118). Nevertheless, its lipid fraction is rich in MUFA, PUFA and lipid-soluble bioactive components including vitamin K, phytosterols, lutein, tocopherol and polyphenols (44, 45). To date, most of the clinical trials on pistachio were performed through inclusion of the pistachio nuts in the regular dietary pattern as intervention study.

Zhang and colleagues investigated the effect of virgin pistachio oil (VPO, *Pistacia vera* L.) on expression of inflammation-related genes in an attempt to elucidate its role in immunology-related pathways, given that inflammation is a major risk factor for many chronic diseases (119). In the study, microarray analyses of RAW 264.7 macrophages revealed that a total of 43 genes were significantly regulated by 5.5 mg/mL of VPO treatment. In addition, 1 mg/mL of VPO and its organic extract significantly reduced inflammatory markers (IL-6, TNF-α, Ifit-2), further implying the potential beneficial effect of VPO to ameliorate cardiovascular disease risk through dietary intervention.

Several other studies have suggested that subclinical hypothyroidism patients were at higher risk of hypertension and dyslipidemia, which may eventually lead to adverse cardiovascular events (120, 121). Nazifi et al. determined the effects of including wild pistachio oil (WPO, *Pistacia* L.) in the diet of animals with experimental hypothyroidism induced through administration of propyl thiouracil (122). The study observed a repression in the increased lipid profiles of the treatment group receiving 5–20% of WPO via the diet. This indicated that WPO might be able to modulate hypothyroidism and lipid metabolism concomitantly.

Similarly, Jamshidi et al. examined the potential preventive effects of WPO (*Pistacia atlanticamutica*) on the elements and risk factors of metabolic syndrome in rats fed fructose solution for ten weeks (123). Consumption of 0.5 mL WPO reduced the levels of IL-6, LDL and total cholesterol in the experimental rats. In addition, experimental rats that received WPO also demonstrated insignificant lower increase in body weight in comparison to the fructose group.

Jebali and colleagues reported the efficacy of pistachio oil (PO, *Pistacia vera L.*) to attenuate inflammatory responses. Proteolipid protein injection was performed to induce autoimmune encephalomyelitis in mice. Administration of 10% v/v PO for one week significantly downregulated the gene expression of *HO-1*, *NOS2*, *MGST1*, *NFκB*, *MAPK* and *AKT1* from NFκB or oxidative stress signaling pathways. Furthermore, injection of 10% v/v PO reduced the elevated Th2($CD8^+$) and number of Th1($CD4^+$) cells in peripheral blood of the experimental mice (124). All these preclinical studies implied the potential health benefits of PO to improve vascular health, modulate glycemic control, as well as reduce oxidative and inflammatory stresses.

Evening primrose oil (EPO)

Evening primrose (*Oenothera biennis* L., *Onagraceae* family) is a biennial plant originated from North and South America. Evening primrose oil (EPO) is generally obtained through cold pressing followed by solvent extraction (such as hexane) from the plant seed (125). Evening primrose seeds can produce around 18–25% of oil; the content varies with factors such as growth conditions, cultivar and age of the seed (126). The seed oil contents include high amounts of essential fatty acid such as γ-linolenic acid (GLA), a compound not commonly found in a normal diet (127). In addition, EPO contains tocopherols (such as α-, γ-, δ-tocopherol), phytosterols (such as β-sitosterol, campesterol) and phenolic compounds (46–49). The oil is commonly used in the management of women's ailments including menopausal hot flashes (128), labor induction (129), cyclical mastalgia (130) and premenstrual syndrome (131). Several clinical studies had evaluated the efficacy of EPO in the treatment of chronic diseases including cardiovascular events, diabetes mellitus, cancer and rheumatologic conditions.

Since the late 1980s, the role of PUFAs has been extensively studied to investigate their activities against established solid and hematological tumors (132, 133). In this regard, the GLA-rich EPO demonstrated the potential to eradicate cancer cells and increase survival rate. Despite the fact that GLA was previously reported to demonstrate cytostatic activity *in vitro*, cytotoxic action of EPO toward cancer cells was still unproven through clinical studies. Based on the results of a double-blind, placebo-controlled trial conducted by van der Merwe, there was no statistically significant difference of survival time or liver size among primary liver cancer patients administered with 36 capsules/day (500 mg EPO/capsule) when compared with placebo (500 mg olive oil/capsule). It was proposed that the findings might be caused by insufficient dosage of GLA supplemented through EPO capsules (134).

Khodeer et al. reported the genoprotective effect against antineoplastic agent cyclophosphamide (CP). EPO showed significant improvement in serum glutamic oxaloacetic transaminase, glutamic pyruvic transaminase and pancreatic amylase in the CP-intoxicated mice. Histopathological scoring further corroborated the protective effects of EPO against CP-induced hepatic and pancreatic toxicity following treatments with either 5 or 10 mg/kg/day of EPO for 14 days. In addition, insulin level in the EPO-treated mice (for 10 mg/kg/day) was also restored to normal (135).

The effect of EPO in reducing the complications of type 2 diabetic mellitus was being investigated by Safaa Hussain et al. in a prospective randomized controlled interventional open label study. Thirteen patients aged between 35 and 60 and diagnosed with type 2 diabetes received 500 mg of metformin and 2 g of EPO capsule (twice daily) for three continual months. The results demonstrated that supplementation of EPO to metformin therapy significantly reduced the serum MDA, TNF-α and blood pressure levels in the patients (136).

Mert et al. studied the effect of EPO on adiponectin and biochemical parameters such as glucose level, total cholesterol, HDL level in a fructose-induced metabolic syndrome rat model. The administration of EPO at a dose of 0.1 mL/rat/day was found to demonstrate anti-inflammatory properties, and significantly increased the adiponectin and HDL levels in the experimental rats after 57 days of high fructose diet. Additionally, EPO was also found to decrease the systolic blood pressure in the rats. These findings implied the potential effectiveness of EPO as antidiabetic agent through regulation of adiponectin and HDL levels in blood (137).

In a randomized, double-blind, placebo-controlled trial performed by Jamilian et al., supplementation of 1000 IUs of vitamin D3 and 1000 mg EPO was also found to improve the glycemia and lipid profiles in 27 women with gestational diabetes mellitus. A significant reduction in serum TAG, VLDL, TC, LDL and TC/HDL was determined in the subjects supplemented with vitamin D and EPO after six weeks of intervention (138). Horrobin hypothesized that administration of GLA could rectify impaired nerve function in diabetic animal models that generally exhibit abnormal fatty acid patterns (139).

The potential cardiovascular protection of EPO toward the cardiovascular risk caused by celecoxib was also investigated. Male Wistar rats treated with 20 mg/kg/day celecoxib were co-treated with either n-3 PUFAs (360 mg EPA and 240 mg DHA/rat/day) or EPO at a dose of 5 g/kg/day over six weeks through gastric gavage. The results suggested that concomitant administration of both n-3 PUFAs or EPO reduced the celecoxib-induced elevation in blood pressure. EPO, but not n-3 PUFAs, was discovered to lower the acceleration of thrombogenesis triggered by celecoxib in the two thrombogenesis models (140). Andjic et al. investigated the effect of EPO on the cardiac function of isolated rat hearts from both male and female Wistar albino rats (141). Their findings suggested that male rats treated with EPO (10 mg/kg/day) for six weeks demonstrated stronger cardiac response compared to female rats. Together, these findings implicated the potential effectiveness of EPO in the secondary prevention of cardiovascular diseases.

EPO may also alleviate systemic chronic inflammation marked by atopic dermatitis and rheumatoid arthritis. A randomized, double-blind, placebo-controlled clinical study was conducted to examine the effect of EPO on patients with mild atopic dermatitis. It was concluded that administration of EPO for four months improved eczema area severity index scores in the mild atopic dermatitis patients (142). Another prospective, randomized controlled clinical trial performed also showed a positive reduction of oxidative stress biomarkers among the rheumatoid arthritis patients who had been taking concentrated fish oil together with EPO. Overall, GLA in EPO is positively associated with anti-inflammatory mechanisms through rapid conversion into dihomo-GLA (DGLA) that function to inhibit production of inflammatory leukotrienes (143). All these findings corroborate the therapeutic effects of EPO in various inflammatory and immunologic pathogeneses.

Melon seed oil (MeSO)

Melon (*Cucumis melo* L.) belongs to the *Cucurbitaceae* family and is cultivated in countries with temperate climates. Melon fruit is rich in cucurbitacin B, cucurbitacin D and cucurbitacin E, which may modulate anti-inflammatory and cytotoxic activities (144). Melon seeds, generally regarded as waste, are also extensively consumed as snacks in certain countries (145). Comprising approximately 91% TAG, melon seed oil (MeSO) possesses an intriguing fatty acids profile similar to soybean and sunflower oils (52, 146). MeSO is reported to be rich in PUFAs. Linoleic acid made up approximately 69% of the total fatty acids present in the oil. However, the content may differ in some cultivars and varies with the use of different extraction methods (50, 51). From a nutritional point view, the rich content of linoleic acid in MeSO reflects its importance and potential usage as a new type of cooking oil and in industrial applications. Significant amounts of bioactive compounds found in MeSO include tocopherol (such as γ- & δ-tocopherol), phytosterols (β-sitosterol & Δ5-avenasterol), phenolic compounds and carotenoids (50–53, 147).

MeSO has been reported to exert free radical-scavenging activity in *in vitro* DPPH assay measured colorimetrically (33, 148). There is, however, a limited number of studies to date investigating the clinical or preclinical efficacy of MeSO in the therapeutic area. Bouazzaoui et al. conducted *in vitro* experiments on the anti-inflammatory and cytotoxicity of MeSO (*Cucumis melo* L. *Inodorus*) extracted using supercritical carbon dioxide or n-hexane. MeSO demonstrated insignificant cytostatic inhibition against the two *IGROV* and *OVAR* human tumor cell lines. In the same studies, nevertheless, MeSO was found to exhibit limited anti-inflammatory activity on soybean lipoxygenase (149).

The study by Hao et al. examined the effects of wild MeSO (*Cucumis melo* var. agrestis) on blood cholesterol and gut microbiota using Golden Syrian hamsters hypercholesterolemia model (150). The researchers observed an upregulation of the expression of CYP7A1 in a dose-dependent manner. An enhanced excretion of fecal bile acids was also observed. The results data also showed that supplementation of 9.5% wild MeSO in the animal diet for six weeks significantly modulated the gut microbiota and reduced plasma cholesterol. The observation may relate to the downregulation of cholesterol synthesis by PUFAs in MeSO via activation of the CYP7A1 gene expression. Based on these studies, it seems fair to claim that MeSO is a promising new vegetable oil that may satisfy demand for human nutrition.

Cucumber seed oil (CSO)

Being a member of the *Cucurbitaceae* family, cucumber (*Cucumis sativus* L.) is an annual climber commonly grown on most continents. As a common ingredient of salads, it is often grown to be eaten fresh. The crisp texture and juiciness of cucumber is due to its high water composition. Containing cucurbitacin

C and cucurbitacin I, these bitter-tasting principles of cucumber were reported to exert anti-proliferative and anti-tumor properties (151).

Cucumber seeds contain more than 53.7% of a nutty flavored oil with a fatty acid profile similar to olive, corn and sunflower oils (152, 153). Cucumber seed oil (CSO) is reported to consist of approximately 3% phospholipid, 62% linoleic acid, 16% oleic acid, 11% stearic acid and 11% palmitic acid (54, 55). The presence of lipid-soluble bioactive compounds such as phytosterols and tocopherols (such as δ-, α-, γ-tocopherol) are also being confirmed by various studies (56, 57).

The therapeutic properties of cucumber fruits and seed extracts are well documented; the extracts are found to be effective in diminishing blood glucose levels, reducing pain related to moderate knee osteoarthritis and exhibiting desirable effects on serum lipid profiles (154–156). From the CSO perspective, Achu et al. investigated the atherogenicity of CSO in female Wistar albino rats after administration of 5% oil incorporated into the animal's diet for three months. There was, however, no significant difference being reported for the CSO experimental rat group in terms of liver weight, weight gain and total serum TAG level compared with palm oil, corn oil and *Cucumeropsis mannii* oil. Interestingly, LDL cholesterol level and the atherogenic ratio of CSO group were shown to be significantly higher when compared to both corn oil- and palm oil-based diets (157). Despite the fact that the biological activities of CSO were less exciting in the current reported studies, the high PUFA content of CSO may still warrant further investigations into its usage.

Kenaf seed oil (KSO)

Kenaf has received considerable attention for industrial usage owing to its high fiber content. Being a side product from kenaf plantation processes, kenaf seeds are often discarded. However, the seeds contain a valuable lipid composition and bioactive ingredients that pose potential for a new, safe edible oil (158). The major fatty acids found in kenaf seed oil (KSO) are palmitic, oleic and linoleic acids. The lipid fraction is rich in phytosterols and tocopherols, where sitosterol and γ-tocopherol are the two major components (58, 59).

The potential application of KSO in chronic diseases management were previously reported. Anti-hypercholesterolemic activity of KSO was studied using high-fat diet-induced dyslipidemia Sprague–Dawley rats. KSO significantly lowered TC, TG, LDL and MDA levels in the rats, with comparable efficacy to the clinically used simvastatin (159). Another study conducted with a different experimental design reported similar results. The efficacy of KSO in lowering the lipid profile of the animals was compared between the high cholesterol diet (HCD) and post-HCD period. Treatment of KSO during the 14-day HCD treatment period significantly improved the cholesterol profile, weight control and liver fat of the rats in comparison to the untreated HCD group. More significant improvement was observed at the post-HFD period (160). Based on the aforementioned two studies, it can be presumed that long-term consumption of KSO may lower the risk of cardiovascular diseases. To the best of our knowledge, however, none of these preclinical studies had been further translated for investigation at the clinical level.

The anticancer properties of KSO were evaluated using both *in vitro* and *in vivo* models. KSO demonstrated cytotoxicity against cervical, lung cancer, leukemic (161–163) and colon cancer (164) cell lines. The promising results from *in vitro* prompted further evaluation using the animal model. For the study on chemoprevention of colon cancer, the rats were pretreated with azoxymethane prior to KSO treatment. Results from the study showed that aberrant crypt foci (earliest identifiable neoplastic lesions) in the colon was reduced by 45, 51 and 53% in rats fed with 0.5, 1 and 1.5 g/kg of KSO, respectively (165), indicating the potential of KSO in reducing the risk of precancerous formation in the colon. In a separate anti-leukemic study, a reduction of the population of immature granulocytes and monocytes were recorded in WEHI-3B/BALB/c mice peripheral blood with a concomitant increased in T cells population. A concurrent reduction of infiltration of leukemic cells into the splenic red pulp was also reported. The findings demonstrated the potential use of KSO in managing leukemia (166). Nonetheless, the application of KSO for oncological therapy remains to be validated at the clinical level.

Conclusions

As outlined in this chapter, the present scientific result findings clearly demonstrated that some of these selected noncommercial plant-based oils possess cardioprotective efficacy, with accumulating data for use in the prevention and treatment of cancer, diabetes, neurodegenerative diseases and cardiovascular diseases (Figure 10.1). While some of these findings are supported by both preclinical (Table 10.2) and clinical (Table 10.3) result data, most of the findings cease at the preclinical level due to the challenge in translating the result for clinical evaluation. It is proposed that more robust and well-designed future preclinical and translational studies must be in place to provide clinically relevant data while deriving an in-depth understanding on the scientific basis and mechanisms of action. For development into functional food or nutraceutical and pharmaceutical applications, it is also crucial to ensure consistency of the oil product. To ensure the standardization of lipid, phytosterol and tocopherol content, proper mitigation must be conducted as early as the plantation stage. Proper choice and selection of plant species, cultivar, seed and planting conditions to the use of different extraction technologies and storage conditions may all affect lipid content and its efficacy.

Table 10.2 Preclinical Studies of Selected Noncommercial Plant-based Edible Oils on Chronic Diseases

Oil	Disease	Experimental Design	Dose and Treatment Duration	Outcomes	References
Pumpkin seed oil (PSO)	Inflammation	Methotrexate-induced nitrosative stress model in rat	Oral gavage; PSO (40 mg/kg BW); daily for 5 days	↓ intestinal damage ↓ serum prostaglandin E_2, malondialdehyde, & nitric oxide levels ↓ myeloperoxidase, xanthine oxidase, & adenosine deaminase activities ↑ reduced glutathione level	(11)
	Inflammation	Acute and chronic chemical-induced topical inflammation model in mice	Topical application; PSO (25, 50 & 100%; 20 μL/ear); acute treatment for 1 & 4 hours; chronic treatment for 6–96 hours	↓ acute and chronic chemical-induced ear edema (dose dependent) for both skin thickness and weight ↓ intense dermal edema and congestion, epidermal hyperplasia and number of infiltrating inflammatory cells	(15)
	Inflammation & genotoxicity	Freund's complete adjuvant (FCA)-induced adjuvant arthritis in rat	Oral gavage; PSO (40 & 500 mg/kg BW); daily for 21 days	↓ edema thickness, plasma tumor necrosis factor-α, erythrocyte sedimentation rate & malondialdehyde ↑ plasma total antioxidant capacity ↓ chromosomal aberration, sperm shape abnormalities & DNA fragmentations	(16)
	Genotoxicity	Azathioprine-induced genotoxicity model in mice	Oral gavage; PSO (4 mL/kg BW); daily for 10 days	↓ the frequencies of micronucleated polychromatic erythrocytes, DNA fragmentation & total sperm abnormalities ↑ increased sperm count, polychromatic erythrocytes % & polychromatics/normochromatics ratio	(12)

(Continued)

Table 10.2 Preclinical Studies of Selected Noncommercial Plant-based Edible Oils on Chronic Diseases *(Continued)*

Oil	Disease	Experimental Design	Dose and Treatment Duration	Outcomes	References
	Cytotoxicity	Bisphenol A-induced cytotoxicity model in mice	Oral gavage; PSO (1 mL/kg BW); daily for 28 days	↓ DNA damage ↓ micronucleated polychromatic erythrocytes and polychromatics/normochromatics ratio in bone marrow cells ↑ protection against hepatic and testes tissues histopathological alterations	(13)
	Cytotoxicity	Nanoalumina-induced cytotoxicity model in pregnant rat	Oral gavage; PSO (4 mL/kg BW); daily from 5th until 19th day of gestation	↑ pregnancy outcomes, fetal growth parameters, antioxidant defences, ↓ DNA damage, histopathological changes of the liver and brain for both pregnant rats and fetuses	(14)
	Hypertension & cardiovascular disease	N^G-Nitro-L-arginine-methyl ester (L-NAME)–induced hypertensive model in rat	Oral gavage; PSO (40 & 100 mg/kg BW); daily for 6 weeks	↓ systolic blood pressure & normalized the ECG changes (including prolongation of the RR interval, increased P wave duration, & ST elevation) ↓ plasma malondialdehyde ↑ plasma nitric oxide ↓ pathological alterations in heart and aorta	(17)
	Cardiovascular disease	Ovariectomy-induced dyslipidemia model in rat	Oral gavage; PSO (40 mg/kg BW); 5 days weekly for 12 weeks	↓ total cholesterol, triglycerides, & LDL-C ↑ HDL-C ↓ systolic and diastolic blood pressures	(18)
	Cardiovascular disease	High fat diet-induced dyslipidemia model in rat	Diet; PSO (50 g/kg BW); 14 weeks	↓ triglyceride, total cholesterol, LDL-C, glutathione S-transferase ↑ superoxide dismutase ↓ endoplasmic reticulum stress & unfolded protein response	(19)

(Continued)

Table 10.2 Preclinical Studies of Selected Noncommercial Plant-based Edible Oils on Chronic Diseases *(Continued)*

Oil	Disease	Experimental Design	Dose and Treatment Duration	Outcomes	References
	Benign prostatic hyperplasia	Testosterone-induced benign prostatic hyperplasia model in rat	Oral gavage; PSO (20–40 mg/kg BW); daily for 20 days	↔ weight gain ↓ prostate size ratio	(21)
	Benign prostatic hyperplasia	Testosterone /Prazosin induced benign prostatic hyperplasia model in rat	Oral gavage; PSO (2.5 mL/kg BW/day); daily for 20 days	↓ weight ratio for ventral prostate ↓ total protein levels in both ventral and dorsolateral lobes	(22)
Watermelon seed oil (WSO)	Neurotoxic disease	Cadmium-induced neurotoxicity model in mice	Oral gavage; WSO (0.25–0.5 g/kg BW); daily for 4 weeks	↑ Acetylcholinesterase activity (dose dependent) in the brain ↑ levels of glutathione, superoxide dismutase, catalase & glutathione peroxidase ↓ lipid peroxidation ↓ cerebral edema and prevented hippocampal damage	(62)
	Neurotoxic disease	Mercury chloride-induced neurotoxicity model in rat	Oral gavage; WSO (200 mg/kg BW); daily for 14 days	↓ mercury chloride-induced degeneration of frontal cerebral cortical neurons	(66)
	Liver disease	Carbon tetrachloride-induced hepatotoxicity model in rat	Oral gavage; WSO (125-250 mg/kg BW); daily for 10 days	↓ serum alkaline phosphatase, aspartate transaminase, & alanine transaminase levels ↓ hepatic sinusoidal dilation with normalization of the cells and presence of mild inflammogens	(65)
	Cardiovascular disease	Diet-induced hypercholesterolemic rat model	Diet; WSO (5% in the diet); 6 weeks	↓ serum triglycerides, total, free and esterified cholesterol ↓ foam cell formation, smooth muscle cell migration in the blood vessel and severe atherosclerosis in the aorta	(67)

(Continued)

Table 10.2 Preclinical Studies of Selected Noncommercial Plant-based Edible Oils on Chronic Diseases *(Continued)*

Oil	Disease	Experimental Design	Dose and Treatment Duration	Outcomes	References
	Cardiovascular disease	Diet-induced hypercholesterolemic rat model	Diet; WSO (5% in the diet); 6 weeks	↓ both hepatic and serum levels of triglycerides, total cholesterol, non-high-density lipoprotein cholesterol ↓ both hepatic and serum levels of malondialdehyde ↓ serum glutathione peroxidase and glutathione reductase levels ↑ hepatic glutathione peroxidase and glutathione reductase levels	(68)
	Diabetes	Alloxan-induced diabetes model in mice	Oral gavage; WSO (150, 200 and 250 mg/kg BW); treatment duration not mentioned	↓ blood glucose level (dose dependent)	(71)
	Inflammation	*Candida albicans* infection-induced inflammation model in rat	Oral gavage; WSO (50, 100 and 150 mg/kg BW); daily for one week	↑ hind paw oedema after 24 hours ↓ hind paw oedema after 48 hours ↓ $CD4^+$ T-lymphocytes, IFN-γ and TNF-α from day 7 to 14 ↑ IL-10 and IgA from day 7 to 14	(74)
Citrus seed fixed oil (CSFO)	Non-small cell lung cancer	Non-small cell human pulmonary-carcinoma A549 cell line & orthotopic lung cancer model in mice	*In vitro*: CSFO (0–400 μg/mL); 24 hours *In vivo*: CSFO (1.75 or 5.25 mg/mice/day); 3 weeks	↓ A549 cell proliferation in culture (dose dependent) ↓ tumor growth & membrane-bound Ras protein in the group receiving 5.25 mg CSFO ↑ apoptosis in both *in vitro* and *in vivo* models	(83)
	Skin cancer	Murine B16F10 melanoma cell line	CSFO (50–1000 μg/mL); 24 hours	↓ growth of murine B16F10 melanoma cells	(35)
	Diabetes & cardiovascular disease	Alloxan-induced diabetes model in mice	Intraperitoneal injection; CSFO emulsion (1000 mg/kg BW); daily for 28 days	↓ fasting blood glucose level ↓ serum triglyceride, total cholesterol, LDL-C, VLDL-C ↑ HDL-C	(81)

(Continued)

Table 10.2 Preclinical Studies of Selected Noncommercial Plant-based Edible Oils on Chronic Diseases *(Continued)*

Oil	Disease	Experimental Design	Dose and Treatment Duration	Outcomes	References
Citrus peel fixed oil (CPFO)	Depression	Chronic unpredictable mild stress (CUMS) model in mice	Inhalation; CPFO (1 mL inside the cage with size of $28 \times 21 \times 17.5$ cm); 1.5 or 24 hours for 5 days	↓ CUMS-induced depressive behavior, hyperactivity of hypothalamic–pituitary–adrenal axis ↓ monoamine neurotransmitter levels, brain-derived neurotrophic factor and its receptor expression in the hippocampus ↓ total cholesterol, triglycerides, HDL-C, LDL-C	(89)
Nigella sativa seed oil (NSO)	Diabetes	Streptozotocin-induced diabetes model in rat	Intraperitoneal injection; 5% *N. sativa* aqueous extract (2 mL/kg BW); NSO (0.2 mL/kg BW); Thymoquinone (3 mg/mL); 6 days per week for 30 days	↓ tissue MDA and glucose ↑ serum insulin and tissue SOD	(93)
	Diabetes	Streptozotocin-induced diabetes model in rat	Intragastric intubation; NSO (0.2 mL/kg BW); isotonic NaCl; daily for 4 weeks	↑ intensity of staining for insulin and preservation of β-cell numbers ↑ (moderately) in the lowered secretory vesicles with granules and slight destruction (with loss of cristae) within the mitochondria of β-cell	(94)
	Multi-organ cancer	Multi-organ carcinogenesis model in rat	Powdered basal diet (1000 or 4000 mg/L NSO); normal basal diet; 0.9% saline injections; NSO (4000 mg/L); daily for 30 weeks	↓ malignant and benign colon tumor sizes ↓ incidences and multiplicities in the lungs, as well as different parts of the alimentary canal (oesophagus and forestomach) ↓ cell proliferation in various organs and lesions	(97)
	Colon Cancer	1,2-dimethylhydrazine dihydrochloride–induced colonic tumorigenesis model in rat	Intragastric intubation; NSO (200 mg/kg BW) in 0.5 mL 5% carboxymethyl cellulose; daily for 14 weeks	↓ occurrence of the DMH-induced colonic aberrant crypt foci at both the initiation and post-initiation stages	(98)

(Continued)

Table 10.2 Preclinical Studies of Selected Noncommercial Plant-based Edible Oils on Chronic Diseases *(Continued)*

Oil	Disease	Experimental Design	Dose and Treatment Duration	Outcomes	References
	Breast cancer	Dimethylbenz(a)anthracene (DMBA)-induced mammary carcinogenesis model in rat	Oral gavage; 400 mg NSO/100 g BW; daily for 3 months	$\downarrow$ the frequency of mammary carcinoma $\downarrow$ the markers of tumorigenicity endocrine derangement $\downarrow$ oxidative stress $\uparrow$ apoptotic activity	(99)
Moringa Seed oil (MSO)	Diabetes	Alloxan–induced diabetes model in rat	Oral gavage; dichloromethane-extracted MSO (2.0 mL/kg BW); petroleum ether-extracted MSO (2.0 mL/kg BW)	$\downarrow$ blood glucose level, cholesterol, triglyceride, low-density lipoprotein, serum total bilirubin, total protein urea and creatinine when compared to diabetic control	(113)
	Diabetes	Streptozotocin–induced diabetes model in rat	Standard diet containing MSO (1.8 mg/kg BW)	$\downarrow$ serum urea nitrogen, creatinine and uric acid levels $\uparrow$ serum glutathione level	(114)
	Colon and breast cancer	*In vitro* cell culture (HepG2, MCF7, HCT 116, Caco-2 and BHK-21 cell lines)	MSO (0–100 μg/mL); MSO nano-micelle (0–100 μg/mL); 24 hours	Anti-proliferative activities against colorectal cell lines (Caco-2 and HCT 116 cells) and breast cancer cell line (MCF-7 cells)	(115)
Pistachio oil (PO)	Inflammation	*In vitro* cell culture (RAW 264.7 *Mus musculus* macrophage cell line)	Oil extract (0.125–1 mg/mL)	Modulation of 43 genes by PO treatment at 5.5 mg/mL $\downarrow$ Ifit expression $\downarrow$ inflammatory markers (Ifit-2, TNF-α and IL-6)	(119)
	Hypothyroidism	Propylthiouracil-induced hypothyroidism model in rat	Standard diet containing PO (5, 10 & 20%); 30 days	$\downarrow$ total cholesterol, triglyceride, HDL and LDL in the experimental groups when compared to experimental group that had PTU only	(122)
	Metabolic syndrome	Fructose-induced metabolic syndrome model in rat	Oral gavage; sunflower oil (0.5 mL); PO (hull + kernel) (0.5 mL); PO (hull only) (0.5 mL); PO (kernel only) (0.5 mL); twice daily for 10 weeks	$\downarrow$ total cholesterol, LDL and IL-6 levels	(123)

(Continued)

Table 10.2 Preclinical Studies of Selected Noncommercial Plant-based Edible Oils on Chronic Diseases *(Continued)*

Oil	Disease	Experimental Design	Dose and Treatment Duration	Outcomes	References
	Multiple Sclerosis	Proteolipid protein (PLP)-induced experimental autoimmune encephalomyelitis (EAE) model in mice	Oral gavage; PO (1, 5 10%; v/v); one week	↓ expression level of *AKT1*, *MAPK*, *NFKB*, *MGST1*, *NOS2* and *HO-1* genes from NFKB or oxidative stress signaling pathways ↓ the number of Th1(CD4+) and ↑ Th2(CD8+) cells in peripheral blood	(124)
Evening primrose oil (EPO)	Genotoxicity	Cyclophosphamide-induced genotoxicity model in mice	EPO pretreatment (5 or 10 mg/kg/day) for 14 days	↓ oxidative, inflammatory, histopathological and genotoxic changes ↑ insulin secretion (return to normal)	(135)
	Metabolic syndrome	Fructose-induced metabolic syndrome model in rat	Oral gavage; EPO (0.1 mL/rat/day); 57 days	↓ systolic blood pressure ↓ uric acid, triglyceride, LDL, VLDL, ALT, AST and LDH levels ↑ HDL level ↓ total oxidant status and TNF-α levels	(137)
	Cardiovascular disease	Celecoxib-induced cardiovascular risk model in rat	Intragastric intubation; n3-PUFAs (360 mg EPA & 240 mg DHA); EPO (5 g/kg BW/day); daily for 6 weeks	↓ blood pressure ↓ the number of thrombi in lung vessels	(140)
	Cardiovascular disease	Adult Wistar albino rats	Intragastric intubation; linseed oil (300 mg/kg BW/day); EPO (10 mg/kg BW/day); daily for 6 weeks	Slightly changed the cardiac function compared to control in male rats but not in female rats. Treated male rats demonstrated higher cardiodynamic values.	(141)
Melon seed oil (MeSO)	Inflammation & cytotoxicity	*In vitro* cell culture (OVAR & IGROV human tumour cell lines)	MeSO (50 mg/L)	↓ inflammation ↓ cancer cell activities	(149)

(Continued)

Table 10.2 Preclinical Studies of Selected Noncommercial Plant-based Edible Oils on Chronic Diseases *(Continued)*

Oil	Disease	Experimental Design	Dose and Treatment Duration	Outcomes	References
	Hypercholesterolemia	Diet-induced atherosclerosis model in Golden Syrian hamsters	Noncholesterol diet; high-cholesterol diet (HCD); HCD containing 4.75 and 9.5% MeSO; 6 weeks	↓ plasma total cholesterol, triacylglycerol and non-HDL-cholesterol (dose-dependent) ↓ formation of high cholesterol-induced atherosclerotic plaque ↑ gene expression of CYP7A1 (dose-dependent) Modified intestinal microbial community	(150)
Cucumber seed oil (CSO)	Atherogenesis	Female Wistar albino rats	Specially formulated rat diet containing either 5% of corn oil, palm oil, *Cucumeropsis mannii* melon seed oil or CSO; 3 months	↑ LDL cholesterol when compared to control, corn and palm oil treatment groups. ↑ atherogenic ratio when compared to corn and palm oil treatment groups.	(157)
Kenaf seed oil (KSO)	Hypercholesterolemia	High fat diet-induced dyslipidemia model in rat	Normal diet; HFD; 90% HFD + 10% KSO; daily for 32 days	↓ Total cholesterol, triglycerides, LDL and MDA level comparable to simvastatin (40 mg/kg/day)	(159)
	Hypercholesterolemia	High cholesterol diet-induced dyslipidemia model in rat	KSO (1 mL/kg BW); daily for 14 days	↓ Total cholesterol and LDL in stage 3 when compared to the HCD group.	(160)
	Colon cancer	Azoxymethane-induced colonic aberrant crypt foci model in rat	Oral gavage; KSO (0.5, 1 and 1.5 g/kg BW); daily for 13 weeks	↓ aberrant crypt foci in colon	(165)
	Leukemia	WEHI-3 leukemia animal model in BALB/c mice	Oral gavage; KSO (0.5, 1 and 1.5 g/kg BW); daily for 2 weeks	↑ T cells population ↓ immature monocytes and granulocytes population in the peripheral blood ↓ infiltration of leukemic cells into the splenic red pulp.	(166)

Notes: High cholesterol diet (HCD), high fat diet (HFD), high-density lipoprotein (HDL), low-density lipoprotein (LDL), malondialdehyde (MDA), body weight (BW).

Table 10.3 Clinical Studies of Selected Noncommercial Plant-based Edible Oils on Chronic Diseases

Oil	Treated Disease	Trial Design and Subjects	Dose and Treatment Duration	Outcomes	References
Pumpkin seed oil (PSO)	Cardiovascular disease & hypertension	**Trial Design:** Randomized, double-blind, placebo-controlled trial **Subjects:** Postmenopausal women (n=35)	**Intervention:** PSO (2 g/day) **Placebo:** Wheat germ oil (3 g/day) **Duration:** 12 weeks	↑ HDL-C ↓ diastolic blood pressure ↓ severity of hot flushes, headaches, joint pains (improved menopausal symptom scores)	(20)
	Benign prostatic hyperplasia	**Trial Design:** Multicentric surveillance study **Subjects:** Benign prostate hyperplasia patients (n=2245)	**Intervention:** PSO (1–2 capsules) **Duration:** 12 weeks	↓ international prostate symptom score ↑ quality of life	(23)
	Benign prostatic hyperplasia	**Trial Design:** Randomized, double-blind, placebo-co ntrolled trial **Subjects:** Benign prostate hyperplasia patients (n=47)	**Intervention:** PSO (320 mg/day) **Placebo:** sweet potato starch (320 mg/day) **Duration:** 12 months	↓ urinary dysfunction & international prostate symptom score ↑ quality of life & maximal urinary flow rate	(24)
	Benign prostatic hyperplasia	**Trial Design:** Prospective randomized clinical trial **Subjects:** Benign prostate hyperplasia patients (n=92)	**Intervention:** PSO (2 tablets) **Placebo:** Prazosin (2 tablets) **Duration:** 6 months	↔ prostate specific antigen level ↓ international prostate symptom score ↑ quality of life & maximal urinary flow rate	(25)
	Urinary disorders	**Trial Design:** Surveillance study **Subjects:** Healthy volunteers (n=45)	**Intervention:** PSO (10g/day) **Duration:** 12 weeks	↓ Overactive bladder symptom score (OABSS)	(26)

(Continued)

Table 10.3 Clinical Studies of Selected Noncommercial Plant-based Edible Oils on Chronic Diseases *(Continued)*

Oil	Treated Disease	Trial Design and Subjects	Dose and Treatment Duration	Outcomes	References
Nigella sativa seed oil (NSO)	Type 2 diabetes mellitus	**Trial Design:** Randomized, double-blind, placebo-controlled trial **Subjects:** T2DM patients aged 30–60 years old (n=72)	**Intervention:** NSO (3 g/day) **Placebo:** Sunflower oil (3 g/day) **Duration:** 12 weeks	↓ body weight and body mass index ↓ fasting blood sugar and glycated hemoglobin ↓ triglyceride and LDL-cholesterol	(95)
	Type 2 diabetes mellitus	**Trial Design:** Single-blind trial **Subjects:** T2DM patients aged 30–60 years old (n=41)	**Intervention:** NSO (equivalent to 0.7 g of *N. sativa* seeds) **Placebo:** Wheat bran oil **Duration:** 40 days	↓ fasting blood glucose level ↑ insulin level	(96)
	Obesity	**Trial Design:** Randomized, double-blind, placebo-controlled trial **Subjects:** Obese females aged 25–50 years old (n=49)	**Intervention:** Low-calorie diet supplemented with NSO (3 g/day) **Placebo:** Low-calorie diet supplemented with sunflower oil (3 g/day) **Duration:** 8 weeks	↓ body weight ↑ level of red blood cell superoxidase dismutase	(100)
	Obesity	**Trial Design:** Randomized, double-blind, placebo-controlled trial **Subjects:** Obese females aged 25–50 years old (n=84)	**Intervention:** Low-calorie diet supplemented with NSO (3 g/day) **Placebo:** Low-calorie diet supplemented with sunflower oil (3 g/day) **Duration:** 8 weeks	↓ body weight and waist circumference ↓ triglyceride and VLDL	(101)
	Rheumatoid arthritis	**Trial Design:** Placebo-controlled trial **Subjects:** Rheumatoid arthritis patients (n=40)	**Intervention:** NSO (1 g/day) **Placebo:** Starch-filled capsule **Duration:** 1 month	↓ disease activity score ↓ number of swollen joints ↓ duration of morning stiffness	(102)

(Continued)

Table 10.3 Clinical Studies of Selected Noncommercial Plant-based Edible Oils on Chronic Diseases *(Continued)*

Oil	Treated Disease	Trial Design and Subjects	Dose and Treatment Duration	Outcomes	References
	Rheumatoid arthritis	**Trial Design:** Randomized, double-blind, placebo-controlled trial **Subjects:** Rheumatoid arthritis patients (n=42)	**Intervention:** NSO (1 g/day) **Placebo:** Paraffin capsule **Duration:** 8 weeks	↑ serum IL-10 level ↓ serum malondialdehyde and nitric oxide when compared with baseline	(103)
	Metabolic syndrome	**Trial Design:** Controlled trial **Subjects:** Patients with metabolic syndrome, characterized by abdominal obesity, dyslipidemia, high blood pressure and high fasting blood glucose (n=60)	**Intervention:** NSO (2.5 mL; twice daily), metformin (500 mg; twice daily) & atorvastatin (10 mg; once daily) **Control:** Metformin (500 mg; twice daily) & atorvastatin (10 mg; once daily) **Duration:** 6 weeks	↓ fasting blood glucose level ↓ total cholesterol ↓ LDL cholesterol	(104)
	Hypertension	**Trial Design:** Randomized, double-blind, placebo-controlled trial **Subjects:** Normal to hypertension stage I patients (n=70)	**Intervention:** NSO (2.5 mL; twice daily) **Placebo:** Mineral oil (2.5 mL; twice daily) **Duration:** 8 weeks	↓ systolic and diastolic blood pressures No other adverse effect	(105)
Evening primrose oil (EPO)	Primary liver cancer	**Trial Design:** Randomized, double-blind, placebo-controlled trial **Subjects:** Patients with a histological diagnosis of primary cancer and an expected survival of more than 14 days (n=62)	**Intervention:** EPO (500 mg; 36 capsules daily) **Placebo:** Olive oil (500 mg; 36 capsules daily) **Duration:** Not available	↔ in the survival time and liver size of EPO supplemented group in comparison to placebo	(134)

(Continued)

Table 10.3 Clinical Studies of Selected Noncommercial Plant-based Edible Oils on Chronic Diseases *(Continued)*

Oil	Treated Disease	Trial Design and Subjects	Dose and Treatment Duration	Outcomes	References
	Type 2 diabetic mellitus	**Trial Design:** Prospective randomized, controlled, interventional open label study **Subjects:** Patients newly diagnosed with type 2 diabetes mellitus (n=26) & healthy control subjects (n=14)	**Intervention:** Metformin (500 mg; twice daily) & EPO (2 g; twice daily) **Active Comparator:** Metformin (500 mg; twice daily) **Control:** No specific intervention **Duration:** 3 months	↓ serum malondialdehyde, hd-CRP & TNF-α levels ↓ systolic and diastolic blood pressure	(136)
	Gestational diabetes	**Trial Design:** Randomized, double-blind, placebo-controlled trial **Subjects:** Pregnant participants diagnosed with gestational diabetes	**Intervention:** Vitamin D (1000 IU) & EPO (1000 mg); once daily **Placebo:** Placebo with color, shape, size and packaging identical to intervention; once daily **Duration:** 6 weeks	↓ fasting plasma glucose, serum insulin, HOMA insulin resistance and HOMA-B cell function indexes ↑ quantitative insulin sensitivity check index ↓ serum total cholesterol, triglyceride, VLDL, LDL and TC/HDL levels	(138)
	Atopic dermatitis	**Trial Design:** Randomized, double-blind, placebo-controlled trial **Subjects:** Mild atopic dermatitis patients with Eczema Area Severity Index (EASI) score equal to or lower than 10 (n=50)	**Intervention:** EPO (450 mg/capsule) **Placebo:** Soybean oil (450 mg/capsule) **Dose:** 8 capsules/day; for patients between ages of 2 and 12 (4 capsules/day) **Duration:** 4 months	↓ EASI score at the end of treatment period ↑ (not significant) transepidermal water loss (TEWL) and skin hydration	(142)

(Continued)

Table 10.3 Clinical Studies of Selected Noncommercial Plant-based Edible Oils on Chronic Diseases *(Continued)*

Oil	Treated Disease	Trial Design and Subjects	Dose and Treatment Duration	Outcomes	References
	Rheumatoid arthritis	**Trial Design:** Randomized, controlled trial **Subjects:** Postmenopausal women with stable rheumatologic therapy (n=60)	**Intervention:** Regular rheumatologic therapy, concentrated fish oil supplement (2 gel capsules daily) & EPO (1.3 g; 2 capsules daily) **Control 1:** Regular rheumatologic therapy only **Control 2:** Regular rheumatologic therapy & concentrated fish oil supplement (5 gel capsules daily) **Duration:** 3 months	↑ plasma levels of lipid peroxidation (thiobarbituric acid reactive substances) & nitric oxide ↑ activities of superoxide dismutase (SOD)	(143)

Declaration of competing interest

The authors declare that there are no conflicts of interest.

Acknowledgments

This work was funded by Fundamental Research Grant Scheme (FRGS/1/2019/STG05/MAHSA/02/2), Ministry of Higher Education Malaysia.

References

1. Organisation, W.H. *Noncommunicable diseases: key facts*. 2021 [cited 2021 18 May 2021]; Available from: https://www.who.int/news-room/fact-sheets/detail/noncommunicable-diseases.
2. Alkhatib, A., *Chapter 33 - Olive oil nutraceuticals and chronic disease prevention: more than an offshoot of the Mediterranean diet*, in *The Mediterranean Diet (Second Edition)*, V.R. Preedy and R.R. Watson, Editors. 2020, Academic Press: London. p. 363–370.
3. Visioli, F., M. Franco, E. Toledo, *et al.*, *Olive oil and prevention of chronic diseases: summary of an International conference*. Nutr. Metab. Cardiovasc. Dis., 2018. **28**(7): p. 649–656.
4. Kadandale, S., R. Marten, and R. Smith, *The palm oil industry and noncommunicable diseases*. Bull. W.H.O., 2019. **97**(2): p. 118–128.
5. Ruan, M., Y. Bu, F. Wu, *et al.*, *Chronic consumption of thermally processed palm oil or canola oil modified gut microflora of rats*. Food Sci. Hum. Wellness, 2021. **10**(1): p. 94–102.
6. Patterson, E., R. Wall, G.F. Fitzgerald, R.P. Ross, and C. Stanton, *Health implications of high dietary omega-6 polyunsaturated fatty acids*. Nutr. Metab., 2012. **2012**: p. 539426.
7. Dubois, V., S. Breton, M. Linder, J. Fanni, and M. Parmentier, *Fatty acid profiles of 80 vegetable oils with regard to their nutritional potential*. Eur. J. Lipid Sci. Technol., 2007. **109**(7): p. 710–732.
8. Alasalvar, C. and B.W. Bolling, *Review of nut phytochemicals, fat-soluble bioactives, antioxidant components and health effects*. Br. J. Nutr., 2015. **113**(S2): p. S68–S78.
9. Chen, B., D.J. McClements, and E.A. Decker, *Design of Foods with Bioactive Lipids for Improved Health*. Annu. Rev. Food Sci. Technol., 2013. **4**(1): p. 35–56.
10. Tetali, S.D., *Terpenes and isoprenoids: a wealth of compounds for global use*. Planta, 2019. **249**(1): p. 1–8.
11. El-Boghdady, N.A., *Protective effect of ellagic acid and pumpkin seed oil against methotrexate-induced small intestine damage in rats*. Indian J. Biochem. Biophys., 2011. **48**(6): p. 380–387.
12. Elfiky, S., I. Elelaimy, A. Hassan, H. Ibrahim, and R. Elsayad, *Protective effect of pumpkin seed oil against genotoxicity induced by azathioprine*. J. Basic Appl. Zool., 2012. **65**(5): p. 289–298.
13. Fawzy, E.I., A.I. El Makawy, M.M. El-Bamby, and H.O. Elhamalawy, *Improved effect of pumpkin seed oil against the bisphenol – A adverse effects in male mice*. Toxicol. Rep., 2018. **5**: p. 857–863.
14. Hamdi, H. and M.M. Hassan, *Maternal and developmental toxicity induced by Nanoalumina administration in albino rats and the potential preventive role of the pumpkin seed oil*. Saudi J. Biol. Sci., 2021.
15. de Oliveira, M.L.M., D.C.S. Nunes-Pinheiro, B.M.O. Bezerra, *et al.*, *Topical anti-inflammatory potential of pumpkin (Cucurbita pepo L.) seed oil on acute and chronic skin inflammation in mice*. Acta Sci. Vet., 2013. **41**(1): p. 1–9.

16. Al-Okbi, S., D.A. Mohamed, E. Kandil, *et al.*, *Anti-inflammatory activity of two varieties of pumpkin seed oil in an adjuvant arthritis model in rats.* Grasas Aceites, 2017. **68**(1): p. 180.
17. El-Mosallamy, A.E., A.A. Sleem, O.M. Abdel-Salam, N. Shaffie, and S.A. Kenawy, *Antihypertensive and cardioprotective effects of pumpkin seed oil.* J. Med. Food, 2012. **15**(2): p. 180–189.
18. Gossell-Williams, M., K. Lyttle, T. Clarke, M. Gardner, and O. Simon, *Supplementation with pumpkin seed oil improves plasma lipid profile and cardiovascular outcomes of female non-ovariectomized and ovariectomized Sprague-Dawley rats.* Phytother. Res., 2008. **22**(7): p. 873–877.
19. Zhao, X.J., Y.L. Chen, B. Fu, *et al.*, *Intervention of pumpkin seed oil on metabolic disease revealed by metabonomics and transcript profile.* J. Sci. Food Agric., 2017. **97**(4): p. 1158–1163.
20. Gossell-Williams, M., C. Hyde, T. Hunter, *et al.*, *Improvement in HDL cholesterol in postmenopausal women supplemented with pumpkin seed oil: pilot study.* Climacteric, 2011. **14**(5): p. 558–564.
21. Gossell-Williams, M., A. Davis, and N. O'connor, *Inhibition of testosterone-induced hyperplasia of the prostate of Sprague-Dawley rats by pumpkin seed oil.* J. Med. Food, 2006. **9**(2): p. 284–286.
22. Tsai, Y.-S., Y.-C. Tong, J.-T. Cheng, *et al.*, *Pumpkin seed oil and phytosterol-F can block testosterone/prazosin-induced prostate growth in rats.* Urol. Int., 2006. **77**(3): p. 269–274.
23. Friederich, M., C. Theurer, and G. Schiebel-Schlosser, *Prosta Fink Forte®-Kapseln in der Behandlung der benignen Prostatahyperplasie. Eine multizentrische Anwendungsbeobachtung an 2245 Patienten.* Complement. Med. Res., 2000. **7**(4): p. 200–204.
24. Hong, H., C.-S. Kim, and S. Maeng, *Effects of pumpkin seed oil and saw palmetto oil in Korean men with symptomatic benign prostatic hyperplasia.* Nutr. Res. Pract., 2009. **3**(4): p. 323.
25. Shirvan, M.K., M.R.D. Mahboob, M. Masuminia, and S. Mohammadi, *Pumpkin seed oil (prostafit) or prazosin? Which one is better in the treatment of symptomatic benign prostatic hyperplasia.* J. Pak. Med. Assoc., 2014. **64**(6): p. 683–685.
26. Nishimura, M., T. Ohkawara, H. Sato, H. Takeda, and J. Nishihira, *Pumpkin seed oil extracted from Cucurbita maxima improves urinary disorder in human overactive bladder.* J. Tradit. Complement. Med., 2014. **4**(1): p. 72–74.
27. Guo, Y., M. Gao, X. Liang, *et al.*, *Quantitative trait loci for seed size variation in cucurbits–A review.* Front. Plant Sci., 2020. **11**: p. 304.
28. Biswas, R., S. Ghosal, A. Chattopadhyay, and S. Datta, *A comprehensive review on watermelon seed oil–An underutilized product.* IOSR J Pharm, 2017. **7**(11): p. 01–07.
29. Montesano, D., F. Blasi, M.S. Simonetti, A. Santini, and L. Cossignani, *Chemical and nutritional characterization of seed oil from Cucurbita maxima L. (var. Berrettina) pumpkin.* Foods, 2018. **7**(3): p. 30.
30. Stevenson, D.G., F.J. Eller, L. Wang, *et al.*, *Oil and tocopherol content and composition of pumpkin seed oil in 12 cultivars.* J. Agric. Food Chem., 2007. **55**(10): p. 4005–4013.
31. Fedko, M., D. Kmiecik, A. Siger, *et al.*, *Comparative characteristics of oil composition in seeds of 31 Cucurbita varieties.* J. Food Meas. Charact., 2020. **14**(2): p. 894–904.
32. Atolani, O., J. Omere, C. Otuechere, and A. Adewuyi, *Antioxidant and cytotoxicity effects of seed oils from edible fruits.* J. Acute Dis., 2012. **1**(2): p. 130–134.
33. Jorge, N., A.C. da Silva, and C.R. Malacrida, *Physicochemical characterisation and radical-scavenging activity of Cucurbitaceae seed oils.* Nat. Prod. Res., 2015. **29**(24): p. 2313–2317.
34. de Conto, L.C., M.A.L. Gragnani, D. Maus, *et al.*, *Characterization of crude watermelon seed oil by two different extractions methods.* J. Am. Oil Chem. Soc., 2011. **88**(11): p. 1709–1714.

35. Rosa, A., B. Era, C. Masala, *et al.*, *Supercritical CO2 extraction of waste citrus seeds: chemical composition, nutritional and biological properties of edible fixed oils.* Eur. J. Lipid Sci. Technol., 2019. **121**(7): p. 1800502.
36. Guneser, B.A. and E. Yilmaz, *Bioactives, aromatics and sensory properties of cold-pressed and hexane-extracted lemon (Citrus limon L.) seed oils.* J. Am. Oil Chem. Soc., 2017. **94**(5): p. 723–731.
37. Anwar, F., R. Naseer, M. Bhanger, *et al.*, *Physico-chemical characteristics of citrus seeds and seed oils from Pakistan.* J. Am. Oil Chem. Soc., 2008. **85**(4): p. 321–330.
38. Matthaus, B. and M. Özcan, *Chemical evaluation of citrus seeds, an agro-industrial waste, as a new potential source of vegetable oils.* Grasas Aceites, 2012. **63**(3): p. 313–320.
39. Alrashidi, M., D. Derawi, J. Salimon, and M. Firdaus Yusoff, *An investigation of physicochemical properties of Nigella sativa L. seed oil from Al-Qassim by different extraction methods.* J. King Saud Univ. Sci., 2020. **32**(8): p. 3337–3342.
40. Benkaci-Ali, F., A. Baaliouamer, J.P. Wathelet, and M. Marlier, *Chemical composition and physicochemical characteristics of fixed oils from Algerian Nigella sativa seeds.* Chem. Nat. Compd., 2012. **47**(6): p. 925–931.
41. Rohman, A., E. Lukitaningsih, M. Rafi, N. Ahmad Fadzillah, and A. Windarsih, *Nigella sativa oil: physico-chemical properties, authentication analysis and its antioxidant activity.* Food Res., 2019. **3**: p. 628–634.
42. Leone, A., A. Spada, A. Battezzati, *et al.*, *Moringa oleifera seeds and oil: characteristics and uses for human health.* Int. J. Mol. Sci., 2016. **17**(12): p. 2141.
43. Nadeem, M. and M. Imran, *Promising features of Moringa oleifera oil: recent updates and perspectives.* Lipids Health Dis., 2016. **15**(1): p. 212.
44. Dreher, M.L., *Pistachio nuts: composition and potential health benefits.* Nutr. Rev., 2012. **70**(4): p. 234–240.
45. Salvador, M.D., R.M. Ojeda-Amador, and G. Fregapane, *Virgin Pistachio (Pistachia vera L.) Oil*, in *Fruit Oils: Chemistry and Functionality*, M.F. Ramadan, Editor. 2019, Springer International Publishing: Cham. p. 181–197.
46. Bialek, A., M. Bialek, M. Jelinska, and A. Tokarz, *Fatty acid composition and oxidative characteristics of novel edible oils in Poland.* CYTA J. Food, 2017. **15**(1): p. 1–8.
47. Pan, F., Y. Li, X. Luo, *et al.*, *Effect of the chemical refining process on composition and oxidative stability of evening primrose oil.* J. Food Process. Preserv., 2020. **44**(10): p. e14800.
48. Timoszuk, M., K. Bielawska, and E. Skrzydlewska, *Evening primrose (Oenothera biennis) biological activity dependent on chemical composition.* Antioxidants, 2018. **7**(8): p. 108.
49. Zhao, B., H. Gong, H. Li, *et al.*, *Fatty acid, triacylglycerol and unsaponifiable matters profiles and physicochemical properties of Chinese evening primrose oil.* J. Oleo Sci., 2019. **68**(8): p. 719–728.
50. Mallek-Ayadi, S., N. Bahloul, and N. Kechaou, *Cucumis melo L. seeds as a promising source of oil naturally rich in biologically active substances: compositional characteristics, phenolic compounds and thermal properties.* Grasas Aceites, 2019. **70**(1): p. e284.
51. Azhari, S., Y. Xu, Q. Jiang, and W. Xia, *Physicochemical properties and chemical composition of Seinat (Cucumis melo var. tibish) seed oil and its antioxidant activity.* Grasas Aceites, 2014. **65**(1): p. 008.
52. Hemavatahy, J., *Lipid composition of melon (Cucumis melo) kernel.* J. Food Compost. Anal., 1992. **5**(1): p. 90–95.
53. Petkova, Z. and G. Antova, *Proximate composition of seeds and seed oils from melon (Cucumis melo L.) cultivated in Bulgaria.* Cogent Food Agric., 2015. **1**(1): p. 1018779.
54. Achu, M., E. Fokou, C. Tchiégang, M. Fotso, and M. Tchouanguep. *Chemical characteristics and fatty acid composition of cucurbitaceae oils from Cameroon.* in *13th World Congress of Food Science & Technology 2006.* 2006.

55. Ali, M.A., M.A. Sayeed, S.K. Ghosh, *et al.*, *Comparative study of the characteristics of seed oil and seed nutrient content of three varieties of Cucumis sativus L.* Pak. J. Sci. Ind. Res., 2011. **54**(2): p. 68.
56. Matthaus, B., K. Vosmann, L.Q. Pham, and K. Aitzetmüller, *FA and tocopherol composition of Vietnamese oilseeds.* J. Am. Oil Chem. Soc., 2003. **80**(10): p. 1013–1020.
57. Yuenyong, J., P. Pokkanta, N. Phuangsaijai, *et al.*, *GC-MS and HPLC-DAD analysis of fatty acid profile and functional phytochemicals in fifty cold-pressed plant oils in Thailand.* Heliyon, 2021. **7**(2): p. e06304.
58. Nyam, K.L., C.P. Tan, O.M. Lai, K. Long, and Y.B. Che Man, *Physicochemical properties and bioactive compounds of selected seed oils.* LWT - Food Sci. Technol., 2009. **42**(8): p. 1396–1403.
59. Mohamed, A., H. Bhardwaj, A. Hamama, and C. Webber, *Chemical composition of kenaf (Hibiscus cannabinus L.) seed oil.* Ind. Crops Prod., 1995. **4**(3): p. 157–165.
60. Mariod, A.A., B. Matthäus, and M. Ismail, *Comparison of supercritical fluid and hexane extraction methods in extracting kenaf (Hibiscus cannabinus) seed oil lipids.* J. Am. Oil Chem. Soc., 2011. **88**(7): p. 931–935.
61. Adaramola, B. and O. Adebayo, *Comparative fatty acids profiling and antioxidant potential of pawpaw and watermelon seed oils.* J. Pharm. Res. Int., 2016: p. 1–9.
62. Adnaik, R., P. Gavarkar, and S. Mohite, *Evaluation of antioxidant effect of Citrullus vulgaris against cadmium-induced neurotoxicity in mice brain.* Int. J. Pharm. Sci. Res., 2015. **6**(10): p. 4316.
63. Hassan, L., H. Sirat, S. Dahham, *et al.*, *Phytochemical and toxicity studies of Citrullus lanatus var. citroides (wild melon) on brine shrimps (lethality test).* Aust. J. Basic Appl. Sci., 2015. **9**(25): p. 60–65.
64. Eke, R., E. Ejiofor, S. Oyedemi, S. Onoja, and N. Omeh, *Evaluation of nutritional composition of Citrullus lanatus Linn.(watermelon) seed and biochemical assessment of the seed oil in rats.* J. Food Biochem., 2021. **45**(6): p. e13763.
65. Madhavi, P., V. Kamala, and R. Habibur, *Hepatoprotective activity of Citrullus lanatus seed oil on CCl4 induced liver damage in rats.* Sch. Acad. J. Pharm., 2012. **1**(1): p. 30–33.
66. Owoeye, O., R. Akinbami, and M. Thomas, *Neuroprotective potential of Citrullus lanatus seed extract and vitamin E against mercury chloride intoxication in male rat brain.* Afr. J. Biomed. Res., 2018. **21**(1): p. 43–49.
67. Oluba, O., O. Adeyemi, G. Ojieh, and I. Isiosio, *Fatty acid composition of Citrullus lanatus (egusi melon) and its effect on serum lipids and some serum enzymes.* Inter. J. Cardio. Res., 2008. **5**(2–10).
68. Oluba, O.M., G.O. Eidangbe, G.C. Ojieh, and B.O. Idonije, *Palm and egusi melon oils lower serum and liver lipid profile and improve antioxidant activity in rats fed a high fat diet.* Int. J. Med. Med. Sci., 2011. **3**(2): p. 47–51.
69. Eidangbe, G.O., G. Ojieh, B. Idonije, and O. Oluba, *Palm oil and egusi melon oil lower serum and liver lipid profile and improve antioxidant activity in rats fed a high fat diet.* J. Food Technol., 2010. **8**(4): p. 154–158.
70. Nameni, R.O., C.Y. Woumbo, A.P. Kengne, *et al.*, *Effects of stifled cooking on the quality and lipid-lowering potential of oils extracted from two species of pumpkin seeds (Citrullus lanatus and Cucumeropsis mannii).* Pharmacology, 2021. **4**(1): p. 47.
71. Sani, U.M., *Phytochemical screening and antidiabetic effect of extracts of the seeds of Citrullus lanatus in alloxan-induced diabetic albino mice.* J. Appl. Pharm. Sci., 2015. **5**(3): p. 51–54.
72. Ghosh, T.K., H. Rahman, D. Bardalai, F. Ali, and H. Rahman, *In-vitro antibacterial study of Aquilaria agallocha heart wood oil and Citrullus lanatus seed oil.* Sch. J. Appl. Med. Sci., 2013. **1**(1): p. 13–15.
73. Adewuyi, A., R. Ayodele Oderinde, and A. Ololade Ademisoye, *Antibacterial activities of nonionic and anionic surfactants from Citrullus lanatus seed oil.* Jundishapur J. Microbiol., 2013. **6**(3): p. 205–208.

74. Apeh, V.O., F.I. Chukwuma, F.N. Nwora, O.U. Njoku, and F.O. Nwodo, *Significance of crude and degummed Citrullus lanatus seed oil on inflammatory cytokines in experimental infection induced by Candida albicans*. Acta Pharm. Sci., 2021. **59**(3): p. 363–383.
75. Madhavi, P., K. Vakati, and H. Rahman, *Evaluation of anti-inflammatory activity of Citrullus lanatus seed oil by in-vivo and in-vitro models*. Int. J. Appl. Pharm. Sci. Res., 2012. **2**(4): p. 104–108.
76. Komane, B., I. Vermaak, G. Kamatou, B. Summers, and A. Viljoen, *The topical efficacy and safety of Citrullus lanatus seed oil: a short-term clinical assessment*. S. Afr. J. Bot., 2017. **112**: p. 466–473.
77. Sharma, K., N. Mahato, M.H. Cho, and Y.R. Lee, *Converting citrus wastes into value-added products: economic and environmently friendly approaches*. Nutrition, 2017. **34**: p. 29–46.
78. Burnett, C.L., M.M. Fiume, W.F. Bergfeld, *et al.*, *Safety assessment of citrus-derived peel oils as used in cosmetics*. Int. J. Toxicol., 2019. **38**(2_suppl): p. 33S–59S.
79. Azar, A.P., M. Nekoei, K. Larijani, and S. Bahraminasab, *Chemical composition of the essential oils of Citrus sinensis cv. Valencia and a quantitative structure-retention relationship study for the prediction of retention indices by multiple linear regression*. J. Serbian Chem. Soc., 2011. **76**(12): p. 1627–1637.
80. United States Pharmacopeial Convention Council of Experts & United States Pharmacopeial Convention Food Ingredient Expert Committee, *Food Chemicals Codex*. 2010: US Pharmacopeia Conv.
81. Chilaka, K., E. Ifediba, and J. Ogamba, *Evaluation of the effects of Citrus sinensis seed oil on blood glucose, lipid profile and liver enzymes in rats injected with alloxan monohydrate*. J. Acute Dis., 2015. **4**(2): p. 129–134.
82. Roe, F.J. and W.E. Peirce, *Tumor promotion by citrus oils: tumors of the skin and urethral orifice in mice*. J. Natl. Cancer Inst., 1960. **24**(6): p. 1389–1403.
83. Castro, M.A., B. Rodenak-Kladniew, A. Massone, *et al.*, *Citrus reticulata peel oil inhibits non-small cell lung cancer cell proliferation in culture and implanted in nude mice*. Food Funct., 2018. **9**(4): p. 2290–2299.
84. Kirbaşlar, F.G., A. Tavman, B. Dülger, and G. Türker, *Antimicrobial activity of Turkish citrus peel oils*. Pak. J. Bot, 2009. **41**(6): p. 3207–3212.
85. Chalova, V.I., P.G. Crandall, and S.C. Ricke, *Microbial inhibitory and radical scavenging activities of cold-pressed terpeneless Valencia orange (Citrus sinensis) oil in different dispersing agents*. J. Sci. Food Agric., 2010. **90**(5): p. 870–876.
86. Benelli, P., C.A. Riehl, A. Smânia Jr, E.F. Smânia, and S.R. Ferreira, *Bioactive extracts of orange (Citrus sinensis L. Osbeck) pomace obtained by SFE and low pressure techniques: Mathematical modeling and extract composition*. J. Supercrit. Fluids, 2010. **55**(1): p. 132–141.
87. Yi, F., R. Jin, J. Sun, B. Ma, and X. Bao, *Evaluation of mechanical-pressed essential oil from Nanfeng mandarin (Citrus reticulata Blanco cv. Kinokuni) as a food preservative based on antimicrobial and antioxidant activities*. LWT - Food Sci. Technol., 2018. **95**: p. 346–353.
88. Güneşer, B., N.D. Zorba, and E. Yılmaz, *Antimicrobial activity of cold pressed citrus seeds oils, some citrus flavonoids and phenolic acids*. Riv. Ital. delle Sostanze Grasse, 2018. **95**(2): p. 119–131.
89. Zhang, L.-L., Z.-Y. Yang, G. Fan, *et al.*, *Antidepressant-like effect of Citrus sinensis (L.) Osbeck essential oil and its main component limonene on mice*. J. Agric. Food Chem., 2019. **67**(50): p. 13817–13828.
90. Salehi, B., C. Quispe, M. Imran, *et al.*, *Nigella plants – traditional uses, bioactive phytoconstituents, preclinical and clinical studies*. Front. Pharmacol., 2021. **12**(417): p. 625386.
91. Hussein El-Tahir, K.E.-D. and D.M. Bakeet, *The black seed Nigella sativa linnaeus - a mine for multi cures: a plea for urgent clinical evaluation of its volatile oil*. J. Taibah Univ. Medical Sci., 2006. **1**(1, Supplement C): p. 1–19.

92. El-Dakhakhny, M., *Studies on the chemical constitution of Egyptian Nigella sativa L. seeds. II[1]) the essential oil.* Planta Med., 1963. **11**(4): p. 465–470.
93. Abdelmeguid, N.E., R. Fakhoury, S.M. Kamal, and R.J. Al Wafai, *Effects of Nigella sativa and thymoquinone on biochemical and subcellular changes in pancreatic β-cells of streptozotocin-induced diabetic rats.* J. Diabetes., 2010. **2**(4): p. 256–66.
94. Kanter, M., M. Akpolat, and C. Aktas, *Protective effects of the volatile oil of Nigella sativa seeds on beta-cell damage in streptozotocin-induced diabetic rats: a light and electron microscopic study.* J. Mol. Histol., 2009. **40**(5-6): p. 379–85.
95. Heshmati, J., N. Namazi, M.-R. Memarzadeh, M. Taghizadeh, and F. Kolahdooz, *Nigella sativa oil affects glucose metabolism and lipid concentrations in patients with type 2 diabetes: a randomized, double-blind, placebo-controlled trial.* Food Res. Int., 2015. **70**: p. 87–93.
96. Ahmad, B., T. Masud, A. Uppal, and A. Naveed, *Effects of Nigella sativa oil on some blood parameters in type 2 diabetes mellitus patients.* Asian J. Chem., 2009. **21**: p. 5373–5381.
97. Salim, E.I., *Cancer chemopreventive potential of volatile oil from black cumin seeds, Nigella sativa L., in a rat multi-organ carcinogenesis bioassay.* Oncol. Lett., 2010. **1**(5): p. 913–924.
98. Salim, E.I. and S. Fukushima, *Chemopreventive potential of volatile oil from black cumin (Nigella sativa L.) seeds against rat colon carcinogenesis.* Nutr. Cancer, 2003. **45**(2): p. 195–202.
99. El-Aziz, M.A.A., H.A. Hassan, M.H. Mohamed, *et al.*, *The biochemical and morphological alterations following administration of melatonin, retinoic acid and Nigella sativa in mammary carcinoma: an animal model.* Int. J. Exp. Pathol., 2005. **86**(6): p. 383–396.
100. Namazi, N., R. Mahdavi, M. Alizadeh, and S. Farajnia, *Oxidative stress responses to Nigella sativa oil concurrent with a low-calorie diet in obese women: a randomized, double-blind controlled clinical trial.* Phytother. Res., 2015. **29**(11): p. 1722–1728.
101. Mahdavi, R., N. Namazi, M. Alizadeh, and S. Farajnia, *Effects of Nigella sativa oil with a low-calorie diet on cardiometabolic risk factors in obese women: a randomized controlled clinical trial.* Food Funct., 2015. **6**(6): p. 2041–2048.
102. Gheita, T.A. and S.A. Kenawy, *Effectiveness of Nigella sativa oil in the management of rheumatoid arthritis patients: a placebo controlled study.* Phytother. Res., 2012. **26**(8): p. 1246–1248.
103. Hadi, V., S. Kheirouri, M. Alizadeh, A. Khabbazi, and H. Hosseini, *Effects of Nigella sativa oil extract on inflammatory cytokine response and oxidative stress status in patients with rheumatoid arthritis: a randomized, double-blind, placebo-controlled clinical trial.* Avicenna J. Phytomed., 2016. **6**(1): p. 34–43.
104. Najmi, A., M. Nasiruddin, R.A. Khan, and S.F. Haque, *Effect of Nigella sativa oil on various clinical and biochemical parameters of insulin resistance syndrome.* Int. J. Diabetes Dev. Ctries., 2008. **28**(1): p. 11–14.
105. Huseini, F.H., M. Amini, R. Mohtashami, *et al.*, *Blood pressure lowering effect of Nigella sativa L. seed oil in healthy volunteers: a randomized, double-blind, placebo-controlled clinical trial.* Phytother. Res., 2013. **27**(12): p. 1849–1853.
106. Basuny, A.M. and M.A. Al-Marzouq, *Biochemical studies on Moringa oleifera seed oil.* MOJ Food Process. Technol., 2016. **2**: p. 40–46.
107. Mensink, R.P. and M.B. Katan, *Effect of dietary trans fatty acids on high-density and low-density lipoprotein cholesterol levels in healthy subjects.* N. Engl. J. Med., 1990. **323**(7): p. 439–445.
108. Allman-Farinelli, M.A., K. Gomes, E.J. Favaloro, and P. Petocz, *A diet rich in high-oleic-acid sunflower oil favorably alters low-density lipoprotein cholesterol, triglycerides, and factor VII coagulant activity.* J. Am. Diet. Assoc., 2005. **105**(7): p. 1071–1079.
109. Schwingshackl, L. and G. Hoffmann, *Monounsaturated fatty acids, olive oil and health status: a systematic review and meta-analysis of cohort studies.* Lipids Health Dis., 2014. **13**: p. 154.

110. Ras, R.T., J.M. Geleijnse, and E.A. Trautwein, *LDL-cholesterol-lowering effect of plant sterols and stanols across different dose ranges: a meta-analysis of randomised controlled studies*. Br. J. Nutr., 2014. **112**(2): p. 214–219.
111. Gupta, R., A.K. Sharma, M.P. Dobhal, M.C. Sharma, and R.S. Gupta, *Antidiabetic and antioxidant potential of β-sitosterol in streptozotocin-induced experimental hyperglycemia*. J. Diabetes, 2011. **3**(1): p. 29–37.
112. Genser, B., G. Silbernagel, G. De Backer, *et al.*, *Plant sterols and cardiovascular disease: a systematic review and meta-analysis*. Eur. Heart J., 2012. **33**(4): p. 444–451.
113. Bola, B.M., H.L. Muhammad, E.O. Ogbadoyi, *et al.*, *In vivo evaluation of antidiabetic properties of seed oil of Moringa oleifera lam*. J. Appl. Life Sci. Int., 2015. **2**: p. 160–174.
114. Yousef, S.M., *Evaluation the protective effect of Moringa oleifera seeds oil against diabetic nephropathy in male rats*. Sci. J. Specif. Edu. Appl. Sci., 2018. **1**(1): p. 267–282.
115. Abd-Rabou, A.A., K.M. A Zoheir, M.S. Kishta, A.B. Shalby, and M.I. Ezzo, *Nano-micelle of Moringa oleifera seed oil triggers mitochondrial cancer cell apoptosis*. Asian Pac. J. Cancer Prev., 2016. **17**(11): p. 4929–4933.
116. Salvador, M.D., R.M. Ojeda-Amador, and G. Fregapane, *Chapter 30 - Pistachio nut, its virgin oil, and their antioxidant and bioactive activities*, in *Pathology*, V.R. Preedy, Editor. 2020, Academic Press: London. p. 309–320.
117. U.S., F.D.A. *Letter of Enforcement Discretion – Nuts and Coronary Heart Disease. Qualified Health Claim.*. 2003 [cited 2021 Accessed 18 May]; Available from: https://www.fda.gov/food/food-labeling-nutrition/qualified-health-claims-letters-enforcement-discretion.
118. Bulló, M., M. Juanola-Falgarona, P. Hernández-Alonso, and J. Salas-Salvadó, *Nutrition attributes and health effects of pistachio nuts*. Br. J. Nutr., 2015. **113**(Suppl 2): p. S79–S93.
119. Zhang, J., P.M. Kris-Etherton, J.T. Thompson, and J.P. Vanden Heuvel, *Effect of pistachio oil on gene expression of IFN-induced protein with tetratricopeptide repeats 2: A biomarker of inflammatory response*. Mol. Nutr. Food Res., 2010. **54**(S1): p. S83–S92.
120. Delitala, A.P., G. Fanciulli, M. Maioli, and G. Delitala, *Subclinical hypothyroidism, lipid metabolism and cardiovascular disease*. Eur. J. Intern. Med., 2017. **38**: p. 17–24.
121. Duntas, L.H. and G. Brenta, *A renewed focus on the association between thyroid hormones and lipid metabolism*. Front. Endocrinol., 2018. **9**(511): p. 1–10.
122. Nazifi, S., M. Saeb, M. Sepehrimanesh, and S. Poorgonabadi, *The effects of wild pistachio oil on serum leptin, thyroid hormones, and lipid profile in female rats with experimental hypothyroidism*. Comp. Clin. Path., 2012. **21**(5): p. 851–857.
123. Jamshidi, S., N. Hejazi, M.-T. Golmakani, and N. Tanideh, *Wild pistachio (Pistacia atlantica mutica) oil improve metabolic syndrome features in rats with high fructose ingestion*. Iran J. Basic Med. Sci., 2018. **21**(12): p. 1255–1261.
124. Jebali, A., M. Noroozi Karimabad, Z. Ahmadi, *et al.*, *Attenuation of inflammatory response in the EAE model by PEGlated nanoliposome of pistachio oils*. J. Neuroimmunol., 2020. **347**: p. 577352.
125. Bayles, B. and R. Usatine, *Evening primrose oil*. Am. Fam. Physician, 2009. **80**(12): p. 1405–1408.
126. Christie, W.W., *The analysis of evening primrose oil*. Ind. Crops Prod., 1999. **10**(2): p. 73–83.
127. Hernandez, E.M., *4 - Specialty Oils: Functional and Nutraceutical Properties*, in *Functional Dietary Lipids*, T.A.B. Sanders, Editor. 2016, Woodhead Publishing: Sawston. p. 69–101.
128. Farzaneh, F., S. Fatehi, M.-R. Sohrabi, and K. Alizadeh, *The effect of oral evening primrose oil on menopausal hot flashes: a randomized clinical trial*. Arch. Gynecol. Obstet., 2013. **288**(5): p. 1075–1079.
129. Kalati, M., M. Kashanian, F. Jahdi, *et al.*, *Evening primrose oil and labour, is it effective? A randomised clinical trial*. J. Obstet. Gynaecol., 2018. **38**(4): p. 488–492.
130. Balci, F.L., C. Uras, and S. Feldman, *Clinical factors affecting the therapeutic efficacy of evening primrose oil on mastalgia*. Ann. Surg. Oncol., 2020. **27**(12): p. 4844–4852.

131. Hardy, M.L., *Herbs of special interest to women*. J. Am. Pharm. Assoc., 2000. **40**(2): p. 234–242.
132. D'Eliseo, D. and F. Velotti, *Omega-3 fatty acids and cancer cell cytotoxicity: implications for multi-targeted cancer therapy*. J. Clin. Med., 2016. **5**(2): p. 15.
133. Shaikh, I.A.A., I. Brown, K.W.J. Wahle, and S.D. Heys, *Enhancing cytotoxic therapies for breast and prostate cancers with polyunsaturated fatty acids*. Nutr. Cancer, 2010. **62**(3): p. 284–296.
134. van der Merwe, C.F., J. Booyens, H.F. Joubert, and C.A. van der Merwe, *The effect of gamma-linolenic acid, an in vitro cytostatic substance contained in evening primrose oil, on primary liver cancer. A double-blind placebo controlled trial*. Prostaglandins Leukot. Essent. Fatty Acids, 1990. **40**(3): p. 199–202.
135. Khodeer, D.M., E.T. Mehanna, A.I. Abushouk, and M.M. Abdel-Daim, *Protective effects of evening primrose oil against cyclophosphamide-induced biochemical, histopathological, and genotoxic alterations in mice*. Pathogens, 2020. **9**(2): p. 98.
136. SafaaHussain, M., M.K. Abdulridha, and M.S. Khudhair, *Anti-inflammatory, antioxidant, and vasodilating effect of evening primrose oil in type 2 diabetic patients*. Int. J. Pharmaceut. Sci. Rev. Res., 2016. **2**(39): p. 5.
137. Mert, H., K. İrak, S. Çibuk, S. Yıldırım, and N. Mert, *The effect of evening primrose oil (Oenothera biennis) on the level of adiponectin and some biochemical parameters in rats with fructose induced metabolic syndrome*. Arch. Physiol. Biochem. 2022. 128(6): p. 1539-1547.
138. Jamilian, M., M. Karamali, M. Taghizadeh, *et al.*, *Vitamin D and evening primrose oil administration improve glycemia and lipid profiles in women with gestational diabetes*. Lipids, 2016. **51**(3): p. 349–356.
139. Horrobin, D.F., *Essential fatty acids in the management of impaired nerve function in diabetes*. Diabetes, 1997. **46**(Supplement 2): p. S90–S93.
140. Zaitone, S.A., Y.M. Moustafa, S.M. Mosaad, and N.F. El-Orabi, *Effect of evening primrose oil and omega-3 polyunsaturated fatty acids on the cardiovascular risk of celecoxib in rats*. J. Cardiovasc. Pharmacol., 2011. **58**(1).
141. Andjic, M., N. Draginic, K. Radoman, *et al.*, *Flaxseed and evening primrose oil slightly affect systolic and diastolic function of isolated heart in male but not in female rats*. Int. J. Vitam. Nutr. Res., 2019. **8**: p. 1–9.
142. Chung, B.Y., S.Y. Park, M.J. Jung, H.O. Kim, and C.W. Park, *Effect of evening primrose oil on Korean patients with mild atopic dermatitis: a randomized, double-blinded, placebo-controlled clinical study*. Ann. Dermatol., 2018. **30**(4): p. 409–416.
143. Vasiljevic, D., M. Veselinovic, M. Jovanovic, *et al.*, *Evaluation of the effects of different supplementation on oxidative status in patients with rheumatoid arthritis*. Clin. Rheumatol., 2016. **35**(8): p. 1909–1915.
144. Yuan, R.-Q., L. Qian, W.-J. Yun, *et al.*, *Cucurbitacins extracted from Cucumis melo L. (CuEC) exert a hypotensive effect via regulating vascular tone*. Hypertens. Res., 2019. **42**(8): p. 1152–1161.
145. Rabadán, A., M.A. Nunes, S.M.F. Bessada, *et al.*, *From by-product to the food chain: melon (Cucumis melo L.) seeds as potential source for oils*. Foods, 2020. **9**(10): p. 1341.
146. Silva, M.A., T.G. Albuquerque, R.C. Alves, M.B.P.P. Oliveira, and H.S. Costa, *Melon (Cucumis melo L.) by-products: potential food ingredients for novel functional foods?* Trends Food Sci. Technol., 2020. **98**: p. 181–189.
147. Rezig, L., M. Chouaibi, K. Msaada, and S. Hamdi, *Chapter 54 - Cold pressed Cucumis melo L. seed oil*, in *Cold Pressed Oils*, M.F. Ramadan, Editor. 2020, Academic Press: London. p. 611–623.
148. Górnaś, P., A. Soliven, and D. Segliņa, *Seed oils recovered from industrial fruit by-products are a rich source of tocopherols and tocotrienols: rapid separation of α/β/γ/δ homologues by RP-HPLC/FLD*. Eur. J. Lipid Sci. Technol., 2015. **117**(6): p. 773–777.

149. Bouazzaoui, N., J. Bouajila, S. Camy, J.K. Mulengi, and J.-S. Condoret, *Fatty acid composition, cytotoxicity and anti-inflammatory evaluation of melon (Cucumis melo L. Inodorus) seed oil extracted by supercritical carbon dioxide.* Sep. Sci. Technol., 2018. **53**(16): p. 2622–2627.
150. Hao, W., H. Zhu, J. Chen, *et al.*, *Wild melon seed oil reduces plasma cholesterol and modulates gut microbiota in hypercholesterolemic hamsters.* J. Agric. Food Chem., 2020. **68**(7): p. 2071–2081.
151. Lim, T.K., *Cucumis sativus*, in *Edible Medicinal And Non-Medicinal Plants: Volume 2, Fruits.* 2012, Springer Netherlands: Dordrecht. p. 239–249.
152. Mariod, A.A., M.E. Saeed Mirghani, and I. Hussein, *Chapter 16 - Cucumis sativus Cucumber, in Unconventional Oilseeds and Oil Sources*, A.A. Mariod, M.E. Saeed Mirghani, and I. Hussein, Editors. 2017, Academic Press: London. p. 89–94.
153. Sharma, V., L. Sharma, and K.S. Sandhu, *Cucumber (Cucumis sativus L.),* in *Antioxidants in Vegetables and Nuts - Properties and Health Benefits*, G.A. Nayik and A. Gull, Editors. 2020, Springer Singapore: Singapore. p. 333–340.
154. Minaiyan, M., B. Zolfaghari, and A. Kamal, *Effect of hydroalcoholic and buthanolic extract of Cucumis sativus seeds on blood glucose level of normal and streptozotocin-induced diabetic rats.* Iran J Basic Med Sci, 2011. **14**(5): p. 436–442.
155. Nash, R.J., B.K. Azantsa, H. Sharp, and V. Shanmugham, *Effectiveness of Cucumis sativus extract versus glucosamine-chondroitin in the management of moderate osteoarthritis: a randomized controlled trial.* Clin. Interv. Aging, 2018. **13**: p. 2119–2126.
156. Soltani, R., M. Hashemi, A. Farazmand, *et al.*, *Evaluation of the effects of Cucumis sativus seed extract on serum lipids in adult hyperlipidemic patients: a randomized double-blind placebo-controlled clinical trial.* J. Food Sci., 2017. **82**(1): p. 214–218.
157. Achu, M.B., E. Fokou, C. Tchiégang, M. Fotso, and F.M.J.A.J.F.S. Tchouanguep, *Atherogenicity of Cucumeropsis mannii and Cucumis sativus oils from Cameroon.* Afr. J. Food Sci., 2008. **2**: p. 21–25.
158. Abd Ghafar, S.A., L. Saiful Yazan, S. Fakhurazi, and M. Ismail, *Toxicological evaluation of kenaf seed supercritical carbon dioxide-oil in male Sprague Dawley rats.* ASM Sci., 2021. **14**: p. 7.
159. Kai, N.S., T.A. Nee, E.L.C. Ling, *et al.*, *Anti–hypercholesterolemic effect of kenaf (Hibiscus cannabinus L.) seed on high–fat diet Sprague Dawley rats.* Asian Pac. J. Trop. Med., 2015. **8**(1): p. 6–13.
160. Cheong, A.M., J.X. Jessica Koh, N.O. Patrick, C.P. Tan, and K.L. Nyam, *Hypocholesterolemic effects of kenaf seed oil, macroemulsion, and nanoemulsion in high-cholesterol diet induced rats.* J. Food Sci., 2018. **83**(3): p. 854–863.
161. Foo, J., L. Yazan, K. Chan, P. Tahir, and M. Ismail, *Kenaf seed oil from supercritical carbon dioxide fluid extraction induced G1 phase cell cycle arrest and apoptosis in leukemia cells.* Afr. J. Biotechnol., 2011. **10**(27): p. 5389–5397.
162. Yazan, L., J. Foo, K. Chan, P. Tahir, and M. Ismail, *Kenaf seed oil from supercritical carbon dioxide fluid extraction shows cytotoxic effects towards various cancer cell lines.* Afr. J. Biotechnol., 2011. **10**(27): p. 5381–5388
163. Wong, Y.H., W.Y. Tan, C.P. Tan, K. Long, and K.L. Nyam, *Cytotoxic activity of kenaf (Hibiscus cannabinus L.) seed extract and oil against human cancer cell lines.* Asian Pac. J. Trop. Biomed., 2014. **4**: p. S510–S515.
164. Abd Ghafar, S.A., M. Ismail, L. Saiful Yazan, *et al.*, *Cytotoxic activity of kenaf seed oils from supercritical carbon dioxide fluid extraction towards human colorectal cancer (HT29) cell lines.* Evid. Based Complement. Alternat. Med., 2013. **2013**: p. 549705.
165. Ghafar, S.A.A., L.S. Yazan, P.M. Tahir, and M. Ismail, *Kenaf seed supercritical fluid extract reduces aberrant crypt foci formation in azoxymethane-induced rats.* Exp. Toxicol. Pathol., 2012. **64**(3): p. 247–251.
166. Foo, J.B., L. Saiful Yazan, S.M. Mansor, *et al.*, *Kenaf seed oil from supercritical carbon dioxide fluid extraction inhibits the proliferation of WEHI-3B leukemia cells in vivo.* J. Med. Plant Res., 2012. **6**(8): p. 1429–1436.

Oral Health Challenges during Chronic Diseases

Prevention and Treatment Using Functional Foods and Nutraceuticals

Sangeeta Jayant Palaskar, Rasika Balkrishna Pawar, and Darshana Rajesh Shah

Contents

Introduction 293
Oral lesions 294
References 314

Introduction

The oral cavity has been described as "the window to general health" (1). According to Seymour, a statement such as "You cannot have good general health without good oral health" is now considered obvious. The oral cavity is an integral part of the body, and the entry gate of general health; thus it acts as an intersection of dentistry and medicine. Both should focus on the quality of life by improving the health of individuals (1).

Approximately 500 medications and more than 100 systemic diseases show oral manifestations. These are common in the geriatric population. It is known that infected teeth were extracted for the cure of systemic diseases by Hippocrates. However, the oral manifestation of systemic conditions is a topic of neglect. The impact of systemic and oral conditions on each other needs more attention and increased collaborations.

Common chronic conditions that have an impact on oral health are atherosclerotic disease, pulmonary disease, diabetes, obesity, arthritis, cancer, Alzheimer's disease, smoking and alcohol-related diseases (1). Growth, development, maintenance and repair of dentition and oral tissues require adequate nutrients (2).

DOI: 10.1201/9781003220053-13

Nutraceuticals are "products, which other than nutrition are also used as medicine". A nutraceutical product may be defined as "a substance, which has a physiological benefit or provides protection against chronic disease" (3).

Functional foods are ingredients that offer health benefits that extend beyond their nutritional value. Examples are foods fortified with vitamins, minerals, probiotics, or fibers. Fruits, vegetables, nuts, seeds and grains are often considered to be functional foods (4).

Chronic diseases are defined broadly as "conditions that last one year or more, limit activities of daily living and require ongoing medical attention or both" (5).

The typical chronic diseases with oral manifestations are cardiovascular diseases, respiratory diseases, Alzheimer's disease, diabetes, arthritis, obesity, cancer, and smoking and alcohol-related chronic diseases (5, 6).

Smoking is a public health problem showing adverse effects on the whole body. It causes a wide range of systemic diseases, including various cancers, coronary heart disease, stroke and peripheral vascular diseases (7). The common reasons for death in developed countries caused by smoking are respiratory heart disease, pneumonia, aortic aneurysm and ischemic heart disease (7). The oral effects of smoking range from stains on teeth and dental restorations, bad breath, periodontal diseases, reduced ability to taste and smell, and impaired wound healing. It also leads to smoker's melanosis and smoker's palate, to potentially malignant lesions such as leukoplakia/erythroplakia and to serious diseases such as oral cancer (8).

Alcohol consumption affects not only the general health but also the dental and oral health of a person. It brings changes in the salivary gland function. Heavy drinkers are at great danger of having dental caries, gingival health alteration, periodontal diseases, tooth disintegration and increased risk of oral cancer. Alcohol abuse with smoking has diverse effects on oral health (9, 10) (Table 11.1).

Oral mucosal lesions associated with smoking, in general, are caused by various toxins and carcinogens produced from burning tobacco. Various carcinogens from burning tobacco, that is, tars, carbon monoxide, benzopyrene, Cd complex-nitrogen oxide, and so on, that are released are harmful. Among all nicotine metabolites, cotinine is the most important; it is detected in blood, urine and gingival fluid (10).

Oral lesions

Smoking and excessive alcohol have increased the risk of the development of cancer, as carcinogens increase the permeability of oral epithelium. Carcinogens in tobacco smoke cause p53 tumor suppressor gene mutation, leading to DNA damage in the cell. This change plays a significant role in the development of cancer. Oral cancer development in a smoking person can also be due to confounding factors like low intake of fresh vegetables and

Table 11.1 Oral Lesions and Conditions Associated with Tobacco Use

Oral precancerous lesions
Leukoplakia
Erythroplakia
Smokeless tobacco keratosis
Oral cancers
Squamous cell carcinomas of the oral mucous membrane
Verrucous carcinomas of the:
Buccal mucosa
Gingiva
Alveolar ridge, Vestibule
Periodontal diseases
Increased plaque and calculus depositions
Gingivitis
Periodontitis
Periodontal pockets
Gingival recession
Alveolar bone loss
Root caries
Peri-implantitis
Halitosis
Taste derangement
Stained teeth and restorations
Black hairy tongue syndrome

fruits, social status, alcohol abuse and so on. (7, 11). The carcinogenic effect of alcohol is possible due to acetaldehyde, which causes damage to oral epithelial cells and oncogene expression of oral keratinocytes. The alcohol also causes a dehydrating effect on the cell wall, enhancing mucosal permeability to toxins, and reduction in epithelial thickness. Oral cancer manifests as ulceration, exophytic or nodular growth. Other symptoms include bleeding, mobility of teeth, difficulty in speech and mastication, dysphagia, trismus and paresthesia (12, 13).

Tobacco smoke and alcohol are strongly associated with oral potentially malignant disorders, predominantly leukoplakia and erythroplakia. Leukoplakia appears as a gray or grayish-white non-scrapable lesion, which could be homogenous or nonhomogenous. Erythroplakia is a red lesion seen on the mucous membrane and has the highest risk of turning into malignancy (7).

Smoking mainly affects the oral microbial flora, among which Candida species is more commonly affected, which causes oral candidiasis (7).

Another lesion associated with heavy smoking is black hairy tongue. This lesion is characterized by hypertrophy of filiform papillae on the dorsum of the tongue, giving it black or brown discoloration (7).

Aloe vera is an indigenous medicinal plant found throughout India. This plant is effective in the treatment of oral ulceration; it also has antiseptic, antibacterial, anti-inflammatory, antioxidant and wound-healing properties (14).

Centella asiatica (gotu kola) is distributed throughout the plains of India. This plant is effective in the treatment of mouth ulcers (14).

Chamomile contains flavonoids, coumarins and essential oils that are effective against mucosal ulcers (14).

Jasminum grandiflorum (Spanish jasmin) leaves are widely used to treat ulcerative stomatitis and oral wounds, due to their antioxidant properties (14).

Spirulina platensis (spirulina) is the most widely used nutraceutical food supplement. Its extract contains a potent pigment, astaxanthin, which not only has a beneficial role in regression of precancerous lesions but is also effective on oral cancer (15).

Curcumin has a very significant role in preventing oral carcinogenesis and has anti-inflammatory properties, thus it is helpful in treating leukoplakia, oral cancer and periodontal diseases (14).

Antioxidants such as β-carotene, provitamin A, vitamin C, vitamin E, zinc, selenium and spirulina have a preventive role against oral cavity cancer. An effective combination of antioxidants like vitamins A, E and C are also valuable for the treatment of leukoplakia. Antioxidant nutrients help in inhibiting cancer cell development and destroy cancer cells through apoptosis by various mechanisms such as stimulation of cytotoxic cytokines, action on gene expression, preventing the development of a tumor's blood supply or by cellular differentiation. A report has also shown a reduction in adverse effects of chemotherapy when given concurrently with antioxidants. Tocopherol (AT) is the commonest and most active form of vitamin E. It is found in plant oil, margarine and green leaves. Tocopherol is an effective antioxidant at high levels of oxygen, protecting cellular membranes from lipidic peroxidation (16).

Xianhuayin, a decoction consisting of *Phellodendron amurense* (Amur cork tree), *Amomum villosum* (Wurfbainia villosa), a plant from ginger family, *Sclerotium price* (poria), Helianthus (sunflower) and *Glycyrrhiza glabra* (licorice), is useful in treating premalignant lesions. The main ingredient in *Glycyrrhiza glabra* is licorice, which helps in epithelial cell regeneration and thus is helpful in treating premalignant lesions and cancer (14).

Caffeinated coffee intake is inversely related to cancer of the oral cavity and pharynx (14).

Palatal leukokeratosis, also known as a smoker's palate, is associated with heavy pipe and cigar smoking. It shows white changes on the hard palate, combined with multiple red dots located centrally in the small elevated nodule. One study on the Indian population showed that smokers have more tendency to develop smoker's melanosis compared to other lesions (7).

Smoking contributes to the discoloration of teeth, dental restorations and dentures. Halitosis is most commonly associated with smoking and alcohol, thus affecting the function of smell and worsening taste perception (7, 12).

Chemical products and toxins in tobacco smoke and alcohol are risk factors for periodontal disease. Smoking and alcohol-related periodontal disease include gingival bleeding, periodontitis, deeper periodontal pockets, increased attachment loss and bone loss. Periodontal degradation is related to increase levels of periodontal pathogens and modulation of host immune response. Alcohol drinkers have nutritional deficiencies leading to poor immunity, which further allows harmful chemicals to penetrate. Dehydration from alcohol consumption causes bacteria and plaque build-up resulting in early symptoms of gingival diseases and further progression to more severe conditions like periodontal diseases (7, 11, 13).

Vitamins (vitamins A, D, E and B-complex) help in the prevention and treatment of some pathological conditions, particularly periodontal disease. Vitamin D plays an essential role in suppressing inflammation by inhibiting inflammatory cells and alters the immune system, thus playing a role in treating periodontitis. Vitamin C maintains the integrity of connective tissue structures such as the gingiva, periodontal ligament and cementum. It also improves the immune system and helps in wound healing. Vitamin E maintains the proper health of periodontal tissue by increasing the nitric oxide synthetase levels and preventing oxidative stress. Vitamin B-complex has a significant role in preventing gingival inflammation. Several minerals such as calcium and magnesium also help in maintaining periodontal health, showing a positive correlation in the suppression of periodontal tissue inflammation (17).

Food supplements such as probiotics, resveratrol and cranberry inhibit biofilm formation, thereby reducing gingival inflammation. They also decrease the pathogenic bacteria load that causes periodontal disease and prevent the adherence of *P. gingivalis*; thus they are valuable for preventing periodontal diseases (17).

Lycopene significantly reduces gingivitis, bleeding index and noninvasive measures of plaque; thus lycopene is helpful in treating gingivitis and periodontitis (14).

Smoking causes delayed wound healing in the mouth due to increased levels of adrenaline and noradrenaline, leading to peripheral vasoconstriction and impaired polymorphonuclear neutrophil function. It also increases the chances of alveolar osteitis, also known as dry socket (7).

Achyranthes bidentata and *Achyranthes aspera* (chaff flower) alcoholic extracts are often used in the treatment of aphthous ulcers (gargled) since they have demonstrated significant wound healing effects (14).

Centella asiatica (gotu kola) remarkably affects wound healing and promotes connective tissue growth (14).

Curcuma longa (turmeric), another old spice in Ayurvedic medicine, has prominent anti-inflammatory, antibacterial and wound-healing effects, and is effective in treating various oral lesions (14).

Smoking is the predisposing factor for dental implant failure due to poor healing or osseointegration. Implant failure is also strongly correlated with exposure of peri-implant tissue to tobacco smoke causing marginal bone loss around implants (7, 11).

Smoking is associated with high caries index, but the direct etiological relationship is lacking. It is known that smoking causes an increase in lactobacillus and *Streptococcus mutans* counts as compared to non-smokers (7, 11).

Spirulina platensis possesses good antibacterial activity against *S. mutans*, *E. faecalis* and *S. aureus* and is effective for prevention of dental caries (15).

Tea is derived from the dried leaves of the plant *Camellia sinensis* (family Theaceae). Its chemical composition contains polyphenols, catechins, caffeine, amino acids, carbohydrates, protein, chlorophyll, volatile compounds, fluoride, minerals and other undefined compounds. The fluoride content in tea is the major contributor for anticarcinogenic activity rather than the polyphenols (18).

Cranberry also possesses anticaries properties, as it inhibits acid production, attachment and biofilm formation by *S. mutans* (18).

Propolis is a resinous mixture that contains numerous flavonoids. It exhibits good antimicrobial activity against a range of oral bacteria and good anti-adhesion properties by inhibiting the adherence of *S. mutans*, thus reducing the carcinogenicity (18).

Nidus vespae, is a traditional Chinese medicine with numerous pharmacological properties. It is similar to propolis, but it also contains waxes and aromatic oils. The extracts and fractions of N. vespae exert antimicrobial activity toward oral bacteria, particularly *S. mutans*, showing significant anti-acidogenic activity (18).

Green tea is helpful in treating oral diseases like chronic periodontitis, dental caries and oral cancer (19).

Red wine stain is mainly associated with staining of teeth. The dietary chromogens in the drink get adsorbed to pellicle, an essential layer of stained material, which cannot be removed easily (12).

Regular and prolonged use of acidic drinks like alcohol makes the oral cavity as well as teeth more acidic in nature. This makes teeth susceptible to mechanical damages like tooth brushing, tooth clenching and so on. Alcohol consumption also causes frequent vomiting, which results in acidic content entering the mouth and causing erosion of the teeth. Thus, acidification causes a reduction in salivary secretion and buffering capacity, increasing the risk of enamel erosion. Erosion is more common on the palatal surface of maxillary teeth, followed by the occlusal surface of posterior teeth (9, 13).

Oral manifestations of cancer are primarily due to side effects of chemotherapy and radiation. Chemotherapeutic agents are cytostatic or cytotoxic in nature; they not only target the malignant cells but also have an adverse effect on the normal cells. Fortunately, many normal adult cells are less sensitive to chemotherapeutic drugs, while normal cells of the oral and gastrointestinal mucosa, hemopoietic system and hair follicles are more sensitive and rapidly dividing. Thus, chemotherapeutic drugs show various adverse effects, which include mucositis, dysgeusia, oral and dental infections, hemorrhage, saliva changes, and neurologic and nutritional problems, impacting quality of life (20, 21).

Mucositis is a painful and debilitating condition in patients undergoing chemotherapy. It is commonly seen on various parts of the oral cavity as inflammation of mucosa or ulcers. It begins five to ten days after initiation of chemotherapy and resolves within two to three weeks, and can be correlated with a standard white blood cell count. Leukoedema is the initial sign of mucositis, which disappears upon stretching of the mucosa (21, 22).

Salvia officinalis (sage), *Matricaria chamomilla* (chamomile), aloe vera and *Gentiana lutea* (yellow gentian) are useful herbal formulations for oral ulceration and chemotherapy-induced mucositis either through gargling or topical application (14).

Plantago major (common plantain), an effective wound healer, has various properties such as anti-ulcerative, anti-inflammatory, antibacterial, antiviral and antioxidant (23).

Hangeshashinto, the traditional medicine of the Japanese, is helpful against chemotherapy-induced oral mucositis by antioxidation and anti-inflammation properties or by suppression of inflammatory cell chemotaxis and cyclooxygenase-2 (COX2) expression (23).

Turmeric is a well-known spice used traditionally for many health benefits. The active element in turmeric is curcumin, which has antioxidant and anti-inflammatory properties and is helpful in healing patients with chemotherapy- and radiotherapy-induced oral mucositis (23).

Zingiber officinale (ginger) active ingredients like 6-gingerol and 6-shogaol are helpful in relieving oral ulcerative mucositis (23).

Quercetin, a natural flavonoid, has antioxidative and anti-inflammatory effects and effectively reduces chemotherapy-related oral mucositis (23).

Honey helps reduce mucositis caused by chemotherapy, radiotherapy and bacterial infection (23).

Nutrients such as vitamin E, zinc and glutamines are effective in preventing and treating oral mucositis due to their antioxidant and immunomodulatory properties (24).

Zinc, an essential trace element, is helpful for tissue-repairing processes and severe oral mucositis in cancer patients (25).

Emblica officinalis (amla) has an exciting antioxidant and astringent property, and it has been demonstrated to be effective in the treatment of aphthous stomatitis and other types of mouth ulcers (14).

Tinospora cordifolia (guduchi) has anti-inflammatory, antioxidant and immunomodulatory effects. This plant can reduce mucositis severity in radiotherapy patients (14).

Oral infections after chemotherapy contribute to 25–50%, with further increase in mortality and morbidity rate. Teeth, gingiva, salivary glands and mucosa are the most common anatomical sites for these complications. The fungal infection associated with chemotherapy is candidiasis caused by Candida. The lesion appears as curd-like or patchy white lesions or red lesions and can be acute or chronic type. Few viral infections caused by herpes simplex virus (HSV), varicella-zoster virus (VZV) and cytomegalovirus (CMV) are seen in patients undergoing chemotherapy (21, 22).

Nutraceuticals such as resveratrol, quercetin, curcumin, epigallocatechin gallate (EGCG), N-acetyl cysteine (NAC) and palmitoylethanolamide (PEA) are helpful in preventing and treating viral infections due to their antiviral, anti-inflammatory and immunomodulatory effects (26).

A nutrient-rich diet reduces the possibility of chronic diseases. Vitamins (vitamins A, D, C, E, B6 and B12), minerals (calcium and magnesium), trace elements (zinc, copper, selenium, etc.), carbohydrates, proteins, fats and water help in making many viral infections less severe (26).

Naturally, bioactive compounds and nutraceuticals such as garlic, green tea, propolis, curcumin, licorice root, cinnamon, resveratrol, ginger and berberine are helpful in the treatment of oral candidiasis (27).

Intraoral bleeding clinically manifests as gingival bleeding or sometimes as submucosal bleeding, which possibly is spontaneous, induced traumatically or from preexisting pathology. Thrombocytopenia, which may be secondarily induced by chemotherapeutic drugs, is the common cause of hemorrhage (21).

Xerostomia is the most common complaint in patients undergoing radiotherapy. It causes irreversible damage to salivary gland tissue, including a reduction in salivary outflow, glandular atrophy, fibrosis of tissue with infiltration of lymphocytes and plasma cells (22). The most common complaint by the patient is discomfort, pain and dryness of the mucosa, increasing the chances for oral infection leading to further difficulties and overall affecting quality of life (22). Xerostomia also increases periodontal disease and dental caries (28, 29).

Another complaint with xerostomia is taste loss, also called hypogeusia. Loss of taste is usually associated with pain, dysgeusia, hyposalivation, loss of pleasure in eating, loss of appetite, weight loss and malnutrition (22, 28).

The most severe and undesirable problem associated with radiotherapy is osteoradionecrosis (ORN). Irradiation typically causes the formation of hypovascular-hypocellular-hypoxic tissue, disrupting the barrier of the oral mucosa, resulting in a nonhealing process. Clinically, necrotic bone tissue is seen, which is related to various other symptoms like paresthesia, secondary infection, fistula formation, bone destruction and pathological fractures (28).

Radiation caries (rampant caries) develop after radiotherapy at a much faster rate than conventional caries. It is mainly caused because of diminished salivary flow and its qualitative alteration. Radiation also has a direct effect on teeth, like changes in odontoblasts, reducing the production of dentin, changes in the organic and inorganic composition of enamel and dentin, making them vulnerable to decalcification. Clinically, the teeth are more brittle with easy enamel chipping (28. 29).

Soft tissue necrosis is a painful condition and can be managed with the help of painkillers and sometimes antibiotics (29).

Trismus (lockjaw) is a common complication after irradiation of head and neck cancer. It arises from hypovascularization and fibrosis of the muscle tissue, manifesting from three to six months after radiation therapy. The muscles affected by trismus during cancer treatment are the masticatory or temporomandibular muscles. Tonic muscle spasms with or without fibrosis of the mastication muscles and TMJ (temporal-mandibular joint) can be minimized or prevented with jaw-opening exercises (28).

Diabetes mellitus (DM) is a chronic metabolic disorder affecting 2.8% of the population worldwide and is expected to affect 4.4% by 2030. This increase in the prevalence of DM is due to lifestyle change and an increase in obesity (30). There are multiple risk factors for the incidence and progression of type 2 diabetes mellitus, of which diet is an important modifiable factor. The majority of literature emphasizes that a diet with high phytochemicals, anti-oxidant capacity and polyphenolic compounds might lower the risk of diabetes (31).

DM is characterized by hyperglycemia because of insulin secretion deficiency or resistance to the action of insulin or both (32). Chronic hyperglycemia leads to various long-term pathologies affecting the oral cavity, so blood glucose control is critical (33). The complications of DM include neuropathy, nephropathy, retinopathy, microvascular and macrovascular complications, decrease in quality of life and also increased rate of mortality (30).

The oral manifestations and complications related to DM are dry mouth (xerostomia), dental caries (including root caries), periapical lesions, gingival and periodontal disease, oral candidiasis, burning mouth (especially glossodynia), altered taste, geographic tongue, coated and fissured tongue, recurrent aphthous stomatitis, increased tendency to infections and delayed wound healing (33).

Possible pathogenesis behind the oral complications of diabetes could be impaired neutrophil function, increased collagenase activity and a reduction in collagen synthesis, microangiopathy and neuropathy (34). The severity of complications depends upon the degree and duration of hyperglycemia (30).

Various pharmacological and non-pharmacological interventions have been developed to date. They aimed at improving glycemic control and preventing or minimizing complications. Based on the currently available knowledge of the pathophysiology of insulin resistance, the use of functional foods and their bioactive components have been considered as a new approach in the prevention and management of diabetes and its complications (31).

Nutraceuticals Used in the Treatment of Diabetes and Its Oral Manifestations: In recent years, there is growing evidence that plant-food polyphenols are proven to be unique nutraceuticals and supplementary treatments for type 2 diabetes mellitus. Their compounds can also prevent the development of various diabetic complications.

Tea has anti-cariogenic properties. Tea components have been shown to inhibit dental caries (35).

Acacia arabica (babul) is also known as prickly acacia. It is from the genus *Acacia* and the family *Fabaceae*. It has been used to treat various diabetic complications such as gingivitis, stomatitis (mouth sores) and pharyngitis. The powdered seeds of *A. arabica* and roots of *Caralluma edulis* have demonstrated a hypoglycemic effect in the normal experimental rabbits by initiating insulin release from pancreatic β cells (30).

Allium cepa (onion) have been known for ages to have many putative health benefits. It is rich in flavonoids and sulfur compounds, which are beneficial in everything from the common cold to diabetes and osteoporosis (31).

Allium sativum (garlic) is best known for its cardiovascular health benefits, hypoglycemic and lipid-lowering effects. Various studies showed that the allium family members contain mild hypoglycemic activity. The active constituents are sulfur-containing compounds, including allicin, allyl propyl disulfide, S-allyl cysteine and S-allyl mercaptocysteine. These compounds decrease the rate of insulin degradation, thus increasing circulating insulin levels (36).

Cucurbita maxima (pumpkin) was reported to cause significant reductions in plasma glucose. Pumpkin extract was also demonstrated to reduce or eliminate the administration of injectable insulin in Type 1 diabetic patients, thereby increasing insulin production and regenerating the damaged pancreatic cells (30, 31).

Eugenia jambolana (jamun), from the family *Myrtaceae*, also known as jamun or black plum, is an Indian plant. Jamun is known for its ethnomedicinal uses. It is used as an adjuvant in the treatment of diabetes. It does have known hypoglycemic action (30, 31).

Gymnema sylvestre (gurmarbooti, gurmar) is widely used in Ayurveda. Nonrandomized controlled clinical trials have confirmed improved glycemic control over those who received conventional treatment alone. Patients on insulin also have been observed to decrease their insulin doses (30, 31).

Lagerstroemia speciosa (banaba) leaves contain corosolic acid, which has anti-diabetic properties (30, 32).

Momordica charantia (bitter melon) has been studied in animal models, and observations showed a reduction in blood glucose levels. Similar results were noted in humans (30, 32).

Since ancient times, *Ocimum sanctum* (tulsi) has been used in Ayurveda. It has been known for the treatment of diabetes and minimizes various complications of diabetes by decreasing blood glucose (30, 32).

Pterocarpus marsupium (vijayasar) is a traditional anti-diabetic Ayurvedic medicinal plant. The extracts of the heartwood and the bark of the plant have shown significant benefits in uncontrolled Type 2 diabetics. In advanced diabetes cases, 2–4 g/day of dried extract decreases fasting blood glucose levels up to ~30 mg/dL and post-prandial blood glucose levels up to ~45 mg/dL. Animal model studies have suggested that it may also affect lipid levels and gastrointestinal glucose absorption, improve beta-cell function and act like insulin (30, 32).

Pycnogenol (French maritime pine bark) is an extract with antioxidant properties. In a multi-centric study Pycnogenol, used along with standard anti-diabetic treatment, showed a significant reduction in blood glucose levels when compared with placebo (30, 36).

Silybum marianum (milk thistle) is from the aster family *Asteraceae*. It shows hepatoprotective effects and has been used to treat alcoholic cirrhosis, viral hepatitis and medication poisonings. It also has the ability to alter insulin resistance in hepatic damage patients (30, 36).

Salacia reticulata (salacia, kotalahimbatu) belongs to the *Celastraceae* family. Aqueous extract from *Salacia reticulata* stem in animal studies and clinical trials demonstrated suppression of increased serum glucose levels after administration with sucrose, maltose and starch in a dose-dependent manner (30, 36).

Trigonella Foenum-graecum (fenugreek) is used as a herb as well as spices, a member of the Fabaceae family. Many studies have confirmed the hypoglycaemic effect of fenugreek. This high fibre content forms a gel in the stomach, thus leisures gastric emptying and delaying glucose absorption. One study concluded that fenugreek seeds' adjunct use significantly improved glycaemic control and decreased insulin resistance (30, 36).

Tinospora cordifolia (guduchi) plant has well-documented antidiabetic effects in traditional medicine. Recently, researchers found that the aqueous extract of guduchi enhances insulin secretion and significantly improves glucose metabolism, thus lowering blood glucose levels (31).

Vaccinium myrtillus (bilberry), also referred to as European blueberry, is a close relative of North American huckleberry and blueberry. The leaves have been used in traditional tea for diabetics and are a rich source of chromium. Animal studies showed decreased blood glucose and triglyceride levels with it. The fruit berry has been used in the improvement of visual acuity and night vision. The animal model proved it as a treatment for diabetic retinopathy (31).

Bilberry fruit is a significant source of anthocyanosides; a class of bioflavonoids is an active therapeutic component. These extracts typically contain 25% anthocyanosides, which are reported to increase vascular permeability, decrease platelet aggregation, improve microvascular circulation and improve retinal regeneration. It has also been proposed as a promising hypoglycemic agent (31).

Polyherbal formulations: various polyherbal formulations have been studied for their antidiabetic action. An animal model study was done to establish the hypoglycemic effect of the polyherbal formulation prepared from *Piper nigrum*, *Tribulus terrestris* and *Ricinus communis* in alloxan-induced diabetic rats. After four weeks of treatment, a significant reduction was noticed in elevated blood glucose levels. Polyherbal formulation performs a maximum decrease in blood sugar level (BSL) at a concentration of 300 mg/kg compared to the standard drug, that is, glibenclamide (37).

Arthritis is a common disease in which the end-point result is joint replacement surgery. Nutraceuticals are an alternative treatment for pathological manifestations of arthritic disease. The efficacy of fish oils (e.g., cod liver oil) in the diet has been demonstrated in several clinical trials, animal feeding experiments and *in vitro* models that mimic cartilage destruction in arthritic disease. Other than this, there is some evidence that other nutraceuticals are beneficial, such as green tea, herbal extracts, chondroitin sulphate and glucosamine (35).

Alzheimer's disease (AD) is a neurodegenerative disease of unknown etiology with progressive memory and cognitive impairment. The oral health of elderly people with Alzheimer's disease and dementia can be poor due to negligence in oral hygiene routine. They may have gingival bleeding, periodontitis, attachment loss and so on. Xerostomia, stomatitis and Candidiasis are also common. These patients need to maintain good oral health, as oral health is an integral part of their general health (38).

Pathogens from the oral cavity enter the bloodstream and cross the blood-brain barrier more readily as a person ages, which causes inflammation and increases the formation of senile plaques, which are conducive to Alzheimer's disease. Also, chronic periodontitis results in an increased level of inflammatory products in the bloodstream and thus may cause cerebral inflammation and neurodegeneration. Periodontal disease and Alzheimer's disease both potentiate each other. A dentist should try to prevent or control oral diseases to preserve function and enhance the quality of life in these patients (38).

Caries prevention can be done in these patients with fluoride therapy (38).

Many natural dietary chemical substances have protective functions against some neurogenerative diseases. The Mediterranean diet has a high content of nutraceuticals. This diet has plant foods, fish and olive oil as primary sources of monounsaturated fat, and moderate intake of wine may have synergistic mechanisms between food components and has neuroprotective effects. Blueberries and strawberries are known to reduce rates of cognitive decline in subjects older than 70 years. Curcumin has an antioxidant, anti-inflammatory and anti-amyloidogenic mechanism and is known to reduce the risk of AD (39, 40).

Resveratrol is a phytoalexin molecule and is a polyphenolic found in the seeds and skin of grapes. It is found to be effective against cancer, diabetes, and inflammation, as it has antioxidant, cardio-protective, reno-protective and hepato-protective properties. This phytoestrogen molecule is a good option for neurodegenerative disorders as per the studies reported (41).

Genistein is one of the simplest isoflavonoids found in many plants of the *Leguminosae* family, such as soy plants. It is known to have numerous prophylactic health effects against cancer, cardiovascular diseases, diabetes and postmenopausal disorders. Genistein also exhibits potent antioxidant properties. Results of some studies have revealed that it is also advantageous in the improvement of learning and memory. N-acetyl-5-methoxytryptamine or commonly known as "melatonin", is a prime metabolite of the amino acid tryptophan. It has roles like scavenging of the free radicals, reduction of bio-oxidation, maintenance of the circadian rhythms and so on (41).

Patients with AD have a decreased level of melatonin in the blood serum and cerebrospinal fluid, causing impaired circadian rhythm. Tannic acid is a plant-derived polyphenolic compound found in plants like legumes, sorghum, bananas, berries and tea. It has antioxidant, anti-inflammatory, radical scavenging, antiviral and antibacterial properties (41).

Tannic acid has a neuroprotective role in AD. Vanillic acid is used as a flavoring agent in different foods and drugs. It is obtained from the plant *Angelica sinensis* and ameliorates spatial learning and memory deficits by checking oxidative stress (41).

Milk thistle (*Silybum marianum*), which has a flavonoid molecule, silibinin, exhibits antioxidative and anti-inflammatory properties. It is reported to be effective against ethanol-induced brain injury and lipopolysaccharide-mediated neurotoxicity and helps in improving memory function (41).

Caffeic acid (3,4-hydroxycinnamic acid; CA) is present in fruits, vegetables, coffee and tea. It exhibits anti-inflammatory, antiviral, anticancer, antihypertensive, antithrombosis, antifibrosis activities and others. CA helps to recover spatial, cognitive and memory functions (41).

Berberine is a yellow-colored plant alkaloid isolated from the stem, bark, roots and rhizomes of many plants like *Hydrastis canadensis* (goldenseal), *Berberis aquifolium* (Oregon grape), *Coptis chinensis* (coptis or golden thread), *Berberis vulgaris* (barberry) and *Berberis aristata* (tree turmeric).

Recent studies indicate that berberine has antioxidative properties and has been beneficial in AD (41).

Icariin is a natural product extracted from the Chinese herb *Herba Epimedii* (bishop's hat, fairy wings) and shows aphrodisiac and antirheumatic properties. It exhibited considerably increased memory retention and spatial learning ability in animal models (41).

Cardiovascular disease (CVD) comprises ischemic heart disease, cerebrovascular disease and peripheral vascular disease (42).

CVD patients have a higher prevalence of periodontal disease. Periodontitis and myocardial infarction have multiple common risk factors (42).

Periodontitis contributes to the development of atheroma plaque and cardiovascular disease. This could be due to cross-reactivity between antibodies against periodontal microorganisms and heat shock proteins HSP60, which cause atheroma. Bacteremia of periodontal origin may release inflammatory mediators such as IL-1α, IL-1ß and TNF, which play a role in atherogenesis. These microorganisms induce platelet aggregation and thromboembolic phenomena (42).

Patients with cardiovascular diseases and metabolic syndrome (MetS) have a poorer periodontal condition than patients without coronary disease and respond less favorably to periodontal treatment. Nonsurgical treatment of periodontal disease in patients with coronary disease reduces the serum C reactive protein levels and levels of lipids in patients with dyslipidemia (42).

MetS, according to the International Diabetes Federation (IDF), is a "complex series of symptoms defined by the presence of three or more of the five characteristic components such as hypertriglyceridemia, hypertension, obesity, low HDL-cholesterol, and elevated blood glucose" (42).

The relationship between diabetes and periodontal disease has been established. It is an example of how a systemic disorder can predispose to oral infections and of how the latter, in turn, can exacerbate the systemic disease (42).

Diabetes and periodontal disease influence each other negatively. There is evidence of a statistically significant decrease in the glycosylated hemoglobin (HbA1c) levels after periodontal interventions (42).

Studies have demonstrated the association between obesity and oral diseases, particularly periodontitis. Adipose tissue secretes a range of hormones and cytokines, which may contribute to systemic inflammation and insulin resistance, leading to type 2 diabetes and cardiovascular diseases (42).

Circulating TNF-α in obese people exacerbates periodontal inflammation (42).

Plasma cholesterol reduction and the antiatherosclerotic effect of lupin proteins have been studied, which also reported reducing glycemia in the postprandial condition. In animal studies, lupin proteins also displayed hypolipidemic and antiatherosclerotic effects (43).

Soy proteins are widely evaluated for metabolic control. Proteins from Glycine max are the prototype plant proteins and are helpful in MetS (43).

Blood pressure changes and weight loss are found following plant protein sources, compared with animal proteins (43).

Insulin resistance, type 2 diabetes, obesity and metabolic syndrome result from the interaction between environment, host genetics, diet and the gut microbiota (43).

Since probiotics, prebiotics and synbiotic diet shape the composition of gut microbiota, restoring its diversity and activity, nutrients and particularly nutraceuticals can be promising to improve metabolic features of MetS (43).

Vitamin D has generally been shown to exert a role in controlling some cases of MetS (43).

Curcumin acts as an insulin sensitizer and has the potential to modify features of MetS (43).

Berberine is a natural plant alkaloid isolated from the Chinese herb *Coptis chinensis* (huanglian), which has potential glucose- and cholesterol-lowering effect. Red yeast rice (RYR) documented lipid-lowering effects (43).

RYR is currently used predominantly in combination with other nutraceuticals for cholesterol reduction. Interestingly, unlike statins, this nutraceutical combination improves the leptin-to-adiponectin ratio (–17.8%) without changing adiponectin levels and endothelial function. In addition, RYR (200 mg), in combination with berberine (500 mg) and policosanols (10 mg), has been shown to improve insulin sensitivity in patients with insulin resistance (43).

Management of MetS patients requires dietary counseling and nutritional supplements of nutraceuticals, which can offer significant help (43).

Mediterranean diet components – olive oil and its antioxidants components, legumes and cereals – can be recommended to these patients (43).

Plant proteins (e.g., soy and lupin), probiotics and prebiotics provide significant benefits by modifying the intestinal microbiome to natural compounds, frequently achieving a drug-like status, for example, RYR, berberine and curcumin, as well as vitamin D, are helpful for the management of common and high-risk metabolic and vascular abnormalities (43).

Many pulmonary disorders like obstructive respiratory diseases, systemic diseases with pulmonary involvement, lung cancer, cystic fibrosis or tuberculosis all have clinical and/or therapeutic involvement of the oral cavity, which signifies regular dental service and oral cavity examination and collaboration between dental practitioners and pulmonologists (44).

Many systemic diseases manifest initially in the mouth. Timely diagnosis by a dental professional can lead to an early referral to the appropriate health professional. The oral cavity serves as a potential reservoir of respiratory pathogens as there is an anatomical connection between the oral cavity and lungs. These pathogens could defeat the immune and mechanical defense mechanisms to reach the lower respiratory tract (44).

The body's defense mechanism defeats pathogens from entering the respiratory tract. Despite the heavy bacterial load found in the oral cavity and upper respiratory tract, healthy individuals are not affected. The oral cavity examination will help in the clinical diagnosis of pulmonary diseases (44).

The American thoracic society defines COPD as "a disease state characterized by the presence of airflow obstruction either due to chronic bronchitis or emphysema". Smoking is the cause of about 87% to 91% of cases of COPD. In recent years, many researchers have proposed an association between COPD and periodontitis (44).

The other common manifestations are oral thrush, the most common mucosal ailment, dental plaque, gingival bleeding, increased periodontal pocket depth, tooth loss or, sometimes, toothless-ness (44).

Primary oral tuberculosis is common in young individuals. The secondary form can involve any age group; however, middle and old age groups are commonly involved with oral manifestations. Ulcers, nodules, fissures, or tuberculomas, which can be single or multiple, painless or painful, affecting tongue and hard palate, are common (44).

Oral tuberculosis may result due to self-inoculation from infected sputum. When the patient coughs, the infected sputum gets lodged into the oral cavity. When the protective layer is lost, the bacteria easily enter and cause oral manifestations. Reactivation and hematogenous spread from the primary lung infection may cause mouth involvement in secondary tuberculosis. Tuberculous ulcers are common on the tongue and palate. Cervical lymphadenopathy, osteomyelitis of jawbones are also found in oral tuberculosis (44).

Oral manifestations of cystic fibrosis (CF) include tooth discoloration, salivary gland enlargement, increased risk of caries, colonization of bacteria and dental malocclusions. The discoloration and the defects can be due to tetracycline intake during the period of development of the permanent dentition. Cystic fibrosis usually affects the mucous glands (44).

CF patients could carry *Pseudomonas aeruginosa* in the pharynx, dorsum of the tongue, buccal mucosa and saliva. This bacterium serves as the source of pulmonary infection in high-risk patients with resistance against chlorhexidine. Patients with cystic fibrosis tend to have chronic nasal and sinus obstruction and develop mouth-breathing habits. Mouth-breathing habits lead to open bite, xerostomia, caries, and malocclusion (44).

Asthma is the most common respiratory disorder. It is often associated with dental cavities, erosions, periodontal disease and oral candidiasis. Beta-agonists cause reduced salivary secretions in asthmatic patients, which will hamper the defensive barrier of the oral environment. Asthmatic patients may have mouth-breathing habits, causing dryness of the alveolar mucosa and gingival inflammation (44).

The third National Health and Nutrition Examination Survey (NHANES) has reported that *P. gingivalis* is associated with a decrease in the prevalence of asthma (44).

Pneumonia is an infection in the lung parenchyma caused by a wide variety of microorganisms, which includes bacteria, viruses, fungi and parasites. Sarcoidosis is a systemic disease with the presence of granulomas in the lungs. In the oral cavity, it manifests as non-tender ulcerations, gingival inflammation, hyperplasia or recession, and localized swelling or nodules. In 6% of patients, the parotid gland impairment is seen, causing xerostomia. In females, it has a tumor-like appearance and xerostomia. Heerfordt Waldenstrom syndrome includes systemic sarcoidosis, xerostomia and parotid gland swelling with bilateral uveitis and facial nerve palsy. Maxillary and the mandibular jawbones are affected with tooth loosening, radiating pain, mandibular tumefaction and maxillary bone loss (44).

In lung carcinomas, the mandible is the most common site of metastases and may be the first manifestation. There is progressive localized swelling, pain, paresthesia of the lower lip due to tumor invasion of the inferior alveolar nerve. Metastases in the soft tissue present as a submucosal mass, highly vascularized, often hemorrhagic, ulcerations can be seen (44).

Patients affected by pulmonary diseases often take inhaled medications, and a more significant portion of these are usually present in the oropharyngeal region. These patients exhibit xerostomia, dental cavities, halitosis, ulcerations, candidiasis, mucosal changes, gingivitis, periodontitis, gastroesophageal reflux, dysphonia, tongue hypertrophy and perioral dermatitis (44).

Lycopene, a carotene found in tomatoes and carrots, has a protective effect against asthma development in a murine model. The deficiency of vitamin A has increased the risk of developing respiratory diseases in children. In a clinical study, ascorbic acid supplementation attenuated exercise-induced bronchoconstriction in asthmatic patients (45).

Supplementation with vitamins C and E was helpful in asthma patients. It was reported that tocotrienol has a vital role in managing oxidative stress in COPD patients. Children with bronchial asthma were reported to have benefited from fish oil. A 50% reduction in asthma prevalence and clinical improvement in patients was found when selenium levels were increased in smoke-exposed individuals (45).

Zinc was beneficial in airway hyper-responsiveness and inflammation of the airway in mice, indicative of a potential treatment strategy in asthmatic patients (45).

Supplementation of iron also showed reduced airway hyper-responsiveness and eosinophilia in a murine model of allergic asthma (45).

The leaves, roots, flowers and bark of *Adhatoda vasica* (adulasa) are helpful in treating gastroesophageal reflux responsible for decreased pH of the saliva and enamel demineralization. This plant is also used in cough, colds, asthma to liquefy sputum, bronchitis and tuberculosis and is used as a bronchodilator (45).

Deciduous woody tree *Albizia lebbeck* (lebbek tree) is known to be effective for colds, coughs, bronchitis and asthma (45).

Boswellia serrata (Indian frankincense, salai guggul, or shallaki) is a traditional medicine used to treat asthma. *Kalanchoe integra* (Never Die) is anthelmintic, immunosuppressive, helps in wound healing, hepatoprotective, anti-inflammatory, antidiabetic, nephroprotective, antioxidant, antimicrobial, analgesic, anticonvulsant and antipyretic. In respiratory conditions, the boiled leaf extract of this plant is helpful in managing acute and chronic bronchitis, pneumonia, bronchial asthma and palpitations. Flavonoids and tannins in leaves of *K. integra* have beneficial effects in the treatment of bronchial asthma (45).

Curcumin has anti-asthmatic effects in both *in vivo* and *in vitro* studies. Herbaceous perennial plant *Glycyrrhiza glabra*, commonly known as licorice, has been used as a flavoring agent in foods and medicines for decades. The root of this plant is used for cough, colds, asthma and COPD (45).

Tulsi, known as *Ocimum sanctum*, is an annual herb used in the traditional Indian system of medicine. The leaves of this plant are traditionally used for cough, colds, asthma and bronchitis (45).

Piper longum (Indian long pepper) is used in Asia and the Pacific islands as traditional medicine. *Piper longum* has been a good remedy for treating tuberculosis and respiratory tract infections. Childhood asthma has been treated with the fruits and roots of this plant (45).

Tylophora indica (Antamoo) is important in the Ayurvedic system of medicine. The leaves of this plant are used to treat inflammatory and allergic disorders like bronchial asthma, bronchitis and whooping cough (45).

Kantkari or *Solanum xanthocarpum* is used in traditional Indian medicine for the treatment of bronchitis and asthma. Sore throat has been treated with the juice of berries (45).

In the Siddha system of medicine, *Solanum xanthocarpum* has been used to treat respiratory diseases. Asthmatic subjects showed improvement in different parameters of pulmonary function tests when treated with powder of whole plants of Solanum xanthocarpum. Zingiber officinale is a dietary component that is known as ginger. The rhizome of this plant is used to treat colds, asthma and bronchitis. The antibacterial activity of ethanolic extracts of ginger rhizome against respiratory pathogens, such as *Streptococcus pyogenes*, *Streptococcus pneumonia*, *Staphylococcus aureus* and *Haemophilus influenzae*, is known. The antiviral activity of fresh ginger has been known against virus-induced plaque formation on airway epithelium (45).

Quercetin is an essential dietary flavonoid found in plants such as onions, apples, tea, berries and broccoli, with anti-asthmatic potential (45).

Colchicine is an alkaloid obtained from the plant *Colchicum autumnale* (lirium) and has been reported to treat pulmonary fibrosis and bronchial asthma (45).

An alkaloid fraction of the *Peganum harmala* plant (harmala) is a beneficial traditional medicinal herb for the treatment of cough and asthma with potent antitussive, expectorant and bronchodilating effects (45).

Resveratrol, which is anti-asthmatic, is a dietary polyphenol found in the skin and seeds of grapes. The efficacy of a polyphenol-rich tomato extract containing naringenin chalcone was tested for sneezing, nasal discharge and obstruction and improved the quality of life of the subjects. Various phyto-constituents obtained from several plant sources have also been used to treat respiratory diseases (45) (Table 11.2).

Table 11.2 Summary of Chronic Systemic Diseases Having Oral Manifestations and Nutraceuticals/Functional Foods for Its Prevention and Treatment

Sr No	Chronic Diseases	Oral Manifestations	Nutraceuticals and Functional Foods
1	Cancer	• Mucositis • Loss of taste • Xerostomia • Dysgeusia • Oral infections • Intraoral bleeding • Osteoradionecrosis • Radiation caries • Soft tissue necrosis • Trismus	• Salvia officinalis (sage) • Matricaria chamomilla (chamomile) • Aloe vera and Gentiana lutea (yellow gentian) • Plantago major • Hangeshashinto • Chamomile • Turmeric • Quercetin • Honey • Zingiber officinale (ginger) • Vitamin E, zinc and glutamines • Zinc • Emblica Officinalis (amla) • Tinospora cordifolia (guduchi) • Resveratrol • Quercetin epigallocatechin gallate (EGCG) • N-acetyl cysteine (NAC) • Palmitoylethanolamide (PEA) • Minerals (calcium and magnesium) • Trace elements (zinc, copper, selenium, etc.) • Garlic • Green tea • Propolis

(Continued)

Table 11.2 Summary of Chronic Systemic Diseases Having Oral Manifestations and Nutraceuticals/Functional Foods for Its Prevention and Treatment *(Continued)*

Sr No	Chronic Diseases	Oral Manifestations	Nutraceuticals and Functional Foods
			• Licorice root, cinnamon • Berberine • Spirulina plantensis (spirulina) • Tea • Nidus vespae
2	Smoking- and alcohol-related chronic diseases	• Oral cancer • Potentially malignant disorders • Discoloration of teeth • Halitosis • Altered taste sensations • Smoker's palate • Hairy tongue • Periodontitis • Dental caries • Tooth wear • Staining • Delayed wound healing • Implant failure	• Aloe vera • Centella asiatica (gotu kola) • Chamomile • Jasminum grandiflorum (Spanish jasmin) • Spirulina plantensis (spirulina) • Curcumin • Antioxidants (β-carotene, provitamin A, vitamin C, vitamin E, zinc, selenium, spirulina) • Vitamins A, E and C • Xianhuayin, a decoction contains Phellodendron amurense (Amur cork tree) • Amomum villosum (Wurfbainia villosa) Sclerotium price (poria) • Helianthus (sunflower) • Glycyrrhiza glabra (licorice) • Caffeinated coffee • Vitamins (A, D, E and B-complex) • Probiotics • Resveratrol • Cranberry • Lycopene • Achyranthes bidentata • Achyranthes aspera (chaff flower) • Centella asiatica (gotu kola) • Curcuma longa (turmeric) • Spirulinaplantensis (Spirulina) • Tea • Nidus vespae • Propolis • Green tea

(Continued)

Table 11.2 Summary of Chronic Systemic Diseases Having Oral Manifestations and Nutraceuticals/Functional Foods for Its Prevention and Treatment *(Continued)*

Sr No	Chronic Diseases	Oral Manifestations	Nutraceuticals and Functional Foods
3	Diabetes	• Xerostomia • Dental caries • Periapical lesions • Gingivitis • Periodontitis • Oral candidiasis • Glossodynia • Altered taste • Geographic tongue • Coated and Fissured tongue • Recurrent aphthous stomatitis • Increased tendency to infections • Delayed wound healing	• Tea • Acacia Arabica (Babul) • Allium Cepa (Onion) • Allium Sativum (Garlic) • Cucurbita Maxima (Pumpkin) • Eugenia jambolana (jamun) • Gymnema sylvestre(gurmar booti, gurmar) • Lagerstroemia speciosa (banaba) • Momordica charantia (bitter melon) • Ocimum sanctum (tulsi) • Pterocarpus marsupium (vijayasar) • Pycnogenol (French maritime pine bark) • Silybum marianum (milk thistle) • Salacia reticulata (salacia, kotalahimbatu) • Trigonella foenum-graecum (fenugreek) • Tinospora Cardifolia (Guduchi) • Vaccinium Myrtillus (bilberry) • Polyherbal formulations (Piper nigrum, Tribulus terrestris and Ricinus communis)
4	Cardiovascular diseases	Periodontitis Oral infections	• Lupin proteins • Soy proteins • Probiotics, prebiotics and synbiotic diet • Vitamin D • Curcumin • Berberine • Red yeast rice • Mediterranean diet
5	Respiratory diseases	Dental caries Gingivitis Periodontitis Tooth loss Oral Infections Ulcers Tuberculoma Cervical lymphadenopathy Osteomyelitis of Jawbones Tooth Discoloration Salivary gland enlargement Malocclusions Erosions Dysphonia Tongue hypertrophy Gastro-Oesophagal reflux	• Lycopene • Fish oil • Selenium, zinc, iron • Adhatoda vasica (Adulasa) • Albizia lebbeck (lebbek tree) • Boswellia serrata (salai guggul or shallaki) • Kalanchoe integra (Never Die) • Curcumin • Glycyrrhiza glabra (Yashtimadhu) • Ocimum sanctum (tulsi) • Piper longum (Indian long pepper) • Tylophora indica (antamool) • Solanum xanthocarpum • Zingiber officinale (ginger) • Colchium autumnale (lirium) • Resveratrol • Naringenin chalcone • Peganum harmala (harmala)

(Continued)

Table 11.2 Summary of Chronic Systemic Diseases Having Oral Manifestations and Nutraceuticals/Functional Foods for Its Prevention and Treatment *(Continued)*

Sr No	Chronic Diseases	Oral Manifestations	Nutraceuticals and Functional Foods
6	Arthritis	Nerve parasthesia Pain in temporo-mandibular joint	• Cod liver oil • Green tea • Herbal extracts • Chondroitin sulfate • Glucosamine
7	Obesity	Periodontitis Poor oral hygiene Dental caries Xerostomia Tooth erosion Dentinal hypersensitivity	
8	Alzheimer's disease	Chronic periodontitis Gingivitis Dental caries	• Mediterranean diet • Blueberries and strawberries • Curcumin • Resveratrol • Genistein • Melatonin • Legumes, sorghum, bananas, berries and tea • Vanillic acid • Milk thistle (Silybum marianum) • Silibinin • Hydrastis canadensis (goldenseal) • Berberis aquifolium (Oregon grape) • Coptis chinensis (coptis or golden thread) • Berberis vulgaris (barberry) • Berberis aristata (tree turmeric) • Caffeic acid • Berberine • Icariin • Epimedium herb (bishop's hat, fairy wings)

References

1. Kane SF. The effects of oral health on systemic health. Gen Dent. 2017 Nov-Dec;65(6):30–34. PMID: 29099363.
2. Pflipsen M, Zenchenko Y. Nutrition for oral health and oral manifestations of poor nutrition and unhealthy habits. Gen Dent. 2017 Nov-Dec;65(6):36–43. PMID: 29099364.
3. Nasri H, Baradaran A, Shirzad H, Rafieian-Kopaei M. New Concepts in Nutraceuticals as Alternative for Pharmaceuticals. Int J Prev Med 2014;5(12):1487–1499.
4. Link, R. What are functional foods? All you need to know. Healthline 2020; January 17.
5. Goodman RA, Posner SF, Huang ES, Parekh AK, Koh HK. Defining and measuring chronic conditions: imperatives for research, policy, program, and practice. Prev Chronic Dis. 2013 Apr 25;10:E66. doi: 10.5888/pcd10.120239. PMID: 23618546; PMCID: PMC3652713.

6. Raghupathi W, Raghupathi V. An Empirical Study of Chronic Diseases in the United States: A Visual Analytics Approach. Int J Environ Res Public Health. 2018 Mar 1;15(3):431. doi: 10.3390/ijerph15030431. PMID: 29494555; PMCID: PMC5876976.
7. Komar K, Glavina A, Boras VV, Verzak Z, Brailo V. Impact of Smoking on Oral Health: Knowledge and Attitudes of Dentists and Dental Students. Acta Stomatal Croat 2018; 52(2):148–155.
8. Fahad AH, Mohamed RA, Hadi Layedh NM. Effect of Alcohol Consumption Severity on Oral Health Status in Relation to Salivary Parameters, Smoking and Tooth Wear in Baghdad, Iraq. Medico-legal Update 2020;20(4):1227–1233.
9. Ozturka O, Fidancib I, Unalc M. Effects of Smoking on Oral Cavity. J Exp Clin Med 2017;34(1):3–7.
10. Reibel J. Tobacco and Oral Diseases. Med Princ Pract 2003;12(1):22–32.
11. Grocock R. The Relevance of Alcohol to Dental Practice. Br Dent J 2017;223:895–899.
12. Khairnar MR, Wadgave U and Khairnar SM. Effect of Alcoholism on Oral Health: A Review. J Alcohol Drug Depend 2017;5(3):1–4.
13. Salehi B, Jornet PL, López EP, Calina D, Sharifi-Rad M, Ramírez-Alarcón K et al. Plant-Derived Bioactives in Oral Mucosal Lesions: A Key Emphasis to Curcumin, Lycopene, Chamomile, Aloe vera, Green Tea and Coffee Properties. Biomolecules 2019;9(3):1–23.
14. Ferrazzano GF, Papa C, Pollio A, Ingenito A, Sangianantoni G and Cantile T. Cyanobacteria and Microalgae as Sources of Functional Foods to Improve Human General and Oral Health. Molecules 2020;25(21):5164.
15. Iqubal A, Khan M, Kumar P, Kumar A, and Ajai K. Role of Vitamin E in Prevention of Oral Cancer - A Review. J Clin Diagn Res 2014;8(10):ZE05–07.
16. Swaminathan PG, Joshi SR, Vidyasagar M, Amit M, Priti K, Neeta P et al. Nutraceutical Basis for Drug Delivery in Periodontal Disease. Pravara Med Rev 2020;12(3):87–90.
17. Loveren CV, Broukal Z, Oganessian E. Functional Foods/Ingredients and Dental Caries. Eur J Nutr 2012;51(2):S15–S25.
18. Gaur S, Agnihotri R. Green Tea: A Novel Functional Food for the Oral Health of Older Adults. Geriatr Gerontol Int 2014;14(2):238–250.
19. Vozza I, Caldarazzo V, Polimeni A and Ottolenghi L. Periodontal Disease and Cancer Patients Undergoing Chemotherapy. Int Dent J 2015;65(1):45–48.
20. Toscano N, Holtzclaw D, Hargitai IA, Shumaker N, Richardson H, Naylor G et al. Oral Implications of Cancer Chemotherapy. J Implant Adv Clin Dent 2009;1:51–69.
21. Wong HM. Oral Complications and Management Strategies for Patients Undergoing Cancer Therapy. ScientificWorldJournal 2014;2014:1–14.
22. Zhang QY, Wang FX, Jia KK, Kong LD. Natural Product Interventions for Chemotherapy and Radiotherapy-Induced Side Effects. Front Pharmacol 2018;9:1–25.
23. Sousa Melo A de, Lima Dantas JB de, Medrado A, Lima HR, Martins GB, Carrera M. Nutritional Supplements in the Management of Oral Mucositis in Patients with Head and Neck Cancer: Narrative Literary Review. Clin Nutr 2021;43:31–38.
24. Yarom N, Ariyawardana A, Hovan A, Barasch A, Jarvis V, Jensen SB et al. Systematic Review of Natural Agents for the Management of Oral Mucositis in Cancer Patients. Support Care Cancer 2013;21:3209–3221.
25. Singh S, Kola P, Kaur D, Singla G, Mishra V, Panesar PS et al. Therapeutic Potential of Nutraceuticals and Dietary Supplements in the Prevention of Viral Diseases: A Review. Front Nutr 2021;8:1–16.
26. Gharibpour F, Bagherniya M, Nosouhian, Shirban F. The Effect of Nutraceuticals and Herbal Medicine on Candida Albicans in Oral Candidiasis: A Comprehensive Review. Adv Exp Med Biol 2021;1308:228–248.
27. Sari J, Nasiloski KS, Gomes AP. Oral Complications in Patients Receiving Head and Neck Radiation Therapy: A Literature Review. Rev Gaúch Odontol 2014;62(4): 395–400.
28. Jham BC, da Silva Freire AR. Oral Complications of Radiotherapy in the Head and Neck. Braz J Otorhinolaryngol 2006;72(5):704–708.
29. Ashwlayan VD, Nimesh S. Nutraceuticals in the Management of Diabetes Mellitus. Pharm Pharmacol Int J 2018;6(2):114–120.

30. Derosa G, Limas CP, Maciás PC, Estrella A, Maffioli P. Dietary and Nutraceutical Approach to Type 2 Diabetes. Arch Med Sci 2014;10(2):336–344.
31. Mirmiran P. Functional Foods-based Diet as a Novel Dietary Approach for Management of Type 2 Diabetes and its Complications: A Review. World J Diabetes 2014;5(3):267.
32. Rohani B. Oral Manifestations in Patients with Diabetes Mellitus. World J Diabetes 2019;10(9):485–489.
33. Garima V, Manoj KM. A Review on Nutraceuticals: Classification and its Role in Various Diseases. Int J Pharm Ther. 2016;7(4):152–160.
34. Khan RA, Elhassan GO, Qureshi KA. Nutraceuticals: In the Treatment & Prevention of Diseases - An Overview. The Pharma Innovation Journal 2014; 3(10): 47-50.
35. Kumari, Mamta. (2015). Nutraceutical-Medicine of Future. Journal of Global Biosciences. 4. 2790-2794.
36. Baldi A, Kumar S. Nutraceuticals as Therapeutic Agents for Holistic Treatment of Diabetes. Int J Green Pharm 2013;7:278–287.
37. Gao SS, Chu CH, Young FY. Oral Health and Care for Elderly People with Alzheimer's Disease. Int J Environ Res Public Health 2020 Jan;17(16):5713.
38. Brennan LJ, Strauss J. Cognitive Impairment in Older Adults and Oral Health Considerations: Treatment and Management. Dental Clinics 2014 Oct 1;58(4):815–828.
39. Mecocci P, Tinarelli C, Schulz RJ, Polidori MC. Nutraceuticals in Cognitive Impairment and Alzheimer's Disease. Front Pharmacol 2014 Jun 23;5:147.
40. Sadhukhan P, Saha S, Dutta S, Mahalanobish S, Sil PC. Nutraceuticals: An Emerging Therapeutic Approach Against the Pathogenesis of Alzheimer's Disease. Pharmacol Res 2018 Mar 1;129:100–114.
41. Carramolino-Cuéllar E, Tomás I, Jiménez-Soriano Y. Relationship Between the Oral Cavity and Cardiovascular Diseases and Metabolic Syndrome. Medicina oral, patologia oral y cirugia bucal 2014 May;19(3):e289.
42. Sirtori CR, Pavanello C, Calabresi L, Ruscica M. Nutraceutical Approaches to Metabolic Syndrome. Ann Med 2017 Nov 17;49(8):678–697.
43. Aishwarya J. Oral Manifestation of Respiratory Disorder - A Review. J Med Sci Clin Res 2016;4(8):12035–12044.
44. Gulati K, Rai N, Chaudhary S, Ray A. Nutraceuticals in respiratory disorders. In Nutraceuticals 2016 Jan 1 (pp. 75–86). Academic Press.

Urogenital System Disorders and Functional Foods and Nutraceuticals

Manish P. Patel, Arya S. Vyas, Praful D. Bharadia, Jayvadan K. Patel, and Dipti H. Patel

Contents

Introduction 317
Urogenital system disorders 320
Urinary tract infection 320
Causative agent 321
Pathophysiology of urinary tract infection 322
Signs and symptoms of urinary tract infection 323
Treatment of urinary tract infection 324
Non-pharmacological treatments (nutraceuticals and functional foods) 325
Dysmenorrhea 329
Pathophysiology of dysmenorrhea 329
Signs and symptoms of dysmenorrhea 330
Treatment of dysmenorrhea 330
Non-pharmacological treatment 331
Premenstrual syndrome and premenstrual dysphoric disorder 332
Signs and symptoms of premenstrual syndrome 333
Pathophysiology of premenstrual syndrome 333
Treatment of premenstrual syndrome 333
Non-pharmacological treatment 334
Conclusion 336
References 336

Introduction

The organs of the genital (reproductive) and urinary systems make up the genitourinary system or urogenital system. Just because of their similar embryological origin (intermediate mesoderm), closeness to one another, and utilization of common structural routes, these organs are frequently grouped together (1).

DOI: 10.1201/9781003220053-14

The urinary tract is a contiguous hollow organ system whose major job is to collect, transport, store, and eliminate urine in a highly coordinated manner on a regular basis. The urinary system guarantees that metabolic products and toxic wastes created in the kidneys are eliminated in this way (2). Through the urinary route, the urinary system is involved in the creation, reabsorption, storage, and elimination of urine (Figure 12.1). These actions assist the body in maintaining a healthy balance of excretory, regulatory, secretory, and homeostatic functions (3).

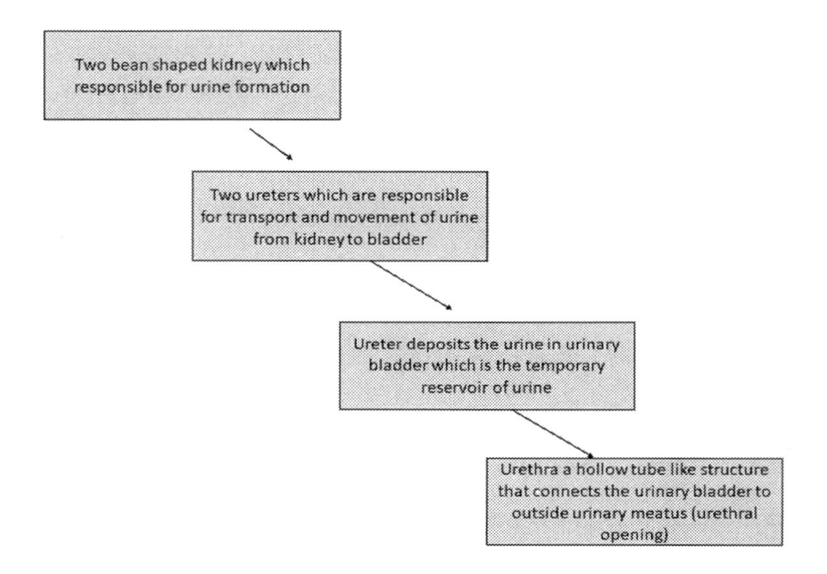

Figure 12.1 *The urinary pathway. (From (3).)*

The fundamental function of the urinary system, in cooperation with the integumentary (skin), digestive, and respiratory systems, is to eliminate water and waste from the body. It plays a role in homeostatic mechanisms that control water and electrolyte balance, acid–base balance, blood pressure, erythrocyte production, 1,25-dihydroxyvitamin D3 production, glucose synthesis, and waste excretion, including foreign chemicals, drugs, hormonal metabolites, and metabolic waste products (3).

Traditionally, the urinary system has been separated into upper and lower urinary tracts. The kidneys and ureters are in the upper tract, while the bladder, urethral sphincter, and urethra are in the lower tract (4). Two kidneys, two ureters, a urinary bladder, and the urethra make up the renal-urologic system (5). The oval kidneys are found in the right and left upper quadrants of the abdomen, deep within the retroperitoneum, near the spine (4). Urine is transported from the kidneys to the bladder by the ureters, which are cylindrical tubes. Each ureter is 22–30 cm in length, with the right ureter being roughly 1 cm shorter than the left. The urinary bladder is a hollow, collapsible chamber that stores urine under low pressure and discharges it at regular intervals (3). The female urethra, a transporter for urine for excretion, is about 4 cm in length, whereas the male urethra, which transports urine for excretion, is longer in length, about 20 cm (3) (Figure 12.2).

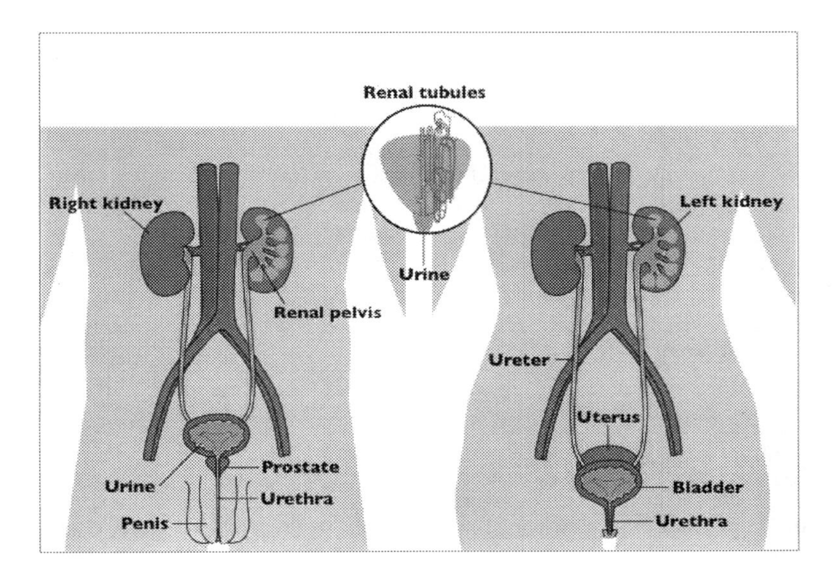

Figure 12.2 *Male urinary tract system (left) and female urinary tract system (right). (From (3).)*

In persons aged 65 and up, genitourinary diseases are widespread, and pharmaceutical therapy is critical to their quality of life. Neurogenic dysfunction, impotence, prostatism, urinary tract infections, and prostate cancer are all frequent genitourinary issues in the elderly, and most of the symptoms may be managed with medication. It's also vital to remember that many urological problems can only be treated with symptom relief rather than a cure (6). Urinary tract infections (UTIs) are one of the most prevalent bacterial diseases, impacting 150 million individuals globally each year (7). UTIs are one of the most prevalent bacterial illnesses found in both the community and in hospitals. UTIs are normally self-limiting in people with no anatomical or functional abnormalities, although they do have a tendency to reoccur (8).

Various drugs can be used for treatment and symptomatic relief of some urinary tract disorders like quinolones (ciprofloxacin, levofloxacin), sulfonamides, amino glycosides, trimethoprim, oxybutynin, flavoxate, prazosin, Ceftibuten, Cefixime, Gentamicin, Tobramycin, Nalidixic acid, and so on (6, 9). Drugs, as just mentioned, are used for the treatment of various urogenital disorders, but sometimes in individuals they may cause adverse drug reactions that may cause harm, and the widespread usage of antibiotics in recent decades has led to the creation of antibiotic-resistant microorganisms and the spread of antibiotic resistance so the use of medicinal plants in the prevention and treatment of many ailments has expanded in recent years. Complementary medicine with medicinal plants is an area of study that deserves special attention. The use of antibiotics in combination with therapeutic plants has mostly synergistic benefits. Herbal treatments have been shown in several trials to significantly diminish bacterial resistance to antibiotics (10). Many nutraceutical products and functional foods like cranberry juice, blueberry, Arctostaphylos uva-ursi, Juniperus communis, Agathosma betulina, Armoracia rusticana (horseradish), and so on, can be used for treatment and symptomatic relief of some urogenital disorders (9, 10).

The male reproductive system can be broadly classified into two parts: internal structure and external structure. Internal organs such as the testes, epididymis, vas deferens, and prostate, as well as exterior tissues such as the scrotum and penis, make up the male reproductive system. These tissues are well vascularized, with numerous glands and ducts to aid in sperm generation, storage, and ejaculation for fertilization, as well as the production of key androgens for male growth. Testosterone, which is generated by Leydig cells in the testes, is the most important male androgen. Testosterone can be transformed to dihydrotestosterone or estradiol in the peripheral via 5-alpha-reductase or aromatase (11). Benign prostatic hyperplasia, prostate cancer, and prostatitis are some examples of disorders of the male reproductive system (12).

The female reproductive system is a complex network of organs that may be divided into external and internal genitalia. The external genitalia include the labia majora and minora, vestibule, Bartholin glands, Skene glands, clitoris, mons pubis, perineum, urethral meatus, and periurethral region, which are all located outside of the true pelvis. The vagina, cervix, uterus, fallopian tubes, and ovaries are all part of the internal genitalia, which are located within the true pelvis (13). Hormones released by the hypothalamus, anterior pituitary, and ovaries govern the female reproductive system, and these hormones interact with one another in a dynamic fashion (14). Disorders such as polycystic ovary syndrome, dysmenorrhea, premenstrual syndrome and premenstrual dysphoric disorder, and infertility are some of disorders of the female reproductive system (15, 16). Therapeutic agents such as progestogens, bromocriptine, danazol, and NSAIDs may be utilized in the treatment and symptomatic relief in some of the aforementioned disorders. Some non-pharmacologic treatments, such as calcium, vitamins, high potassium foods, magnesium, lifestyle, and dietary modification, may help in the treatment or in providing some symptomatic relief to the patient (12).

Urogenital system disorders

Urinary tract infection

The term "urinary tract infection" (UTI) refers to an infection that can occur anywhere throughout the urinary system, from the urethral meatus to the perinephric fascia, and is typically caused by bacteria. The urethra, bladder, ureters, and renal pelvis and parenchyma are all parts of this route. The prostate, epididymis, and perinephric fascia are associated tissues that can get infected and act as sites of recurrent UTIs. Urethritis, which affects only the urethra; cystitis, which affects the bladder; and pyelonephritis, which affects the upper urinary tract structures, are some examples of urinary tract infections (17).

A UTI can be broadly classed as either complicated or uncomplicated. An uncomplicated UTI is the most common kind of infection and occurs when there are no anatomical or functional abnormalities inside the urinary system. A complicated UTI arises when there is an abnormal urinary tract or another condition that enhances infection susceptibility (18).

In outpatient and inpatient settings, urinary tract infections (UTIs) are one of the most prevalent illnesses. Asymptomatic bacteriuria, acute uncomplicated cystitis, recurrent cystitis, severe UTI, catheter-associated asymptomatic bacteriuria, catheter-associated UTI (CAUTI), prostatitis, and pyelonephritis are all clinical entities included in the "UTI." The most frequent manifestations of UTI is acute cystitis, which is more common in women (19).

Although a UTI can lead to life-threatening sepsis, the majority of infections are mild. UTI, on the other hand, causes great misery to the individual and is linked to expensive healthcare and social expenses. UTIs account for seven million clinic visits in the United States each year, costing more than $1.6 billion (18). The microbiological etiology of urinary infections has long been thought to be well established and reliable. In acute community-acquired uncomplicated infections, *Escherichia coli* remains the most common uropathogen (80%), followed by *Staphylococcus saprophyticus* (10% to 15%). Uncomplicated cystitis and pyelonephritis are seldom caused by *Klebsiella*, *Enterobacter*, and *Proteus* species, as well as *enterococci*. Antimicrobial resistance is modifying many of the characteristics of the bacteria typically linked with UTI (20).

Causative agent

Urine is normally sterile. It contains fluids, salts, and waste materials but is normally devoid of bacteria, viruses, and fungus. When small organisms, mainly bacteria from the digestive tract, adhere to the opening of the urethra and grow, an infection arises. Bacteria can also enter the bladder through the urethra, and infections can spread through the blood and lymph. E. coli is the most common bacteria responsible for UTI, accounting for 80% of cases, followed by *Staphylococcus saprophyticus*, which accounts for 5–10% of cases. E. coli is a Gram-negative commensal of the distal colon that also hosts Bacteroides and Bifidobacteria, as well as other anaerobic bacteria. In terms of virulence factors, uropathogenic E. coli varies from intestine pathogenic E. coli. In addition to the microorganisms just described, UTI is linked to *Klebsiella*, *Proteus*, *Pseudomonas*, and *Enterobacter* (21–23) (Figure 12.3).

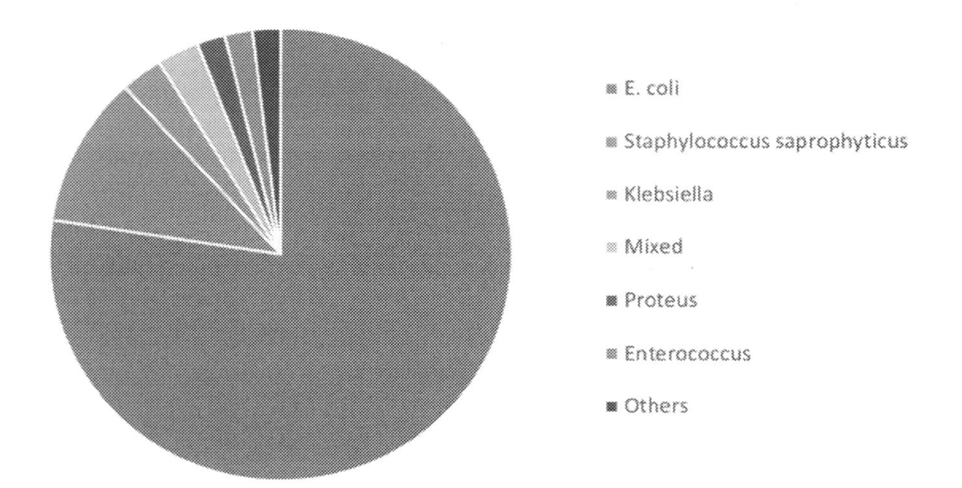

Figure 12.3 *Contribution of microbes for causing urinary tract infection in percentage. (From (21).)*

Pathophysiology of urinary tract infection

UTIs are caused by the combination of bacterial pathogenicity with host biologic and behavioral variables, which overrides the host's very effective defense systems. The ascending, hematogenous, and lymphatic pathways are three potential routes for germs to infiltrate and propagate throughout the urinary system (24).

In healthy patients most uropathogens originate from rectal flora and enter the urinary tract via the urethra into the bladder. Uropathogens first attach to and colonize the urothelium of the distal urethra through the ascending pathway. Up to 50% of infections in individuals with established cystitis may ascend into the upper urinary tracts, and most bouts of pyelonephritis are caused by bacteria ascending from the bladder via the ureter and into the renal pelvis. Bacterial ascent is helped by situations that impede ureteral peristalsis, such as pregnancy and ureteral blockage. Bacteria that enter the renal pelvis through the collecting ducts can infiltrate the renal parenchyma and damage the renal tubules. Infection of the kidney via the hematogenous pathway is uncommon in healthy people. In immunocompromised individuals with *Staphylococcus aureus* bacteremia or *Candida fungemia* from oral sources, the renal parenchyma may be breached on rare occasions. Upper UTIs can occur as a result of bacteria spreading through the bloodstream, such as in persistent bacteraemia, and are frequently linked with a deep source of infection, such as endocarditis. Bacteria from nearby organs can occasionally enter the urinary tract via the lymphatics. Retroperitoneal abscesses and severe bowel infections are two conditions linked to the lymphatic system (25, 26) (Figure 12.4).

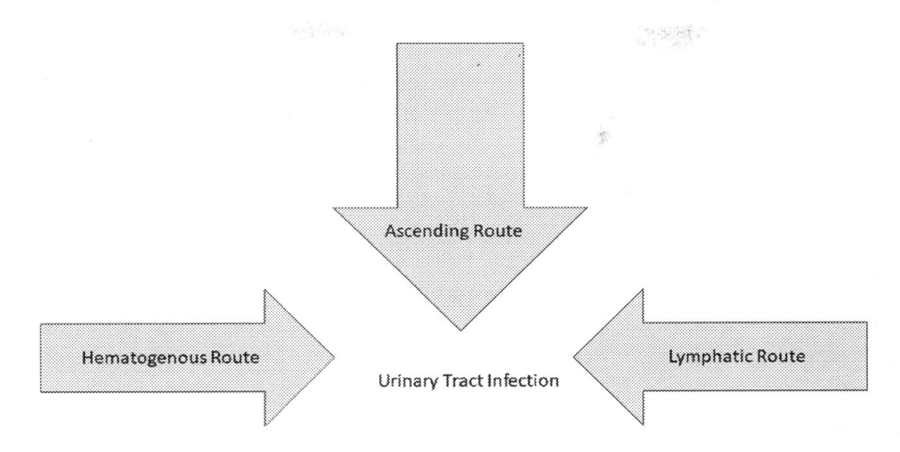

Figure 12.4 *Routes of injection that cause urinary tract infections. (From (25).)*

Lower urinary tract infections, often known as cystitis, are far more common in women than in men. This is due to anatomical variations in women, such as shorter urethral length and a moist periurethral environment. Urinary tract infections usually begin with periurethral contamination by a uropathogen resident in the gut, then colonization of the urethra, and ultimately migration of the pathogen to the bladder or kidney via the flagella and pili of the pathogen. In the pathophysiology of UTI, bacterial adhesion to the uroepithelium is crucial. Infections happen when bacteria's pathogenicity systems outsmart the host's defences. Upper urinary tract infections, commonly known as pyelonephritis, occur when uropathogens pass through the ureters and into the kidneys. When germs cling to a urinary catheter, a kidney, or a bladder stone, or when they are trapped in the urinary system by a physical blockage, infections can ensue. The affected kidney may be enlarged in severe instances of pyelonephritis, with elevated abscesses on the surface. Bacteremia or endocarditis caused by *Staphylococcus aureus* can cause hematogenous seeding of the bacteria in the kidneys, resulting in suppurative necrosis or abscess development within the renal parenchyma. Gram-negative bacteria, on the other hand, seldom induce kidney illness through the hematogenous pathway. The principal renal defect described in an experimental model of pyelonephritis is the failure to optimally concentrate urine. This concentration deficiency appears early in the illness and can be quickly reversed with antibiotic treatment. An obstruction can cause the damaged kidney to deteriorate over time, resulting in renal insufficiency (27).

Signs and symptoms of urinary tract infection

Based on the existence and absence of symptoms, UTI can be classified as asymptomatic or symptomatic infection. Acute cystitis, acute urethritis, or vaginitis are the most common causes of acute dysuria in young women. In healthy women, recurrent UTI (more than three infections in a year) is linked to underlying genetic variables as well as acquired behavioral host factors. The majority of simple infections in women have no long-term consequences or kidney impairment. Acute cystitis is characterized by frequent urination, urgency, and dysuria. The urine is discolored, cloudy, and perhaps unpleasant. The presence of microscopic hematuria is common. Because cystitis is not a systemic infection, patients should not be feverish or have an increased C-reactive protein (CRP) or erythrocyte sedimentation rate (ESR). Except in pregnant women, newborns, and before urological surgery, asymptomatic UTIs are significant bacteriuria that is not always associated with symptoms and does not require therapy. Fever, malaise, loin discomfort (which may be bilateral), bacteriuria, and pyuria are all symptoms of acute pyelonephritis. Some symptoms of cystitis may also be present. The most frequent nosocomial infection is catheter-associated UTI, which accounts for 40% of all hospital-acquired infections. Bacteriuria develops in 1–2% of healthy people after catheterization, and the chance of infection rises with the length of catheterization; after 17 days, 90% of patients have bacteriuria. Approximately 10% of individuals who have a short-term catheterization acquire symptomatic infections. Antibiotics are less effective against microbes that have been lodged in a biofilm. Patients with substantial concomitant conditions, such as diabetes or sickle-cell anemia, as well as those who are prone to stone disease or papillary necrosis and subsequent obstruction, are at a higher risk of more serious upper tract infections and abscesses (22, 28, 29).

Treatment of urinary tract infection

Although UTI treatment guidelines have been established, adherence to them in terms of antibiotic selection, dosage, and duration is low. Noncompliance can lead to reduced antibiotic efficacy (due to the development of resistance) and higher healthcare expenses. Although amoxicillin has long been the first-line antibiotic for urinary tract infections, rising E. coli resistance has made it a less desirable option. Since the 1940s, when sulfonamides were first introduced, antimicrobial medicines have been utilized to treat urinary tract infections. The treatment of an acute uncomplicated UTI (cystitis) is usually easy, as the uropathogens detected have a predicted distribution. A three-day regimen of trimethoprim sulfamethoxazole (TMP-SMX; 160/800 mg twice daily) or TMP (200 mg twice daily) for patients with sulfa allergies has been the first-line therapy for acute uncomplicated UTI for decades. TMP-SMX resistance is rising among commonly acquired E. coli over the world. As a result, the most suitable empiric treatment for these infections has been re-evaluated. First-line antibiotics for acute, uncomplicated UTIs include nitrofurantoin monohydrate/macrocrystals, TMP-SMX, and fosfomycin trometamol. Low collateral damage (selection for drug-resistant organisms), strong effectiveness, and favorable resistance profiles characterize these drugs. Because of increased resistance in some populations, TMP-SMX is only used as a first-line of defense if local resistance rates are not greater than 20%. Fluoroquinolones and beta-lactam drugs are examples of second-line medications. These medications have a higher risk of collateral damage, future infections such *Clostridium difficile*, and poorer effects on normal fecal flora than first-line treatments. Amoxicillin/clavulanate (Augmentin) or cephalosporins such cefixime (Suprax), cefpodoxime, cefprozil (Cefzil), or cephalexin (Keflex) are more options available for treatment of urinary tract infections.

First-line antibiotics have a high efficacy rate. In areas with less than 20% resistance, a three-day TMP-SMX regimen has 90–100% clinical effectiveness and a five- to seven-day nitrofurantoin regimen has an estimated 93% (84–95%) clinical effectiveness. Fosfomycin is thought to have a clinical effectiveness of 91%. Second-line antibiotics such as fluoroquinolones have an estimated clinical efficacy of 90% (85–98%), whereas beta lactam medicines have a clinical efficacy of 89%. (79–98%). Efficacy is determined by bacterial resistance rates, with more resistance resulting in lower treatment efficacy. When choosing an antibiotic therapy, several considerations must be taken into account. These include the duration of therapy, the antimicrobial agent's range of activity, community resistance prevalence, the risk of side effects, and pharmacokinetics. To reach high urinary drug levels, ideal antimicrobial medicines for UTI therapy have primary excretion pathways through the urinary tract (30–32).

Non-pharmacological treatments (nutraceuticals and functional foods)

Stephen DeFelice, MD, founder and chairman of the Foundation for Innovation in Medicine (FIM) in Cranford, New Jersey, invented the phrase "nutraceutical" in 1989, combining the words "nutrition" with "pharmaceutical". DeFelice proceeded to define nutraceutical as, "a food (or part of a food) that provides medical or health benefits, including the prevention and/or treatment of a disease". When functional food aids in the prevention and/or treatment of disease(s) and/or disorder(s) other than anemia, it is called a nutraceutical (33). It can also be defined as "A product isolated or purified from foods that is generally sold in medicinal forms not usually associated with the food (34). A nutraceutical is demonstrated to have a physiological benefit or provide protection against chronic disease." It's worth noting that the term nutraceutical, as it's widely used in marketing, doesn't have a legal definition. Nutraceuticals differ from dietary supplements in the following ways: (1) nutraceuticals must not only enhance the diet but also help in the prevention and/or treatment of illness and/or disorder; and (2) nutraceuticals are utilized as traditional foods or as the only component of a meal or diet. Dietary components include a variety of functions that go beyond basic nutrition, which has led to the creation of the functional food concept and nutraceuticals (33). A nutraceutical, or dietary supplement, has the potential to supply a concentrated form of a postulated bioactive ingredients from a food, given in a non-food matrix, and utilized to improve health in doses greater than those available from typical foods. These are supplied in drug-like packaging: pills, extracts, tablets, and so on (35).

According to the International Food Information Council (IFIC), functional foods are "foods or dietary components that may provide a health benefit beyond basic nutrition" (36). Functional foods are classified as dietary supplements that enhance a regular diet and must be sold in calibrated dosages, such as pills or tablets, in South Korea (35). Functional foods resemble conventional foods in appearance, with the former being ingested as part of a regular diet. Functional foods, in contrast to traditional diets, have established physiological advantages and can lower the risk of chronic illness beyond fundamental nutritional functions, such as gut health maintenance. Food is dubbed "functional food" when it is cooked or prepared with "scientific intelligence," with or without understanding of how or why it is being utilized. As a result, functional food provides the body with the necessary vitamins, lipids, proteins, carbs, and other nutrients for healthy survival (33). Broccoli, carrots, and tomatoes, for example, are called functional foods because they contain physiologically active components such as sulforaphane, beta carotene, and lycopene; other examples include garlic, flax seed, fish, black tea, green tea, soybean, and so on (35).

The distinction between nutraceuticals and functional foods is blurry, and most consumers and businesses use the terms interchangeably. Functional foods are broadly defined as foods that are designed to be consumed as part of a regular diet yet include biologically active components that have the potential to improve health or reduce the risk of disease (35).

Examples of functional foods and nutraceuticals in urinary tract infection are Arctostaphylos uva-ursi (bearberry), Juniperus communis (juniper), Vaccinium macrocarpon (cranberry), Vaccinium myrtillus (blueberry), Cinnamomum verum (cinnamon), Agathosma betulina (buchu), Hybanthus enneaspermus, Armoracia rusticana (horseradish), Hydrastis canadensis (goldenseal), Equisetum arvense (horsetail), Urtica dioica (nettle), Plantago major L, Beth root, Dandelion root, Fennel, and Honeysuckle (10, 21). Some of these functional foods and nutraceuticals are discussed in greater depth in the following sections.

Cranberry (Vaccinium macrocarpon) Cranberry has been linked to the prevention of urinary tract infections by a number of studies. The plant, which belongs to the Ericaceae family, may be effective against E. coli, one of the most common causes of bacterium-mediated UTIs, by reducing bacterial adhesion to bladder walls, making germs more likely to be washed out during urination. Consumption of cranberry juice provides considerable protection against both susceptible and resistant E. coli strains (37). Consumption of cranberry juice has a significant anti-adherence effect against several E. coli uropathogenic strains (38). Biofilm development by uropathogenic Escherichia coli has been found to be inhibited by cranberry juice cocktail (CJC) (39).

For the prevention of urinary tract infections, the juice and extracts of the American cranberry (Vaccinium macrocarpon) fruit are frequently used (UTIs). Cranberry products prevent UTI by acting as an antiadhesive, preventing E. coli from adhering to uroepithelial cells. The antiadhesive action of the cranberry fruit may be related to its proanthocyanidins, flavanol glycosides, anthocyanins, and organic and phenolic acids. Cranberry may help preserve urinary tract health by increasing both in vitro and urine bacterial antiadhesion activities, which have been linked to proanthocyanidins. Cranberry extracts inhibit bacterial growth and can help reduce harmful biofilm development (40).

The active principle of cranberries, which is the total of all chemicals that contribute to cranberry effects, has yet to be discovered. There's no denying that Proanthocyanidins, particularly type-A proanthocyanidins, play a crucial role (41). Because of these health advantages, cranberry extract is available in a variety of commercial forms. The normal dose of the juice in acute conditions is 250–500 mL two to three times a day, while for prevention, 250–500 mL per day is sufficient. There are other solid dose forms available, such as pills with concentrated cranberry extract. In acute cases, taking two to three capsules two to four times per day is suggested, as is taking one capsule two to three times per day for prevention. In conclusion, existing data show that cranberry preparations have beneficial effects against UTIs; however, these effects are primarily related to prophylactic activities, such as preventing the development of infections, or in combination with conventional antibiotics, and sole intake of the herb is not recommended for UTI treatment (10).

Blueberry Blueberries have also been used to treat and prevent urinary tract infections in the past. Blueberries, like the cranberry, have bioactive chemicals that prevent E. coli from adhering to the bladder's walls (42). The most active elements in blueberry extracts against UTI are tannins (10).

Arctostaphylos uva-ursi (uva-ursi) Arctostaphylos uva-ursi is a shrub indigenous to North America's hilly regions, however it has spread to other parts of the globe. It belongs to the Ericaceae family. The therapeutic component of the plant is the leaves, which contain the glycoside arbutoside. Arbutoside is hydrolyzed to glucose and the aglycone hydroquinone in the intestines after consumption. In the liver, hydroquinone is absorbed and subsequently glucuronidated. The glucuronide of hydroquinone is subsequently transported to the kidneys and eliminated in the urine. The hydroquinone glucuronide will breakdown spontaneously if the pH of the urine is sufficiently alkaline (>7), releasing the hydroquinone to serve as a direct antimicrobial agent. This study was unique in that uva-ursi is often used to treat lower urinary tract infections (UTIs), not as a preventative measure. The German Commission E has approved the use of uva-ursi solely for the treatment of urinary tract infections, not for the prevention of them. Part of the rationale for this is fear that long-term exposure to synthetic hydroquinone may be carcinogenic, as evidenced by data from industrial exposures and laboratory studies. The Commission E's suggested dose of uva-ursi is 3 g leaves extracted in 150 ml water by either hot or cold infusion up to four times a day, resulting in 400–840 mg arbutoside. The tannins in the uva-ursi leaf are extracted by hot infusions (as well as other standardized extracts), which might induce stomach distress; cool infusions do not have this problem. Corilagin, a tannin derived from uva-ursi, has recently been found to increase the efficacy of beta-lactam antibiotics against methicillin-resistant Staphylococcus aureus in vitro. Uva-ursi should be avoided during pregnancy and lactation, as well as for those with renal failure and dyspepsia (tannin-containing extracts only). It is also not recommended for minors (43).

Juniperus communis (juniper) Juniperus communis (juniper), a member of the Cupressaceae family, as well as other closely related species such as Juniperus monosperma (Engelm) Sarg and Juniperus osteosperma (Utah juniper), have antibacterial properties (10). Some Juniperus species have antibacterial qualities that have been used in traditional folk medicine to treat ailments like TB, pneumonia, and urinary tract infections (44).

Antimicrobial terpenoids found in juniper leaves may also have diuretic properties (43). It's worth noting that juniper's volatile oil includes nephrotoxic chemicals, particularly hydrocarbon terpenoids. These negative effects, on the other hand, may appear only after taking massive dosages that much surpass the therapeutic amount (10). Juniperus branch extracts had strong antibacterial activity against both S. aureus strains, however they had a different effect on biofilm production depending on the biofilm producing strain's ability (44).

Cinnamomum verum (cinnamon) Cinnamon is a member of the Lauraceae family and has antibacterial and antioxidant properties. It includes bioactive phytochemical substances that have been used to treat UTIs, including as trans-cinnamaldehyde, eugenol, trans-cinnamyl acetate, and proanthocyanidins (10).

Cinnamon bark extract contains a significant amount of trans-cinnamaldehyde (TC). The Food and Drug Administration has certified it for use in foods as a generally regarded as safe (GRAS) molecule (FDA). Trans-cinnamaldehyde was also shown to have no genotoxic or mutagenic effects, according to the analysis. TC has previously been shown to have antibacterial action against Clostridium botulinum, Staph aureus, E. coli, and Salmonella typhimurium. Although cinnamon or cinnamon oil has been used to treat UTIs for a long time, Amalaradjou et al. (2010) were the first to show that trans-cinnamaldehyde might inactivate and limit the growth of UPEC biofilms on urinary catheters. Trans-cinnamaldehyde prevented the adhesion and invasion of uroepithelial cells by UPEC by downregulating important virulence genes in the pathogen, according to follow-up research by the same group (Amalaradjou et al., 2011). These findings suggest that trans-cinnamaldehyde might be used as an antibiotic to treat urinary tract infections. Trans-cinnamaldehyde's antibacterial impact can be due to a variety of methods. The hydrophobicity of essential oils or their components, such as trans-cinnamaldehyde, is an important feature that allows them to target the lipid-rich bacterial cell membrane and mitochondria. This makes these membranes more permeable, allowing ions and other cell contents to flow out. Trans-cinnamaldehyde is thought to kill bacteria by reducing energy synthesis and glucose absorption, in addition to its action on cell membranes (45).

Armoracia rusticana (horseradish) A. rusticana (synonyms: Cochlearia armoracia, Radic ula armoracia), a member of the Brassi caceae family, has historically been used to treat urinary tract infections. It showed promising effects in the prevention of recurrent UTI in children (10). It was discovered that horseradish isothiocyanates are responsible for the herb's antibacterial properties. These bioactive chemicals were proven to inhibit the pathogenic process of uropathogenic E. coli penetrating human cells (46).

Barberry (Berberries vulgaris) Berberine, found in barberry, has outstanding anti-infection capabilities. It has been shown in studies to destroy bacteria that cause urinary tract infections (E. coli, Streptococci) (42).

Marshmallow root (Althaea officinalis) Marshmallow root inhibits the growth of germs in the urinary tract while also strengthening and cleaning the bladder. It's utilized as a diuretic, emollient, and demulcent. It calms the urinary tract and aids in the treatment of kidney and bladder irritation. It efficiently prevents urine bleeding (42).

Chebulic myrobalan (Terminalia chebula Retz.) Chebulic myrobalan was found to have antibacterial action against multidrug-resistant uropathogenic E. coli, with the phenolic chemicals found in it being primarily responsible for this activity (42).

Dysmenorrhea

Dysmenorrhea, or unpleasant cramping associated with the start and early days of menstruation, is classified as either primary or secondary, depending on whether or not there is underlying uterine illness. Secondary dysmenorrhea can be caused by uterine disorders such as endometriosis, uterine polyps, or fibroids, as well as IUD-related problems and pelvic inflammatory disease (15). Dysmenorrhea is quite common in adolescent females and has been studied extensively. In this demographic, it has been recognized as a primary cause of sickness, leading to school absenteeism and activity nonparticipation. Pain during menstruation without an identified pathologic lesion is referred to as primary dysmenorrhea. Nausea, vomiting, diarrhea, and headache are all symptoms of menstruation discomfort. The exact etiology of dysmenorrhea is unknown (47). Primary dysmenorrhea is a widespread and frequent ailment among young women, with severe forms causing work or school absenteeism in 15% and moderate forms requiring no prescription or only occasional over-the-counter (OTC) analgesics in approximately 30% (48).

Pathophysiology of dysmenorrhea

The pathophysiology of primary dysmenorrhea has been debated practically since the dawn of modern medicine, and several ideas have been suggested as to why the disorder exists. Myometrial activity, uterine blood flow, prostaglandins (PG), ovarian steroids, posterior pituitary hormones, cervical factors, nerves, and psychological variables are the most usually mentioned (49).

Primary dysmenorrhea is caused by an overproduction of prostaglandins during ovulatory menses. The withdrawal of progesterone from the typically involuting corpus luteum in the second half of an ovulatory cycle promotes the production of phospholipids, particularly omega-6 fatty acids, which are subsequently converted to arachidonic acid and then to prostaglandins. Increased intrauterine pressure and aberrant uterine contractions result from the generation of prostaglandins. Furthermore, uterine vascular constriction causes reduced blood flow, ischemia of the uterine muscles, and increased sensitivity of pain receptors, all of which produce pelvic discomfort. Endometrial blood flow has been demonstrated to diminish during uterine contractions, implying that the ischemia that results is to blame for the discomfort. Prostaglandins are also converted to leukotrienes, which, together with prostaglandin F2-alpha, cause systemic symptoms such as nausea, vomiting, headache, and disorientation, which can occur as a result of menstruation cramps. The necessity for ovulatory cycles to occur for primary dysmenorrhea to occur partly explains why most adolescent girls do not have dysmenorrhea at the time of menarche, but may experience discomfort several months later once they have established more regular menses (50).

Signs and symptoms of dysmenorrhea

While lower abdominal cramps is the most common symptom of dysmenorrhea, many teenagers have additional menstruation-related symptoms including headaches and vomiting. Symptoms usually begin a few hours before or after the commencement of menstrual flow, and remain for the first 24 to 48 hours (51). Primary dysmenorrhea is characterized by variable, spasmodic menstrual cramps, sometimes known as "labor-like" sensations, that begin just a few hours before or with the commencement of monthly flow and persist only two to three days. The pains are the most strong on the first or second day of menstrual flow, or more precisely the first 24–36 hours, which coincides with the maximum prostaglandin production into the menstrual fluid (48). Early menarche and increased length and volume of menstrual flow are linked to the severity of dysmenorrhea symptoms. Furthermore, cigarette smoking may lengthen the duration of dysmenorrhea, owing to nicotine-induced vasoconstriction (51).

Treatment of dysmenorrhea

The goal of treatment for primary dysmenorrhea is to stop the generation of PGs, lower uterine tone, or block pain perception by a direct analgesic action (52). Nonsteroidal anti-inflammatory medicines (NSAIDs) are the most popular pharmacological therapies for dysmenorrhea. Traditional NSAIDs reduce prostaglandin synthesis by inhibiting the activity of both COX-1 and COX-2 isoforms. As a result of the reduced amounts of prostaglandins, the uterus contracts less vigorously, causing less pain. In a small group of dysmenorrheic women treated with NSAIDs, Chan and Dawood discovered that $PGF_2 \alpha$ reduced and pain improved. COX-2 medications that target specific enzymes may also help with dysmenorrhea symptoms. These COX-2 inhibitors spare prostaglandins, which are necessary for the integrity of the gastric mucosa and are generated by COX-1. The only COX-2 inhibitor authorized by the US Food and Drug Administration (FDA) for the treatment of primary dysmenorrhea is celecoxib (Celebrex®). Not all dysmenorrhea sufferers, especially teenagers and young adults, react to NSAIDs, and some of those who do report only partial improvement. Most NSAIDs block solely cyclooxygenase and have little effect on the generation of other inflammatory mediators such leukotrienes, according to one theory. However, in teenagers, therapy with the leukotriene receptor antagonist montelukast (Singulair®) in the FDA-approved dose (for asthma) and commencing soon before the menstrual cycle failed to improve dysmenorrhea symptoms (53). Examples of non-steroidal anti-inflammatory drugs include flurbiprofen, tiaprofenic acid, piroxicam, mefenamic acid, ibuprofen, naproxen sodium, celecoxib, and so on (52).

In around 70–80% of women, combined hormonal contraceptives are successful in treating dysmenorrhea. The generation of PG, progesterone, and vasopressin will be reduced if ovulation and preventing endometrial growth are inhibited. The combination oral contraceptive pill (OCP), the contraceptive intravaginal ring, and the patch have all been shown to help with dysmenorrhea. The possibility of deep vein thrombosis is the main worry with OCPs. Primary dysmenorrhea appears to be treated effectively with progesterone-only contraception. Croxatto et al. found that the etonogestrel (68 mg) contraceptive implant helped 85% of women with dysmenorrhea. Unfortunately, research on other types of progesterone-only contraceptives is scarce, but they remain viable alternatives since they prevent ovulation and menstrual periods (52).

Non-pharmacological treatment

In certain trials, herbal remedies, transcutaneous nerve stimulation, acupuncture, and heat treatment have all been shown to help with dysmenorrhea. In young adult women, a low-fat vegetarian diet was linked to a reduction in the duration and severity of dysmenorrhea (53). Dietary therapy such as magnesium, vitamin B6, vitamin B1, vitamin E, and omega-3 fatty acids made up the majority of the interventions (fish oil). Toki shakuyaku- san, a Japanese herbal combination, was one of them (54).

Magnesium One small research found that magnesium therapy was more beneficial than placebo for other outcomes, with those receiving magnesium needing much less time off work and less medication. Both the magnesium and placebo groups had minor side effects, with the meta-analysis revealing no meaningful difference. Women who received magnesium therapy had significantly lower levels of PGF2 alpha in their menstrual blood than those who received placebo ($p < 0.05$), which paralleled the participants' therapeutic pain reduction. This points to a possible biological reason for magnesium therapy in the treatment of dysmenorrhoea. It plays a function in muscular relaxation and vasodilation as well as inhibiting PGF2 alpha production (54).

Omega-3 fatty acids The use of omega-3 fatty acids on a regular basis significantly decreased dysmenorrheal symptoms and discomfort. A number of preliminary studies have also suggested that fish oil supplements may be useful in the treatment of primary dysmenorrhea. The first group received fish oil daily for two months, followed by placebo for two months. The second group received placebo daily for the first two months, followed by fish oil for two months. The Cox Menstrual Symptom Scale did not differ significantly across the groups at baseline or after two months of placebo. However, omega-3 treatment resulted in a substantial decrease ($P<0.0004$). Moghadamnia et al. found that omega-3 fatty acid supplementation decreased abdominal discomfort, low-back pain, and the requirement for rescue NSAID medication in a group of teenage females (55). According to the findings, nutritional supplementation with omega-3 very long polyunsaturated fatty acids is useful in reducing dysmenorrhea symptoms in teenagers. The significant effect of omega-3 fatty acid intake on the body is most likely due to a development of less powerful prostaglandins and leukotrienes (56). According to the findings, supplementing the diet with omega-3 fatty acids can help young women with primary dysmenorrhea experience less pain. Furthermore, the usage of omega-3 fatty acids lowered the amount of ibuprofen needed as a rescue dosage to treat severe menstrual pain (55).

Premenstrual syndrome and premenstrual dysphoric disorder

Premenstrual syndrome or premenstrual dysphoric disorder (PMDD), which affects 3% to 40% of women depending on the severity, is characterized by persistent and troublesome physical and mental problems during the premenstrual period (57). During the luteal phase of their menstrual cycles, up to 80% of women experience one or more physical, psychological, or behavioral symptoms without suffering significant interruption in their regular functioning. Premenstrual syndrome affects 20% to 32% of premenopausal women, with mild to moderate symptoms affecting many aspects of their lives; PMDD affects 3% to 8% of premenopausal women with more severe symptoms (58).

Premenstrual syndrome has no established cause; however, the onset of symptoms is linked to ovarian hormone levels. Premenstrual syndrome causes women to be hypersensitive to the typical hormonal changes that occur during the menstrual cycle (59). Several studies have shown that the symptoms are caused by cyclical fluctuations in estrogen and progesterone levels (60). Treatment with ovulation inhibitors, throughout pregnancy, and after menopause, generally improves symptoms. This supports the theory of cyclic ovarian activity. The precise involvement of ovarian function in the onset, manifestation, and remission of symptoms, however, has yet to be identified. Single nucleotide polymorphisms in the ESR1 (oestrogen receptor alpha) gene have been found to be linked to PMDD. Furthermore, antidepressant or anxiolytic medicines might relieve premenstrual symptoms by stabilizing neurotransmitters like serotonin and affecting aminobutyric acid (GABA) imbalance. As a result, it appears that these neurotransmitters are important in the development of premenstrual symptoms (59).

Signs and symptoms of premenstrual syndrome

Premenstrual syndrome symptoms range from minor to severe, interfering with everyday personal and professional life. Physical, psychological, and behavioral symptoms are the three most prevalent types. Tiredness, edema, breast fullness, headache, weight gain, bodily pains, and swelling of the limbs are some of the physical symptoms. Irritability, anxiousness, mood swings, sorrow, depression, poor attention, hypersomnia/insomnia, and withdrawal from typical activities are examples of emotional or behavioral symptoms (59).

Pathophysiology of premenstrual syndrome

The lack of symptoms before puberty, during pregnancy, after menopause, and during therapy with gonadotrophin-releasing hormone (GnRH) analogues supports the hypothesis that the etiology of premenstrual syndrome is centered around the ovarian cycle. Sex hormones easily travel across the blood–brain barrier, and receptors for them may be found in many parts of the brain, including the amygdala and hypothalamus. Progesterone is thought to be metabolized in the brain into allopregnanolone and pregnanolone, which activates the GABA-inhibitory neurotransmitter system. $GABA_A$ receptors are linked to mood, cognition, and affective changes. Pregnanolone and allopregnanolone have anxiolytic, sedative, and anesthetic effects at high concentrations, while at lower levels, allopregnanolone can provoke anxiety, bad mood, and hostility. After exposure to high quantities of allopregnanolone, GABA receptors become less responsive to it, resulting in increased symptoms during the luteal phase. Estrogen and progesterone have an effect on serotonergic activity in the brain. Progesterone raises monoamine oxidase (MAO), which reduces the availability of 5-hydroxytryptamine (5-HT) and causes depression. Estrogen, on the other hand, causes MAO to be degraded, increasing the availability of free tryptophan in the brain, which improves serotonin transport and so activates 5-HT binding sites in the brain, resulting in an antidepressant effect (61).

Treatment of premenstrual syndrome

Treatment and management of premenstrual syndrome includes therapeutic agents such as oral contraceptives, gonadotropin releasing hormone agonist (GnRHa), estradiol, and selective serotonin reuptake inhibitors (62).

Selective serotonin reuptake inhibitors (SSRIs) For women with severe symptoms and for women with milder symptoms who have failed to respond to previous medications, SSRIs can be considered first-line therapy. In comparison to the placebo, 14 RTCs exhibited an improvement. Nausea, vomiting, diarrhea, dry mouth, anxiety, headache, palpitations, dizziness, and a decrease in libido are some of the side effects. To decrease adverse effects, researchers are now focused on utilizing modest dosages during the luteal phase (63).

Oral contraceptives Oral contraceptives (OCs) have been found to alleviate menstrual-related discomfort by regulating hormonal swings. For the treatment of premenstrual syndrome/premenstrual dysphoric condition, ovulation suppression is frequently effective (PMDD). Premenstrual syndrome/PMDD requires long-term treatment since symptoms reoccur when medication is stopped, and the illness generally lasts until ovarian function declines with menopause. The lack of placebo-controlled trials limits conclusions about the treatment of premenstrual syndrome/PMDD with oral contraceptives. Many premenstrual syndrome/PMDD symptoms, such as breast tenderness, headache, bloating, and depression, can also be adverse effects of various oral contraceptive tablets, further aggravating the situation (63).

Gonadotropin releasing hormone agonist The use of a gonadotropin releasing hormone agonist (GnRH agonist) causes a "medical oopherectomy" and menopausal plasma estrogen and progesterone concentrations. In the treatment of premenstrual syndrome and PMDD, GnRH agonists have been demonstrated to be beneficial. Hot flashes, vaginal dryness, and, on rare occasions, depression, headaches, and muscular pains are all side effects of GnRH agonists' hypoestrogenic condition. Long-term GnRH therapy can also lead to osteoporosis and a higher risk of cardiovascular disease. Women with severe premenstrual syndrome and depression, whether primary or secondary, may not react to GnRH agonists (64).

Non-pharmacological treatment

Chasteberry The dried, mature fruits of the chaste tree (Vitex agnus-castus), a Mediterranean plant, are used to make this herb. Chasteberry has a long history of usage as a women's medicine, with Hippocrates recording the first medical descriptions in the 4th century B.C. Chasteberry is now widely used and acknowledged in Europe as a therapy for premenstrual syndrome, dysmenorrhea, mastodynia (painful breast swelling), and menopause in women. Chasteberry includes a number of active chemicals that have a balancing or normalizing impact on several parts of the reproductive system. Essential oils, iridoid glycosides (agnuside and aucubin), and flavonoids (casticin and iso-orientin) are some of the active chemicals (65).

Chasteberry's efficacy in the treatment of menstruation diseases, as well as its high tolerance, have been proven in a number of studies (66). In a three-month randomized double-blind, placebo-controlled experiment, 37 women with monthly irregularities and latent prolactinemia were shown to benefit from chasteberry. Compared to placebo, women who took the chasteberry extract (20 mg/day) exhibited a substantial drop in prolactin release, a significant average increase in the luteal phase of five days, a rise to normal levels of progesterone during the mid-luteal phase, and a reduction in premenstrual syndrome symptoms (65).

Evening primrose oil (EPO) Evening primrose oil is one of the treatments that has lately been suggested to be useful to premenstrual syndrome sufferers. The oil is derived from the seeds of the evening primrose (Oenothera biennis), and it is the sole commonly available source of y-linolenic acid (GLA) for medicinal purposes. The GLA level of seed oils from this plant varies greatly, however one with a constant amount of 9% GLA (equal to 45 mg per capsule) is marketed under the brand name Efamol (Vita-Glow) (67).

The argument for this application is that women with premenstrual syndrome have an unbalanced essential fatty acid profile, which can be restored with EPO supplementation. According to Brush et al., women with premenstrual syndrome had high amounts of n6 essential fatty acids but low levels of all linoleic acid metabolites, including arachidonic acid. These researchers discovered that gamma-linolenic acid levels in premenstrual syndrome patients were below detectable levels, and they hypothesized that premenstrual syndrome is linked to a malfunction in the conversion of linoleic acid to gamma-linolenic acid. Brush investigated activity of EPO worked for premenstrual syndrome symptoms in 68 women. The women were given EPO (1–2 g per day) for three days before to the beginning of premenstrual syndrome symptoms and until menses began. After at least three months of therapy, 41 women (61%) experienced total alleviation of their symptoms and 16 (23%) had partial relief, according to a patient self-report scale (63).

Calcium The metabolism of calcium, magnesium, and vitamin D is influenced by ovarian steroids. Estrogen regulates calcium metabolism, intestinal absorption, and parathyroid gene expression and release, resulting in a variation in calcium levels during the menstrual cycle. Hypocalcemia has been linked to a variety of mood disorders that are comparable to premenstrual syndrome symptoms. According to some data, women with premenstrual syndrome have underlying calcium dysregulation, secondary hyperparathyroidism, and vitamin D insufficiency. Calcium carbonate 1200 mg/day in split dosages has been demonstrated to reduce premenstrual syndrome symptoms in two controlled studies. Calcium, 500 mg twice daily, decreased fatigability, appetite fluctuations, and depression in women with premenstrual syndrome, according to Ghanbari et al. (68).

Conclusion

The health of the urogenital system is an important aspect to maintain human well-being. From this chapter we can conclude that there are many researches done as well as going on for nutraceuticals and functional foods as complementary and alternative therapy for treating various genitourinary tract diseases and disorders. They are demonstrating positive efficacy by showing curative effect in comparison with placebo, but alone it may not give desired effects in severe cases. For better and proper curative purpose, nutraceuticals and functional foods need to be administered co-currently with pharmacologically classified therapeutic agents.

References

1. Abolbashari M, Atala A, Yoo JJ. Genitourinary System [Internet]. Translational Regenerative Medicine. Elsevier Inc.; 2015. 495–505 p. Available from: http://dx.doi.org/10.1016/B978-0-12-410396-2.00036-0
2. Hickling DR, Sun TT, Wu XR. Anatomy and physiology of the urinary tract: relation to host defense and microbial infection. Urinary tract infections: Molecular pathogenesis and clinical management. 2017 Feb 15:1–25.
3. Velho AM, Velho RM. Anatomy and Physiology Series: The Kidney and Lower Urinary Tract. Journal of Renal Nursing. 2013 Mar;5(2):76–80.
4. Linsenmeyer TA. Urologic Anatomy and Physiology. Phys Med Rehabil Clin N Am [Internet]. 1993;4(2):221–47. Available from: https://doi.org/10.1016/S1047-9651(18)30579-5
5. Wallace MA. Anatomy and Physiology of the Kidney. AORN Journal. 1998 Nov 1;68(5):799–820.
6. Atala A, Amin M. Current Concepts in the Treatment of Genitourinary Tract Disorders in the Older Individual. Drugs & Aging. 1991 May;1(3):176–93.
7. Flores-Mireles AL, Walker JN, Caparon M, Hultgren SJ. Urinary Tract Infections: Epidemiology, Mechanisms of Infection and Treatment Options. Nat Rev Microbiol [Internet]. 2015;13(5):269–84. Available from: http://dx.doi.org/10.1038/nrmicro3432
8. Foxman B. The Epidemiology of Urinary Tract Infection. Nat Rev Urol [Internet]. 2010;7(12):653–60. Available from: http://dx.doi.org/10.1038/nrurol.2010.190
9. Monroy-Torres R, Medina-Jiménez AK. Cranberry Juice and Other Functional Foods in Urinary Tract Infections in Women: A Review of Actual Evidence and Main Challenges Frontiers in Clinical Drug Research - Anti Infectives: Volume 4. 2019;(June):183–211.
10. Fazly Bazzaz BS, Darvishi Fork S, Ahmadi R, Khameneh B. Deep Insights into Urinary Tract Infections and Effective Natural Remedies. African J Urol [Internet]. 2021;27(1). Available from: https://doi.org/10.1186/s12301-020-00111-z
11. Gurung P, Yetiskul E, Jialal I. Physiology, Male Reproductive System. StatPearls [Internet]. 2020; Available from: https://www.ncbi.nlm.nih.gov/books/NBK538429/
12. Walker R. Clinical Pharmacy and Therapeutics. E-Book. Elsevier Health Sciences; 2011 Oct 24. 753, 715 p.
13. Hoare BS, Khan YS. Anatomy, Abdomen and Pelvis, Female Internal Genitals. StatPearls [Internet]. 2020;(August). Available from: http://www.ncbi.nlm.nih.gov/pubmed/32119488
14. Vogazianou A. and Physiology of the Female Reproductive System. Adv Pract Endocrinol Nurs. 2019;739–52.
15. Koda-Kimble MA. Koda-Kimble and Young's Applied Therapeutics: The Clinical Use of Drugs. Lippincott Williams & Wilkins; 2012 Feb 1. 47–1, 47–9 p.

16. Smith S, Pfeifer SM, Collins JA. Diagnosis and Management of Female Infertility. J Am Med Assoc. 2003;290(13):1767–70.
17. Barnett BJ, Stephens SD. Urinary Tract Infection: An Overview. Am J Med Sci [Internet]. 1997; Available from: http://dx.doi.org/10.1016/S0002-9629(15)40208-3
18. Sheerin NS. Urinary Tract Infection. Medicine (Baltimore) [Internet]. 2011;39(7):384–9. Available from: http://dx.doi.org/10.1016/j.mpmed.2011.04.003
19. Gupta K, Grigoryan L, Trautner B. In the Clinic® Urinary Tract Infection. Ann Intern Med. 2017;167(7):ITC49–64.
20. Ronald A. The Etiology of Urinary Tract Infection: Traditional and Emerging Pathogens. Disease-a-Month. 2003;49(2):71–82.
21. Komala M, Bhowmik D, Sampath Kumar KP. Urinary Tract Infection: Causes, Symptoms, Diagnosis and it's Management. J Chem Pharm Sci. 2013;6(1):22–8.
22. Vasudevan R. Urinary Tract Infection: An Overview of the Infection and the Associated Risk Factors. J Microbiol Exp. 2014;1(2):42–54.
23. Minardi D, d'Anzeo C, Conti M. Urinary Tract Infections in Women: Etiology and Treatment Options. Int J Gen Med. 2011;4:333.
24. Sobel JD, Kaye D. Urinary Tract Infections [Internet]. Eighth Edition. Vol. 1, Mandell, Douglas, and Bennett's Principles and Practice of Infectious Diseases. Elsevier Inc.; 2014. 886–913.e3 p. Available from: http://dx.doi.org/10.1016/B978-1-4557-4801-3.00074-6
25. Davis N, Flood H. The pathogenesis of urinary tract infections. In: Clinical management of complicated urinary tract infection. 2011. p. 978–53.
26. Walsh C, Collyns T. The Pathophysiology of Urinary Tract Infections. Surg (United Kingdom) [Internet]. 2017;35(6):293–8. Available from: http://dx.doi.org/10.1016/j.mpsur.2017.03.007
27. Le HSLJ. PSAP 2018 Book 1 Urinary Tract Infections. PSAP 2018 B 1- Infect Dis. 2018;(Sobel 2014):7–28.
28. Lee JBL, Neild GH. Urinary tract infection. Medicine. 2007 Aug;35(8):423–8.
29. McLaughlin SP, Carson CC. Urinary Tract Infections in Women. Med Clin North Am. 2004;88(2):417–29.
30. White B. Diagnosis and Treatment of Urinary Tract Infections in Children. Am Fam Physician. 2011;1:265–79.
31. Nicolle LE. Urinary Tract Infection: Traditional Pharmacologic Therapies. Am J Med. 2002;49(2):111–28.
32. Chu CM, Lowder JL. Diagnosis and Treatment of Urinary Tract Infections Across Age Groups. Am J Obstet Gynecol [Internet]. 2018;219(1):40–51. Available from: https://doi.org/10.1016/j.ajog.2017.12.231
33. Cencic A, Chingwaru W. The Role of Functional Foods, Nutraceuticals, and Food Supplements in Intestinal Health. Nutrients. 2010;2(6):611–25.
34. Trottier G, Boström PJ, Lawrentschuk N, Fleshner NE. Nutraceuticals and Prostate Cancer Prevention: A Current Review. Nat Rev Urol [Internet]. 2010;7(1):21–30. Available from: http://dx.doi.org/10.1038/nrurol.2009.234
35. Gul K, Singh AK, Jabeen R. Nutraceuticals and Functional Foods: The Foods for the Future World. Crit Rev Food Sci Nutr. 2016;56(16):2617–27.
36. Keservani RK, Kesharwani RK, Vyas N, Jain S, Raghuvanshi R, Sharma AK. Nutraceutical and Functional Food As Future Food: A Review. Der Pharm Lett. 2010;2(1):106–16.
37. Howell AB. Cranberry Juice and Adhesion of Antibiotic-Resistant Uropathogens. JAMA. 2002;287(23):3082.
38. Di Martino P, Agniel R, David K, Templer C, Gaillard JL, Denys P, et al. Reduction of Escherichia Coli Adherence to Uroepithelial Bladder Cells After Consumption of Cranberry Juice: A Double-blind Randomized Placebo-controlled Cross-over Trial. World J Urol. 2006;24(1):21–7.
39. Tao Y, Pinzón-Arango PA, Howell AB, Camesano TA. Oral Consumption of Cranberry Juice Cocktail Inhibits Molecular-scale Adhesion of Clinical Uropathogenic Escherichia Coli. J Med Food. 2011;14(7–8):739–45.

40. Laplante KL, Sarkisian SA, Woodmansee S, Rowley DC, Seeram NP. Effects of Cranberry Extracts on Growth and Biofilm Production of Escherichia Coli and Staphylococcus Species. Phyther Res. 2012;26(9):1371–4.
41. Davidson E, Zimmermann BF, Jungfer E, Chrubasik-Hausmann S. Prevention of Urinary Tract Infections with Vaccinium Products. Phyther Res. 2014;28(3):465–70.
42. Rev P, Bag A, Bhattacharyya SK, Chattopadhyay RR. Medicinal Plants and Urinary Tract Infections : An update. Pharmacognosy Review. 2008;2(4):277–84.
43. Yarnell E. Botanical Medicines for the Urinary Tract. World J Urol. 2002;20(5):285–93.
44. Marino A, Bellinghieri V, Nostro A, Miceli N, Taviano MF, Güvenç AŞ, et al. In Vitro Effect of Branch Extracts of Juniperus Species from Turkey on Staphylococcus Aureus Biofilm. FEMS Immunol Med Microbiol. 2010;59(3):470–6.
45. Roshni Amalaradjou MA, Venkitanaray K. Natural Approaches for Controlling Urinary Tract Infections. In: Urinary Tract Infections. InTech; 2011. p. 227–44.
46. Mutters NT, Mampel A, Kropidlowski R, Biehler K, Günther F, Bălu I, et al. Treating Urinary Tract Infections Due to MDR E. coli with Isothiocyanates – a Phytotherapeutic Alternative to Antibiotics? Fitoterapia. 2018;129(July):237–40.
47. Davis AR, Westhoff CL. Primary Dysmenorrhea in Adolescent Girls and Treatment with Oral Contraceptives. J Pediatr Adolesc Gynecol. 2001;14(1):3–8.
48. Dawood Yusoff M. Primary Dysmenorrhea Advances in Pathogenesis and Management. Obstet Gynecol. 2006;108(2):428–41.
49. Åkerlund M. Pathophysiology of Dysmenorrhea. Acta Obstet Gynecol Scand. 1979; 58(87 S):27–32.
50. Ryan SA. The Treatment of Dysmenorrhea. Pediatr Clin North Am [Internet]. 2017; 64(2):331–42. Available from: http://dx.doi.org/10.1016/j.pcl.2016.11.004
51. Harel Z. Dysmenorrhea in Adolescents and Young Adults: Etiology and Management. J Pediatr Adolesc Gynecol. 2006;19(6):363–71.
52. Ferries-Rowe E, Corey E, Archer JS. Primary Dysmenorrhea: Diagnosis and Therapy. Obstet Gynecol. 2020;136(5):1047–58.
53. Harel Z. Dysmenorrhea in Adolescents and Young Adults: From Pathophysiology to Pharmacological Treatments and Management Strategies. Expert Opin Pharmacother. 2008;9(15):2661–72.
54. Proctor M, Murphy PA . Herbal and Dietary Therapies for Primary and Secondary Dysmenorrhoea. Cochrane Database Syst Rev. 2002;2001(2): 1–23.
55. Rahbar N, Asgharzadeh N, Ghorbani R. Effect of Omega-3 Fatty Acids on Intensity of Primary Dysmenorrhea. Int J Gynecol Obstet [Internet]. 2012;117(1):45–7. Available from: http://dx.doi.org/10.1016/j.ijgo.2011.11.019
56. Harel Z, Biro FM, Kottenhahn RK, Rosenthal SL. Supplementation with Omega-3 Polyunsaturated Fatty Acids in the Management Of Dysmenorrhea in Adolescents. Am J Obstet Gynecol. 1996;174(4):1335–8.
57. Milewicz A, Jedrzejuk D. Premenstrual Syndrome: From Etiology to Treatment. Maturitas. 2006;55(SUPPL. 1):47–54.
58. Biggs WS, Demuth RH. Premenstrual Syndrome and Premenstrual Dysphoric Disorder. Am Fam Physician. 2011;84(8):918–24.
59. Ryu A, Kim TH. Premenstrual Syndrome: A Mini Review. Maturitas [Internet]. 2015; 82(4):436–40. Available from: http://dx.doi.org/10.1016/j.maturitas.2015.08.010
60. Hofmeister S, Bodden S Premenstrual Syndrome Premenstrual Dysphoric Disorder. Am Fam Physician. 2016;94(3):236–240.
61. Walsh S, Ismaili E, Naheed B, O'Brien S. Diagnosis, Pathophysiology and Management of Premenstrual Syndrome. Obstet Gynaecol. 2015;17(2):99–104.
62. Imai A, Ichigo S, Matsunami K, Takagi H. Premenstrual Syndrome: Management and Pathophysiology. Clin Exp Obstet Gynecol. 2015;42(2):123–8.
63. Verma RK, Chellappan DK, Pandey AK. Review on Treatment of Premenstrual Syndrome: From Conventional to Alternative Approach. J Basic Clin Physiol Pharmacol. 2014;25(4):319–27.

64. Rapkin A. A Review of Treatment of Premenstrual Syndrome & Premenstrual Dysphoric Disorder. Psychoneuroendocrinology. 2003;28(SUPPL. 3):39–53.
65. JL M. Black Cohosh and Chasteberry : Clin Nutr Insights. 1998;6(15):15–8.
66. Loch EG, Selle H BN. Treatment of Premenstrual Syndrome with a Phytopharmaceutical Formulation Containing Vitex Agnus Castus. J Womens Health Gend Based Med [Internet]. 2000; Available from: http://www.ncbi.nlm.nih.gov/pubmed/13546749
67. Khoo SK, Munro C, Battistutta D. Evening Primrose Oil and Treatment of Premenstrual Syndrome. Med J Aust. 1990;153(4):189–92.
68. Rapkin AJ, Akopians AL. Pathophysiology of Premenstrual Syndrome and Premenstrual Dysphoric Disorder. Menopause Int. 2012;18(2):52–9.

13

Functional Foods and Nutraceuticals and Respiratory Diseases

Azadeh Dehghani and Mehran Rahimlou

Contents

Introduction 341
Nutrition and respiratory tract overview 342
Functional foods and nutraceuticals 343
Dietary patterns 344
Fruits and vegetables and fiber 345
Vitamins and minerals 346
Fish oil...348
Probiotics 348
Phenolic compounds 349
Herbal tea...352
Conclusions..352
References..353

Introduction

The human lung contains 100,000 small airways and 200 million alveoli, with an area of more than 130 m^2 (1). The surface of the lungs is responsible for absorbing oxygen, which is also exposed to inhaled particles such as allergens, germs, pollutants, and tobacco smoke (2, 3). Respiratory diseases (RSDs) cause a significant burden of illness in primary and secondary care, including death, worldwide (4, 5). The incidence of acute respiratory infections (ARI) varies considerably with age, season, and year, and the mortality rate is subject to annual and seasonal fluctuations (4, 6).

In 2008, the World Health Organization (WHO) announced that 9.5 million people had died of lung disease, accounting for one-sixth of global deaths (7). Some pathogenic viruses (influenza, respiratory syncytial virus (RSV), coronavirus, adenovirus, and rhinovirus) and bacteria can cause significant respiratory infections. Early and accurate diagnosis of these viruses makes it possible

DOI: 10.1201/9781003220053-15

to treat infections properly (8). Tuberculosis (TB), asthma, and lung cancer are classified as primary pulmonary system disorders, but conditions associated with infection, cardiovascular disease, obesity, and sickle cell disease are secondary. Furthermore aspiration pneumonia, airway obstruction caused by allergic food reactions are acute conditions, and cystic fibrosis (CF) and chronic obstructive pulmonary disease (COPD) are chronic conditions (9, 10). The chronic respiratory disease between adults is commonly divided into obstructive and restrictive conditions, with obstructive disorders being further divided into two – reversible and irreversible – disease groups (11).

Complications and mortality from pulmonary diseases are staggering. Hundreds of millions of people around the world suffer from chronic respiratory problems, and approximately four million people die each year from chronic respiratory diseases (12). Tuberculosis has infected about 8.6 million people and killed 1.3 million in 2012, mostly in sub-Saharan Africa, where the immunodeficiency virus epidemic continues unmitigated (13). Many physiological and anatomical conversions also affect the respiratory system in pregnant women (14). It is estimated that COPD will soon be among the top three reasons for death, and 235 million people globally suffer from asthma. The prevalence of COPD and obstructive asthma worldwide is estimated at 8 to 10 percent, depending on the country, while the prevalence of asthma varies between 2 and 33 percent (11, 15, 16). The results of the researchers found that the risk of respiratory complications in premature infants is higher (17).

In December 2019, a new virus called SARS-CoV-2 was found in all countries of the world and caused unusual respiratory diseases in many people. The virus is contagious to humans and has caused a pandemic around the world: Covid-19. However, previous studies and research have shown that respiratory diseases are 5 of the 30 most common causes of death worldwide and are among the leading causes of disability (15, 16). Chronic obstructive pulmonary disease (COPD) is the third leading cause of death in the United States, with the highest rate of death reported in the elderly, according to a 2017 report (18).

Nutrition and respiratory tract overview

Lifestyle and diet are important risk factors for the development, progression, and control of various chronic diseases, including obstructive pulmonary disease (COPD) and asthma (19). Many studies demonstrated that cigarette smoking is the main risk factor for lung abnormalities. Results of studies also showed that the rate of lung diseases, especially interstitial lung diseases in males and patients with hepatitis C, likewise the elderly with a history of pneumonia, has been reported more (20, 21). Effective drugs for a wide range of lung diseases have been identified in science. This knowledge is taught in the academic curriculum among the pharmaceutical and medical professions (22).

A healthy diet and adequate exercise play an essential role in health and disease and reduce the risk of several chronic diseases. High-energy diets and inadequate nutrients, and inactivity, cause weight gain and obesity, which can be linked to chronic diseases such as cardiovascular disease, diabetes, metabolic syndrome, and kidney and lung disease (23). Diets containing large amounts of fruits and vegetables and antioxidants have a positive effect on conditions such as asthma; food interventions and diets are considered an influential key factor in creating and managing many chronic diseases (22, 24, 25). Stress, air pollution, cigarette smoke, illness, and even exercise can increase free radicals. Hence, the need to regularly consume foods that are naturally rich in antioxidants such as fruits, vegetables, nuts and seeds, whole grains, legumes, and vegetable seasonings is vital and thus fights degenerative diseases. Consumption of more functional foods, nutraceuticals, and dietary supplements is critical as a form of food therapy in various human diseases (26, 27).

Nutritional factors, nutrients, fruits, vegetables; antioxidant vitamins such as vitamin A, vitamin C, vitamin E, beta-carotene and carotenoids; valuable fatty acids such as omega-3; and some minerals such as sodium, magnesium, and selenium have potential protective roles involved in the oxidative process and inflammatory response in the emergence or development of lung diseases (28). Micronutrients include vitamins A, C, D, E, B6, folate, B12, iron, zinc, selenium, magnesium, and copper. They act as antioxidants and are essential for biochemical reactions and support the immune functions found in the diet (29).

Functional foods and nutraceuticals

Foods of natural origin can be used to prevent and treat diseases, increase patient life expectancy, and maintain good health. Foods that have health-promoting effects are considered functional. Extra nutritional elements that occur naturally in plants and that can have a biological impact are known as functional food bioactive chemicals. Natural dietary bioactive substances have been linked to a lower risk of several chronic diseases (30, 31). Functional food is used in the treatment and cure of diseases such as coronary heart disease, cancer, respiratory diseases, viral infections, and inflammation, and functional food inhibits the aging process and improves physical and mental well-being (32, 33).

The natural ingredients in food that gives food flavor and aroma are very useful and can act as expectorants and anti-cough, anti-congestion, antimicrobial agents; useful for respiratory diseases. They reduce inflammation and increase the secretion of mucous membranes, eliminate pathogens, reduce mucosal viscosity by weakening hydrogen bonds in the mucosa, and simplify the flow of air into the lungs (32). Aging, oxidative stress, and inflammatory actions reduce pulmonary function. Antioxidant or anti-inflammatory activities of some micronutrients or phytochemicals help maintain lung function.

Fruits, vegetables, herbal teas, and seafood are all good for lung health (34–37). Studies show that fatty acids, vitamins, minerals, and some antioxidant phytochemicals; consumption of magnesium, selenium, zinc, folate, niacin, vitamins A, D, E, K; Eicosapentaenoic acid (EPA), Docosahexaenoic acid (DHA), and fiber in the diet affect lung function and diseases (38).

Dietary patterns

Studies revealed that a healthy/cautious diet could reduce the risk of lung diseases such as COPD, and an unhealthy/Western-style diet may increase the risk (39). Increased consumption of vegetables, fruits, and fiber has been shown to help prevent lung disease, while red meat, desserts, and refined grains are potential risk factors for these diseases (40).

Excessive consumption of red meat is associated with an increased risk of lung disease, especially COPD, which is attributed to nitrates, which act as a preservative or color stabilizer during meat processing (41, 42). Excessive consumption of unhealthy foods such as red/processed meats, concentrated refined and high-energy foods, fast foods, added sugar, all stimulate inflammation and oxidative stress in lung disease, but healthy diets rich in fruits, vegetables, whole grains, coffee, nuts, fish, wine and legumes have been suggested to reduce the inflammatory response (43).

Low energy intake, as well as unbalanced intake of macronutrients like low protein and insufficient intake of some micronutrients and vitamins such as iron, zinc, calcium, potassium, folate, vitamin B6, retinol, and niacin in pulmonary patients, increase the risk of malnutrition and adverse outcomes (44).

Mediterranean diets or Western diets, have different effects on lung disease. The Mediterranean diet includes high consumption of fruits, vegetables, olive oil, bread, whole grains, nuts, and seeds, low to medium consumption of dairy products, fish, and poultry, and low consumption of red meat (45). It has protective effects for respiratory diseases, atopic symptoms, persistent wheezing, atopic wheezing, allergies, and asthma. The Western diet is characterized by high consumption of refined grains, red and cooked meats, fast foods such as hamburgers, French fries, desserts and sweets, salty snacks, and high-fat dairy products. The results of various studies have demonstrated that this consumption pattern is associated with an increased risk of asthma in children, airway overreaction, wheezing, COPD, asthma, and allergies (19, 46, 47).

A ketogenic diet is a diet that contains a very high amount of fat (about 50 to 60 percent) and a low amount of carbohydrates (5 to 10 percent). With the lack of carbohydrates (glucose) in this diet, ketones are used as the primary source of energy (48). The ketogenic diet is an adjunct treatment in many diseases such as polycystic ovary syndrome, diabetes mellitus, neurological disorders, acne, respiratory and cardiovascular diseases, and cancer (49, 50).

One of the effects of the ketogenic diet is reducing the amount of respiratory exchange ratio (RER), which is due to oxidation of fats more than usual (51).

Because the ketogenic diet is an adjunct therapy and reduces the partial pressure of arterial carbon dioxide (PETCO2), it reduces RER and the respiratory quotient (RQ), the production of metabolic carbon dioxide and pulmonary ventilation, and increased fat oxidation (51, 52).

Fruits and vegetables and fiber

Fruits and vegetables have a positive effect on diseases and lung capacity due to their antioxidant and anti-inflammatory properties. For example, apples, pears, and flavonoids have been shown to be associated with lung disease (53, 54). Excessive consumption of fruits, vegetables, and whole grains in the diet is associated with a reduced risk of chronic diseases, including lung disease (55–57). Articles demonstrated that fruit consumption of more than 180 g/day is linked with a lowering prevalence of respiratory symptoms (53). High fruit and vegetable intake is inversely correlated to COPD risk. High intake of apples, pears, and grapes, which are non-citrus fruits, were independently related to low cough with phlegm (58). Also, apples and pears, which are in the solid fruits group, were oppositely connected with chronic cough and breathlessness in COPD (53). Broccoli, cabbage, onions, garlic, tomatoes, grapes, carrots, whole wheat bread, beans, legumes, raspberries, cherries, strawberries, and soy foods are popular sources of phytochemicals (59).

Vegetables and fruits, as well as whole grains, contain high levels of vitamin C, vitamin E, flavonoids, phenolic acids, phytic acid, avanthamide, and selenium (60). Dietary components, including lutein, lycopene, zeaxanthin, carotene, and cryptoxanthin, have protective effects and are positively correlated with lung function indices (61, 62). Red, purple, yellow, orange, and green fruits and vegetables such as mangoes, citrus fruits, blackberries, cherries, tomatoes, carrots, spinach, broccoli, and other dark green leafy vegetables and spices, as well as fat-soluble vitamins A, E, C, beta-carotene, and selenium, are important and valuable sources of antioxidants (63–65).

Today's diets have shifted to high-consumption ready-to-eat foods that are high in saturated fat, low in fiber and antioxidants, and low in fruits and vegetables (66). The results showed that allergic asthma and rhinitis in children and adults are reduced by consuming antioxidants in fruits and vegetables (67). Dietary fiber intake is beneficial for health and the prevention of chronic diseases such as CVD, some cancers, diabetes, respiratory diseases, and COPD (55). The results of studies showed an inverse relationship between total fiber intake on the immune system and the development of allergies and asthma. There is also an inverse association between soluble fiber intake and inflammation of the respiratory tract (68, 69). Dietary fiber balances the gut microbiome, lower levels of C-reactive protein (CRP), and lower plasma levels of IL-6 and the TNF-α2 receptor, and is thus associated with asthma (70, 71). A high intake of dietary soluble fiber was associated with lower levels of sputum macrophages, lymphocytes, and neutrophils (72).

Vitamins and minerals

Immuno-nutrition is the modulation of the immune system by modifying the nutrients consumed in a person's diet (73). Different aspects of lung maturity are affected by micronutrients (74).

Vitamin A has two sources in diets, which are retinol as a pre-formed vitamin A and carotenoids as a pro-vitamin A (75). Vitamin A retinoids regulate the expression of extracellular matrix proteins that promote airway development in individuals (74). Dietary retinol is consumed from animal sources and orange and yellow fruits and vegetables. Carotenoids, including dietary alpha-carotene, beta-carotene, lycopene, and beta-cryptoxanthins, are antioxidants that prevent the proliferation of pro-inflammatory cytokines and downregulate oxidative stress (76). It also increases the protective immune responses of Th2, for example, the expression of interleukin (IL) 4 (77). Studies in lung diseases such as asthma showed that increasing serum vitamin A promotes good lung function, as retinoic acid can reverse the airway overreaction, and protect against respiratory tract disease by reducing oxidative stress (78). Retinol deficiency causes squamous epithelium metaplasia, the first respiratory defense barrier (79).

Vitamin D has anti-inflammatory properties and is involved in processes other than the functions of calcium and phosphate homeostasis. Low serum 25 (OH) D levels are related to several diseases, for example, autoimmune disorders and infectious diseases. Vitamin D modulates the immune system by autocrine and paracrine methods. Different cytokines, protease/antiprotease levels, oxidative stress, and cellular elements affect fibro proliferation, regeneration, and lung function. Vitamin D levels may also be involved in these processes (80, 81). Vitamin D has immunomodulatory properties. Vitamin D may also play a role in lung growth, and there is a strong association between serum vitamin D and basal lung function (FEV_1 and FVC), which may explain the association between vitamin D deficiency and lung disease (82).

Vitamin C protects immune cells against oxidative stress. Vitamin C is the main antioxidant in the extracellular respiratory fluid. Lung growth and development and proper lung function in children and adults are affected by various vitamins, including vitamin C (83).

Consumption of vitamin E in the diet also has a protective effect against cancer, especially cancer of the respiratory system. Every 2 mg daily increase in vitamin E intake in the diet reduces the risk of lung cancer by 5 percent (84). Vegetable oils such as soybean oil, sunflower oil, corn, cottonseed, nuts, especially walnuts, wheat germs, and dates are the main sources of vitamin E (85). α-tocopherol can reduce inflammation, and its levels are low in adults or children with asthma, so increasing α-tocopherol through diet can be helpful in preventing or controlling allergic diseases and asthma (86, 87). Taking vitamin A, vitamin D, vitamin C, the antioxidant lycopene, and genistein may be helpful for asthma patients (88).

Vitamin K plays a crucial role in the pathology of Covid-19, and its deficiency is associated with cytokine storm, multiple organ damage, thrombotic complications, and high mortality seen in severe and fatal Covid-19 patients (89).

The fourth most abundant metal in the human body is magnesium. The results of various studies have shown a significant association between hypomagnesemia and respiratory distress syndrome (RDS), respiratory failure, interstitial pulmonary emphysema in infants, and the development of BPD and asthma. Magnesium plays a role in inflammatory and oxidative responses (90).

Selenium is a trace element. Low serum selenium levels are associated with increased systemic inflammatory responses, poor prognosis, and malnutrition in patients. Selenium plays a vital role in supporting viral infections such as coronavirus 2019 (Covid-19) and patients with critical illnesses. Selenium can minimize disease severity and reduce oxidative stress, cytokine pathology, and thus boost immunity. Selenium is an antioxidant in acute stress conditions, and protects cells against oxidative stress. Serum CRP levels also had a significant negative correlation with serum selenium. Selenium is present in selenoprotein, and by reducing the rate of peroxidation, it suppresses the metabolism of inflammatory cells and protects cells against oxidation. Selenium supplements enhancement total T cells, exclusively $CD4^+$, T cell and percentage gain in NK cells, followed by propagation in NK cell cytotoxicity (91–96).

Zinc is an essential metal that is involved in regulating the metabolism of carbohydrates and fats and also controls the function of the cardiovascular and nervous systems. Zinc participates in various biological processes as a cofactor, signal molecule, and structural element (97). Zinc has antibacterial properties. Zinc supplementation improves acute lower respiratory infections in children, and zinc deficiency is associated with reduced phagocytic activity in pneumococcal infections (98, 99). It is involved in many metabolic and chronic diseases, such as cancer, diabetes, and neurological disorders, and several infectious diseases, such as tuberculosis and pneumonia (100). Its deficiency causes thymus atrophy and T cell lymphopenia, and a decrease in immature and premature B cells. It also increases the production of pro-inflammatory cytokines (interleukins IL-1, IL-6, and tumor necrosis factor (TNF)) and thus reduces the production of antibodies. Also, it enhances the adhesion of circulating monocytes to endothelial cells in the target tissue (101). Besides, it is essential for the growth and function of all cells and for achieving an optimal immune response to various stimuli, and preventing damage to tissues and organs (98).

Fish oil

Fish oil contains omega-3 fatty acids, docosahexaenoic acid (DHA), and eicosapentaenoic acid (EPA). It is an anti-inflammatory nutrient that is involved in inflammation and chronic diseases such as COPD and pulmonary fibrosis and helps to manage different types of lung disease progression and prevent pulmonary fibrosis (102–104). Consumption of fish oil and PUFAs is also associated with the development and control of respiratory disease (40).

Pieces of evidence suggested that regular consumption of a Western inflammatory diet often contains disproportionately high amounts of omega-6 fatty acids compared to omega-3 fatty acids, often associated with an increased incidence of pulmonary distress (105, 106). Eating omega-3 rich fish is associated with improved lung function and reduced risk of asthma and asthma-like symptoms (107). The results of studies have shown that oleic acid and hexadecanoic acid have anti-inflammatory effects in pulmonary patients such as those with asthma. A diet rich in functional foods fortified with omega-3 fatty acids can be used to treat respiratory illnesses such as allergic asthma (108).

Probiotics

The lung is a sterile organ that contains a variety of microbiomes. In the pathogenesis of COPD, the lung microbiome is altered (2, 109). Impaired lung microbiome balance and recurrent infections lead to a rapid decline in lung function. Nutrition can balance the lung microbiome and is promising for the prevention of decreased lung function (110). Manipulation and modification of the intestinal microbiota of infants for the treatment and prevention of infectious diseases by probiotics have been considered (111). Probiotics are considered to be promising foods to reduce the risk or control of infectious diseases in the human body, and have also been used as a supplement to reduce the dangers of widespread use of antibiotics such as diarrhea (111–113). The human microbiome is a community of microorganisms that live in symbiotic communication in various parts of the human body, like the gastrointestinal tract, skin, and respiratory tract (114, 115). Probiotics are beneficial microorganisms that, when consumed in optimal amounts, have boosting effects on the functioning of the digestive and respiratory systems (112, 116).

Numerous studies have shown the effect of diet and nutrition on the microbiome and interaction with the immune system to improve human health. They have potential benefits, especially in the prevention or treatment of many different chronic inflammatory diseases, and the results suggest that the composition of the microbiome plays an essential role in intestinal inflammation (117–120).

The results of various studies have shown that the use of probiotics is associated with a reduced risk of systemic diseases such as cancer, allergies, asthma, and ear and urinary tract infections (121). Probiotics directly and indirectly stimulate and modulate the immune system (122). They have a functional role in the gastrointestinal tract and its intestinal epithelium and are also associated with the function of the immune system (112). Accelerated lymphatic maturation, Th1/Th2 balance, epithelial repair (via endotoxin signaling), and mucosal healing are performed by resident microbes (123).

Probiotics have beneficial effects in severe and critical diseases such as Covid-19 by manipulating the intestinal microbiota, activating mucosal immunity, modulating the innate and adaptive immune response, reducing the displacement of opportunistic organisms, and suppressing opportunistic pathogens in the gut (124).

Gut–lung interaction is bilateral, intestinal–lung axis maintains host homeostasis. Intestinal microbiome affects lung diseases. In cases of dysbiosis, the transfer of microbial metabolites and endotoxins from the gut through the bloodstream affects the lungs. In cases of inflammation in the lungs, intestinal–pulmonary axis disorder occurs (125–127).

Covid-19 patients usually have intestinal microbiome dysbiosis, hyperinflammation, cytokine storm, and immunosuppressive disorders, for which probiotics are an important strategy (124, 125, 127). Fermented products, such as fermented dairy products, contain lactobacilli, which lower interleukin-12 and increase interleukin-10 (derived from CD4 + T-helper type 2), thereby having an anti-inflammatory effect. Interleukin-10 inhibits monocyte/macrophage function and pro-inflammatory cytokines (128).

Probiotics strengthen the immune system against infections and induce AMP, IgA, and IgG, and restore the activity of natural killer cells (NK), which shortens the duration of illness and improves recovery (129–131). Probiotics have an immune-stimulating effect, leading to the production of innate and acquired immune peptides and the activation of immune cells. Neutrophils, Paneth cells, and epithelial cells are active cells that produce lactoferrin, lysozyme, and defensins, which are antimicrobial peptides (AMPs) against pathogens (128, 132). Various studies have shown that the use of probiotics can enhance the activity of natural killer cells (NK), a number of innate immune cells, and increase host immune defenses, reduce the duration of infection, and improve chronic lung disease (130, 133).

Phenolic compounds

Phenolic compounds (such as anthocyanins, phenolic acids, anthocyanins, flavonoids, tannins, coumarin, and acetylbene) are products of secondary metabolism produced in plants with aromatic rings and hydroxyl groups (134, 135). Cereals, spices, tea, coffee, and fruits contain phenolic acids (136).

Bioactive and natural beneficial chemical compounds in plants are called phytochemicals. They contribute to the aroma, color, and taste of a plant and protect plants from disease, damage, and environmental hazards such as stress, pollution, drought, and pathogenic attack (59).

Polyphenols are found in many plants such as tea, coffee, cocoa, cereals, and vegetables. The anti-pulmonary activity of polyphenols may be due to the modulation of apoptosis, autophagy, oxidative stress, and inflammation (137). Dietary intake of total flavonoids, catechins, anthocyanins, and proanthocyanidins has a robust protective inverse relationship with lung function diminution and age-related lung function decline (138–140). Significant correlations were found between higher levels of lutein/zeaxanthin and catechins and improved lung function and forced expiratory volume in one second (FEV1) and forced vital capacity (FVC) (110, 138).

Many biological compounds (more than 8000 compounds) have been identified in plants; they help control and treat many chronic diseases. Some of these compounds include polyphenols, carotenoids, alkaloids, terpenoids, polyunsaturated fatty acids, omega-3 fatty acids, and so on (59). Phytochemicals are also available in supplement form today, but there is no evidence that they have the same health benefits as dietary phytochemicals derived from natural foods (59). Many phytochemicals can protect humans from disease. These compounds are called plant secondary metabolites and they have antioxidant properties, stimulate the immune system, reduce platelet aggregation, have an antimicrobial effect, and modulate detoxification enzymes. They also modulate hormone metabolism, have anti-cancer properties, neutralize free radicals, reduce low-density lipoprotein (LDL) and cholesterol, normalize blood pressure and clotting, and improve arterial elasticity (59, 141).

In studies on functional foods and plant extracts, some (for example, Mimosa pigra (mimosa), Nigella sativa L. (black cumin), Euphorbia hirta L., Allium sativum (garlic), Zingiber officinale (ginger), Allium ascolanicum (shallot), Thymus vulgaris L. (thyme), curcumin, and ginseng) are considered to be anti-inflammatory, anti-asthma, anti-bronchitis, and anti-emphysema treatments. They may have antitussive effects, cause dilation of the bronchi, and act as muscle relaxants. They are well-known to boost the antigen-specific antibodies located in the respiratory mucus and increment the function of the respiratory tract, and moreover to diminish inflammation. The activity of the NF-KB factor, which is involved in the formation of inflammatory processes in the respiratory system, and endothelin-induced tracheal contraction is inhibited by extracts of some plants (142). Turmeric polyphenol curcumin is rich in antioxidants and has significant anti-inflammatory properties that are associated with improved lung function. Lung function levels in smokers are improved by consuming curcumin (143).

Phenolic acids are associated with inhibiting the transcriptional activity of specific proteins that help control inflammation. Caffeic acid is one of the phenolic compounds that blocks the biosynthesis of leukotrienes and has anti-tumor properties, and is involved in allergic diseases, asthma, and immunodeficiency diseases (26).

Flavonoids belong to six groups (flavonols, flavones, flavanols, flavanones, anthocyanidins, isoflavonoids). Kaempferol, quercetin, myricetin, and galangin are classified in the flavonol group; apigenin, chrysin, and luteolin are in the flavones group. Flavanols mainly contain catechins (epicatechin, epicatechin gallate [ECG], epigallocatechin [EGC], and [EGCG]). Hesperitin, naringenin, and eriodictyol are in the group of flavanones. Malvidin, cyanidin, pelargonidin, peonidine, and delphinidin are in the anthocyanidins group. Daidzein, genistein, formononetin, and glycitein are common dietary isoflavonoids (136, 144–146). Flavonoids are found in various fruits, vegetables, and whole grains, and in oranges and orange juice, apples, African star apples, tomatoes, avocados, guava, pineapples, watermelons, carrots, cashews, pineapples, cocoa, chocolate, and tea (26).

Flavonoids are compounds in plants and vegetables that have anti-inflammatory and antioxidant properties and are helpful in treating inflammatory diseases such as lung disease. In addition, flavonoids play a protective role in cells by inhibiting enzymes involved in cell proliferation and modulating the expression of apoptosis-related proteins (147, 148). Macrophages and tumor necrosis factor (TNF-α), interleukin-1 (IL-1), and IL-6 stimulate the inflammatory cascade (149). The results of studies showed that flavonols, flavonoids and flavonoids, quercetin, and hesperetin, and naringin were negatively associated with the development and progression of some lung diseases such as asthma (150). Flavonoids reduce inflammatory responses in many conditions. These mechanisms are via the MAPK signaling pathway and downregulating the TLR4/NF-kB signaling pathway (151–153).

Alkaloids are found in microorganisms, plants, and animals. Alkaloids are produced in the processes of transamination reaction or amino acid biosynthesis (154, 155). Alkaloids have anti-inflammatory, antidepressant, antihypertensive, anti-vomiting, antioxidant, anti-tumor, diuretic, sympathomimetic, antiviral, sedative, and antimicrobial properties (154, 156–159).

Lycopene, alpha carotene, beta carotene, xanthine, zeaxanthin, beta cryptoxanthin, and lutein are among the carotenoids found in fruits and vegetables and are very good for health (160). A diet rich in carotenoids acts as an antioxidant and has been linked to various diseases and cancers (161).

Lycopene is the most potent carotenoid found in tomatoes, tomato products, red fruits such as watermelon, pink grapefruit, and apricots. Oxidative stress and airway smooth muscle contraction, and excessive mucus secretion in pulmonary patients are reduced by adequate lycopene intake (162).

Resveratrol is a polyphenolic compound found in grapes, peanuts, and berries (163). Resveratrol has protective effects on aging, fibrosis, cancer, cardiovascular diseases, autoimmune diseases, and lung disease by inhibiting oxidative stress and inflammation. It is considered a beneficial treatment option for the treatment of lung and respiratory diseases (163–165). Resveratrol is helpful for relieving pulmonary function. It is considered polyphenolic phytoalexin, anti-inflammatory, anti-apoptotic, antioxidant, anticancer, antihypertensive, and antifibrotic agent (166). Resveratrol significantly increases apoptosis in activated T cells, induces enhanced induction of myeloid-derived suppressor cells, and also reduces the production of inflammatory cytokines (167). Resveratrol modulates dysbiosis and increases beneficial bacteria such as lactobacillus. In this way, it improves the immune response and prevents the production of cytokine storms (168).

The results of various studies have shown that soy consumption is positively associated with improved lung function and is inversely associated with the risk of shortness of breath and chronic respiratory symptoms, and COPD and productive coughs. Tofu (soybean curd), miso (fermented soybean paste), natto (fermented soybeans), bean sprouts, soup, and soy milk are in the group of soy products that are related to COPD and respiratory symptoms. Flavonoids in soy foods act as a potent anti-inflammatory agent in the lungs and can have anti-inflammatory and anticancer effects, especially in people who use tobacco (58, 169).

Herbal tea

Phytomedicine is plant-based and uses various herbal ingredients to prevent and treat disease. Green tea has antioxidant and anti-inflammatory effects and is helpful for lung cancer patients. Epigallocatechin gallate (EGCG), which is involved in pulmonary fibrosis, is the most critical nutrient in green tea (170). EGCG has the effect of regulating inflammatory responses in several fibrotic diseases and improving the well-being of patients with respiratory diseases. Thealfavin is a natural phenol extracted from black tea and green tea fermentation. Patients can enjoy the nutritional benefits of EGCG and taflavin while drinking green tea (171, 172).

The use of herbal teas such as anise, fennel, chamomile, saffron, cardamom and black seed, cumin, and licorice improves the frequency and intensity of cough. It increases the percentage of forced exhalation volume in 1 second/ mandatory vital capacity in patients with lung diseases, especially allergic asthma (173). Mint leaves, which contain menthol, menthone, menthyl acetate, limonene, and neomenthol, are often used as raw materials for cold medicines and can reduce shortness of breath (174).

Conclusions

Respiratory diseases are among the most essential and common fatal diseases in the present century. These diseases affect many population groups. Lifestyle, physical activity, consumption of some diets, some nutrients, and dietary patterns seem to be important in the development and management of respiratory diseases in humans. The relationship between dietary patterns, diet, nutrient intake, and weight status in lung disease at different stages of life has been studied. Nutrients in some dietary components have positive and beneficial effects on the function of the pulmonary system. Vitamins, minerals, polyphenols, fish oils, and healthy foods such as these can boost the body's immunity. These compounds are antioxidants and anti-inflammatory and have a positive effect on the lung mucosa. Probiotics and fermented foods also increase immunity and reduce the severity of diseases, including lung diseases, by replacing beneficial microbiota in the intestine and lungs through the intestinal–lung axis.

Many studies have shown the dramatic effect of consuming fruits and vegetables and their nutrients on respiratory diseases. Nutritionists evaluate the diets of patients with lung diseases such as asthma and COPD and make helpful recommendations for patients.

References

1. Hauptmann, M. and U.E. Schaible, *Linking microbiota and respiratory disease*. FEBS letters, 2016. **590**(21): p. 3721–3738.
2. Sze, M.A., J.C. Hogg, and D.D. Sin, *Bacterial microbiome of lungs in COPD*. International journal of chronic obstructive pulmonary disease, 2014. **9**: p. 229.
3. Bassis, C.M., et al., *Analysis of the upper respiratory tract microbiotas as the source of the lung and gastric microbiotas in healthy individuals*. MBio, 2015. **6**(2): p. e00037-15.
4. Pishgar, E., et al., *Mortality rates due to respiratory tract diseases in Tehran, Iran during 2008–2018: a spatiotemporal, cross-sectional study*. BMC public health, 2020. **20**(1): p. 1–12.
5. Beckham, J.D., et al., *Respiratory viral infections in patients with chronic, obstructive pulmonary disease*. Journal of infection, 2005. **50**(4): p. 322–330.
6. Steppuhn, H., et al., *Time trends in incidence and mortality of respiratory diseases of high public health relevance in Germany*. Journal of Health Monitoring, 2017.**2**(3): p. 231–240.
7. Mendis, S., D. Bettcher, and F. Branca, World Health Organization Global Status Report on Noncommunicable Diseases 2014. 2014.
8. Zhang, N., et al., *Recent advances in the detection of respiratory virus infection in humans*. Journal of medical virology, 2020. **92**(4): p. 408–417.
9. Raymond, J.L. and K. Morrow, *Krause and Mahan's Food and the Nutrition Care Process E-Book*. 2020: Elsevier Health Sciences.
10. Ferkol, T. and D. Schraufnagel, *The global burden of respiratory disease*. Annals of the American thoracic society, 2014. **11**(3): p. 404–406.
11. Burney, P., D. Jarvis, and R. Perez-Padilla, *The global burden of chronic respiratory disease in adults*. The international journal of tuberculosis and lung disease, 2015. **19**(1): p. 10–20.
12. Organization, W.H., Global surveillance, prevention and control of chronic respiratory diseases: a comprehensive approach. 2007. p. vii, 146.
13. Organization, W.H., *Global tuberculosis report 2013*. 2013: World Health Organization.
14. Mehta, N., et al., *Respiratory disease in pregnancy*. Best practice & research clinical obstetrics & gynaecology, 2015. **29**(5): p. 598–611.
15. Yuki, K., M. Fujiogi, and S. Koutsogiannaki, *COVID-19 pathophysiology: A review*. Clinical immunology, 2020. **215**: p. 108427.
16. Paul, E., G.W. Brown, and V. Ridde, *COVID-19: time for paradigm shift in the nexus between local, national and global health*. BMJ global health, 2020. **5**(4): p. e002622.

17. Greenough, A., et al., *Respiratory outcomes in early childhood following antenatal vitamin C and E supplementation*. Thorax, 2010. **65**(11): p. 998–1003.
18. Budinger, G.S., et al., *The intersection of aging biology and the pathobiology of lung diseases: a joint NHLBI/NIA workshop*. Journals of gerontology series A: biomedical sciences and medical sciences, 2017. **72**(11): p. 1492–1500.
19. Berthon, B.S. and L.G. Wood, *Nutrition and respiratory health—feature review*. Nutrients, 2015. **7**(3): p. 1618–1643.
20. Washko, G.R., et al., *Lung volumes and emphysema in smokers with interstitial lung abnormalities*. New England journal of medicine, 2011. **364**(10): p. 897–906.
21. Choi, W.-I., et al., *Risk factors for interstitial lung disease: a 9-year nationwide population-based study*. BMC pulmonary medicine, 2018. **18**(1): p. 1–7.
22. Semen, K.O., O.P. Yelisyeyeva, and A. Bast, *Interaction of diet and drugs in lung disease*. Current opinion in pulmonary medicine, 2020. **26**(4): p. 359–362.
23. https://www.cdc.gov/chronicdisease/about/index.htm.
24. Nishida, C., P. Shetty, and R. Uauy, *Diet, nutrition and the prevention of chronic diseases. scientific background papers of the Joint WHO/FAO Expert Consultation (Geneva, 28 January–1 February 2002)-Introduction*. Public health nutrition, 2004. **7**(1): p. 99–100.
25. Julia, V., L. Macia, and D. Dombrowicz, *The impact of diet on asthma and allergic diseases*. Nature reviews immunology, 2015. **15**(5): p. 308–322.
26. Adefegha, S.A., *Functional foods and nutraceuticals as dietary intervention in chronic diseases; novel perspectives for health promotion and disease prevention*. Journal of dietary supplements, 2018. **15**(6): p. 977–1009.
27. Adefegha, S.A. and G. Oboh, *Phytochemistry and mode of action of some tropical spices in the management of type-2 diabetes and hypertension*. African journal of pharmacy and pharmacology, 2013. **7**(7): p. 332–346.
28. Romieu, I., *Nutrition and lung health [State of the Art]*. The international journal of tuberculosis and lung disease, 2005. **9**(4): p. 362–374.
29. Gasmi, A., et al., *Micronutrients as immunomodulatory tools for COVID-19 management*. Clinical immunology, 2020: p. 108545.
30. Lima, G.P.P., et al., *Polyphenols in fruits and vegetables and its effect on human health*. Food and nutrition sciences, 2014: p. 1065–1082.
31. Zhang, Y.-J., et al., *Antioxidant phytochemicals for the prevention and treatment of chronic diseases*. Molecules, 2015. **20**(12): p. 21138–21156.
32. Lim, S.-L. and S. Mohamed, *Functional food and dietary supplements for lung health*. Trends in food science & technology, 2016. **57**: p. 74–82.
33. Pearson, W., et al., *Pilot study investigating the ability of an herbal composite to alleviate clinical signs of respiratory dysfunction in horses with recurrent airway obstruction*. Canadian journal of veterinary research, 2007. **71**(2): p. 145.
34. Reyfman, P.A., et al., *Defining impaired respiratory health. A paradigm shift for pulmonary medicine*. American journal of respiratory and critical care medicine, 2018. **198**(4): p. 440–446.
35. Garcia-Larsen, V., et al., *Dietary antioxidants and 10-year lung function decline in adults from the ECRHS survey*. European respiratory journal, 2017. **50**(6).
36. Siedlinski, M., et al., *Dietary factors and lung function in the general population: wine and resveratrol intake*. European respiratory journal, 2012. **39**(2): p. 385–391.
37. Kelly, Y., A. Sacker, and M. Marmot, *Nutrition and respiratory health in adults: findings from the health survey for Scotland*. European respiratory journal, 2003. **21**(4): p. 664–671.
38. Leng, S., et al., *Dietary nutrients associated with preservation of lung function in Hispanic and non-Hispanic white smokers from New Mexico*. International journal of chronic obstructive pulmonary disease, 2017. **12**: p. 3171.
39. Zheng, P.-F., et al., *Dietary patterns and chronic obstructive pulmonary disease: a meta-analysis*. COPD: Journal of chronic obstructive pulmonary disease, 2016. **13**(4): p. 515–522.

40. Annesi-Maesano, I. and N. Roche, Healthy behaviours and COPD. European respiratory review. 2014. **23**(134): p. 410–415.
41. McKeever, T.M., et al., *Patterns of dietary intake and relation to respiratory disease, forced expiratory volume in 1 s, and decline in 5-y forced expiratory volume.* The American journal of clinical nutrition, 2010. **92**(2): p. 408–415.
42. De Batlle, J., et al., *Cured meat consumption increases risk of readmission in COPD patients.* European respiratory journal, 2012. **40**(3): p. 555–560.
43. Scoditti, E., et al., *Role of diet in chronic obstructive pulmonary disease prevention and treatment.* Nutrients, 2019. **11**(6): p. 1357.
44. Laudisio, A., et al., *Dietary intake of elderly outpatients with chronic obstructive pulmonary disease.* Archives of gerontology and geriatrics, 2016. **64**: p. 75–81.
45. Arvaniti, F., et al., *Adherence to the Mediterranean type of diet is associated with lower prevalence of asthma symptoms, among 10–12 years old children: the PANACEA study.* Pediatric Allergy and Immunology, 2011. **22**(3): p. 283–289.
46. De Batlle, J., et al., *Mediterranean diet is associated with reduced asthma and rhinitis in Mexican children.* Allergy, 2008. **63**(10): p. 1310–1316.
47. Wood, L.G., M.L. Garg, and P.G. Gibson, *A high-fat challenge increases airway inflammation and impairs bronchodilator recovery in asthma.* Journal of allergy and clinical immunology, 2011. **127**(5): p. 1133–1140.
48. Masood, W., P. Annamaraju, and K.R. Uppaluri, *Ketogenic diet.* StatPearls [Internet], 2020.
49. Cross, J., et al., *The ketogenic diet in childhood epilepsy: where are we now?* Archives of disease in childhood, 2010. **95**(7): p. 550–553.
50. Paoli, A., et al., *Beyond weight loss: a review of the therapeutic uses of very-low-carbohydrate (ketogenic) diets.* European journal of clinical nutrition, 2013. **67**(8): p. 789–796.
51. Tagliabue, A., et al., *Effects of the ketogenic diet on nutritional status, resting energy expenditure, and substrate oxidation in patients with medically refractory epilepsy: A 6-month prospective observational study.* Clinical nutrition, 2012. **31**(2): p. 246–249.
52. Alessandro, R., et al., *Effects of twenty days of the ketogenic diet on metabolic and respiratory parameters in healthy subjects.* Lung, 2015. **193**(6): p. 939–945.
53. Tabak, C., et al., *Diet and chronic obstructive pulmonary disease: independent beneficial effects of fruits, whole grains, and alcohol (the MORGEN study).* Clinical & experimental allergy, 2001. **31**(5): p. 747–755.
54. Hirayama, F., A.H. Lee, and C.W. Binns, *Dietary factors for chronic obstructive pulmonary disease: epidemiological evidence.* Expert review of respiratory medicine, 2008. **2**(5): p. 645–653.
55. Andrianasolo, R.M., et al., *Association between dietary fibre intake and asthma (symptoms and control): results from the French national e-cohort NutriNet-Santé.* British journal of nutrition, 2019. **122**(9): p. 1040–1051.
56. Slavin, J.L. and B. Lloyd, *Health benefits of fruits and vegetables.* Advances in nutrition, 2012. **3**(4): p. 506–516.
57. Hosseini, B., et al., *Effects of fruit and vegetable consumption on risk of asthma, wheezing and immune responses: a systematic review and meta-analysis.* Nutrients, 2017. **9**(4): p. 341.
58. Butler, L.M., et al., *Dietary fiber and reduced cough with phlegm: a cohort study in Singapore.* American journal of respiratory and critical care medicine, 2004. **170**(3): p. 279–287.
59. Saxena, M., et al., *Phytochemistry of medicinal plants.* Journal of pharmacognosy and phytochemistry, 2013. **1**(6).
60. Slavin, J., *Whole grains and human health.* Nutrition research reviews, 2004. **17**(1): p. 99–110.
61. Ochs-Balcom, H., et al., *Antioxidants, oxidative stress, and pulmonary function in individuals diagnosed with asthma or COPD.* European journal of clinical nutrition, 2006. **60**(8): p. 991–999.

62. Guenegou, A., et al., *Serum carotenoids, vitamins A and E, and 8 year lung function decline in a general population.* Thorax, 2006. **61**(4): p. 320–326.
63. Adefegha, S. and G. Oboh, *Enhancement of total phenolics and antioxidant properties of some tropical green leafy vegetables by steam cooking.* Journal of food processing and preservation, 2011. **35**(5): p. 615–622.
64. Adefegha, S.A. and G. Oboh, *Acetylcholinesterase (AChE) inhibitory activity, antioxidant properties and phenolic composition of two Aframomum species.* Journal of basic and clinical physiology and pharmacology, 2012. **23**(4): p. 153–161.
65. Adefegha, S.A. and G. Oboh, *Inhibition of key enzymes linked to type 2 diabetes and sodium nitroprusside-induced lipid peroxidation in rat pancreas by water extractable phytochemicals from some tropical spices.* Pharmaceutical biology, 2012. **50**(7): p. 857–865.
66. Wood, L.G. and P.G. Gibson, *Dietary factors lead to innate immune activation in asthma.* Pharmacology & therapeutics, 2009. **123**(1): p. 37–53.
67. Nadeem, A., et al., *Increased oxidative stress and altered levels of antioxidants in asthma.* Journal of allergy and clinical immunology, 2003. **111**(1): p. 72–78.
68. Berthon, B.S., et al., *Investigation of the association between dietary intake, disease severity and airway inflammation in asthma.* Respirology, 2013. **18**(3): p. 447–454.
69. Halnes, I., et al., *Soluble fibre meal challenge reduces airway inflammation and expression of GPR43 and GPR41 in asthma.* Nutrients, 2017. **9**(1): p. 57.
70. Ma, Y., et al., *Association between dietary fiber and serum C-reactive protein.* The American journal of clinical nutrition, 2006. **83**(4): p. 760–766.
71. Ma, Y., et al., *Association between dietary fiber and markers of systemic inflammation in the Women's Health Initiative Observational Study.* Nutrition, 2008. **24**(10): p. 941–949.
72. Benisi-Kohansal, S, P. Saneei, M. Salehi-Marzijarani, B. Larijani, and A. Esmaillzadeh, *Whole-grain intake and mortality from all causes, cardiovascular disease, and cancer: a systematic review and dose-response meta-analysis of prospective cohort studies.* Advances in nutrition, 2016. **7**(6): p. 1052–1065.
73. Grimble. R.F., *Immunonutrition.* Current opinion in gastroenterology. 2005, **21**(2): p. 216–222.
74. Massaro, D. and G. DeCarlo Massaro, Lung development, lung function, and retinoids. The New England journal of medicine, 2010, **362**(19): p. 1829–1831.
75. Litonjua, A.A., *Fat-soluble vitamins and atopic disease: what is the evidence?* Proceedings of the Nutrition Society, 2012. **71**(1): p. 67–74.
76. Johnson, E.J., *The role of carotenoids in human health.* Nutrition in clinical care, 2002. **5**(2): p. 56–65.
77. Mucida, D., et al., *Reciprocal TH17 and regulatory T cell differentiation mediated by retinoic acid.* Science, 2007. **317**(5835): p. 256–260.
78. Riccioni, G., et al., *Antioxidant vitamin supplementation in asthma.* Annals of clinical & laboratory science, 2007. **37**(1): p. 96–101.
79. Brown, C.C. and R.J. Noelle, *Seeing through the dark: new insights into the immune regulatory functions of vitamin A.* European journal of immunology, 2015. **45**(5): p. 1287–1295.
80. Gilbert, C.R., S.M. Arum, and C.M. Smith, *Vitamin D deficiency and chronic lung disease.* Canadian respiratory journal, 2009. **16**(3): p. 75–80.
81. Cantorna, M.T., J. Zhao, and L. Yang, *Vitamin D, invariant natural killer T-cells and experimental autoimmune disease.* Proceedings of the nutrition society, 2012. **71**(1): p. 62–66.
82. Black, P.N. and R. Scragg, *Relationship between serum 25-hydroxyvitamin d and pulmonary function in the third national health and nutrition examination survey.* Chest, 2005. **128**(6): p. 3792–3798.
83. Gilliland, F.D., et al., *Children's lung function and antioxidant vitamin, fruit, juice, and vegetable intake.* American journal of epidemiology, 2003. **158**(6): p. 576–584.

84. Zhu, Y.-J., et al., *Association of dietary vitamin E intake with risk of lung cancer: a dose-response meta-analysis.* Asia pacific journal of clinical nutrition, 2017. **26**(2): p. 271–277.
85. Traber, M.G. and J. Atkinson, *Vitamin E, antioxidant and nothing more.* Free radical biology and medicine, 2007. **43**(1): p. 4–15.
86. Al-Abdulla, N.O., L.M. Al Naama, and M.K. Hassan, *Antioxidant status in acute asthmatic attack in children.* JPMA-Journal of the Pakistan medical association, 2010. **60**(12): p. 1023.
87. SCHÜNEMANN, H.J., et al., *The relation of serum levels of antioxidant vitamins C and E, retinol and carotenoids with pulmonary function in the general population.* American journal of respiratory and critical care medicine, 2001. **163**(5): p. 1246–1255.
88. Gupta, A., et al., *Relationship between serum vitamin D, disease severity, and airway remodeling in children with asthma.* American journal of respiratory and critical care medicine, 2011. **184**(12): p. 1342–1349.
89. Ali, A.M., et al., *Vitamin K in COVID-19—potential anti-covid-19 properties of fermented milk fortified with bee honey as a natural source of vitamin K and probiotics.* Fermentation, 2021. **7**(4): p. 202.
90. Fridman, E. and N. Linder, *Magnesium and bronchopulmonary dysplasia.* Harefuah, 2013. **152**(3): p. 158–61, 182.
91. Forceville, X., *Seleno-enzymes and seleno-compounds: the two faces of selenium.* Critical care, 2006. **10**(6): p. 1–2.
92. Forceville, X., et al., *Selenoprotein P, rather than glutathione peroxidase, as a potential marker of septic shock and related syndromes.* European surgical research, 2009. **43**(4): p. 338–347.
93. Costa, N.A., et al., *Erythrocyte selenium concentration predicts intensive care unit and hospital mortality in patients with septic shock: a prospective observational study.* Critical care, 2014. **18**(3): p. 1–7.
94. Lee, Y.-H., et al., *Serum selenium levels in patients with respiratory diseases: A prospective observational study.* Journal of thoracic disease, 2016. **8**(8): p. 2068.
95. Khatiwada, S. and A. Subedi, *A Mechanistic Link Between Selenium and Coronavirus Disease 2019 (COVID-19).* Current nutrition reports, 2021: p. 1–12.
96. Wood, S.M., et al., *β-Carotene and selenium supplementation enhances immune response in aged humans.* Integrative medicine, 2000. **2**(2-3): p. 85–92.
97. Prasad, A.S., Discovery of zinc for *h*uman *h*ealth and *b*iomarkers of zinc deficiency. Molecular, genetic, and nutritional aspects of major and trace minerals. 2017: p. 241–260.
98. Gammoh, N.Z. and L. Rink, *Zinc in infection and inflammation.* Nutrients, 2017. **9**(6): p. 624.
99. Eijkelkamp, B.A., et al., *Dietary zinc and the control of Streptococcus pneumoniae infection.* PLoS pathogens, 2019. **15**(8): p. e1007957.
100. Maret, W., *Zinc and human disease.* Interrelations between essential metal ions and human diseases, 2013: p. 389–414.
101. Haase, H. and L. Rink, *Zinc signals and immune function.* Biofactors, 2014. **40**(1): p. 27–40.
102. Steinemann, N., et al., *Associations between dietary patterns and post-bronchodilation lung function in the SAPALDIA Cohort.* Respiration, 2018. **95**: p. 454–463.
103. Zhao, H., et al., *Pulmonary delivery of docosahexaenoic acid mitigates bleomycin-induced pulmonary fibrosis.* BMC pulmonary medicine, 2014. **14**(1): p. 1–10.
104. Schwartz, J., *Role of polyunsaturated fatty acids in lung disease.* The American journal of clinical nutrition, 2000. **71**(1): p. 393s–396s.
105. Simopoulos, A.P., *The importance of the omega-6/omega-3 fatty acid ratio in cardiovascular disease and other chronic diseases.* Experimental biology and medicine, 2008. **233**(6): p. 674–688.
106. Galland, L., *Diet and inflammation.* Nutrition in clinical practice, 2010. **25**(6): p. 634–640.

107. Laerum, B., et al., *Relationship of fish and cod oil intake with adult asthma.* Clinical & experimental allergy, 2007. **37**(11): p. 1616–1623.
108. Ruxton, C., et al., *The impact of long-chain n-3 polyunsaturated fatty acids on human health.* Nutrition research reviews, 2005. **18**(1): p. 113–129.
109. Hilty, M., et al., *Disordered microbial communities in asthmatic airways.* PloS one, 2010. **5**(1): p. e8578.
110. Zhai, T., et al., *Potential micronutrients and phytochemicals against the pathogenesis of chronic obstructive pulmonary disease and lung cancer.* Nutrients, 2018. **10**(7): p. 813.
111. Wolvers, D., et al., *Guidance for substantiating the evidence for beneficial effects of probiotics: prevention and management of infections by probiotics.* The journal of nutrition, 2010. **140**(3): p. 698S–712S.
112. Rijkers, G.T., et al., *Guidance for substantiating the evidence for beneficial effects of probiotics: current status and recommendations for future research.* The journal of nutrition, 2010. **140**(3): p. 671S–676S.
113. Kekkonen, R.A., et al., *The effect of probiotics on respiratory infections and gastrointestinal symptoms during training in marathon runners.* International journal of sport nutrition and exercise metabolism, 2007. **17**(4): p. 352–363.
114. Kumpitsch, C., et al., *The microbiome of the upper respiratory tract in health and disease.* BMC biology, 2019. **17**(1): p. 1–20.
115. Lloyd-Price, J., et al., *Strains, functions and dynamics in the expanded Human Microbiome Project.* Nature, 2017. **550**(7674): p. 61–66.
116. Pineiro, M. and C. Stanton, *Probiotic bacteria: legislative framework—requirements to evidence basis.* The journal of nutrition, 2007. **137**(3): p. 850S–853S.
117. Borody, T.J., et al., *Bacteriotherapy using fecal flora: toying with human motions.* Journal of clinical gastroenterology, 2004. **38**(6): p. 475–483.
118. D'souza, A.L., et al., *Probiotics in prevention of antibiotic associated diarrhoea: meta-analysis.* Bmj, 2002. **324**(7350): p. 1361.
119. Salminen, S., *Gastrointestinal physiology and function: the role of probiotics and probiotics.* Br. J. Nutr, 1998. **80**(1): p. 147–171.
120. Rousseaux, C., et al., *Lactobacillus acidophilus modulates intestinal pain and induces opioid and cannabinoid receptors.* Nature medicine, 2007. **13**(1): p. 35–37.
121. Lenoir-Wijnkoop, I., et al., *Probiotic and prebiotic influence beyond the intestinal tract.* Nutrition reviews, 2007. **65**(11): p. 469–489.
122. Prescott, S.L. and B. Björkstén, *Probiotics for the prevention or treatment of allergic diseases.* Journal of allergy and clinical immunology, 2007. **120**(2): p. 255–262.
123. Hooper, L.V., D.R. Littman, and A.J. Macpherson, *Interactions between the microbiota and the immune system.* Science, 2012. **336**(6086): p. 1268–1273.
124. Angurana, S.K. and A. Bansal, *Probiotics and COVID-19: Think about the link.* British journal of nutrition, 2020: p. 1–26.
125. Dhar, D. and A. Mohanty, *Gut microbiota and Covid-19-possible link and implications.* Virus research, 2020. **285**: p. 198018.
126. Dumas, A., et al., *The role of the lung microbiota and the gut–lung axis in respiratory infectious diseases.* Cellular microbiology, 2018. **20**(12): p. e12966.
127. Keely, S., N.J. Talley, and P.M. Hansbro, *Pulmonary-intestinal cross-talk in mucosal inflammatory disease.* Mucosal immunology, 2012. **5**(1): p. 7–18.
128. Rashidi, K., et al., *Effect of probiotic fermented dairy products on incidence of respiratory tract infections: a systematic review and meta-analysis of randomized clinical trials.* Nutrition journal, 2021. **20**(1): p. 1–12.
129. Trebichavský, I. and I. Šplíchal, *Probiotics manipulate host cytokine response and induce antimicrobial peptides.* Folia microbiologica, 2006. **51**(5): p. 507–510.
130. Reale, M., et al., *Daily intake of Lactobacillus casei Shirota increases natural killer cell activity in smokers.* British journal of nutrition, 2012. **108**(2): p. 308–314.
131. Dong, H., et al., *Selective effects of Lactobacillus casei Shirota on T cell activation, natural killer cell activity and cytokine production.* Clinical & experimental immunology, 2010. **161**(2): p. 378–388.

132. Diamond, G., et al., *The roles of antimicrobial peptides in innate host defense.* Current pharmaceutical design, 2009. **15**(21): p. 2377–2392.
133. Alberda, C., et al., *Effects of probiotic therapy in critically ill patients: a randomized, double-blind, placebo-controlled trial.* The American journal of clinical nutrition, 2007. **85**(3): p. 816–823.
134. Mazid, M., T. Khan, and F. Mohammad, *Role of secondary metabolites in defense mechanisms of plants.* Biology and medicine, 2011. **3**(2): p. 232–249.
135. Ohri, P. and S.K. Pannu, *Effect of phenolic compounds on nematodes-A review.* Journal of applied and natural science, 2010. **2**(2): p. 344–350.
136. Habauzit, V. and C. Morand, *Evidence for a protective effect of polyphenols-containing foods on cardiovascular health: an update for clinicians.* Therapeutic advances in chronic disease, 2012. **3**(2): p. 87–106.
137. He, Y.-Q., et al., *Natural product derived phytochemicals in managing acute lung injury by multiple mechanisms.* Pharmacological research, 2021. **163**: p. 105224.
138. Garcia-Larsen, V., et al., *Ventilatory function in young adults and dietary antioxidant intake.* Nutrients, 2015. **7**(4): p. 2879–2896.
139. Mehta, A.J., et al., *Dietary anthocyanin intake and age-related decline in lung function: longitudinal findings from the VA Normative Aging Study–3.* The American journal of clinical nutrition, 2016. **103**(2): p. 542–550.
140. Garcia-Larsen, V., et al., *Dietary intake of flavonoids and ventilatory function in European adults: A GA2LEN study.* Nutrients, 2018. **10**(1): p. 95.
141. Rao, B.N., *Bioactive phytochemicals in Indian foods and their potential in health promotion and disease prevention.* Asia Pacific journal of clinical nutrition, 2003. **12**(1).
142. Widelska, G., et al., *The impact of functional food on the prevention and treatment of respiratory diseases.* Current issues in pharmacy and medical sciences, 2020. **33**(4): p. 228–232.
143. Ng, T.P., et al., *Curcumins-rich curry diet and pulmonary function in Asian older adults.* PLoS One, 2012. **7**(12): p. e51753.
144. Temidayo, A.R., *Extraction and isolation of flavonoids present in the methanolic extract of leaves of Acanthospermum hispidium Dc.* Global journal of medicinal plant research, 2013. **1**(1): p. 111–23.
145. Matias, I., A.S. Buosi, and F.C.A. Gomes, *Functions of flavonoids in the central nervous system: astrocytes as targets for natural compounds.* Neurochemistry international, 2016. **95**: p. 85–91.
146. Liu, R.H., *Dietary bioactive compounds and their health implications.* Journal of food science, 2013. **78**(s1): p. A18–A25.
147. Lee, K.-H. and C.-G. Yoo, *Simultaneous inactivation of GSK-3β suppresses quercetin-induced apoptosis by inhibiting the JNK pathway.* American journal of physiology-lung cellular and molecular physiology, 2013. **304**(11): p. L782–L789.
148. Lago, J.H.G., et al., *Structure-activity association of flavonoids in lung diseases.* Molecules, 2014. **19**(3): p. 3570–3595.
149. Ricciardolo, F.L., et al., *Nitric oxide in health and disease of the respiratory system.* Physiological reviews, 2004. **84**(3): p. 731–765.
150. Knekt, P., et al., *Flavonoid intake and risk of chronic diseases.* The American journal of clinical nutrition, 2002. **76**(3): p. 560–568.
151. Zhao, Y., et al., *Flavonoid VI-16 protects against DSS-induced colitis by inhibiting Txnip-dependent NLRP3 inflammasome activation in macrophages via reducing oxidative stress.* Mucosal immunology, 2019. **12**(5): p. 1150–1163.
152. Zhao, F., et al., *Apigenin attenuates acrylonitrile-induced neuro-inflammation in rats: Involved of inactivation of the TLR4/NF-κB signaling pathway.* International immunopharmacology, 2019. **75**: p. 105697.
153. Zhai, K.-f., et al., *Liquiritin from Glycyrrhiza uralensis attenuating rheumatoid arthritis via reducing inflammation, suppressing angiogenesis, and inhibiting MAPK signaling pathway.* Journal of agricultural and food chemistry, 2019. **67**(10): p. 2856–2864.

154. Demirgan, R., et al., *In vitro anticancer activity and cytotoxicity of some papaver alkaloids on cancer and normal cell lines.* African journal of traditional, complementary and alternative medicines, 2016. **13**(3): p. 22–26.
155. Aniszewski, T., *Alkaloids - Secrets of Life: Aklaloid Chemistry, Biological Significance, Applications and Ecological Role.* 2007: Elsevier.
156. Kaur, R. and S. Arora, *Alkaloids-important therapeutic secondary metabolites of plant origin.* Journal of critical reviews, 2015. **2**(3): p. 1–8.
157. Adefegha, S.A., G. Oboh, and T.A. Olasehinde, *Alkaloid extracts from shea butter and breadfruit as potential inhibitors of monoamine oxidase, cholinesterases, and lipid peroxidation in rats' brain homogenates: a comparative study.* Comparative clinical pathology, 2016. **25**(6): p. 1213–1219.
158. Hayfaa, A.A.-S., A.M.A.-S. Sahar, and M.A.-S. Awatif, *Evaluation of analgesic activity and toxicity of alkaloids in Myristica fragrans seeds in mice.* Journal of pain research, 2013. **6**: p. 611.
159. Nesterova, Y.V., et al., *Antidepressant activity of diterpene alkaloids of Aconitum baicalense Turcz.* Bulletin of experimental biology and medicine, 2011. **151**(4): p. 425–428.
160. Yeum, K.-J. and R.M. Russell, *Carotenoid bioavailability and bioconversion.* Annual review of nutrition, 2002. **22**(1): p. 483–504.
161. Michaud, D.S., et al., *Intake of specific carotenoids and risk of lung cancer in 2 prospective US cohorts.* The American journal of clinical nutrition, 2000. **72**(4): p. 990–997.
162. Wood, L.G., et al., *Airway and circulating levels of carotenoids in asthma and healthy controls.* Journal of the American college of nutrition, 2005. **24**(6): p. 448–455.
163. Salehi, B., et al., *Resveratrol: A double-edged sword in health benefits.* Biomedicines, 2018. **6**(3): p. 91.
164. Gülçin, İ., *Antioxidant properties of resveratrol: a structure–activity insight.* Innovative food science & emerging technologies, 2010. **11**(1): p. 210–218.
165. Meng, X., et al., *Health benefits and molecular mechanisms of resveratrol: A narrative review.* Foods, 2020. **9**(3): p. 340.
166. Zhu, X.-d., X.-p. Lei, and W.-b. Dong, *Resveratrol as a potential therapeutic drug for respiratory system diseases.* Drug design, development and therapy, 2017. **11**: p. 3591.
167. Rieder, S.A., P. Nagarkatti, and M. Nagarkatti, *Multiple anti-inflammatory pathways triggered by resveratrol lead to amelioration of staphylococcal enterotoxin B-induced lung injury.* British journal of pharmacology, 2012. **167**(6): p. 1244–1258.
168. Alghetaa, H., et al., *Resveratrol-mediated attenuation of superantigen-driven acute respiratory distress syndrome is mediated by microbiota in the lungs and gut.* Pharmacological research, 2021. **167**: p. 105548.
169. Hirayama, F., et al., *Soy consumption and risk of COPD and respiratory symptoms: a case-control study in Japan.* Respiratory research, 2009. **10**(1): p. 1–7.
170. Sriram, N., S. Kalayarasan, and G. Sudhandiran, *Epigallocatechin-3-gallate exhibits anti-fibrotic effect by attenuating bleomycin-induced glycoconjugates, lysosomal hydrolases and ultrastructural changes in rat model pulmonary fibrosis.* Chemico-biological interactions, 2009. **180**(2): p. 271–280.
171. Oh, C.-M., et al., *Consuming green tea at least twice each day is associated with reduced odds of chronic obstructive lung disease in middle-aged and older Korean adults.* The journal of nutrition, 2018. **148**(1): p. 70–76.
172. Hwang, Y.-Y. and Y.-S. Ho, *Nutraceutical support for respiratory diseases.* Food science and human wellness, 2018. **7**(3): p. 205–208.
173. Haggag, E.G., et al., *The effect of a herbal water-extract on histamine release from mast cells and on allergic asthma.* Journal of herbal pharmacotherapy, 2003. **3**(4): p. 41–54.
174. Kaur, G., A. Gaurav, T. Lamb, M. Perkins, T. Muthumalage, and I., Rahman, *Current perspectives on characteristics, compositions, and toxicological effects of e-cigarettes containing tobacco and menthol/mint flavors.* Frontiers in physiology, 2020. **11**: p. 613948.

14

Strategies Using Functional Food and Nutraceuticals to Prevent and Treat Arthritis

Shikha Sharma, Ramesh Bhonde, and Kalpana Joshi

Contents

Introduction 361
Fish oil 362
Glucosamine sulfate, chondroitin sulfate and hyaluronic acid 364
Ayurvedic formulations 365
Ginger 365
Olive oil 366
Curcumin 367
Vitamin C 368
Avocado/soy unsaponifiables 369
Conclusion and future direction 369
Acknowledgment 370
References 370

Introduction

Osteoarthritis (OA) is a degenerative disease characterized by the degeneration of chondrocytes, prominently due to inflammation that affects joint function, leading to decreased lifestyle. OA accounts for 85% of the disease burden worldwide (Żęgota et al., 2021). Other factors associated with the development of OA through the disintegration of synovial fluid include mechanical factors and metabolic factors (Żęgota et al., 2021). OA was considered to be a noninflammatory disease in earlier studies, but later inflammation in synovial fluid was observed in a certain subset of patients, implicating its role in enhancing the pathogenesis of OA (Sokolove and Lepus, 2013). Mostly, inflammation in synovial fluid is regarded as the early step for the initiation of OA, which eventually progresses to cartilage damage (Sokolove and Lepus, 2013). OA leads to a higher risk for the development of obesity, diabetes mellitus, joint trauma and aging (Felson, 2006). In addition, OA is associated with an active immune response and not with an adaptive immune response (Sokolove and Lepus, 2013).

DOI: 10.1201/9781003220053-16

Rheumatoid arthritis (RA) is an inflammatory autoimmune disease characterized by synovitis, joint inflammation and pannus formation (Chen et al., 2021; Wang et al., 2021a). Poor diet quality, obesity and gut microbiota may lead to the progression of RA through enhancing pain, stress and inflammation at the joint (McGarrity-Yoder et al., 2021; Somers et al., 2021; Chen et al., 2021). RA was reported to affect 0.5% to 1% of the population and was found to be three times more prevalent in women (Mannucci et al., 2021). Immune cells including B cells, T cells, macrophages and synoviocytes play a prominent role in evoking an immune response in RA through the release of cytokines that leads to the generation of reactive oxygen species (ROS) and tissue damage (Mannucci et al., 2021), suggesting the requirement of antioxidants to counterbalance the oxidative stress response. Vitamins, flavonoids, carotenoids and phytochemicals are the rich constituents of antioxidants that play a role in ameliorating chronic illness (Mannucci et al., 2021). Nutraceuticals are the substances present in food that exhibit anti-inflammatory and antioxidative effects that help in improving and treating diseases and disorders (Mannucci et al., 2021). They have been implicated in improving the health status of patients with inflammation, obesity, diabetes, cancer, cardiovascular diseases and atherosclerosis (Nasri et al., 2014). Several clinical trials on nutraceuticals support their general safety and efficacy in treating various diseases. Descriptions of a few beneficial nutraceuticals implicated in the treatment of OA and RA follow (Figure 14.1).

Fish oil

Fish oil has shown a beneficial effect in patients with OA with the decrease in induced inflammatory cytokines in a dose-dependent manner (Castrogiovanni et al., 2016). Omega-3 fatty acids exert an anti-inflammatory response either by themselves or by their metabolites, namely, resolvins, protectins and maresins. Among these, resolvins are widely studied metabolites, divided into two classes, class D and class E resolvins. Class D resolvins are derived from docosahexaenoic acid (DHA), while class E resolvins are derived from eicosapentaenoic acid (EPA) (Serhan et al., 2006; Serhan et al., 2000; Hong et al., 2003). Omega-3 fatty acid metabolite competes with omega-6 fatty acid metabolite to mediate the inflammatory cycle and plays a vital role in the reduction of inflammation and regulation of autoimmunity (Neuhofer et al., 2013; Watson et al., 2019). Research studies have shown that omega-3-fatty acids including EPA and DHA from fish oil decrease pro-inflammatory mediators and increase joint lubrication under in vitro conditions (Castrogiovanni et al., 2016). Another group revealed that the supplementation of omega-3 fatty acids including EPA and DHA from fish oil improved pain and stiffness (Gruenwald et al., 2009). Hill et al. found that low-dose fish oil 15 mL/day (blend of fish oil and sunola oil 1:9 ratio, 0.45 g omega-3 fatty acids) is superior to high-dose fish oil (15 mL/day) (4.5 g omega-3 fatty acids) in reducing pain at two years in OA treatment, suggesting the effective utility of this compound (Hill et al., 2016).

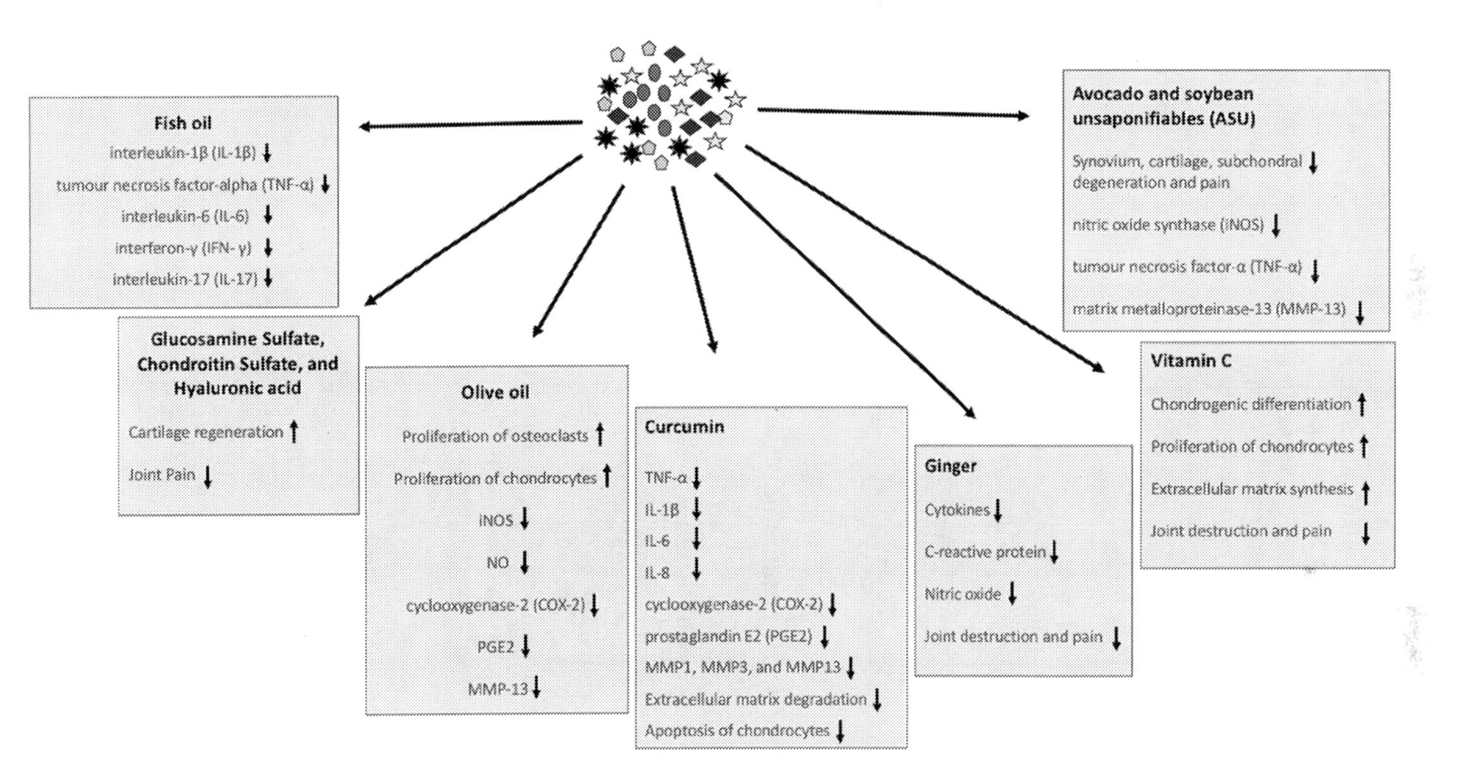

Figure 14.1 *Nutraceuticals for the treatment of arthritis.*

However, the lack of clinical trials on the utility of fish oil for OA treatment hinders its recommendation for routine use (Castrogiovanni et al., 2016). Various studies have shown that fish oil consumption by RA patients resulted in a decrease in the level of interleukin-1β (IL-1β), tumor necrosis factor-alpha (TNF-α), interleukin-6 (IL-6), interferon-γ (IFN-γ) and interleukin-17 (IL-17) (Espersen et al., 1992; Caughey et al., 1996; Ye et al., 2012; Novak et al., 2003; Babcock et al., 2002; Zhao et al., 2004; Trebble et al., 2003). In other studies, EPA and DHA suppresses the proliferation of T cells and subsequently the formation of IL-2. Omega-3 fatty acid is also known to regulate the Th1 and Th2 ratio in RA, suggesting its role in the proliferation and differentiation of T cells (Pompos and Fritsche, 2002; Bi et al., 2017; Mizota et al., 2009; Sierra et al., 2006). Studies performed in various animal models have reported the beneficial effect of omega-3-fatty acid in reducing experimental induced arthritis (Kostoglou-Athanassiou et al., 2020). Various clinical studies on supplementation of omega-3 fatty acids on human RA patients resulted in a significant decrease in the number of tender and swollen joints, DAS28, VAS score, cytokine, neutrophil leukotriene B4 production and macrophage interleukin-1 production (Kostoglou-Athanassiou et al., 2020).

Glucosamine sulfate, chondroitin sulfate and hyaluronic acid

Glycosaminoglycans (GAGs) include hyaluronic acid, glucosamine sulfate and chondroitin sulfate, which are the basic components of the extracellular matrix, synthesized by chondrocytes and synoviocytes (Kirkham and Samarasinghe, 2009). GAGs can also be supplemented with food, which raises interest to evaluate the therapeutic benefit of these molecules to protect from OA (Kirkham and Samarasinghe, 2009). Clinical and preclinical data showed that supplementation of glucosamine sulfate exerted a beneficial effect on the treatment of OA by stimulating cartilage regeneration and reducing joint pain and ameliorating joint function (Kirkham and Samarasinghe, 2009). Glucosamine treatment exhibits no adverse effects, thus it is advantageous for the management of OA. However, there are also clinical trial data that state no benefit with glucosamine treatment (Wang et al., 2021c).

Glucosamine sulfate and chondroitin sulfate exerted their anti-inflammatory and antioxidant properties in two randomized controlled trials to mitigate the catabolic and degenerative process (Michel et al., 2005; Kahan et al., 2009). Converse to this, a pharmacokinetic study revealed competition between glucosamine and chondroitin sulfate in intestinal absorption, indicating no synergic effect of both molecules (Jackson et al., 2010). For availing the therapeutic benefit of these molecules, it would be advisable to consume these molecules at different times according to the individual clinical cases (Lopez, 2012). A randomized controlled trial suggests the effectiveness of both the molecules for the treatment of moderate to severe knee pain (Clegg et al., 2006). Another GAG, hyaluronic acid (HA) showed a beneficial effect on the joints by improving the mechanical properties of synovial fluid and exerting biochemical control on joint tissues (Maneiro et al., 2004). HA has been commonly used in the clinical context through local injection in joints to improve pain and function. Other authors suggested the oral use of HA, which can be absorbed by the intestine and released into joint tissue (Adams et al., 2000). Western Ontario and McMaster universities osteoarthritis index (WOMAC) scores for improvement are encouraging, suggesting the requirement of long-term studies (Balogh et al., 2008). A study demonstrated that oral combination of HA, chondroitin sulfate and keratin matrix ameliorates pain, joint function and stiffness in early symptomatic knee OA (Galluccio et al., 2015). Matsuno et al. revealed that the combination of glucosamine-chondroitin-quercetin (GCQG) was effective for the treatment of OA while exerting no effect on RA treatment (Matsuno et al., 2009).

Ayurvedic formulations

Ayurvedic formulation comes under the traditional medicine system of India and is also popular worldwide, based on its treatment regime on diet, exercise, herbals and a healthy lifestyle (Chopra et al., 2013; Sharma et al., 2013; Kessler et al., 2018). Various studies have reported the beneficial effect of Ayurvedic medicine for the treatment of knee OA (Chopra et al., 2013; Sharma et al., 2013; Kessler et al., 2018). In 2013, Chopra et al. (2013) published a randomized controlled trial on Ayurvedic formulation SGCG derived from Emblica officinalis and Boswellia serrata and showed its efficacy for the treatment of knee OA comparative to glucosamine and celecoxib. However further studies are needed for accessing the safety of these formulations (Chopra et al., 2013). Herbs may offer a complementary or alternative method for effective and safe treatment of OA and RA. Extensive literature is available on Ayurveda-based herbal therapy for OA and RA and the mechanisms of action and safety are explored in detail (Lindler et al., 2020). Sumantran et al. demonstrated that an aqueous extract of W. somnifera showed a significant chondroprotective effect on damaged human OA cartilage via diminishing the gelatinase activity of collagenases (Sumantran et al., 2007).

Ginger

Ginger is described as an ancient herb of India and is mostly used for cooking along with it exerting various beneficial effects for the treatment of various diseases. The anti-inflammatory and antioxidant property of ginger holds its potential for the treatment of OA through reducing the expression of cytokines, C-reactive protein and nitric oxide (Naderi et al., 2015; Mozaffari-Khosravi et al., 2016). In addition, other studies also showed that ginger treatment reduced pain and disability in patients suffering from OA and improved their quality of life (Bartels et al., 2015; Amorndoljai et al., 2015; Rondanelli et al., 2017). The anti-inflammatory activities of ginger have been widely investigated in patients as well as in vitro and in vivo models. Treatment with ginger has reduced the production of PGE2, NO, IL-1β, IL-12, TNF-α, monocyte chemoattractant protein-1 (MCP-1), and regulated on activation, normal T cell expressed and secreted (RANTES) (Tripathi et al., 2008). Ginger has also inhibited the expression of MHC class II molecules, interferon-gamma (IFN-γ) and IL-2, resulting in inhibition of the antigen-presenting activity of macrophages and T cell function (Tripathi et al., 2008). Chondroprotective activity of Amla water extract was demonstrated in cartilage explant from arthritis patients (Sumantran et al., 2008).

Olive oil

Olive oil and its polyphenols including tyrosol, hydroxytyrosol, oleuropein and oleocanthal exhibit antioxidant and anti-inflammatory properties that are vital for the treatment of OA. Gong et al. showed that with supplementation of extract of olive leaf in drinking water (500 mg/kg/day), 16-week-old rabbits averted the proliferation of osteoclasts, the bone-resorbing cells, and observed matured cartilage tissue, proliferating chondrocytes and proteoglycans were present at the injured sites, suggesting a joint protective property of olive leaf extract (Gong et al., 2013). Extra-virgin olive oil (EVOO) commonly found in Mediterranean diets constitutes a high level of polyphenols such as tyrosol, hydroxytyrosol and oleuropein as well as unsaturated fatty acids including oleic and linoleic (Chin and Pang, 2017). Using the rodent model, Musumeci et al. found that a combination of physical activity (treadmill training) and EVOO-enriched diet ameliorated the cartilage damage by enhancing synovial lubricin levels and reducing interleukin-1 levels (Musumeci et al., 2013). Another study using mice and rabbit models of OA showed that supplementation of olive and grape seed extract exhibited a chondroprotective effect and showed decreased cartilage abrasion at the knee (Mével et al., 2016). A study revealed that oleuropein, a polyphenol found in olives, reduced joint lesions at the medial, lateral, tibial and femoral compartments concomitant with the decrease in inflammatory marker PGE2 and cartilage degradation marker collagen-2, suggesting its chondroprotective effect (Horcajada et al., 2015). Various randomized controlled trials conducted in OA patients using topical and oral supplementation of olive extracts, virgin olive oil and hydroxytyrosol showed reduced joint pain, edema and improved mobility (Chin and Pang, 2017). Mével et al. observed that pre-treatment of rabbit chondrocytes with olive and grape seed extract and then exposing them to IL-1β, decreased the expression of iNOS, NO, cyclooxygenase-2 (COX-2), PGE2 and MMP-13 involved in inflammation and cartilage degradation (Chin and Pang, 2017). It was shown that oleocanthal polyphenol exerted a protective effect on ATDC-5 chondrocytes, and macrophages, from LPS-induced cell death through decreasing the expression of macrophage inflammatory protein 1α, iNOS, IL-1β, IL-6, TNF-α and p38 phosphorylation (Chin and Pang, 2017). Another olive polyphenol, hydroxytyrosol, was shown to exhibit antioxidant and anti-inflammatory properties on chondrocytes treated with H_20_2 by reducing the oxygen radical species, and inflammatory molecules including COX-2 and iNOS. It also prevents cartilage degradation by decreasing MMP-13 expression, and terminal chondrocyte differentiation by reducing runt-related transcription factor 2 (RUNX2) and vascular endothelial growth factor (VEGF) (Chin and Pang, 2017). Hydroxytyrosol has also been known to promote the expression of autophagy marker (LC3II) through SIRT-1 and reduce the expression of p62 in chondrocytes with or without the presence of H_2O_2. A recent study showed that EVOO-derived polyphenol ligstroside aglycone and acetylated ligstroside exerted anti-inflammatory effects by reducing the expression of pro-inflammatory genes including MMP-13 and NOS2 at both transcript and protein levels in human OA cartilage explant. They also lessen the release of nitric oxide and prevent the loss of proteoglycan (de Andres et al., 2020). A study showed that EVOO polyphenol oleocanthal (OC) exhibited an anti-inflammatory effect in LPS treated chondrocytes through the downregulation of IL-6, IL-8, CCL3, LCN2 and TNF-α. It also decreased the expression of NOS2 and COX-2 protein as well as NO production. OC exerted its mechanism of action through MAPK/P38/NF-kB pathways (Scotece et al., 2018). Szychlinska et al. revealed an increase in the lubricin expression and a decrease in the IL-6 level in rat articular cartilage supplemented with EVOO diet along with physical exercise (Szychlinska et al., 2019).

Curcumin

Curcumin is an important phytochemical in turmeric. It imparts a chondroprotective effect through its anti-inflammatory, antioxidative and anti-catabolic activity, thus exerting a beneficial effect for OA pathogenesis treatment (Zhang et al., 2016). Curcumin decreases the production of inflammatory mediators including TNF-α, IL-1β, IL-6, IL-8, cyclooxygenase-2 (COX-2) and prostaglandin E2 (PGE2) leading to decreased extracellular matrix degradation and apoptosis of chondrocytes (Zhang et al., 2016). In addition, curcumin inhibits the expression matrix metalloproteinases (MMPs) through regulating activator protein 1 (AP-1) and nuclear factor kappa B (NF-kB) pathway (Zhang et al., 2016). Curcumin promotes anti-oxidative effects by reducing the production of reactive oxygen and nitrogen species (Zhang et al., 2016). Studies showed that curcumin oral or intraperitoneal injection in OA animal models reduced the expression of inflammatory cytokines such as IL-1β and TNF-α (Moon et al., 2010; Colitti et al., 2012). Using the mouse model of the medical meniscus (DMM), it was found that the oral administration of curcumin reduced the progression of OA without exerting pain relief through preventing proteoglycan loss, cartilage erosion, synovitis and subchondral plate thickness as well as inflammatory mediators. Indeed, this study also reported that topical administration of curcumin OA pain is observed by decreased tactile hypersensitivity and refined locomotor behavior. Using primary chondrocytes, they showed that curcumin suppressed the expression of inflammatory mediators including TNF-α, IL-1β, MMP1, MMP3, and MMP13, and aggrecanase ADAMTS5, and increased the expression of CITED2 (chondroprotective transcriptional regulator) (Zhang et al., 2016). Wang et al. demonstrated that curcumin improved the cell viability and extracellular matrix protection in IL-1β treated chondrocytes by increasing the expression of SOX9, Col2α and AGG, possibly through NF-κB/HIF-2α pathway (Wang et al., 2021b). Another report also depicted that curcumin reduces inflammation in OA conditions through the NF-κB-SOX9 mechanism (Buhrmann et al., 2021). A study suggested that a combination of curcumin and probucol exhibit chondroprotective effect from inflammatory cytokines through inhibiting apoptosis and autophagy-related PI3K/Akt/mTOR pathway (Han et al., 2021). The chondroprotective effect of curcumin was also reported by the suppression of miR-34a through reducing apoptosis by E2F1/PITX1and increasing autophagy by Akt/mTOR pathway (Yao et al., 2021). A study showed that the combination of palmitoyl-glucosamine with curcumin was implicated in the reduction of cartilage damage and improvement of locomotor function through reducing the level of inflammatory cytokines including TNF-α, IL-1β, NGF and metalloproteases 1, 3 and 9 leading to improved OA condition (Gugliandolo et al., 2020). Another group showed the beneficial effect of topical administration of 5% curcumin in reducing knee OA (Jamali et al., 2020). Atabaki et al. observed the immunomodulatory effect of curcumin in OA patients through the decrease in the expression of C-reactive protein (CRP), B cells, CD4+ and CD8+ T cells, and Th17 cells along with reduced Visual Analog Score (VAS) (Atabaki et al., 2020).

Vitamin C

Vitamin C, also named L-ascorbic acid, is a water-soluble nutrient that cannot be synthesized by most mammals, including humans (Dunlap, 2021). The last step of vitamin C biosynthesis requires the enzyme gulonolactone oxidase (GULO), which is absent in most mammals, thus it requires supplementation from the diet (Dunlap, 2021). The main source of vitamin C includes citrus fruits, potatoes and vegetables (Dunlap, 2021). Vitamin C act as a cofactor in many metabolic pathways implicated in musculoskeletal development and repair in humans (Padayatty and Levine, 2016). It is also an important cofactor for enzymes including prolyl hydroxylase and lysyl hydroxylase, essential for collagen synthesis. Vitamin C plays a vital role in chondrogenic differentiation and extracellular matrix synthesis (Takamizawa et al., 2004). In addition, it also promotes the differentiation of mesenchymal stem cells to various cell types, including adipocytes, osteoblasts, myoblasts and chondrocytes (Fulzele et al., 2013). Vitamin C is also essential for the enhancement of iron absorption in the gut, which is important for collagen synthesis for musculoskeletal health (Chambial et al., 2013). Vitamin C has also been implicated in preventing chondrocyte dysfunction by reducing oxidative stress. Various clinical trials have also demonstrated the beneficial effect of vitamin C for the prevention and treatment of OA. A recent study showed that vitamin C supplementation increased the proliferation of chondrocytes (Lindsey et al., 2019). Another report suggested the increase in the expression of inflammatory marker TGF-β within marginal osteophytes using Guinea pigs as a model system (Kraus et al., 2004). Using a mice model, it was shown that intra-articular injection of vitamin C/FeCl3 delayed OA formation (Liao et al., 2018). Yao et al. reported that a combination of magnesium and vitamin C reduced joint destruction and pain in mice models with OA (Yao et al., 2020). Recently, a study investigated the effect of four nutraceuticals – catechin hydrate, gallic acid, α-tocopherol and ascorbic acid – on humans for articular cartilage formation and inflammation. It was found that younger but obese females responded better to these nutraceuticals than older but leaner females through the enhanced formation of the extracellular matrix and reduced inflammation (Amr et al., 2021).

Avocado/soy unsaponifiables

Avocado and soybean unsaponifiables (ASU) are the vegetable extracts derived from the fruits and seeds of avocado and soybean oil (Salehi et al., 2020). Various studies revealed the potential of ASU for the treatment of OA in reducing synovium, cartilage, subchondral degeneration and pain through decreasing the expression of antioxidant anti-inflammatory markers including nitric oxide synthase (iNOS), tumor necrosis factor-α (TNF-α) and matrix metalloproteinase-13 (MMP-13) (Jangravi et al., 2021; Al-Afify et al., 2018).

Conclusion and future direction

OA treatment involves anti-inflammatory factors, analgesics, surgical procedures, lifestyle modification and physical activity. Recently, nutraceuticals have been proposed as an alternative treatment of OA. Nutraceuticals including fish oil, olive oil, curcumin, vitamin C and glycosaminoglycans (GAGs) defined in this chapter have shown their efficacy in cartilage regeneration, extracellular matrix formation, anti-inflammatory and antioxidant properties in reducing pain and increasing joint lubrication for OA treatment. However, there is a lack of human clinical trial data on nutraceuticals to validate their efficacy in OA treatment. Recently, Grzanna et al. showed that a combination of (ASU), glucosamine (GLU) and chondroitin sulfate (CS) is effective in reducing pain in OA treatment through regulating the expression of PGE2, IL-6, IL-8 and MCP-1 (Grzanna et al., 2020). It would be interesting to explore whether nutraceuticals alone or in combination would work best for the management of metabolic, inflammatory and oxidative pathways to drive the regeneration of cartilage for the amelioration of various stages of OA. Moreover, further studies are required to validate if supplementation of these nutraceuticals in a regular diet can suppress the onset of OA in patients prone to it.

Acknowledgment

RB, KJ and SS contributed to the concept, data collection and analysis of the manuscript. SS wrote the manuscript.

References

Adams ME, Lussier AJ, Peyron JG. A risk-benefit assessment of injections of hyaluronan and its derivatives in the treatment of knee osteoarthritis. Drug Saf; 2000;23:115–130.

Al-Afify ASA, El-Akabawy G, El-Sherif NM, El-Safty FEA, El-Habiby MM. Avocado soybean unsaponifiables ameliorates cartilage and subchondral bone degeneration in mono-iodoacetate-induced knee osteoarthritis in rats. Tissue Cell; 2018;52:108–115.

Amorndoljai P, Taneepanichskul S, Niempoog S, et al. Improving of knee osteoarthritic symptom by the local application of ginger extract nanoparticles: a preliminary report with short term follow-up. J Med Assoc Thai; 2015;98(9):871–877.

Amr M, Mallah A, Abusharkh H, et al. *In vitro* effects of nutraceutical treatment on human osteoarthritic chondrocytes of females of different age and weight groups. J Nutr Sci; 2021;24(10):e82.

Atabaki M, Shariati-Sarabi Z, Tavakkol-Afshari J, Mohammadi M. Significant immunomodulatory properties of curcumin in patients with osteoarthritis; a successful clinical trial in Iran. Int Immunopharmacol; 2020;85:106607.

Babcock TA, Novak T, Ong E, Jho DH, Helton WS, Espat NJ. Modulation of lipopolysaccharide-stimulated macrophage tumor necrosis factor-alpha production by omega-3 fatty acid is associated with differential cyclooxygenase-2 protein expression and is independent of interleukin-10. J Surg Res; 2002;107(1):135–139.

Balogh L, Polyak A, Matheetal D. Absorption, uptake and tissue affinity of high-molecular-weight hyaluronan after oral administration in rats and dogs. J Agric Food Chem; 2008;56:10582–10593.

Bartels EM, Folmer VN, Bliddal H, Altman RD, Juhl C, Tarp S, Zhang W, Christensen R. Efficacy and safety of ginger in osteoarthritis patients: a meta-analysis of randomized placebo-controlled trials. Osteoarthritis Cartilage; 2015;23(1):13–21.

Bi X, Li F, Liu S, Jin Y, Zhang X, Yang T, et al. ω-3 polyunsaturated fatty acids ameliorate type 1 diabetes and autoimmunity. J Clin Invest; 2017;127(5):1757–71.

Buhrmann C, Brockmueller A, Mueller AL, Shayan P, Shakibaei M. Curcumin attenuates environment-derived osteoarthritis by Sox9/NF-kB signaling axis. Int J Mol Sci; 2021; 22(14):7645.

Castrogiovanni P, Trovato FM, Loreto C, et al. Nutraceutical supplements in the management and prevention of osteoarthritis. Int J Mol Sci; 2016;17(12):2042.

Caughey GE, Mantzioris E, Gibson RA, Cleland LG, James MJ. The effect on human tumor necrosis factor alpha and interleukin 1 beta production of diets enriched in n-3 fatty acids from vegetable oil or fish oil. Am J Clin Nutr; 1996;63(1):116–122.

Chambial S, Dwivedi S, Shukla KK, John PJ, Sharma P. Vitamin C in disease prevention and cure: an overview. Indian J Clin Biochem; 2013;28(4):314–328.

Chen Y, Ma C, Liu L, He J, Zhu C, Zheng F, Dai W, Hong X, Liu D, Tang D, Dai Y. Analysis of gut microbiota and metabolites in patients with rheumatoid arthritis and identification of potential biomarkers. Aging (Albany NY); 2021;20:13(undefined).

Chin KY, Pang KL. Therapeutic effects of olive and its derivatives on osteoarthritis: from bench to bedside. Nutrients; 2017;9(10):1060.

Chopra A, Saluja M, Tillu G, Sarmukkaddam S et al. Ayurvedic medicine offers a good alternative to glucosamine and celecoxib in the treatment of symptomatic knee osteoarthritis: a randomized, double-blind, controlled equivalence drug trial. Rheumatology; 2013;52:1408.

Clegg DO, Reda DJ, Harris CL et al. Glucosamine, chondroitin sulfate, and the two in combination for painful knee osteoarthritis. N Engl J Med; 2006;354:795–808.

Colitti M, Gaspardo B, Della Pria A, Scaini C, Stefanon B. Transcriptome modification of white blood cells after dietary administration of curcumin and non-steroidal anti-inflammatory drug in osteoarthritic affected dogs. Vet Immunol Immunopathol; 2012; 147(3-4):136–146.

de Andrés MC, Meiss MS, Sánchez-Hidalgo M, González-Benjumea A, Fernández-Bolaños JG, Alarcón-de-la-Lastra C, Oreffo RO. Osteoarthritis treatment with a novel nutraceutical acetylated ligstroside aglycone, a chemically modified extra-virgin olive oil polyphenol. J Tissue Eng. 2020 May 27;11:2041731420922701.

Dunlap B, Patterson GT, Kumar S, et al. Vitamin C supplementation for the treatment of osteoarthritis: perspectives on the past, present, and future. Ther Adv Chronic Dis; 2021;12:20406223211047026.

Espersen GT, Grunnet N, Lervang HH, Nielsen GL, Thomsen BS, Faarvang KL, et al. Decreased interleukin-1 beta levels in plasma from rheumatoid arthritis patients after dietary supplementation with n-3 polyunsaturated fatty acids. Clin Rheumatol; 1992;11(3):393–395.

Felson DT. Clinical practice. Osteoarthritis of the knee. N Engl J Med; 2006;354(8):841–848.

Fulzele S, Chothe P, Sangani R, et al. Sodium-dependent vitamin C transporter SVCT2: expression and function in bone marrow stromal cells and in osteogenesis. Stem Cell Res; 2013;10(1):36–47.

Galluccio F, Barskova T, Cerinic MM. Short-term effect of the combination of hyaluronic acid, chondroitin sulfate, and keratin matrix on early symptomatic knee osteoarthritis. Eur J Rheumatol; 2015;2(3):106–108.

Gong DZ, Geng CY, Jiang LP, Wang LH, Yoshimura H, Zhong LF. Repair effect of olive leaf extract on experimental cartilaginous injuries in rabbits. Chin J Pharmacol Toxicol; 2013;27:200–204.

Gruenwald J, Petzold E, Busch R, Petzold HP, Graubaum HJ. Effect of glucosamine sulfate with or without omega-3 fatty acids in patients with osteoarthritis. Adv Ther; 2009;26:858–871.

Grzanna MW, Secor EJ, Fortuno LV, Au AY, Frondoza CG. Anti-inflammatory effect of carprofen is enhanced by avocado/soybean unsaponifiables, glucosamine and chondroitin sulfate combination in chondrocyte microcarrier spinner culture. Cartilage; 2020;11(1):108–116.

Gugliandolo E, Peritore AF, Impellizzeri D, et al. Dietary supplementation with palmitoyl-glucosamine co-micronized with curcumin relieves osteoarthritis pain and benefits joint mobility. Animals (Basel); 2020;10(10):1827.

Han G, Zhang Y, Li H. The combination treatment of curcumin and probucol protects chondrocytes from TNF-α induced inflammation by enhancing autophagy and reducing apoptosis via the PI3K-Akt-mTOR pathway. Oxid Med Cell Longev; 2021;2021:5558066.

Hill CL, March LM, Aitken D, Lester SE, Battersby R, Hynes K, Fedorova T, Proudman SM, James M, Cleland LG, et al. Fish oil in knee osteoarthritis: a randomised clinical trial of low dose versus high dose. Ann Rheum Dis; 2016;75:23–29.

Hong S, Gronert K, Devchand PR, Moussignac RL, Serhan CN. Novel docosatrienes and 17S-resolvins generated from docosahexaenoic acid in murine brain, human blood, and glial cells. Autacoids in anti-inflammation. J BiolChem; 2003;278(17): 14677–14687.

Horcajada MN, Sanchez C, MembrezScalfo F, Drion P, Comblain F, Taralla S, Donneau AF, Offord EA, Henrotin Y. Oleuropein or rutin consumption decreases the spontaneous development of osteoarthritis in the Hartley guinea pig. Osteoarthritis Cartilage; 2015;23(1):94–102.

Jackson CG, Plaas AH, Sandy JD, Hua C, Kim-Rolands S, Barnhill JG, Harris CL, Clegg DO. The human pharmacokinetics of oral ingestion of glucosamine and chondroitin sulfate taken separately or in combination. OsteoarthrCartil; 2010;19:297–302.

Jamali N, Adib-Hajbaghery M, Soleimani A. The effect of curcumin ointment on knee pain in older adults with osteoarthritis: a randomized placebo trial. BMC Complement Med Ther; 2020;20(1):305.

Jangravi Z, Basereh S, ZareeMahmoudabadi A, Saberi M, Alishiri GH, Korani M. Avocado/soy unsaponifiables can redress the balance between serum antioxidant and oxidant levels in patients with osteoarthritis: a double-blind, randomized, placebo-controlled, cross-over study. J Complement Integr Med; 2021 Apr 2. doi: 10.1515/jcim-2020-0265. ePub ahead of print.

Kahan A, Uebelhart D, de Vathaire F, Delmas PD, Reginster JY. Long-term effects of chondroitin sulfate on knee osteoarthritis: The study on osteoarthritis progression prevention, a two-year, randomized, double- blind, placebo-controlled trial. Arthritis Rheum; 2009;60:524–533.

Kessler CS, Dhiman KS, Kumar A, et al. Effectiveness of an Ayurveda treatment approach in knee osteoarthritis-a randomized controlled trial. Osteoarthritis Cartilage; 2018; 26(5):620–630.

Kirkham S.G., Samarasinghe R.K. Review article: Glucosamine. J Orthop Surg; 2009;17:72–76.

Kostoglou-Athanassiou I, Athanassiou L, Athanassiou P. The effect of omega-3 fatty acids on rheumatoid arthritis. Mediterr J Rheumatol; 2020;31(2):190–194.

Kraus VB, Huebner JL, Stabler T, Flahiff CM, Setton LA, Fink C, Vilim V, Clark AG. Ascorbic acid increases the severity of spontaneous knee osteoarthritis in a guinea pig model. Arthritis Rheum; 2004;50(6):1822–1831.

Liao Z, Xing Z, Chen Y, et al. Intra-articular injection of ascorbic acid/ferric chloride relieves cartilage degradation in rats with osteoarthritis. Nan Fang Yi Ke Da Xue Bao; 2018;38:62–68.

Lindler BN, Long KE, Taylor NA, and Lei W. Use of herbal medications for treatment of osteoarthritis and rheumatoid arthritis. Medicine; 2020;7:67

Lindsey RC, Cheng S, Mohan S. Vitamin C effects on 5-hydroxymethylcytosine and gene expression in osteoblasts and chondrocytes: Potential involvement of PHD2. PLoS One; 2019;14(8):e0220653.

Lopez HL. Nutritional interventions to prevent and treat osteoarthritis. Part II: focus on micronutrients and supportive nutraceuticals. PM R; 2012;4:S155–S168.

Maneiro E, de Andres MC, Ferández-Sueiro JL, Galdo F, Blanco FJ. The biological action of hyaluronan on human osteoarthritic articular chondrocytes: the importance of molecular weight. Clin Exp Rheumatol; 2004;22:307–312.

Mannucci C, Casciaro M, Sorbara EE, et al. Nutraceuticals against oxidative stress in autoimmune disorders. Antioxidants (Basel); 2021;10(2):261.

Matsuno H, Nakamura H, Katayama K, Hayashi S, Kano S, Yudoh K, Kiso Y. Effects of an oral administration of glucosamine-chondroitin-quercetin glucoside on the synovial fluid properties in patients with osteoarthritis and rheumatoid arthritis. Biosci Biotechnol Biochem; 2009;73(2):288–292.

McGarrity-Yoder ME, Insel KC, Crane TE, Pace TWW. Diet quality and disease activity in rheumatoid arthritis. Nutr Health; 2021 Oct 20:2601060211044311.

Mével E, Merceron C, Vinatier C, Krisa S, Richard T, et al., Olive and grape seed extract prevents post-traumatic osteoarthritis damages and exhibits in vitro anti IL-1β activities before and after oral consumption. Sci Rep; 2016;6:33527.

Michel BA, Stucki G, Frey D, de Vathaire F, Vignon E, Bruehlmann P, Uebelhart D. Chondroitins 4 and 6 sulfate in osteoarthritis of the knee: a randomized, controlled trial. Arthritis Rheum; 2005;52:779–786.

Mizota T, Fujita-Kambara C, Matsuya N, Hamasaki S, Fukudome T, Goto H, et al. Effect of dietary fatty acid composition on Th1/Th2 polarization in lymphocytes. JPEN J Parenter Enteral Nutr; 2009;33(4):390–396.

Moon DO, Kim MO, Choi YH, Park YM, Kim GY. Curcumin attenuates inflammatory response in IL-1beta-induced human synovial fibroblasts and collagen-induced arthritis in mouse model. Int Immunopharmacol; 2010;10(5):605–610.

Mozaffari-Khosravi H, Naderi Z, Dehghan A, et al. Effect of ginger supplementation on proinflammatory cytokines in older patients with osteoarthritis: outcomes of a randomized controlled clinical trial. J Nutr Gerontol Geriatr; 2016;35(3):209–218.

Musumeci G, Trovato FM, Pichler K, et al. Extra-virgin olive oil diet and mild physical activity prevent cartilage degeneration in an osteoarthritis model: an in vivo and in vitro study on lubricin expression. J NutrBiochem; 2013;24(12):2064–2075.

Naderi Z, Mozaffari-Khosravi H, Dehghan A, Nadjarzadeh A, Huseini HF. Effect of ginger powder supplementation on nitric oxide and C-reactive protein in elderly knee osteoarthritis patients: a 12-week double-blind randomized placebo-controlled clinical trial. J Tradit Complement Med; 2015;6(3):199–203.

Nasri H, Baradaran A, Shirzad H, Rafieian-Kopaei M. New concepts in nutraceuticals as alternative for pharmaceuticals. Int J Prev Med; 2014;5(12):1487–1499.

Neuhofer A, Zeyda M, Mascher D, Itariu BK, Murano I, Leitner L, et al. Impaired local production of proresolving lipid mediators in obesity and 17-HDHA as a potential treatment for obesity-associated inflammation. Diabetes; 2013;62(6):1945–1956.

Novak TE, Babcock TA, Jho DH, Helton WS, Espat NJ. NF-kappa B inhibition by omega-3 fatty acids modulates LPS-stimulated macrophage TNF-alpha transcription. Am J Physiol Lung Cell Mol Physiol; 2003;284(1):L84–89.

Padayatty SJ, Levine M. Vitamin C: the known and the unknown and Goldilocks. Oral Dis; 2016;22(6):463–493.

Pompos LJ, Fritsche KL. Antigen-driven murine CD4+ T lymphocyte proliferation and interleukin-2 production are diminished by dietary (n-3) polyunsaturated fatty acids. J Nutr; 2002;132(11):3293–3300.

Rondanelli M, Riva A, Morazzoni P, Allegrini P, et al. The effect and safety of highly standardized Ginger (Zingiber officinale) and Echinacea (Echinacea angustifolia) extract supplementation on inflammation and chronic pain in NSAIDs poor responders. A pilot study in subjects with knee arthrosis. Nat Prod Res; 2017;31(11):1309–1313.

Salehi B, Rescigno A, Dettori T, Calina D, Docea AO, Singh L, Cebeci F, Özçelik B, Bhia M, DowlatiBeirami A, Sharifi-Rad J, Sharopov F, Cho WC, Martins N. Avocado-soybean unsaponifiables: a panoply of potentialities to be exploited. Biomolecules; 2020;10(1):130.

Scotece M, Conde J, Abella V, López V, Francisco V, Ruiz C, Campos V, Lago F, Gomez R, Pino J, Gualillo O. Oleocanthal Inhibits Catabolic and Inflammatory Mediators in LPS-Activated Human Primary Osteoarthritis (OA) Chondrocytes Through MAPKs/NF-κB Pathways. Cell Physiol Biochem. 2018;49(6):2414–2426.

Serhan CN, Clish CB, Brannon J, Colgan SP, Chiang N, Gronert K. Novel functional sets of lipid-derived mediators with antiinflammatory actions generated from omega-3 fatty acids via cyclooxygenase 2-nonsteroidal antiinflammatory drugs and transcellular processing. J Exp Med; 2000;192(8):1197–1204.

Serhan CN, Gotlinger K, Hong S, Lu Y, Siegelman J, Baer T, et al. Anti-inflammatory actions of neuroprotectin D1/protectin D1 and its natural stereoisomers: assignments of dihydroxy-containing docosatrienes. J Immunol; 2006;176(3):1848–1459.

Sharma MR, Mehta CS, Shukla DJ, Patel KB, Patel MV, Gupta SN. Multimodal Ayurvedic management for Sandhigatavata (Osteoarthritis of knee joints). Ayu; 2013;34(1):49–55.

Sierra S, Lara-Villoslada F, Comalada M, Olivares M, Xaus J. Dietary fish oil n-3 fatty acids increase regulatory cytokine production and exert anti-inflammatory effects in two murine models of inflammation. Lipids; 2006;41(12):1115–1125.

Sokolove J, Lepus CM. Role of inflammation in the pathogenesis of osteoarthritis: latest findings and interpretations. Ther Adv Musculoskelet Dis; 2013;5(2):77–94.

Somers TJ, Blumenthal JA, Dorfman CS, Huffman KM, Edmond SN, Miller SN, Wren AA, Caldwell D, Keefe FJ. Effects of a weight and pain management program in patients with rheumatoid arthritis with obesity: a randomized controlled pilot investigation. J Clin Rheumatol. 2022 Jan 1;28(1):7–13.

Sumantran VN, Kulkarni A, Boddul S, Chinchwade T, et al. Chondroprotective potential of root extracts of Withania somnifera in osteoarthritis. J Biosci; 2007;32:299–307.

Sumantran VN, Kulkarni A, Chandwaskar R, Harsulkar A, Patwardhan B. Chondroprotective potential of fruit extracts of Phyllanthus emblica in osteoarthritis. Evid Based Complement Alternat Med; 2008;5(3):329–335.

Szychlinska MA, Di Rosa M, Castorina A, Mobasheri A, Musumeci G. A correlation between intestinal microbiota dysbiosis and osteoarthritis. Heliyon. 2019 Jan 12;5(1):e01134.

Takamizawa S, Maehata Y, Imai K, Senoo H, Sato S, Hata R. Effects of ascorbic acid and ascorbic acid 2-phosphate, a long-acting vitamin C derivative, on the proliferation and differentiation of human osteoblast-like cells. Cell Biol Int. 2004;28(4):255–265.

Trebble T, Arden NK, Stroud MA, Wootton SA, Burdge GC, Miles EA, et al. Inhibition of tumour necrosis factor-alpha and interleukin 6 production by mononuclear cells following dietary fish-oil supplementation in healthy men and response to antioxidant cosupplementation. Br J Nutr; 2003;90(2):405–412.

Tripathi S, Bruch D, Kittur DS. Ginger extract inhibits LPS induced macrophage activation and function. BMC Complement Altern Med; 2008;8:1.

Wang L, Zhu L, Jiang J, Wang L, Ni W. Decision tree analysis for evaluating disease activity in patients with rheumatoid arthritis. J Int Med Res; 2021a;49(10):3000605211053232.

Wang P, Ye Y, Yuan W, Tan Y, Zhang S, Meng Q. Curcumin exerts a protective effect on murine knee chondrocytes treated with IL-1β through blocking the NF-κB/HIF-2α signaling pathway. Ann Transl Med; 2021b;9(11):940.

Wang SJ, Wang YH, Huang LC. Liquid combination of hyaluronan, glucosamine, and chondroitin as a dietary supplement for knee osteoarthritis patients with moderate knee pain: a randomized controlled study. Medicine (Baltimore); 2021c;100(40):e27405.

Watson JE, Kim JS, Das A. Emerging class of omega-3 fatty acid endocannabinoids & their derivatives. Prostaglandins Other Lipid Mediat; 2019;143:106337.

Yao H, Xu J, Wang J, et al. Combination of magnesium ions and vitamin C alleviates synovitis and osteophyte formation in osteoarthritis of mice. Bioact Mater; 2020;6:1341–1352.

Yao J, Liu X, Sun Y, Dong X, Liu L, Gu H. Curcumin-alleviated osteoarthritic progression in rats fed a high-fat diet by inhibiting apoptosis and activating autophagy via modulation of MicroRNA-34a. J Inflamm Res; 2021;14:2317–2331.

Ye P, Li J, Wang S, Xie A, Sun W, Xia J. Eicosapentaenoic acid disrupts the balance between Tregs and IL-17+ T cells through PPARgamma nuclear receptor activation and protects cardiac allografts. J Surg Res; 2012;173(1):161–170.

Żęgota Z, Goździk J, Głogowska-Szeląg J. Efficacy of herbal and naturally-derived dietary supplements for the management of knee osteoarthritis: a mini-review. Wiad Lek; 2021;74(8):1975–1983.

Zhang Z, Leong DJ, Xu L, et al. Curcumin slows osteoarthritis progression and relieves osteoarthritis-associated pain symptoms in a post-traumatic osteoarthritis mouse model. Arthritis Res Ther; 2016;18(1):128.

Zhao Y, Joshi-Barve S, Barve S, Chen LH. Eicosapentaenoic acid prevents LPS-induced TNF-alpha expression by preventing NF-kappaB activation. J Am Coll Nutr; 2004; 23(1):71–78.

Functional Foods of Polyphenolics for Alzheimer's Disease

Hanish Singh Jayasingh Chellammal and Dhani Ramachandran

Contents

Introduction 378
Current treatments of AD 380
Functional foods of polyphenolics for Alzheimer's disease 380
Polyphenol mechanisms and therapeutic target pathways in AD.................381
Resveratrol 381
Clinical studies and clinical pharmacology of resveratrol 383
Curcumin 383
Clinical studies and clinical pharmacology of curcumin 384
Quercetin 385
Clinical studies and clinical pharmacology of quercetin 385
α-Mangostin..386
Clinical studies and clinical pharmacology of α-mangostin 386
Catechins 386
Clinical studies and clinical pharmacology of catechins 387
Hesperidin and hesperetin..388
Clinical studies and clinical pharmacology of hesperidin and hesperetin 388
Naringenin..389
Clinical studies and clinical pharmacology of naringenin....................390
Luteolin..390
Clinical studies and clinical pharmacology of luteolin........................391
Myricetin..392
Mangiferin..392
Clinical studies and clinical pharmacology of mangiferin....................393
Baicalein and baicalin 394
Clinical studies and clinical pharmacology of baicalin and baicalein 395
Vanillic acid 395
Clinical studies and clinical pharmacology of vanillic acid..................396

DOI: 10.1201/9781003220053-17

Fisetin 396
Clinical studies and clinical pharmacology of fisetin 397
Rutin 397
Clinical studies and clinical pharmacology of rutin 398
Bioavailability and pharmacokinetics of polyphenols 398
Conclusion 399
References 400

Abbreviations

3xTg-AD	triple-transgenic "AD" mouse
8OHdG	8-hydroxyl-2-deoxyguanosine
8OHG	8-hydroxyguanosine
Ach	acetylcholine
AChE	acetylcholinesterase
ACTH	adrenocorticotropic hormone
AD	Alzheimer's disease
ADAS	Alzheimer disease assessment scale
ADCS-ADL	Alzheimer's Disease Cooperative Study – Activities of Daily Living
Akt	serine/threonine-specific protein kinases
AMPK	5-AMP activated protein kinases
APE1	apurinic/apyrimidinic endonuclease 1
ApoE	apolipoprotein E
APP/PS1	amyloid precursor protein/presenilin 1
ASD	autism spectrum disorder
Aβ	amyloid beta
BACE-1	beta-site APP-cleaving enzyme 1
Bax	Bcl-2-associated X protein
BBB	blood–brain barrier
BDNF	brain-derived neurotrophic factor
CAT	catalase
CK-1 δ	casein kinase I isoform delta
COX-1/2	cyclooxygenase-1/2
CPT	continuous performance task
CREB	cyclic AMP response element-binding protein
CRMP2	collapsing response mediator protein-2
DSM IV	Diagnostic and Statistical Manual of Mental Disorders
ECG	epicatechin-3-gallate
EGC	epigallocatechin
EGF	epidermal growth factor
EOAD	early onset Alzheimer's disease
ER	endoplasmic reticulum
ERK1-CREB	extracellular signal-regulated kinase1- cAMP Response Element-Binding Protein
FBG	fasting blood glucose

$GABA_A$	γ amino butyric acid
GFAP	glial fibrillary acidic protein
GLUT	glucose transporters
GSH	glutathione
GSH-Px	glutathione peroxidase
GSK3β	glycogen synthase kinase-3β
GST	glutathione-s-transferase
HDL	high density lipoprotein
HPA	hypothalamic pituitary adrenal axis
Hsd	hesperidin
Hst	hesperetin
i.c.v/ICV	Intracerebroventricular
ICAM-1	intercellular adhesion molecule-1
IGFs	insulin like growth factor
IL/s	interleukin/s
IL1β/IL-6	interleukin1β/6
iNOS	inducible nitric oxide synthase
IRE1α	inositol-requiring enzyme 1α
IRs	insulin receptors
IV infusion	intravenous infusion
JAK, ERK, STAT and JNK phosphorylation	(Janus kinase, Extracellular signal-regulated kinases, signal transducer and activator of transcription, the c-Jun N-terminal kinase)
KEAP1	Kelch-like ECH-associated protein 1
LOAD	late onset Alzheimer's disease
LOX	lipooxygenase
LPO	lipid peroxidation
LPS	lipopolysaccharide
LRP1	low-density lipoprotein receptor-related protein 1
LTP	long-term potentiation
MAG	myelin-associated glycoprotein
MAO-A and B	Monoamine oxidase A and B
MAPKs	mitogen activated protein kinase
MAP	mitogen activated protein
MBP	myelin basic protein
M-CSF	macrophage colony-stimulating factor
MDA	malondialdehyde
MEK1	meiotic chromosome-axis-associated kinase 1
MMP	matrix metalloproteinases
MMSE	Mini-Mental State Examination
MOG	myelin oligodendrocyte glycoprotein
Na^+/K^+-ATPase	Sodium potassium adenosine triphosphate
NFkB	nuclear factor kappa B
NFkB-IkB	nuclear factor kappa B-inhibitor of κB
NFTs	neurofibrillary tangles
NGF	nerve growth factor

NLRP3	NOD-LRR-and pyrin domain-containing protein 3
NO	nitric oxide
NR2A	NMDA-subunit NMDAR2(N-methyl-D-aspartate receptor)
Nrf2-ARE	nuclear erythroid 2-related factor 2- antioxidant response element
NSE	Neuron-Specific Enolase
P(+) and P(−)	Effect of polyphenols; "+" activation and "−" inhibition
PERK	Protein kinase RNA-like endoplasmic reticulum kinase
PGC-1α	Peroxisome proliferator-activated receptor gamma
PI3K/Akt-mTOR	phosphatidylinositol-3-kinase (PI3K)/Akt and the mammalian target of rapamycin
PI3K/Akt	phosphatidylinositol 3-kinase/protein kinas B
PI3	phosphoinosital-3
PLGA NPs	Polylactic-co-glycolic acid nanoparticles
PLP	proteolipid protein
PLP	pyridoxal 5′-phosphate
PSEN1	presenilin 1
PSP95	synapse-related proteins
QOL	Quality of life
RAGE	receptor for advanced glycation end products
ROS	reactive oxygen species
sAPPβ (soluble)	soluble amyloid protein precursor
SIRT1/3	member of the sirtuin family
SOD1-mRNA	superoxide dismutase 1 (SOD1) mRNA
SOD	superoxide dismutase
STZ	Streptozotocin
TLR2 and TLR4	toll like receptors 2 and 4
Tm	tunicamycin
TNF-α	tumor necrosis factor-α
TrkB	tyrosine kinase B
TXNIP/TRX	thioredoxin (TRX)-interacting protein
USFDA	United States Food and Drug Administration
VLDL	very low-density lipoprotein
α7-nAChR	α7-nicotinic acetylcholine receptors

Introduction

Alzheimer's disease (AD) is a progressive neurodegenerative disease. Worldwide, 50 million people are existing with dementia, and it is anticipated to reach 82 million by 2030 and 152 million by 2050 (1). Neuropathology of AD is characterized by the accretion of brain amyloid beta-protein (Aβ). Aβ is a proteolytic component of amyloid precursor protein (APP). AD is one among the principal progressive neurodegenerative diseases; it mainly induces reminiscence loss with intra-neuronal fibrillary tangles (NFTs) formation and accumulation of plaques of Aβ protein in cerebral parenchyma. β- and γ-secretase enzymes induce the cleavage of APP and form Aβ peptide. The phosphorylation of tau protein is mediated by glycogen synthetase kinase-3β and produces NFTs within the neurons (2). The foremost indication of AD is failure of short-term remembrance (amnesia). This hinders the habitual capabilities of daily life activities and imparts cognitive dysfunction. Behavioral deficits of AD comprise anxiety, ferocity, deteriorated mood, insomnia and restlessness. Aging is the prospective factor for AD and is of two forms: early onset AD (EOAD) and late onset AD (LOAD). The neurodegeneration and symptoms appear subsequently after 65 years old in LOAD and before 65 years of age in EOAD. The characteristic neuropathological alteration in AD is the escalation of acetylcholinesterase enzyme around the Aβ plaques, promoting the neurofibril tangles formation and cholinergic transmission loss in the projecting neurons of basal forebrain to hippocampus and neocortex (3). Other neurotransmitters of biogenic amines (norepinephrine, dopamine and serotonin) were imbalanced because of the elevation of monoamine oxidase enzyme (MAO) A and B around the plaques. Evidence indicates that principal excitatory neurotransmitter (glutamate) level is reduced in AD, contributing to dysfunction in long-term potentiation. Moreover, neuro-inflammatory process due to the triggering of astrocytes and chemokines leads to elevated level of tumor necrosis factor (TNF-α) and interleukins (ILs) causing cognitive dysfunction (4). The inflammation is accompanied with elevated cyclooxygenase enzyme (5). Neuroinflammation is also well linked with activated levels of toll-like receptors (TLR2 and TLR4) and nuclear factor kappa B (NFƘB) (6). A considerable decrease in neuronal membrane bound Na^+/K^+-ATPase is observed due to the accumulation of Aβ in nucleus basalis and thalamus (7). Stress affects the hippocampal cells and destructs neuronal integrity through hypothalamic-pituitary-adrenal (HPA) axis dysfunction, leading to the loss of spatial performance and declarative memory (8). Synaptic plasticity and the long-term potentiation involve nitric oxide (NO) and accumulation of Aβ peptides causes inhibition of NO, eventually leads to memory dysfunction. In the AD brain, there is a reduced activity of insulin degrading enzyme and there is a loss of insulin receptors (IRs) (9). Moreover, the neuroprotective neurotropic factors (brain derived neurotropic factor (BDNF), epidermal growth factor (EGF) and cerebrolysins) were reduced during cerebral atrophy (10). ApoE is a component of high-density and very low-density lipoproteins (HDL and VLDL). Increased levels of lipids were also associated with stimulation of the NLRP3 inflammasome and enhancement of Aβ aggregation (6). All the multifactorial neuropathological changes are associated with generation of reactive oxygen and nitrogen species. The impact of oxidative stress in AD specifies immeasurable disparity in diverse biomacromolecules such as DNA and RNA oxidation, which elevates the levels of 8-hydroxyl-2-deoxyguanosine (8OHdG) and 8-hydroxyguanosine (8OHG). Oxidative stress is implicated in many chronic diseases and in AD-type neurodegeneration. It impacts very critically and causes elevation of neurotransmitter metabolic enzymes (MAO A & B and AChE), induces inflammation by microglial activation, elevates corticosteroid by dysregulation of HPA axis and also causes oxidative deamination of biogenic amines (11). Glycogen synthase kinase- 3 (GSK) exists in two forms: GSK-3 α and GSK-3 β. GSK-3 β dysregulation leads to AD by promoting Aβ production. NRF2-ARE pathways associated with KEAP1 regulation have significant effect during neurodegeneration and NRF2-NFkB interplays its specific role in major CNS disorders to control the oxidative stress. NRF2 enhancer/NF-κB inhibitor drugs have a potential effect on neuroinflammatory control. Increased oxidative stress leads to phosphorylation of tau protein through GSK-3 β stimulation (12).

Current treatments of AD

AChE-inhibitors recuperate the moderate symptoms of AD. Four categories of molecules are approved by USFDA for therapy of Alzheimer's dementia: (i) AChE inhibitors for mild to moderate AD: Tacrine, Rivastigmine, Donepezil and Galantamine; (ii) Glutamate regulator: Memantine; (iii) Orexin receptor antagonist: Suvorexant for treating insomnia in mild-to-moderate AD; and (iv) Anti-amyloid antibody: IV infusion for changing disease progression (13).

Functional foods of polyphenolics for Alzheimer's disease

Polyphenols are natural products and have influential antioxidant properties. Polyphenolic antioxidant constituents protect the neurons by quenching endogenous oxidants and also scavenge the ROS. Several classes of polyphenols such as flavonoids, phenolic acids, stilbenes, lignans and their subclasses contribute to the protective effect on neurons (14). Several studies support and implicate that the oxidative stress involves as the main pathogenic condition linked to Aβ aggregation and dementia. Neurodegenerative disorder due to oxidative stress responds effectively to treatment with polyphenols and associated therapeutic supplements. It was well reported that the dietary enhancement and consumption of polyphenol-rich fruits affords neuroprotection and delays the incidence of disease (15). Much research on neuroprotective properties has been elucidated in polyphenol-rich fruits and vegetables, which include plums, blueberries, cherries, apples, onions, leafy and root vegetables, and kiwi (8, 14, 16–18). Vitamin E is a vital nutrient having an antioxidant effect and its effect can be markedly intensified by vitamin C co-supplementation. A Mediterranean-style diet reduces the risk for AD and other cognitive dysfunctions (19). Traditional medicines of herbal drugs, which are rich in polyphenols, are among the targeted therapeutic strategies for AD-type dementia. *Citrus natsudaidai, Citrus tamurana, Panax ginseng, Radix glycyrrhizae, Gingko biloba, Tabernaemontana divaricate, Illicium verum, Bacopa monnieri, Alium sativam, Ferulago campestris, Mentha arvensis, Iris pseudopumila, Asparagus adscendens, Radix Angelicae sinensis, Radix Polygalae, Huperzia Saururus, Ptychopetalum olacoides, Thespesia populnea, Salvia lavandulifolia, Celastrus paniculatus, Vaccinium corymbosum, Prunus avium, Evolvulus alsinoides* and *Polygonum multiflorum* are some of the herbal drugs containing polyphenols and other phytoconstituents used in neurodegenerative disease and are cerebroprotective (17, 20–22). The pharmacotherapeutic effects of polyphenols are mediated through various signaling mechanisms and molecular pathways; thereby improving neurological health. Cholinergic hypothesis affirms that acetylcholine concentration in the hippocampal cells preserves cognitive function. Hydrolysis of acetylcholine by acetylcholinesterase enzyme leads to cognitive decline and promotes the accumulation of Aβ peptide. Increase in AChE enzyme and Aβ accumulation cause oxidative stress, further leading to neurodegeneration. AChE enzyme inhibition by polyphenols is reported in various studies (23). Many plant extracts, their fractions and isolated constituents also reveal a promising effect on AChE inhibition. The AChE inhibitory property of polyphenols mediates the neuroprotection and improves cognition. They also primarily act as free radical scavengers and also exert regulations through Nrf2/ARE pathways, improve the biogenic amine (acetylcholine, serotonin and dopamine) turnover with regulated neurotransmitter metabolic enzymes (AChE and MAO), reduce the inflammation by targeting NRLP3 inflammasome receptor and protect the neurotropic factors (15) (Figure 15.1).

Amyloid protein precursor undergoes breakdown by *β-secretase* and *α-secretase* enzymes. The sAPPβ (soluble) produces the Aβ oligomerization and fibrillization. Aβ activates the microglial cells and increases the TLR, NFkB-IkB signaling for neuroinflammation. This triggers the inflammatory components of cytokines (IL-1β, IL-6, TNF-α, iNOS) and COX-2. Oxidative stress signaling induces the ROS and promotes Aβ, cellular oxidative damages and activation of NLRP3, leading to neurodegeneration. NFR2-KEAP1 controls to maintain the homeostasis during oxidative injury by increasing antioxidants. BDNF helps in the neuroprotection by activating ERK1-CREB and maintains the normal physiological neuronal function. P(+) and P(–): Effect of polyphenols; "+" activation and "–" inhibition (Figure 15.1).

Polyphenol mechanisms and therapeutic target pathways in AD

Resveratrol

Resveratrol is a polyphenolic compound and is a phytoalexin present in peels of grapefruit, blueberries, raspberries, mulberries and red wine. It has a potential effect on treating Alzheimer's dementia (24). It also has anti-inflammatory, anti-ischemic, antihyperglycemic, antihypercholesteremic, anticancer and cardioprotective effects (25). Resveratrol preserves the neuroprotective property due to its potential antioxidant effect. In AD-type dementia, resveratrol exhibited a neuroprotective effect in scopolamine and streptozotocin induced neurotoxicity in preclinical models (26). Resveratrol exhibits neuroprotective property by inhibiting TNF-α, IL-1 and Aβ production as well

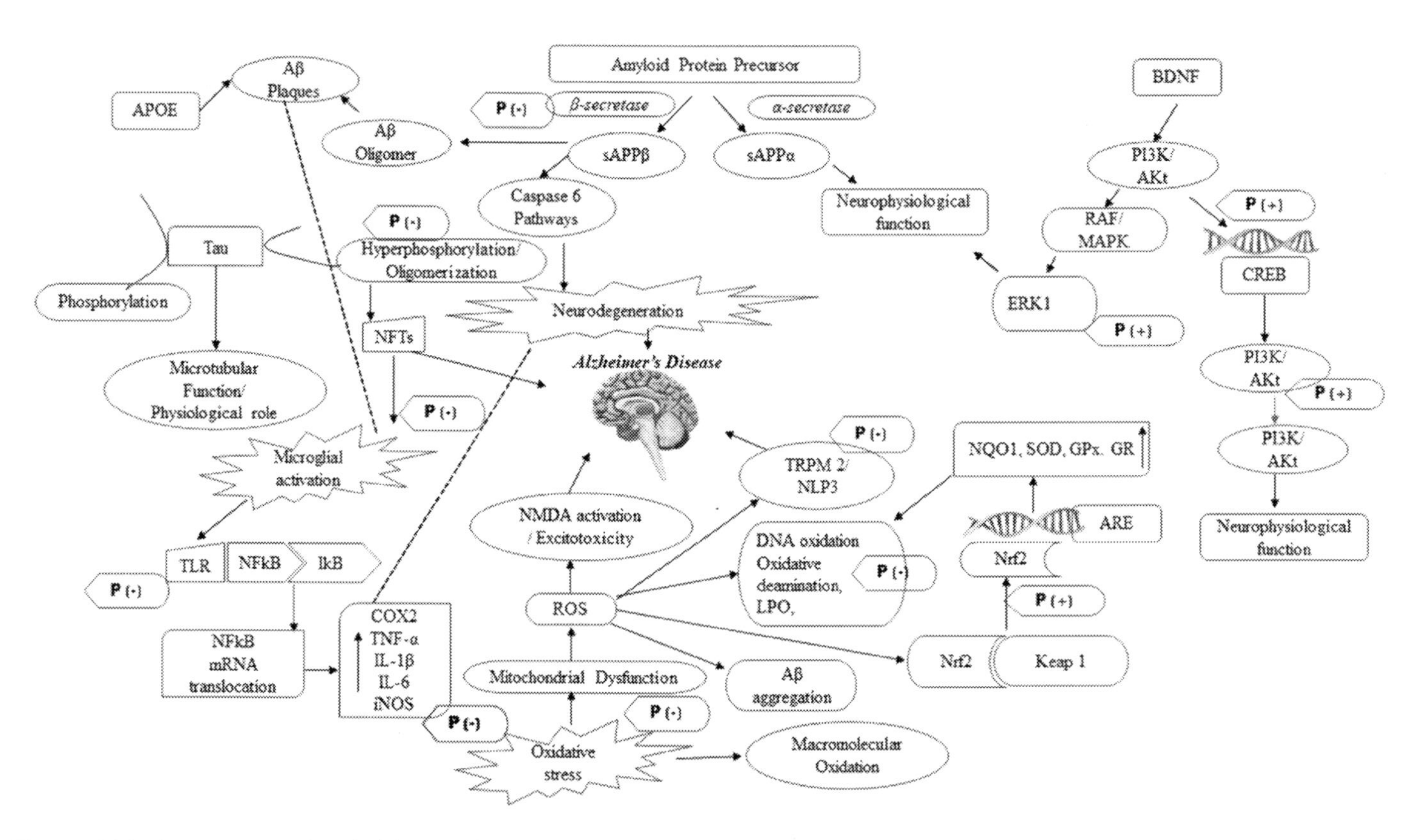

Figure 15.1 *Concept map of probable polyphenolic drug targets on neurodegeneration pathways (simplified).*

as aggregation. Diet supplementation (100 mg/kg, resveratrol) in transgenic (3xTg-AD) Alzheimer's animal model revealed the neuroprotective and cognitive improvement property. Resveratrol results in stimulation of AMPK (5-AMP activated protein kinases), resulting in upregulation of SIRT1. Thereby it signals the coactivator 1-alpha of Peroxisome proliferator-activated receptor gamma (PGC-1α) and AMP response element-binding protein (CREB) to improve cognitive functions (27). Resveratrol inhibits the neuronal inflammation through NF-κB signaling pathway. Resveratrol, 20 mg/kg intraperitoneal treatment in Wistar rats, enhanced the cognitive function through improving habituation memory. It also reduced the AChE enzyme and improved the biomarkers of antioxidant system. During neurodegeneration, the overexpression of COX-1 and COX-2 is noticed and resveratrol potentially inhibits the enzymes, resulting in reduction of neuroinflammation. NR2A of NMDA receptors and α7-nAChR were activated by resveratrol and implicates the enhancement in learning and memory. Resveratrol also improves the neurotrophic factors BDNF and NGF to nourish the neurons. It also protects from the neurodegeneration in conditions where neurodegeneration is possible due to diabetic conditions and insulin resistance by reducing the function of GSK-3β enzyme (10, 28). Resveratrol protects the cognitive function during cerebral ischemic conditions. The activated molecular pathways of JAK, ERK, STAT and JNK phosphorylation were alleviated by its treatment (29). In $A\beta_{(1\text{-}42)}$ induced neurodegeneration, resveratrol treatment downregulates (*in vivo* and *in vitro*) the inflammatory molecular pathways such as NF-κB/IL-1β/NLRP3 and TXNIP/TRX/NLRP3 (30, 31). Its potential effect on activation of NRF2 receptors well assures the antioxidant property for neuroprotection and plays a pivotal role in all types of neurodegeneration (32).

Clinical studies and clinical pharmacology of resveratrol

Acute doses of resveratrol at 250 mg and 500 mg improved the cerebral blood flow. Randomized controlled trial with 150–200 mg/kg enhanced the memory in aged adults and three to six months of treatment enhanced the memory process through improving the hippocampal functional connectivity. Significant memory retention, psychomotor speeding and word retention were exhibited in placebo-controlled double-blind trials. In Alzheimer's patients synthetic resveratrol at escalation doses from 500 mg to 1000 mg (13 weeks incremental) reduced the amyloid β peptide and inflammatory markers (TNF-α and IL-1) in cerebrospinal fluid (33–35).

Curcumin

The source of curcumin is rhizome of turmeric plant, *Curcuma longa*. The component curcumin is abundantly found in southeast Asian countries, including India as a main source. The rhizomes are used in traditional medicinal practice. Curcumin possesses various medicinal properties such as anti-inflammatory, antioxidant, anticancer and cerebroprotective activity. It is also reported to have various medicinal values. The neuroprotective property of curcumin was well correlated with the anti-amyloid and anti-neuroinflammatory effects (36). Natural curcumin in early studies exposed that it has the potential influence to bind with the neurotoxic Aβ plaques and reduces Aβ aggregation to afford neuroprotection (37). Curcumin 80 mg/kg improved the streptozotocin-induced neurodegeneration. This neuroactive effect is due to its potential antioxidant properties and indicates the effectiveness against sporadic Alzheimer's-type dementia. It improved the cognitive behavioral memory and habituation memory. Curcumin also has the property to inhibit the AChE and it improves the cerebral blood flow. The therapeutic effects of cognitive improvement were found at the dosages of 25 and 50 mg/kg in preclinical models. Curcumin improved the turnover of insulin in the brain, increased the cholinergic transmission and enhanced the antioxidant property. The insulin-like growth factor (IGFs) level is elevated by 300 mg/kg of curcumin in Alzheimer's disease–induced rats. Curcumin also reduced the neuroinflammation at 100 mg/kg and improved the neurotrophic BDNF in the rat hippocampus at 300 mg/kg (38–41). Curcumin suppresses the CRMP2 (collapsing response mediator protein-2) and inhibits the hippocampal axonal degeneration (37). Curcumin has strong potential antioxidant properties for neuroprotection and is mediated by NRF2 and apurinic/apyrimidinic endonuclease 1 (APE1) pathways (42). It regulates the AChE genes and controls its overexpression. Curcumin also enhances the synaptic transmission in the hippocampus through expression of synapse-related proteins (PSP95) (43, 44). Curcumin inhibits the neuronal inflammatory markers (TNF-α and IL-1) and reveals the evidence of inhibiting Aβ-induced microglial activation through ERK1/2 and p38 signaling (45). NFkB and COX-2 enzymes were also reduced with the treatment of curcumin (46). At different doses (20, 40, 80, 160 mg/kg), curcumin exhibits a dose-dependent neuroprotective effect on the hippocampal proteins such as Synapsin II, BDNF and CREB. It was found that 160 mg/kg is effectively neuroprotectant (47). Curcumin also enhances the neurogenesis through activation of ERK and MAP kinases pathway. It also enhances the cognitive function by signaling through PAKt/tAkt and PGSk/tGSK pathways. This activation in turn inhibits the GSK-3β function to enhance the neuroprotective property (48).

Clinical studies and clinical pharmacology of curcumin

Clinical trials with curcumin revealed that its different formulations were effective in AD patients. In healthy elderly people it improved the memory. Biocurcumax TM (1500 mg/day) study for 12 months observed memory decline in the placebo control group, indicating that curcumin protected from cognitive decline in the elderly (49). Acute (single dose) and chronic (four weeks) treatment (Longvida 400 mg (containing 80 mg of curcumin), an optimized lipid formulation) improved the working memory and attention (50). Placebo-controlled studies on Theracurmin (90 mg curcumin formulation), two times a day for 18 months treatment in adults with dementia (age range 54 to 84 years) improved the short-term, verbal and visual memory. It also reduced the neurotoxic amyloid and tau proteins in the amygdala (51).

Quercetin

Quercetin is a nutritional diet flavonoid, abundantly found in fruits, vegetables and nuts. As a polyphenolic flavonoid compound it is present in figs, grapes, apples, red onions, lettuce, berries, tomatoes and walnuts. It has potential health benefits as an antioxidant and has the pharmacotherapeutic properties such as cardioprotective, neuroprotective, anti-obesity, antidiabetic, antibacterial and anticancer effects (52). The neuroprotective effect of quercetin is proved by various investigations and it has a potential property to reduce the Aβ aggregation and enhance neurogenesis. It controls the inflammatory cytokines. During neurodegenerative conditions, quercetin protects the neurons through activation of Nrf2-ARE pathways (53). It also promotes the improvement of antioxidant biomarkers (SOD and GSH). Quercetin enhances the cognitive behavioral function, inhibits the AChE enzyme and improves the cholinergic transmission (52). It has a potential effect on suppression of endoplasmic reticulum (ER) stress with controlled phosphorylation of inositol-requiring enzyme 1α (IRE1α) and PKR-like ER-resident kinase (PERK). Thus, it continues to block the functional property of molecular proteins such as thioredoxin interacting protein (TXNIP) and NLRP3 inflammasome. All these effects of quercetin on molecular signaling pathways lead to a reduction in tau phosphorylation (54). In diabetic conditions, quercetin was found to restore glucose transporters (GLUT1 to 4) and low-density lipoprotein receptor-related protein 1 (LRP1), which are the key factors of brain insulin regulation for neurocognition (55). Twenty-five mg/kg of quercetin improved the behavioral memory in transgenic mice (56). It regulated the apolipoprotein (ApoE) and attenuated the Aβ production. Fifty mg/kg of quercetin diminished the radiation-induced brain injury by reducing the oxidative stress. Quercetin 30 mg/kg in mice reduced the stress, anxiety, depression and improved the cognitive behavioral memory (short and long term) by reducing the inflammatory markers (TNF-α, IL-1β, COX-2 and IL-6) (57). It also inhibits GSK3β and reduces the tau phosphorylation. Polyphenolic foods and rejuvenating agents rich in quercetin are highly encouraged for health and well-being. Daily recommended intake of 200 to 1200 mg as a dietary supplement promotes health (58, 59).

Clinical studies and clinical pharmacology of quercetin

Clinical study is conducted in a placebo-controlled, double-blinded trial. Mini-Mental State Examination (MMSE) and assessments on cognition impairment by rating scale were evaluated on the 12th week and the 24th week of the study. Elderly subjects with 60 mg equivalent quercetin were administered and it revealed cognitive improvement (60).

α-Mangostin

α-Mangostin is a polyphenol xanthone molecule extracted from *Garcinia mangostana* L. fruit. It's antioxidant potential exhibited the importance in application of various pharmacotherapeutic properties such as anti-inflammatory, antitumor, antibacterial, antidiabetic, antihypercholesteremic and neuroprotective effects. It inhibits brain inflammation by reducing the interleukins (1β, 6), TNF-α and COX-2 in mice upon a dose of 40 mg/kg (61). α-Mangostin has a property to inhibit the soluble amyloid protein precursor and reduces the conversion of neurotoxic Aβ fragments. It inhibits the AChE enzyme and improves the cholinergic turnover in the brain. It has a property to inhibit mitogen activated protein kinase (MAPKs) and to increase all the antioxidant biomarkers leading to neurocognitive function. In neuroblastoma cells, α-mangostin increased the membrane potential and decreased the level of reactive oxygen species (ROS). In rat brain isolated synaptosomes, glutathione peroxidase and glutathione levels were increased and indicates the free radical scavenging property in neuronal cells (62). It downregulates the expression of NF-κB to attenuate the inflammation (63). α-Mangostin increased the neuropeptide BDNF, reduced the brain tissue apoptosis by reducing p53, Bax and capsase-3 proteins. These pharmacological properties indicate that α-mangostin has the potential effect to improve cognitive functions.

Clinical studies and clinical pharmacology of α-mangostin

Randomized placebo-controlled trial with 245 mL of Verve energy drink (Mangosteen formula) significantly improved the antioxidant property and anti-inflammatory property. The study was conducted in healthy subjects and the evaluations were carried after two weeks of the daily supplement drink (64).

Catechins

Catechins are composed of epigallocatechin-3-Gallate, catechin and epicatechin. The primary source of functional food catechins are apples, cherries, strawberries, apricots, beans, and green tea. Green tea (*Camellia sinensis*) is considered as the major source of catechin for human beings as it is a nutraceutical and widely consumed for health and well-being (65). Green tea consumption reduced the incidence of cognitive dysfunction. Apart from a neuroprotective effect, catechins have wide pharmacotherapeutic properties such as anti-inflammatory, antibacterial, anti-allergic, antidiabetic, anti-obesity and cardioprotective effects (66). Catechins in green tea have preeminent antioxidant property and reduced the neurotoxicity induced by Aβ injection in the hippocampus of preclinical animal model. Green tea treatment at 400 mg/kg body weight in experimental rats for three weeks enhanced the cognitive behavioral function, recognition memory and reduced the biomarkers of AD. The biomarkers of lipid peroxidation, reactive oxygen species and AChE were reduced (67). (–)-Epigallocatechin (EGC) and (–)-epicatechin-3-gallate (ECG) reduced the neurotoxicity induced by Aβ oligomers and fibrils and alleviated the ROS production in APP/PS1 mice brain (68). It also exerts protection on glutamate-induced neurotoxicity by alleviating ROS. The catechin molecule 2S,3R-6 methoxycarbonyl gallocatechin protected the injury in SH-SY5Y cells by inflection of NR2B expression, the signaling by PI3K/Akt and caspase pathway (69). Green tea catechins were also found to alleviate the scopolamine (1 mg/kg) induced memory impairment in mice treated with different concentrations (30, 100 and 300 mg/kg). This improved the passive avoidance memory, spatial learning and upregulated the synaptophysin levels (70). In neuroblastoma cells pretreatment of EGC reduced the ROS and exhibited the antiapoptotic effect. In rats treated with 400 mg/kg for 14 days, it increased the density of neuronal efferent nerve fibers. It possesses strong antioxidant properties by improving SOD, catalase, glutathione and downregulated the AGE receptors. Epicatechin 30 mg/kg decreased the ROS and glial fibrillar acidic protein in hippocampus. In primary cortical neurons induced with Aβ, the ECG pretreatment increased the cell viability, and downregulated the ROS and caspase levels. It also increased the functional actions of nicotinic acetylcholine receptors (α7 nAChR) and phosphatidylinositol 3-kinase/protein kinas B (PI3K/Akt) signaling for neuroprotection (71). EGC also improved the STZ induced neurotoxicity in mice, indicating a potential effect on the neurodegeneration in diabetic conditions. Daily administration of EGC 3 mg/kg for three weeks on LPS induced inflammation in mice, attenuated the inducible nitric oxide (iNO), COX-2 expression and cytokines [ICAM-1 (intercellular adhesion molecule-1) and M-CSF (macrophage colony-stimulating factor)], TNF- α and IL-16 (72). In mice fed with catechin it also enhanced the effect of SIRT1 and SIRT3 thereby reducing the neuroinflammatory signaling by NFkB (4). It is also having an effect on inhibiting NLRP3-inflammasome (73). The catechins also mediate the improved neuronal signals through CREB-BDNF/Bcl-2–dependent pathway and promote the synaptic plasticity (74).

Clinical studies and clinical pharmacology of catechins

The cross-sectional study with the determination of MMSE in more green tea consuming elderly (≥ 70 years old) Japanese showed lower incidence of cognitive dysfunction when compared to the lesser green tea consuming aged peoples. Other studies state that in Chinese population (aged ≥ 50 years), the prevalence of cognitive dysfunction is lowered by frequent green tea consumption. A seven-year follow-up study in Chinese residents (≥ 70 years old) revealed that the people consuming tea have high levels of verbal fluency (75). In a randomized controlled trial, EGC (with dose escalation from 200 to 800 mg) is predicted to have a better cognitive score in Alzheimer disease assessment scale (ADAS) (76). Clinical study (27 healthy adults) with 135 mg epigallocatechin-3-gallate single dose improved the cerebral blood flow, and another study with 300 mg improved the electroencephalogram from central brain areas (77). In a randomized controlled trial with supplement (600 mg/d flavanols, 80 mg (-)-epicatechins, and theobromine 50 mg (Mars Symbioscience) improved the attention speed, verbal fluency and working memory (78).

Hesperidin and hesperetin

Hesperidin (glycone flavonoid) and hesperetin (aglycone flavonoid) are found in citrus species, especially in citrus fruits (Hesperidin) and in citrus peels (hesperetin), and have preeminent pharmacological properties. They are reported to have strong antioxidant capacities and other therapeutic effects. The compounds are said to have an antioxidant effect and neuroprotective, anti-inflammatory, antidiabetic, anticancer and cardioprotective properties (79). Hesperidin (Hsd) and hesperetin (Hst) have good penetrability to blood–brain barrier (BBB), which is an additional advantage in the pharmacotherapeutic effects of neurological disorders. Hst (10 and 20 mg/kg) in rats exhibited an antioxidant effect in STZ injected (i.c.v) neurotoxic model, increased the antioxidant biomarkers in the brain (GPx, SOD, glutathione reductase and catalase) and improved cognitive performance in behavioral tasks (80). Hsd and Hst improved the cognitive function in ischemic reperfusion-induced cognitive defects by signaling through BDNF. Both compounds have a protective effect against cytotoxicity induced by H_2O_2 on PC-12 cells by attenuating caspase-3 activity and calcium concentration. Hst has the ability to prevent neurotoxicity by blocking the excitotoxicity and Hsd exerts the best antioxidant property for neuroprotection. These compounds effectively reduce the nitrosative stress and improve neuronal function. Both the compounds also block the nitrergic pathway. Hsd has the property to activate the neural crest survival and both Hsd and Hst activate the PI3 and MAP kinase cascade for the neuroprotection. These also mediate the antioxidant capacity through ERK/Nrf2 pathway and increase the antioxidant biomarkers in the brain (81). Moreover, Hsd inhibited inflammation induced by LPS and inhibited the microglial activation through regulating the NLRP3 pathway (82). In olfactory bulbectomized mice, Hsd (50 mg/kg) treatment improved the BDNF, NGF and attenuated the interleukins (1β and 6) and AChE quantity in the hippocampus (83). Hsd also protected against the pathological changes induced by aluminum chloride at 100 mg/kg (60 days) by reducing the APP expression and inhibiting AChE in rat cortex and hippocampus (84).

Clinical studies and clinical pharmacology of hesperidin and hesperetin

Randomized placebo-controlled study (double blind) by administration of orange juice (eight weeks) in 37 healthy adults aged 60–81 years was assessed for cognitive function. Two groups were administered with different concentrations such as (i) high flavanone (HF)-549 mg Hsd and 60 mg narirutin/liter, (ii) low flavanone (LF), 64 mg and 10 mg/L Hsd and narirutin respectively. The study found Hsd at higher concentration improved the cognitive and execution function (Mini-Mental State Examination) with good episodic memory (85). Another randomized study (placebo control, single blinded) with two groups administered a dose of 42.15 mg Hsd in 500 mL citrus juice (Tropicana) and 240 mL concentrate (energy (87.7 kcal), fructose (6.38 g), glucose (5.36 g) and sucrose (10.2 g)) as placebo. This study found the improved objective performance and psychomotor speed with reduced errors determined by continuous performance task (CPT) (86). Japanese population cohort study (n=13373) in elderly people found the consumption of flavonoid rich (Hsd) juice indicated proportional risk reduction of dementia in chronological volume consumption (87).

Naringenin

Naringenin and Naringin are natural dietary polyphenolic compounds of citrus fruits (grapefruit, lemons, mandarins and oranges). Consumption of vegetables and fruits with high content of naringenin gives good health benefits. Naringenin is highly antioxidant and has pharmacotherapeutic values in treating several diseases such as diabetes, depression, cardiac disorders, and neurodegenerative disorders. Naringenin has good penetrability towards BBB (88). Naringenin attenuated the scopolamine induced amnesia in experimental animals at 50 and 100 mg dose/kg. In this study it improved the turnover of brain biogenic amines (serotonin, noradrenaline and dopamine) and attenuated the AChE enzyme and free radicals (89). The neuroprotective effect of naringenin improved the signaling of insulin in the brain and enriched cognitive functions in STZ (i.c.v) induced neurotoxicity. It also condensed the inflammatory markers (TNF-α and IL-6) as well as caspases 3 and 9. Naringin reduced the AChE enzyme and improved the antioxidant biomarkers (SOD, glutathione peroxidase and reductase and catalase). Administration of naringenin (100 mg/kg) in rats improved the habituation memory and antioxidant biomarkers in $A\beta_{(1-40)}$ induced neurotoxicity. It also reduced the DNA fragmentation, which clearly indicates the protection against the Aβ toxicity and its free radical scavenging property. Mitochondrial complex enzyme activities and cognitive functions are restored by 40 and 80 mg/kg naringin while treated in galactose induced neurotoxicity. Two weeks of naringin (80 mg/kg) improved the antioxidants and it has improved the Nrf2 protein and reduced the inflammatory markers (iNOS, TNF-α and COX-2) (90–93). In the insulin signaling in diabetic rats naringenin regulated the GLUT1 and GLUT3 expression and reduced the lipoprotein related protein (LRP1). In insulin signaling molecules, naringenin regulated insulin receptors (IRs) PI3K and Akt but was less effective than quercetin. This indicates that naringenin affords neuroprotection during diabetic conditions (55). Naringenin treatment in LPS induced neurotoxicity in rats, improved the behavioral cognitive tasks, hippocampal antioxidants (SOD, catalase, GSH) and lowered the malondialdehyde (MDA) as well as AChE. Naringenin reduced the inflammatory signaling molecule (NF-κB, TLR4, TNFα and COX2), nitric oxide synthase (iNOS) and GFAP. It elevated the Nrf2 domain. This strongly reveals the neuroprotective effect of naringenin through antioxidant and anti-inflammatory pathways (94). Naringenin also has the potential to inhibit the MAO-A and B enzymes, indicating that it could turn over the biogenic amine concentration in the brain and may afford cognitive improvement and treatment for other neurological disorders (95, 96). Moreover, naringenin has the property for reducing tau phosphorylation and neurotransmitter metabolic enzymes *in vitro* (97, 98). Naringenin protects the dopaminergic neurons by activating the Nrf2/ARE signals in SH-SY5Y cells (99). Chinese herbal oral liquid (Shenzhiling, containing 10 poly herbal extracts including naringenin) treated in APP/PS1 mice for three months revealed the neuroprotective effect by decreasing the myelin sheath and attenuated the myelin and oligodendrocyte damages by reducing Aβ accumulation. It also increased the myelin protein expression and other associated signaling molecules (MBP, PLP, MAG, and PI3K/Akt-mTOR) in the hippocampus (100).

Clinical studies and clinical pharmacology of naringenin

Naringenin and Hst are the major active components in orange juice. Naringenin was studied in various clinical studies; however, studies were limited to neuropharmacological research. In a randomized (double blind) study of 12-week duration, orange juice consumption protected from DNA impairment and lipid peroxidation. It also augmented the antioxidant enzymes and abridged obesity. Other studies stated that orange juice consumption improves the vascular function and antioxidant effect. In studies related to plasma pharmacokinetics, a significant amount of naringenin metabolites were observed in endurance trained (4–12 years trained) healthy men treated with 500 mL orange juice (101). However, studies are limited to metabolic and antioxidant effects; it may be predicted that naringenin-containing functional foods rejuvenate health.

Luteolin

The polyphenolic luteolin is a tetrahydroxy flavone in nature and distributed in functional foods such as navel oranges, carrots, rosemary, olive oil, green pepper, broccoli, parsley, chamomile tea, thyme, dandelion and the culinary herb called oregano. It possesses innumerable pharmacological properties such as antioxidant, neuroprotectant, anti-inflammatory, antidiabetic and antitumor activities (102). Luteolin possesses neuroprotective properties and improves the cognitive dysfunction by $A\beta_{(25\text{-}35)}$ peptide in mice. Ten and five mg/kg of luteolin improved the spatial memory and working memory and increased the antioxidant enzymes. It also reduced the AChE enzyme and the signaling of occludin and claudin-5 in the cerebral cortex. Luteolin increased the BDNF and enhanced the TrkB (tyrosine kinase B). It also increased the cerebrovascular blood flow and improved the interactions of neuro-glio-vascular system (103). In STZ-ICV injected rats, luteolin at 10 mg and 20 mg/kg improved the habituation memory, attenuated the neurotoxic effect induced by STZ and the function of hippocampal CA1 region is restored (104). Luteolin treatment in intracerebral hemorrhage induced rats reduced the cerebral inflammation, repressed the proinflammatory cytokines (IL-6 & 1β, TNF-α) and prevented microglial infiltration. The mechanistic influence of luteolin is facilitated through inhibition of TLR4/TRAF6/NF-κB pathways, preventing the p65 translocation to the nucleus (105). In high fat treated mice, luteolin (10 mg/kg) boosted the cognition and inhibited the inflammatory cytokines. In PC12 luteolin is said to enhance the cholinergic function and neurite outgrowth by initiation of ERK1/2 and PI3K/Akt signaling (106). Luteolin 10 mg/kg in high fat diet treated mice for 20 weeks reversed the insulin resistance and restored the normal adipocytokines. Moreover, it increased the BDNF and synapsin 1 in the cortex and hippocampus. These designate the potential benefit of luteolin in the treatment of cognitive dysfunction in metabolic syndrome conditions (107). Luteolin-containing plants were commonly used as functional foods and used in culinary and medicinal purposes; their extracts were shown to inhibit the GSK-3 β, BACE-1 and CK-1 δ. Luteolin also inhibited the hyperglycemic activation and Aβ toxicity induced by high glucose and $A\beta_{(1-42)}$ peptide in hippocampal neurons of rat (108). In Huntington's mouse striatal cells luteolin mediates the antioxidant effects by Nrf2/ARE transcription and enhanced the SOD1-mRNA (109). Luteolin (100 mg/kg) attenuates apoptosis and autophagy by inhibiting the PI3K/Akt pathway in rats in methamphetamine-induced neurotoxicity (110). Intraperitoneal administration of luteolin (20 mg/kg) reduced the Aβ accumulation in Tg2576 mice and is predicted to mediate the effect by inactivation of GSK-α which is considered for the presenilin 1 (PSEN1) phosphorylation and PSEN1 and APP interaction (111). In SD rats 50 mg/kg luteolin reversed the cognitive function and increased the stimulation of PKA/CREB/BDNF pathway on cognitive deficits induced by pentylenetetrazole (112). Luteolin, 10–70 mg/kg in preclinical animal models modulate the brain insulin resistance to protect the cognitive functions, the Aβ deposition may be reduced by regulation of insulin resistance and its signaling by gut microbiota (113). These pharmacological actions and molecular targets of luteolin indicate its therapeutic benefits in treating neurodegenerative diseases.

Clinical studies and clinical pharmacology of luteolin

A randomized, placebo control study was intended to evaluate the effect of luteolin (LuMus-Basel 20) on memory in healthy subjects (adult male/female, aged 18 to 40). The study aimed to evaluate the verbal memory, visual memory, mood state, depressive symptoms and anxiety. The dosage was 500 mg luteolin daily for 7.5 days (2 × 250 mg capsules). However, the study reports are yet to be published (https://clinicaltrials.gov/ct2/show/record/NCT04468854, retrieved on July 26, 2021) (114). Another study (randomized controlled) of luteolin combined with palmitoylethanolamide is designed to determine the behavioral effects, MMSE and Alzheimer's Disease Cooperative Study – Activities of Daily Living (ADCS–ADL) activities in frontotemporal dementia patients (115). Luteolin (100 mg) and quercetin (70 mg), combined in capsule form were administered to children aged 4–10 years having autism spectrum disorders (ASD). It's an open label trial conducted for 26 weeks. The assessment performed with Diagnostic and Statistical Manual of Mental Disorders (DSM-IV). The study results were statistically significant and exhibit the benefit of luteolin formulation in adaptive functioning and behavior. The clinical evaluation revealed that it could be a pharmacotherapeutic agent to regulate the behavioral effects (language, communication and stereotype) (116).

Myricetin

Myricetin is a poly hexahydroxy flavonol present in the bark of *Myrica rubra*. Its commonly found in functional polyphenolic food and vegetables (berries, honey, lady finger (*Abelmoschus esculentus*) and nuts). Myricetin possesses good antioxidant, antidiabetic, anti-inflammatory and neuroprotective effects. It crosses the BBB and has a pharmacological effect in the brain (117). Myricetin was given to mice induced with scopolamine and $FeSO_4$ (induce oxidative stress and dementia). The treatment of myricetin improved the cognitive behavioral parameters and inhibited the oxidative damage in the brain by reducing the brain iron content. It also reduced the AChE enzyme and increased the acetylcholine level. This indicates the cholinergic neurotransmission enhancement in the hippocampus. Myricetin also restored the SOD, GPx, Catalase and reduced the MDA (malondialdehyde) levels. It also indicated the inhibition of TrR1 expression, which mediates the attenuated iron levels in the brain (118). Administration of myricetin (40 mg/kg) in LPS induced mice prevented the variations in mRNA and pro-IL-1β protein levels. These effects were mediated through inhibition NLRP3 driven inflammation. Moreover, myricetin binds to the protein kinases (MEK1, JAK1, Akt and PI3K) and leads to NLRP3 ubiquitination (119). In SH-SY5Y cells induced with high molecular weight Aβo (HMW- Aβo), myricetin increased the cell viability by reducing the membrane disruption induced by HMW- Aβo. It also reduced the membrane LPO and increased the fluidity, resistance of cell membrane and suppressed the mitochondrial dysfunction (120). Myricetin isolated from *Hypericum afrum* have MAO-B inhibitory activity and exhibits its possible effects in AD through biogenic amine regulation (121). Myricetin (40 mg/kg) was treated in mice for 21 consecutive days and induced with restrain stress simultaneously (four hours/day). The study indicated that the stress induced elevation of corticosterone is reduced through regulating HPA axis. This is evident from the behavioral and biochemical changes. Myricetin reduced the ACTH and improved the BDNF, which reveals an effective reduction of stress-induced cognitive impairment (122). Myricetin decreases the glutamate content in synaptosomes through voltage-dependent Ca^{2+} channel blocking.

Mangiferin

Mangiferin is a polyphenolic xanthone having various pharmacotherapeutic benefits. It's a chief phytochemical component in *Mangifera indica* L. (mango tree) and extensively found in fruits and some herbs. Mangiferin exhibits various properties such as antioxidant, antiviral, immunomodulatory, anticancer, antidiabetic, neuroprotective and anti-inflammatory effects. It has the ability to traverse the BBB and it exerts the pharmacotherapeutic effects on CNS disorders (123). Due to its strong antioxidant property, it alleviates the mitochondrial dysfunction and neuroinflammatory cytokines and exerts neuroprotection. The cognitive function is well improved in various preclinical animal models. In murine microglial cells mangiferin administration alleviated the levels of NO, TNF-α, IL-6 & 1β activation. It also condensed the mRNA and the proteins of COX-2 & iNOS, which eventually leads to the reduction of NFkB and NLRP3 signaling. In cultured neuronal cells treated with mangiferin and induced with neurotoxicants (Aβ and glutamate) exhibited the neuroprotection by reducing oxidative stress markers and improved the antioxidants (GSH, GPX, GST, SOD and CAT). In Neuro-2A (N2A) cells treated with Aβ, mangiferin reduced the inflammatory LOX and inhibited the AChE. Swiss albino mice induced with neurotoxic aluminum chloride and treated with mangiferin for 21 days exhibited the inhibition of AChE and oxidative stress. The neurobehavioral parameters were improved with cognitive function. It also reduced the TNF-α and IL1-β (124). APP/PS1 mice treated with 50 mg/kg (mangiferin) for 22 weeks improved the behavioral parameters and reduced the phosphorylation of tau protein. Twenty mg treatment in LPS-induced mice reduced the level of inflammatory marker (TNF-α). Mangiferin treatment (10, 20 and 40 mg/kg) improved the open field exploration and grip strength and improved the antioxidant potential. It also exhibited the inhibitory effect on MAO- B enzyme (125). In scopolamine induced mice, mangiferin (10, 20 and 40 mg/kg) regulated the cholinergic neurotransmission by increasing the Ach and inhibiting the AChE enzyme. It is also found to increase the BDNF, GSH, dopamine and noradrenaline (126). Mangiferin pretreatment in aged and scopolamine induced mice restored brain AChE and increased the noradrenaline and dopamine concentration in brain (127). In corticosterone induced stress mice, mangiferin treatment for seven days attenuated the oxidative stress by reducing MDA levels, IL-1β, TNFα, iNOS, and COX-2 (125). Mangiferin, treated at concentrations of 10, 50, 120 or 100 mg/kg rats improved the novel object recognition test and open field behavior. In glioblastoma (human) cell line U138-MG, it improved the neurotrophic NGF and mediated the synaptic plasticity (128).

Clinical studies and clinical pharmacology of mangiferin

Cognitive effect, stress response and mood were evaluated in a randomized, crossover, double-blinded study conducted in healthy adults. Zynamite (Mango leaf extract) is prepared from leaf extract (300 mg) containing ≥ 60% mangiferin. Seventy-two healthy volunteers (males 50% and females 50%), aged 18–45 years, were recruited in this study. One single acute dose is administered during the test visits. The data was collected using questionnaires assessing mood and cognitive tasks. Blood samples were analyzed after 300 minutes of drug administration to determine the levels of BDNF, cortisol, adrenaline, noradrenaline and prolactin levels. Working memory, episodic memory, attention, executive function, stress visual test, individual task performance (word and picture recognition) and spatial working memory were determined. The single acute dose polyphenol rich drug (Zynamite®) administration broadly improved the cognition, short- and long-term memory, cognitive tasks and executive functions (129).

Baicalein and baicalin

Baicalin and baicalein are two flavonoids extracted from *Scutellaria baicalensis* roots, leaves of *Oroxylum indicum* and *Thymus vulgaris*. It is widely used in traditional medicine and as a food additive (130). Chemically, baicalein is an aglycone derivative from baicalin. Pharmacological studies in baicalein indicated its anticonvulsive, anxiolytic, antioxidant, anticancer, antidiabetic, antithrombotic, anti-inflammatory and neuroprotective properties. It also possesses antiviral and antibacterial properties (131). Various studies revealed that it has a remarkable neuroprotective action and it potentially improved the cognition. In $A\beta_{(25-35)}$ induced (i.c.v) neurotoxicity in mice, pretreatment of baicalein (5 or 10 mg/kg, i.p.) and post treatment for 7 to 13 weeks alleviated the neurotoxicity and exerts neuroprotective effect (132). In $AlCl_3$ induced neurotoxicity in rats treated with baicalein (5, 10 and 20 mg/kg) for four successive weeks improved the memory. It enhanced the behavioral memory performance and improved the brain acetylcholine level with decrease in AChE enzyme (133). In primary cortical neuronal cells it reduces the $A\beta_{(25-35)}$ induced LOX enzyme and reduces the apoptosis. In SH-SY5Y cells, it reduced the aggregation of Aβ and oxidative stress (134, 135). *In vitro* (CHO cells- human APP cells) and *in vivo* transgenic mice (Tg2576) studies indicate that baicalein directly decrease the production of Aβ and diverts to nonamyloidogenic pathway. It improves the spatial learning memory performance and revealed that its effects are also associated with signaling of $GABA_A$ receptor (136). In APP/PS1 mice, baicalein treatment at 40 and 80 mg/kg for two months improved the LTP of hippocampus through activating serine threonine Kinase (Akt) phosphorylation. It also repressed 12/15LO and GSK3β, decreased the BACE1 and prevented the tau phosphorylation. The electrophysiological recording of CA1 hippocampal region showed the improved synaptic plasticity, indicating the remarkable neuroprotective effect of baicalein (137). A formulation (Bushen-Huatan-Yizhi) containing baicalin improved the cognition in $A\beta_{(1-42)}$ induced rats. The treatment with this formulation reduced the hippocampal tau phosphorylation by signaling through GSK-3β/CREB pathway and these molecular components were improved (138). In rats induced with stress and depression, treatment with baicalin (20, 40 mg/kg) improved the behavioral performance, reduced the IL-1β & 6 in prefrontal cortex. It also down streamed the NLRP3 activation, altogether revealing that the baicalin could ameliorate the neuronal inflammation for cognitive improvement (139). Baicalin 10 mg/kg treatment in rats induced with Aβ, indicated an improvement in neuronal stem cell proliferation and differentiation. It was observed that there is a reduction in nuclear pyknosis in hippocampal cells and pyramidal cells. It also increases the proteins such as Nestin, GFAP and NSE for the neuroprotection (140). In male ICR mice induced with chronic unpredictable mild stress and treated with baicalin relatively enhanced the behavioral tasks. In the biochemical aspects, the expression of protein signaling molecules enhanced the molecular cascading involving BDNF/ERK/CREB pathway and this mechanism was the key factor for neuroprotection (141). Baicalin containing granules (Shenqi Yizhi) inhibits the Aβ induced cognitive dysfunction. SD rats injected with neurotoxic Aβ and treated with the granules (9.8 g/kg, 4.9 g/kg and 2.45 g/kg). After the treatment for 60 days, the JAK2/STAT3 signaling was reduced significantly and this exerts neuroprotection (142). Treatment of baicalin in mouse brain vascular cell line (*in vitro*) studies and in mice (*in vivo*), induced with LPS neurotoxicity, attenuated the production of inflammatory cytokines cascading. In the *in vivo* study, the mice were treated with 60 mg/kg of baicalin and *in vitro* cell cultures were treated with multiple dose concentrations starting from 1μg/mL to 32 μg/mL. This treatment attenuated TNF- α and IL1-β, enhanced the antioxidant signaling through improved Nrf2 signaling (143). These multiple targets of baicalin on controlling inflammation, biogenic amine turn over and inhibitory effects on Aβ aggregation as well as tau phosphorylation is due to its antioxidant potential.

Clinical studies and clinical pharmacology of baicalin and baicalein

Extracts of *Scutellaria baicalensis* containing the baicalin (≥60%) and *Acacia catechu* containing catechin (≥10%) were administered in a combined formulation (UP326) to the healthy volunteers (aged 35–65 years, n=83). The study is a randomized placebo control, designed with two groups. UP326 was administered for four weeks and the evaluations were performed weekly. In this, the working memory speed is evaluated with task shifting. Mental flexibility and cognitive processing of higher order were considered as index of cognitive performance. The treatment suggested improved attitude and working memory (144).

Vanillic acid

Vanillic acid (VA) is polyphenolic flavoring agent; it's a benzoic acid derivative and has various pharmacotherapeutic effects. It is abundantly found in the plant *Radix Angelica sinensis* (*Oliv.*) applied in traditional Chinese medicine and also found in some edible plants and fruits. It is reported to hold preeminent antioxidant, antihypertensive, anti-inflammatory and neuroprotective effects. The potential antioxidant effect of vanillic acid plays a pivotal role in various diseases persuaded by oxidative stress. It is possessing the neuroprotective effect and is revealed in mice treated with STZ (i.c.v). VA at 25, 50 and 100 mg/kg treatment for 28 days improved the cognitive behavioral function and antioxidant enzymes (SOD, catalase and Gpx). It also reduced the corticosterone levels in the plasma, AChE and TNF-α levels in the brain homogenate. This evidently indicates the influence on HPA axis regulation. VA reduced the stress hormone corticosterone and the inflammatory mediator and affords neuroprotection (8). VA 30 mg/kg was treated in mice for two weeks and induced with LPS in-between the two weeks VA treatment. The neurobehavioral effects were improved in spatial learning and hippocampal learning. VA treatment significantly attenuated the neuroinflammatory cytokines and reduced the Aβ accumulation. It is revealed that the neuroprotective effect of VA is mediated through the inhibition of LPS/RAGE mediated

JNK signaling (145). The neuroprotective role of VA in bilateral carotid artery occlusion induced rats is evident form its role in improving the behavioral parameter and in reducing the IL-6, TNF-α as well as TUNEL positive cells. It is also shown to increase the IL-10 signaling in rat hippocampus (146). Aβ induced neurotoxic mice treatment with 30 mg/kg vanillic acid for three weeks improved the cognitive function and reduced the neuroinflammation. VA treatment reduced the BACE-1 expression and reduced the Aβ cascading. HT22 cells treated with Aβ and VA (100 μM, 24 h) reduced the NFkB signaling and reduced the inflammatory stream. It is apparent that VA exerts the neuroprotection through increasing the hemeoxygenase-1 mediated by Akt/GSK-3β/Nrf2 signaling (145).

Clinical studies and clinical pharmacology of vanillic acid

Traditional Chinese herbal formulations containing the vanillic acid or the plant containing VA are very commonly used for health, well-being and to treat neurological conditions. Yokukansan, Buqi Huoxue recipe, Modified Didang decoction, Compound *Polygonum multiflorum* extract and Shengui Yizhi recipe are the herbal formulations in which the *Radix Angelica Sinensis* is one ingredient among the different herb compositions (147). Clinical trials in these formulation yielded positive effects in mental function and behavioral tasks. To determine the effect of Yokukansan in neuropsychiatric symptoms in AD, a randomized trial (double-blind) was designed. Patients with AD (144) were treated with 7.5 gm/day formulation (in 75 AD patients) and placebo (70 participants) respectively. The participants' learning and behavioral effects have been improved in MMSE (140). Clinical investigation in compound *Polygonum multiflorum* extract containing *Radix Angelica Sinensis* was investigated in 209 AD patients. The patients (n=120) were assessed for MMSE before and after 12 weeks treatment with compound *Polygonum multiflorum* extract. The group was compared to the control (n=60, Chinese herbal) and standard drug treated group (n=29, Naofukang). The scores on MMSE and Ability of Daily Living Scale of the patients treated with the compound *Polygonum multiflorum* extract significantly improved (147, 148).

Fisetin

Fisetin is a tetrahydroxy flavone contained in strawberries with high content and is profusely found in fruits as well as vegetables (mangoes, apple and strawberries). It has antioxidant, anti-inflammatory and neuroprotective activities. It has an anti-amyloidogenic property and is indicated as a reliable therapeutic molecule for the treatment of neurodegeneration (149). Fisetin 15 mg/kg treatment in mice brain in aluminum chloride induced neurotoxicity exhibited improved cognitive behavioral function and improved the biochemical regulation in the hippocampus and cortex. Fisetin treatment inhibited the Aβ aggregation, reduced protein expressions ASK-1, p-JNK, p53, caspase-9 and 3, cytochrome c, as well as modulated the ratio of Bax/Bcl-2. The reduction in

TUNEL-positive and fluoro-jade C-stained cells indicated the ASK-1 & p-JNK the probable Aβ aggregation mediator, which is regulated by Fisetin treatment (150). Fisetin also enhances the ERK- dependent LTP in the hippocampus and improved the synaptic function. Fisetin (5, 10 and 25 mg/kg) treatment improved the recognition index and cognition. In rat hippocampal slices, it activates the ERK and signals the CREB to phosphorylation (151). Fisetin in catecholaminergic PC12 exhibits protection against tunicamycin (Tm)-mediated cytotoxicity by amplifying the phosphorylation of p38 MAPK and activating the Nrf2 protein. The effect through activation of Nrf2-ARE pathway is considered to be the factor for neuroprotection. It also enhanced the antioxidant enzymes and reduced the stress biomarkers (152). Fisetin in diabetic rats prevented the neuroinflammation by reducing the inflammatory markers (TNF-α and IL-6), eventually COX-2 is reduced with the reduction in NFkB signaling (153). Moreover six weeks treatment with 15 mg/kg of fisetin in aged rats reduced the AChE enzyme, improved the antioxidant enzymes and Na^{2+}/K^{+}ATPase and calcium ATPase. It also upregulated the autophagy gene expression of (Atg-3 and Beclin-1), sirtuin-1 and reduced the inflammatory markers (IL-1β and TNF-α) as well as SIRT-2 genes of aged rats (154). Oxidative stress protection and the potential to inhibit the neuroinflammation are the preeminent factors for cognitive improvement and neuroprotective effects of fisetin.

Clinical studies and clinical pharmacology of fisetin

Some of the clinical trials detailed about fisetin's potential effects on insulin regulation, inflammatory cytokine control and osteoarthritis. These studies suggest that it may attenuate the oxidative stress to improve health. Studies in isolated fisetin related to neurodegeneration are limited; however, strawberries containing fisetin have been proved to improve cognitive function. The study is a longitudinal cohort study with 925 participants (58–98 years old). The study was conducted from 1997 to 2018 with 2152 participants. The participants were administered with a food assessment questionnaire (strawberry consumption and its frequency). They are subjected to annual assessment for AD including clinical judgment and computerized scoring. The study revealed that strawberry consumption diminishes the incidence of Alzheimer's dementia in aged adults and it may be due to the presence of total flavonoid content and other phytoconstituents (155). It is evident that the maximum concentration of fisetin is present in strawberries (160 μg/g) and its adaption in food habits improves the cognitive function and reduces the risk of AD (156).

Rutin

Rutin (rutoside or quercetin-3-O-rutinoside) is chemically quercetin with the combination of glycoside (rutinose). It is found commonly in fruits and vegetables especially in citrus fruits. It has broad pharmacotherapeutic properties such as immunomodulatory, antiproliferative, neuroprotective, anti-inflammatory, anticarcinogenic and cardioprotective effects. Rutin interferes with inflammatory cytokines contributing to neuroprotection and improves the cognitive function in preclinical models (157). In SH-SY5Y and BV-2 cells induced with Aβ, the treatment of rutin inhibited the ROS, NO, MDA, iNOS, MMP, TNF-α, IL-1β and increased the SOD, CAT, GSH and GPx (158). Rutin (25 mg/kg) treatment in STZ (icv) injected animals (rats), improved the antioxidants (SOD, GPx, GR and Catalase) and significantly improved the behavioral memory tasks. It also reduced the NFkB, iNOS, TNF-α, IL-18 and GFAP (159). Rutin treatment at 100 mg/kg in rats for three weeks induced with Aβ exhibited an increase in hippocampal signaling proteins by MAPK (BDNF-ERK-CREB) pathways (160). In cell cultures, rutin improved the cell viability and mediated the neuronal cell differentiation through JNK and p38 MAPK (161). Pomegranate juice containing rutin is treated in Tg2576 mice and is shown to increase the Na^+/K^+ATPase and reduce the AChE. This shows the improvement in synaptic plasticity in AD by rutin (162). In *Hypericum perforatum* containing rutin treatment on aluminum chloride induced rats, the behavioral parameters were improved. It also inhibited the AChE enzyme and improved the noradrenaline (NA) as well as dopamine (DA), exhibiting the preeminent neuroprotection and biogenic neurotransmitter amine regulation (163). A herbal formulation with rutin as a content, called Sanweidoukou decoction, ameliorates Aβ induced neurotoxicity through MAPK/NF-κB pathways (164). These studies on rutin comprehensively exhibit its antioxidant and neuroprotective property with marked inhibition in neuroinflammation.

Clinical studies and clinical pharmacology of rutin

Rutin possesses strong antioxidant effect. It has been studied in a randomized control trial for its effects on oxidative stress and glycemic control. Diabetic patients of age 35 years and above were enrolled and divided into three different groups. Twenty patients in group A were treated with 60 mg of rutin (combined with 160 mg of vitamin C). Twenty patients of group B were treated with their antidiabetic drug and Vitamin C (500 mg). In group C, the patients (n=13) received only their prescribed antidiabetic drug. The treatments were carried for eight weeks. Fasting blood glucose, blood cholesterol, antioxidant stress biomarkers (SOD, MDA) and quality of life (QOL) by means of SF-36 questionnaire determined at baseline and at the end of study. The study revealed substantial decreases in FBG, total cholesterol, LDL and VLDL. It improved the QOL and indicates the improvement in physical working and energy domains in group A. This study states that rutin is a lead natural product for neuroprotection and controls the oxidative stress; thereby rutin possibly will be a therapeutic agent on treating neurodegeneration in diabetic conditions and AD (165).

Bioavailability and pharmacokinetics of polyphenols

Polyphenols are micronutrients beneficial as pharmacotherapeutic agents for treating various diseases including neurodegeneration. The pharmacological effect of polyphenols will be proportional to the consumption and its systemic absorption. They are present in the form of polymers, esters and glycosides, which may require hydrolyzation with intestinal enzymes before absorption and is with the limitation of action by intestinal microflora. However, circulating polyphenols and their metabolites have good tissue penetrability and exert greater therapeutic effect (166). Moreover, polyphenols are susceptible for the environmental conditions that may dampen the pharmacological effect. It is very essential to target the brain by crossing BBB for effective neuroprotection. Many techniques and formulation developments were performed to improve their bioavailability and to preserve the chemical nature. Commonly microencapsulation, ionic gelation, hydrogel emulsions, liposomes, micelles and nanoparticle systems were employed (167). To enhance the delivery of curcumin in the brain, Di-blockPEG-PLANPs were prepared by nanoprecipitation and treated in transgenic mice (Tg2576). This highly improved working memory. Intravenous administration of PLGA NPs in rats improved the delivery and was found to have stable concentration of curcumin in brain. Quercetin in the form of zein NPs, liposomes and PLGA NPs in animal model of AD, improved the cognitive function and ameliorated the biochemical markers in the brain. Polymeric micelles of resveratrol prepared by nanoprecipitation protected the Aβ induced neurotoxicity in PC2 cells (168). Solid lipid nanoparticles of pomegranate extract treatment in aluminum chloride induced animal model of AD, improved the novel object recognition test, behavioral parameters and brain antioxidant enzymes (CAT, GSH) with enhanced total antioxidant capacity (169). The novel developments in polyphenol formulations illuminate the therapeutic insights by improving bioavailability and BBB penetrability for treating AD and other neurological diseases.

Conclusion

The molecular targets by polyphenols in the treatment of AD and other neurological diseases are esteemed with various therapeutic outcomes in preclinical studies as well as clinical trials. Interestingly the established neurological theories escort targeted approach by interacting with various signaling proteins, receptors and enzymes. Scientifically, the functional food polyphenols such as resveratrol, catechins, quercetin, curcumin, luteolin and other polyphenolics were well interpreted for the treatment of oxidative stress and neurological disorders. Even though the neuropathology of AD spindle with multifactorial mechanism, oxidative stress leading to amyloid generation and neuroinflammation are considered prime factors that polyphenols target to exert their therapeutic properties. Polyphenol treatment alone or in combination is established to afford symptomatic improvement; moreover supplemental polyphenols and dietary intake at an early age protects from the incidence of neurodegeneration and elderly dementia. Optimized drug delivery of polyphenols with enhanced bioavailability and BBB penetration impacts the therapeutic strategies of age-related neurological diseases.

References

1. World Health Organization. Global action plan on the public health response to dementia 2017–2025. 2017: https://www.who.int/news-room/fact-sheets/detail/dementia.
2. Nguyen TT, Nguyen TD, Nguyen TKO, Vo TK, Vo VG. Advances in developing therapeutic strategies for Alzheimer's disease. Biomed Pharmacother. 2021;139:111623.
3. Dubey SK, Lakshmi KK, Krishna KV, et al. Insulin mediated novel therapies for the treatment of Alzheimer's disease. Life Sci. 2020;249:117540.
4. Jayasena T, Poljak A, Smythe G, Braidy N, Münch G, Sachdev P. The role of polyphenols in the modulation of sirtuins and other pathways involved in Alzheimer's disease. Ageing Res Rev. 2013;12(4):867–883.
5. Qi Y, Cheng X, Jing H, Yan T, et al. Combination of schisandrin and nootkatone exerts neuroprotective effect in Alzheimer's disease mice model. Metab Brain Dis. 2019;34(6):1689–1703.
6. Pirzada RH, Javaid N, Choi S. The roles of the NLRP3 inflammasome in neurodegenerative and metabolic diseases and in relevant advanced therapeutic interventions. Genes (Basel). 2020;11(2):131.
7. Liguri G, Taddei N, Nassi P, Latorraca S, Nediani C, Sorbi S. Changes in Na+, K+-ATPase, Ca2+-ATPase and some soluble enzymes related to energy metabolism in brains of patients with Alzheimer's disease. Neurosci Lett. 1990;112(2–3):338–342.
8. Singh JCH, Kakalij RM, Kshirsagar RP, Kumar BH, Komakula SSB, Diwan PV. Cognitive effects of vanillic acid against streptozotocin-induced neurodegeneration in mice. Pharm Biol. 2015;53(5):630–636.
9. Lyra SN de M, Gonçalves RA, Boehnke SE, Forny-Germano L, Munoz DP, De Felice FG. Understanding the link between insulin resistance and Alzheimer's disease: Insights from animal models. Exp Neurol. 2019;316:1–11.
10. Drygalski K, Fereniec E, Koryciński K, et al. Resveratrol and Alzheimer's disease. From molecular pathophysiology to clinical trials. Exp Gerontol. 2018;113:36–47.
11. Jayasingh Chellammal HS, Veerachamy A, Ramachandran D, Gummadi SB, Manan MM, Yellu NR. Neuroprotective effects of 1'δ-1'-acetoxyeugenol acetate on Aβ(25-35) induced cognitive dysfunction in mice. Biomed Pharmacother. 2019;109:1454–1461.
12. Sivandzade F, Prasad S, Bhalerao A, Cucullo L. NRF2 and NF-κB interplay in cerebrovascular and neurodegenerative disorders: Molecular mechanisms and possible therapeutic approaches. Redox Biology. 2019;21:101059.
13. Alzheimer's Association. Medications for memory, cognition and dementia-related behaviors. https://www.alz.org/alzheimers-dementia/treatments/medications-for-memory
14. Cassidy L, Fernandez F, Johnson JB, Naiker M, Owoola AG, Broszczak DA. Oxidative stress in Alzheimer's disease: A review on emergent natural polyphenolic therapeutics. Complement Ther Med. 2020;49:102294.
15. Colizzi C. The protective effects of polyphenols on Alzheimer's disease: A systematic review. Alzheimer's Dement Transl Res Clin Interv. 2019;5:184–196.
16. Jayasingh Chellammal HS, Menon BV, Hasan MH, et al. Neuropharmacological studies of ethanolic extract of *Vaccinium corymbosum* on Alzheimer's type dementia and catatonia in Swiss albino mice. J Herbmed Pharmacol. 2021;10(2):241–248.
17. Ebrahimi A, Schluesener H. Natural polyphenols against neurodegenerative disorders: Potentials and pitfalls. Ageing Res Rev. 2012;11(2):329–345.
18. Hanish Singh Jayasingh C, Mohamed Mansor M, Afiq A, Yasothini M, Punitha C, Pavithiraa C. Neurocognitive effects of *Prunus domestica* fruit extract on scopolamine-induced amnesic mice. J Appl Pharm Sci. 2020;10(11):59–66.
19. Kontush A, Mann U, Arlt S, et al. Influence of vitamin E and C supplementation on lipoprotein oxidation in patients with Alzheimer's disease. Free Radic Biol Med. 2001;31(3):345–354.
20. Hanish Singh JC, Muralidharan P, Narsimha Reddy Y, Sathesh Kumar S, Alagarsamy V. Anti-amnesic effects of *Evolvulus alsinoides* Linn. in amyloid β (25–35) induced neurodegeneration in mice. Pharmacologyonline. 2009;1:70–80.

21. Lye S, Aust CE, Griffiths LR, Fernandez F. Exploring new avenues for modifying course of progression of Alzheimer's disease: The rise of natural medicine. J Neurol Sci. 2021;422:117332.
22. Vinitha E, Singh HJC, Kakalij RM, Kshirsagar RP, Kumar BH, Diwan PV. Neuroprotective effect of *Prunus avium* on streptozotocin induced neurotoxicity in mice. Biomed Prev Nutr. 2014;4(4):519–525.
23. Mazumder MK, Choudhury S. Tea polyphenols as multi-target therapeutics for Alzheimer's disease: An *in silico* study. Med Hypotheses. 2019;125:94–99.
24. Griñán-Ferré C, Bellver-Sanchis A, Izquierdo V, et al. The pleiotropic neuroprotective effects of resveratrol in cognitive decline and Alzheimer's disease pathology: From antioxidant to epigenetic therapy. Ageing Res Rev. 2021;67:101271.
25. Komorowska J, Wątroba M, Szukiewicz D. Review of beneficial effects of resveratrol in neurodegenerative diseases such as Alzheimer's disease. Adv Med Sci. 2020;65(2):415–423.
26. Gocmez SS, Gacar N, Utkan T, Gacar G, Scarpace PJ, Tumer N. Protective effects of resveratrol on aging-induced cognitive impairment in rats. Neurobiol Learn Mem. 2016;131:131–136.
27. Corpas R, Grinan-Ferre C, Rodriguez-Farre E, Pallas M, Sanfeliu C. Resveratrol induces brain resilience against Alzheimer neurodegeneration through proteostasis enhancement. Mol Neurobiol. 2018;56(2):1502–1516.
28. Wang R, Zhang Y, Li J, Zhang C. Resveratrol ameliorates spatial learning memory impairment induced by $A\beta_{1-42}$ in rats. Neuroscience. 2017;344:39–47.
29. Chang C, Zhao Y, Song G, She K. Resveratrol protects hippocampal neurons against cerebral ischemia-reperfusion injury via modulating JAK/ERK/STAT signaling pathway in rats. J Neuroimmunol. 2018;315(81):9–14.
30. Feng YS, Tan ZX, Wu LY, Dong F, Zhang F. The involvement of NLRP3 inflammasome in the treatment of neurodegenerative diseases. Biomed Pharmacother. 2021; 138(139)111428.
31. Huang J, Huang N, Xu S, et al. Signaling mechanisms underlying inhibition of neuroinflammation by resveratrol in neurodegenerative diseases. J Nutr Biochem. 2021; 88:108552.
32. Qu Z, Sun J, Zhang W, Yu J, Zhuang C. Transcription factor NRF2 as a promising therapeutic target for Alzheimer's disease. Free Radic Biol Med. 2020;159:87–102.
33. Khorshidi F, Poljak A, Liu Y, Lo JW, Crawford JD, Sachdev PS. Resveratrol: A "miracle" drug in neuropsychiatry or a cognitive enhancer for mice only? A systematic review and meta-analysis. Ageing Res Rev. 2021;65:101199.
34. Huhn S, Beyer F, Zhang R, et al. Effects of resveratrol on memory performance, hippocampus connectivity and microstructure in older adults – A randomized controlled trial. Neuroimage. 2018;174:177–190.
35. Zhu CW, Grossman H, Neugroschl J, Parker S, Burden A, Luo X, et al. A randomized, double-blind, placebo-controlled trial of resveratrol with glucose and malate (RGM) to slow the progression of Alzheimer's disease: A pilot study. Alzheimer's Dement Transl Res Clin Interv. 2018;4(1):609–616.
36. Shende P, Mallick C. Nanonutraceuticals: A way towards modern therapeutics in healthcare. J Drug Deliv Sci Technol. 2020;58:101838.
37. Wang Y, Yin H, Li J, Zhang Y, Han B, Zeng Z, et al. Amelioration of β-amyloid-induced cognitive dysfunction and hippocampal axon degeneration by curcumin is associated with suppression of CRMP-2 hyperphosphorylation. Neurosci Lett. 2013; 557(PB):112–117.
38. Voulgaropoulou SD, van Amelsvoort TAMJ, Prickaerts J, Vingerhoets C. The effect of curcumin on cognition in Alzheimer's disease and healthy aging: A systematic review of pre-clinical and clinical studies. Brain Res. 2019;1725:146476.
39. Tiwari V, Chopra K. Protective effect of curcumin against chronic alcohol-induced cognitive deficits and neuroinflammation in the adult rat brain. Neuroscience. 2013; 244:147–158.

40. Rajasekar N, Dwivedi S, Tota SK, et al. Neuroprotective effect of curcumin on okadaic acid induced memory impairment in mice. Eur J Pharmacol. 2013;715(1–3):381–94.
41. Ghosh S, Banerjee S, Sil PC. The beneficial role of curcumin on inflammation, diabetes and neurodegenerative disease: A recent update. Food Chem Toxicol. 2015; 83:111–124.
42. Abrahams S, Haylett WL, Johnson G, Carr JA, Bardien S. Antioxidant effects of curcumin in models of neurodegeneration, aging, oxidative and nitrosative stress: A review. Neuroscience. 2019;406:1–21.
43. Chen F, He Y, Wang P, et al. Curcumin can influence synaptic dysfunction in APPswe/PS1dE9 mice. J Tradit Chinese Med Sci. 2018;5(2):168–176.
44. Akinyemi AJ, Oboh G, Fadaka AO, Olatunji BP, Akomolafe S. Curcumin administration suppress acetylcholinesterase gene expression in cadmium treated rats. Neurotoxicology. 2017;62:75–79.
45. Shi X, Zheng Z, Li J, et al. Curcumin inhibits Aβ-induced microglial inflammatory responses *in vitro*: Involvement of ERK1/2 and p38 signaling pathways. Neurosci Lett. 2015;594:105–110.
46. Seo EJ, Fischer N, Efferth T. Phytochemicals as inhibitors of NF-κB for treatment of Alzheimer's disease. Pharmacological Research. 2018;129:262–273.
47. Namgyal D, Ali S, Mehta R, Sarwat M. The neuroprotective effect of curcumin against Cd-induced neurotoxicity and hippocampal neurogenesis promotion through CREB-BDNF signaling pathway. Toxicology. 2020;442:152542.
48. SoukhakLari R, Moezi L, Pirsalami F, Ashjazadeh N, Moosavi M. The passive avoidance memory improving effect of curcumin in young adult mice: Considering hippocampal MMP-2, MMP-9 and Akt/GSK3β. Pharma Nutrition. 2018;6(3):95–99.
49. Rainey-Smith SR, Brown BM, Sohrabi HR, et al. Curcumin and cognition: A randomised, placebo-controlled, double-blind study of community-dwelling older adults. Br J Nutr. 2016;115(12):2106–2013.
50. Cox KHM, Pipingas A, Scholey AB. Investigation of the effects of solid lipid curcumin on cognition and mood in a healthy older population. J Psychopharmacol. 2015;29(5): 642–651.
51. Small GW, Siddarth P, Li Z, et al. Memory and brain amyloid and tau effects of a bioavailable form of curcumin in non-demented adults: A double-blind, placebo-controlled 18-month trial. Am J Geriatr Psychiatry. 2018;26(3):266–277.
52. Grewal AK, Singh TG, Sharma D, Sharma V, et al. Mechanistic insights and perspectives involved in neuroprotective action of quercetin. Biomed Pharmacother. 2021; 140:111729.
53. Dong F, Wang S, Wang Y, et al. Quercetin ameliorates learning and memory via the Nrf2-ARE signaling pathway in d-galactose-induced neurotoxicity in mice. Biochem Biophys Res Commun. 2017;491(3):636–641.
54. Chen J, Deng X, Liu N, et al. Quercetin attenuates tau hyperphosphorylation and improves cognitive disorder via suppression of ER stress in a manner dependent on AMPK pathway. J Funct Foods. 2016;22:463–476.
55. Sandeep MS, Nandini CD. Influence of quercetin, naringenin and berberine on glucose transporters and insulin signalling molecules in brain of streptozotocin-induced diabetic rats. Biomed Pharmacother. 2017;94:605–611.
56. Sabogal-Guáqueta AM, Muñoz-Manco JI, Ramírez-Pineda JR, Lamprea-Rodriguez M, Osorio E, Cardona-Gómez GP. The flavonoid quercetin ameliorates Alzheimer's disease pathology and protects cognitive and emotional function in aged triple transgenic Alzheimer's disease model mice. Neuropharmacology. 2015;93:134–145.
57. Mehta V, Parashar A, Udayabanu M. Quercetin prevents chronic unpredictable stress induced behavioral dysfunction in mice by alleviating hippocampal oxidative and inflammatory stress. Physiol Behav. 2017;171:69–78.
58. Zaplatic E, Bule M, Shah SZA, Uddin MS, Niaz K. Molecular mechanisms underlying protective role of quercetin in attenuating Alzheimer's disease. Life Sci.2019;224:109–119.

59. Okamoto T. Safety of quercetin for clinical application (Review). Int J Mol Med. 2005; 16(2):275–278.
60. Nishimura M, Ohkawara T, Nakagawa T, et al. A randomized, double-blind, placebo-controlled study evaluating the effects of quercetin-rich onion on cognitive function in elderly subjects. Funct Foods Heal Dis. 2017;7(6):353.
61. Nava Catorce M, Acero G, Pedraza-Chaverri J, Fragoso G, Govezensky T, Gevorkian G. Alpha-mangostin attenuates brain inflammation induced by peripheral lipopolysaccharide administration in C57BL/6J mice. J Neuroimmunol. 2016;297:20–27.
62. Yang A, Liu C, Wu J, Kou X, Shen R. A review on α-mangostin as a potential multi-target-directed ligand for Alzheimer's disease. Eur J Pharmacol. 2021;897:173950.
63. Jin X, Liu M-Y, Zhang D-F, Zhong X, et al. Natural products as a potential modulator of microglial polarization in neurodegenerative diseases. Pharmacol Res. 2019;145(77): 104253.
64. Xie Z, Sintara M, Chang T, Ou B. Daily consumption of a mangosteen-based drink improves *in vivo* antioxidant and anti-inflammatory biomarkers in healthy adults: a randomized, double-blind, placebo-controlled clinical trial. Food Sci Nutr. 2015;3(4):342–348.
65. Gadkari PV, Balaraman M. Catechins: Sources, extraction and encapsulation: A review. Food Bioprod Process. 2015;93:122–138.
66. Rashidinejad A, Boostani S, Babazadeh A, et al. Opportunities and challenges for the nanodelivery of green tea catechins in functional foods. Food Res Int. 2021;142:110186.
67. Schimidt HL, Carrazoni GS, Garcia A, Izquierdo I, Mello-Carpes PB, Carpes FP. Strength training or green tea prevent memory deficits in a β-amyloid peptide-mediated Alzheimer's disease model. Exp Gerontol. 2021;143:111186.
68. Chen T, Yang Y, Zhu S, et al. Inhibition of Aβ aggregates in Alzheimer's disease by epigallocatechin and epicatechin-3-gallate from green tea. Bioorg Chem. 2020;105:104382.
69. Liu J, Fan Y, Kim D, et al. Neuroprotective effect of catechins derivatives isolated from Anhua dark tea on NMDA-induced excitotoxicity in SH-SY5Y cells. Fitoterapia. 2019;137:104240.
70. Bae HJ, Kim J, Jeon SJ, et al. Green tea extract containing enhanced levels of epimerized catechins attenuates scopolamine-induced memory impairment in mice. J Ethnopharmacol. 2020;258:112923.
71. Rameshrad M, Razavi BM, Hosseinzadeh H. Protective effects of green tea and its main constituents against natural and chemical toxins: A comprehensive review. Food Chem Toxicol. 2017;100:115–137.
72. Lee Y-J, Choi D-Y, Yun Y-P, Han SB, Oh K-W, Hong JT. Epigallocatechin-3-gallate prevents systemic inflammation-induced memory deficiency and amyloidogenesis via its anti-neuroinflammatory properties. J Nutr Biochem. 2013;24(1):298–310.
73. Welcome MO. Neuroinflammation in CNS diseases: Molecular mechanisms and the therapeutic potential of plant derived bioactive molecules. PharmaNutrition. 2020; 11:100176.
74. Rodrigues J, Assunção M, Lukoyanov N, Cardoso A, Carvalho F, Andrade JP. Protective effects of a catechin-rich extract on the hippocampal formation and spatial memory in aging rats. Behav Brain Res. 2013;246:94–102.
75. Ide K, Matsuoka N, Yamada H, Furushima D, Kawakami K. Effects of tea catechins on Alzheimer's disease: Recent updates and perspectives. Molecules. 2018;23(9):1–13.
76. Mecocci P, Polidori MC. Antioxidant clinical trials in mild cognitive impairment and Alzheimer's disease. Biochim Biophys Acta - Mol Basis Dis. 2012;1822(5):631–638.
77. Cascella M, Bimonte S, Muzio MR, Schiavone V, Cuomo A. The efficacy of Epigallocatechin-3-gallate (green tea) in the treatment of Alzheimer's disease: An overview of pre-clinical studies and translational perspectives in clinical practice. Infect Agent Cancer. 2017;12(1):1–7.
78. Baker LD, Rapp SR, Shumaker SA, et al. Design and baseline characteristics of the cocoa supplement and multivitamin outcomes study for the Mind: COSMOS-Mind. Contemp Clin Trials. 2019;83:57–63.

79. Singh B, Singh JP, Kaur A, Singh N. Phenolic composition, antioxidant potential and health benefits of citrus peel. Food Res Int. 2020;132:109114.
80. Kheradmand E, Hajizadeh Moghaddam A, Zare M. Neuroprotective effect of hesperetin and nano-hesperetin on recognition memory impairment and the elevated oxygen stress in rat model of Alzheimer's disease. Biomed Pharmacother. 2018;97:1096–1101.
81. Roohbakhsh A, Parhiz H, Soltani F, Rezaee R, Iranshahi M. Neuropharmacological properties and pharmacokinetics of the citrus flavonoids hesperidin and hesperetin - A mini-review. Life Sci. 2014;113(1–2):1–6.
82. Qiao O, Ji H, Zhang Y, et al. New insights in drug development for Alzheimer's disease based on microglia function. Biomed Pharmacother. 2021;140:111703.
83. Antunes MS, Jesse CR, Ruff JR, et al. Hesperidin reverses cognitive and depressive disturbances induced by olfactory bulbectomy in mice by modulating hippocampal neurotrophins and cytokine levels and acetylcholinesterase activity. Eur J Pharmacol. 2016;789:411–420.
84. Thenmozhi AJ, Raja TRW, Janakiraman U, Manivasagam T. Neuroprotective effect of hesperidin on aluminium chloride induced Alzheimer's disease in Wistar rats. Neurochem Res. 2015;40(4):767–776.
85. Kean R, Lamport D, Ellis J, et al. Chronic consumption of orange juice flavonoids is associated with cognitive benefits: An 8 week randomised double-blind placebo-controlled trial in older adults. Appetite. 2018;130:308.
86. Alharbi MH, Lamport DJ, Dodd GF, et al. Flavonoid-rich orange juice is associated with acute improvements in cognitive function in healthy middle-aged males. Eur J Nutr. 2016;55(6):2021–2029.
87. Zhang S, Tomata Y, Sugiyama K, Sugawara Y, Tsuji I. Citrus consumption and incident dementia in elderly Japanese: The Ohsaki Cohort 2006 Study. Br J Nutr. 2017;117(8):1174–1180.
88. Joshi R, Kulkarni YA, Wairkar S. Pharmacokinetic, pharmacodynamic and formulations aspects of Naringenin: An update. Life Sci. 2018;215:43–56.
89. Zaki HF, Abd-El-Fattah MA, Attia AS. Naringenin protects against scopolamine-induced dementia in rats. Bull Fac Pharmacy, Cairo Univ. 2014;52(1):15–25.
90. Umukoro S, Kalejaye HA, Ben-Azu B, Ajayi AM. Naringenin attenuates behavioral derangements induced by social defeat stress in mice via inhibition of acetylcholinesterase activity, oxidative stress and release of pro-inflammatory cytokines. Biomed Pharmacother. 2018;105:714–723.
91. Sachdeva AK, Kuhad A, Chopra K. Naringin ameliorates memory deficits in experimental paradigm of Alzheimer's disease by attenuating mitochondrial dysfunction. Pharmacol Biochem Behav. 2014;127(2):101–110.
92. August PM, dos Santos BG. Naringin and naringenin in neuroprotection and oxidative stress. In: Oxidative Stress and Dietary Antioxidants in Neurological Diseases. Elsevier; 2020. p. 309–323.
93. Rahigude A, Bhutada P, Kaulaskar S, Aswar M, Otari K. Participation of antioxidant and cholinergic system in protective effect of naringenin against type-2 diabetes-induced memory dysfunction in rats. Neuroscience. 2012;226:62–72.
94. Khajevand-Khazaei M-R, Ziaee P, Motevalizadeh S-A, et al. Naringenin ameliorates learning and memory impairment following systemic lipopolysaccharide challenge in the rat. Eur J Pharmacol. 2018;826:114–122.
95. Olsen HT, Stafford GI, van Staden J, Christensen SB, Jäger AK. Isolation of the MAO-inhibitor naringenin from *Mentha aquatica* L. J Ethnopharmacol. 2008;117(3):500–502.
96. Mohamed EI, Zaki MA, Chaurasiya ND, et al. Monoamine oxidases inhibitors from Colvillea racemosa: Isolation, biological evaluation, and computational study. Fitoterapia. 2018;124:217–223.
97. Md S, Gan SY, Haw YH, Ho CL, Wong S, Choudhury H. *In vitro* neuroprotective effects of naringenin nanoemulsion against β-amyloid toxicity through the regulation of amyloidogenesis and tau phosphorylation. Int J Biol Macromol. 2018;118:1211–1219.

98. Ali MY, Jannat S, Edraki N, et al. Flavanone glycosides inhibit β-site amyloid precursor protein cleaving enzyme 1 and cholinesterase and reduce Aβ aggregation in the amyloidogenic pathway. Chem Biol Interact. 2019;309:108707.
99. Lou H, Jing X, Wei X, Shi H, Ren D, Zhang X. Naringenin protects against 6-OHDA-induced neurotoxicity via activation of the Nrf2/ARE signaling pathway. Neuropharmacology. 2014;79:380–388.
100. Zheng M, Liu Z, Mana L, et al. Shenzhiling oral liquid protects the myelin sheath against Alzheimer's disease through the PI3K/Akt-mTOR pathway. J Ethnopharmacol. 2021;278:114264.
101. Pereira-Caro G, Clifford MN, Polyviou T, et al. Plasma pharmacokinetics of (poly)phenol metabolites and catabolites after ingestion of orange juice by endurance trained men. Free Radic Biol Med. 2020;160:784–795.
102. Kwon Y. Luteolin as a potential preventive and therapeutic candidate for Alzheimer's disease. Exp Gerontol. 2017;95:39–43.
103. Liu R, Gao M, Qiang G-F, et al. The anti-amnesic effects of luteolin against amyloid β25–35 peptide-induced toxicity in mice involve the protection of neurovascular unit. Neuroscience. 2009;162(4):1232–1243.
104. Wang H, Wang H, Cheng H, Che Z. Ameliorating effect of luteolin on memory impairment in an Alzheimer's disease model. Mol Med Rep. 2016;13(5):4215–4220.
105. Yang Y, Tan X, Xu J, et al. Luteolin alleviates neuroinflammation via downregulating the TLR4/TRAF6/NF-κB pathway after intracerebral hemorrhage. Biomed Pharmacother. 2020;126:110044.
106. El Omri A, Han J, Kawada K, Ben Abdrabbah M, Isoda H. Luteolin enhances cholinergic activities in PC12 cells through ERK1/2 and PI3K/Akt pathways. Brain Res. 2012;1437:16–25.
107. Liu Y, Fu X, Lan N, et al. Luteolin protects against high fat diet-induced cognitive deficits in obesity mice. Behav Brain Res. 2014;267:178–188.
108. Elmazoglu Z, Galván-Arzate S, Aschner M, et al. Redox-active phytoconstituents ameliorate cell damage and inflammation in rat hippocampal neurons exposed to hyperglycemia+Aβ1-42 peptide. Neurochem Int. 2021;145.104993
109. Oliveira AM, Cardoso SM, Ribeiro M, Seixas RSGR, Silva AMS, Rego AC. Protective effects of 3-alkyl luteolin derivatives are mediated by Nrf2 transcriptional activity and decreased oxidative stress in Huntington's disease mouse striatal cells. Neurochem Int. 2015;91:1–12.
110. Tan X-H, Zhang K-K, Xu J-T, et al. Luteolin alleviates methamphetamine-induced neurotoxicity by suppressing PI3K/Akt pathway-modulated apoptosis and autophagy in rats. Food Chem Toxicol. 2020;137:111179.
111. Rezai-Zadeh K, Douglas Shytle R, Bai Y, et al. Flavonoid-mediated presenilin-1 phosphorylation reduces Alzheimer's disease β-amyloid production. J Cell Mol Med. 2009; 13(3):574–588.
112. Zhen JL, Chang YN, Qu ZZ, Fu T, Liu JQ, Wang WP. Luteolin rescues pentylenetetrazole-induced cognitive impairment in epileptic rats by reducing oxidative stress and activating PKA/CREB/BDNF signaling. Epilepsy Behav. 2016;57:177–184.
113. Daily JW, Kang S, Park S. Protection against Alzheimer's disease by luteolin: Role of brain glucose regulation, anti-inflammatory activity, and the gut microbiota-liver-brain axis. BioFactors. 2021;47(2):218–231.
114. de Quervain D. Influence of Luteolin on Memory in Healthy Subjects (LuMus-Basel 20). ClinicalTrials.gov Identifier: NCT04468854. 2021. p. 1–8. Available from: https://clinicaltrials.gov/ct2/show/record/NCT04468854
115. I.R.C.C.S. Fondazione Santa Lucia. Trial record 5 of 16 for: Luteolin. Palmitoylethanolamide combined with luteoline in frontotemporal dementia patients. A Randomized controlled trial (PEA-FTD). https://clinicaltrials.gov/ct2/show/NCT04489017.
116. Taliou A, Zintzaras E, Lykouras L, Francis K. An open-label pilot study of a formulation containing the anti-inflammatory flavonoid luteolin and its effects on behavior in children with autism spectrum disorders. Clin Ther. 2013;35(5):592–602.

117. Song X, Tan L, Wang M, et al. Myricetin: A review of the most recent research. Biomed Pharmacother. 2021;134:111017.
118. Wang B, Zhong Y, Gao C, Li J. Myricetin ameliorates scopolamine-induced memory impairment in mice via inhibiting acetylcholinesterase and down-regulating brain iron. Biochem Biophys Res Commun. 2017;490(2):336–342.
119. Chen H, Lin H, Xie S, et al. Myricetin inhibits NLRP3 inflammasome activation via reduction of ROS-dependent ubiquitination of ASC and promotion of ROS-independent NLRP3 ubiquitination. Toxicol Appl Pharmacol. 2019;365:19–29.
120. Kimura AM, Tsuji M, Yasumoto T, et al. Myricetin prevents high molecular weight Aβ1-42 oligomer-induced neurotoxicity through antioxidant effects in cell membranes and mitochondria. Free Radic Biol Med. 2021;171:232–244.
121. Manzoor S, Hoda N. A comprehensive review of monoamine oxidase inhibitors as Anti-Alzheimer's disease agents: A review. Eur J Med Chem. 2020;206:112787.
122. Wang QM, Wang GL, Ma ZG. Protective effects of myricetin on chronic stress-induced cognitive deficits. Neuroreport. 2016;27(9):652–658.
123. Lum PT, Sekar M, Gan SH, Pandy V, Bonam SR. Protective effect of mangiferin on memory impairment: A systematic review. Saudi J Biol Sci. 2021;28(1):917–927.
124. Kasbe P, Jangra A, Lahkar M. Mangiferin ameliorates aluminium chloride-induced cognitive dysfunction via alleviation of hippocampal oxido-nitrosative stress, proinflammatory cytokines and acetylcholinesterase level. J Trace Elem Med Biol. 2015;31: 107–112.
125. Feng ST, Wang ZZ, Yuan YH, Sun HM, Chen NH, Zhang Y. Mangiferin: A multipotent natural product preventing neurodegeneration in Alzheimer's and Parkinson's disease models. Pharmacol Res. 2019;146:104336.
126. Kavitha M, Nataraj J, Essa MM, Memon MA, Manivasagam T. Mangiferin attenuates MPTP induced dopaminergic neurodegeneration and improves motor impairment, redox balance and Bcl-2/Bax expression in experimental Parkinson's disease mice. Chem Biol Interact. 2013;206(2):239–247.
127. Biradar SM, Joshi H, Chheda TK. Neuropharmacological effect of mangiferin on brain cholinesterase and brain biogenic amines in the management of Alzheimer's disease. Eur J Pharmacol. 2012;683(1–3):140–147.
128. Pardo Andreu GL, Maurmann N, Reolon GK, et al. Mangiferin, a naturally occurring glucoxilxanthone improves long-term object recognition memory in rats. Eur J Pharmacol. 2010;635(1–3):124–128.
129. Wightman EL, Jackson PA, Forster J, et al. Acute effects of a polyphenol-rich leaf extract of *Mangifera indica* l. (zynamite) on cognitive function in healthy adults: A double-blind, placebo-controlled crossover study. Nutrients. 2020;12(8):1–16.
130. Lu Y, Joerger R, Wu C. Study of the chemical composition and antimicrobial activities of ethanolic extracts from roots of *Scutellaria baicalensis* Georgi. J Agric Food Chem. 2011;59(20):10934–10942.
131. Sowndhararajan K, Deepa P, Kim M, Park SJ, Kim S. Baicalein as a potent neuroprotective agent: A review. Biomed Pharmacother. 2017;95:1021–1032.
132. Wang SY, Wang HH, Chi CW, Chen CF, Liao JF. Effects of baicalein on β-amyloid peptide-(25-35)-induced amnesia in mice. Eur J Pharmacol. 2004;506(1):55–61.
133. Zhou L, Tan S, Shan YL, et al. Baicalein improves behavioral dysfunction induced by Alzheimer's disease in rats. Neuropsychiatr Dis Treat. 2016;12:3145–3152.
134. Yin F, Liu J, Ji X, Wang Y, Zidichouski J, Zhang J. Baicalin prevents the production of hydrogen peroxide and oxidative stress induced by Aβ aggregation in SH-SY5Y cells. Neurosci Lett. 2011;492(2):76–79.
135. Lebeau A, Esclaire F, Rostène W, Pélaprat D. Baicalein protects cortical neurons from β-amyloid (25-35) induced toxicity. Neuroreport. 2001;12(10):2199–2202.
136. Zhang SQ, Obregon D, Ehrhart J, et al. Baicalein reduces β-amyloid and promotes nonamyloidogenic amyloid precursor protein processing in an Alzheimer's disease transgenic mouse model. J Neurosci Res. 2013;91(9):1239–1246.

137. Gu X, Xu L, Liu Z, et al. The flavonoid baicalein rescues synaptic plasticity and memory deficits in a mouse model of Alzheimer's disease. Behav Brain Res. 2016;311:309–321.
138. Yang S-S, Shi H-Y, Zeng P, Xia J, Wang P, Lin L. Bushen-Huatan-Yizhi formula reduces spatial learning and memory challenges through inhibition of the GSK-3β/CREB pathway in AD-like model rats. Phytomedicine. 2021;90(16):153624.
139. Liu X, Liu C. Baicalin ameliorates chronic unpredictable mild stress-induced depressive behavior: Involving the inhibition of NLRP3 inflammasome activation in rat prefrontal cortex. Int Immunopharmacol. 2017;48:30–34.
140. Zhao J, Lu S, Yu H, Duan S, Zhao J. Baicalin and ginsenoside Rb1 promote the proliferation and differentiation of neural stem cells in Alzheimer's disease model rats. Brain Res. 2018;1678:187–194.
141. Jia Z, Yang J, Cao Z, et al. Baicalin ameliorates chronic unpredictable mild stress-Induced depression through the BDNF/ERK/CREB Signaling Pathway. Behav Brain Res. 2021;414:113463.
142. Li P, Wu Q, Li X, Hu B, Wen W, Xu S. Shenqi Yizhi Granule attenuates Aβ1–42 induced cognitive dysfunction via inhibiting JAK2/STAT3 activated astrocyte reactivity. Exp Gerontol. 2021;151:111400.
143. Wang X, Yu J, Sun Y, Wang H, Shan H, Wang S. Baicalin protects LPS-induced blood–brain barrier damage and activates Nrf2-mediated antioxidant stress pathway. Int Immunopharmacol. 2021;96:107725.
144. Yimam M, Burnett BP, Brownell L, Jia Q. Clinical and preclinical cognitive function improvement after oral treatment of a botanical composition composed of extracts from *Scutellaria baicalensis* and *Acacia catechu*. Behav Neurol. 2017;2017:1–10.
145. Ullah R, Ikram M, Park TJ, et al. Vanillic acid, a bioactive phenolic compound, counteracts LPS-induced neurotoxicity by regulating c-Jun N-terminal kinase in mouse brain. Int J Mol Sci. 2021;22(1):1–21.
146. Khoshnam SE, Sarkaki A, Rashno M, Farbood Y. Memory deficits and hippocampal inflammation in cerebral hypoperfusion and reperfusion in male rats: Neuroprotective role of vanillic acid. Life Sci. 2018;211:126–132.
147. Ting YW, Wei ZX, Chen S, Shuo SC, et al. Chinese herbal medicine for Alzheimer's disease: Clinical evidence and possible mechanism of neurogenesis. Biochem Pharmacol. 2017;141:143–155.
148. Chen L, Huang J, Xue L. Effect of compound *Polygonum multiflorum* extract on Alzheimer's disease in Chinese. J South Med Univ. 2010;35:612–615.
149. Ushikubo H, Watanabe S, Tanimoto Y, et al. 3,3',4',5,5'-Pentahydroxyflavone is a potent inhibitor of amyloid β fibril formation. Neurosci Lett. 2012;513:51–56.
150. Prakash D, Sudhandiran G. Dietary flavonoid fisetin regulates aluminium chloride-induced neuronal apoptosis in cortex and hippocampus of mice brain. J Nutr Biochem. 2015;26(12):1527–1539.
151. Xiao S, Lu Y, Wu Q, et al. Fisetin inhibits tau aggregation by interacting with the protein and preventing the formation of β-strands. Int J Biol Macromol. 2021;178:381–393.
152. Zhang L, Wang H, Zhou Y, Zhu Y, Fei M. Fisetin alleviates oxidative stress after traumatic brain injury via the Nrf2-ARE pathway. Neurochem Int. 2018;118:304–313.
153. Sandireddy R, Yerra VG, Komirishetti P, Areti A, Kumar A. Fisetin imparts neuroprotection in experimental diabetic neuropathy by modulating Nrf2 and NF-κB pathways. Cell Mol Neurobiol. 2016;36(6):883–892.
154. Singh S, Singh AK, Garg G, Rizvi SI. Fisetin as a caloric restriction mimetic protects rat brain against aging induced oxidative stress, apoptosis and neurodegeneration. Life Sci. 2018;193:171–179.
155. Agarwal P, Holland TM, Wang Y, Bennett DA, Morris MC. Association of strawberries and anthocyanidin intake with Alzheimer's dementia risk. Nutrients. 2019;11(12): 1–11.
156. Khan N, Syed DN, Ahmad N, Mukhtar H. Fisetin: A dietary antioxidant for health promotion. Antioxid Redox Signal. 2013;19(2):151–162.

157. Enogieru AB, Haylett W, Hiss DC, Bardien S, Ekpo OE. Rutin as a potent antioxidant: Implications for neurodegenerative disorders. Oxid Med Cell Longev. 2018;2018:1–17.
158. Wang S, Wang Y-J, Su Y, et al. Rutin inhibits β-amyloid aggregation and cytotoxicity, attenuates oxidative stress, and decreases the production of nitric oxide and proinflammatory cytokines. Neurotoxicology. 2012;33(3):482–490.
159. Javed H, Khan MM, Ahmad A, et al. Rutin prevents cognitive impairments by ameliorating oxidative stress and neuroinflammation in rat model of sporadic dementia of Alzheimer type. Neuroscience. 2012;210:340–352.
160. Moghbelinejad S, Nassiri-Asl M, Naserpour Farivar T, et al. Rutin activates the MAPK pathway and BDNF gene expression on beta-amyloid induced neurotoxicity in rats. Toxicol Lett. 2014;224(1):108–113.
161. Sivanantham B, Krishnan UM, Rajendiran V. Amelioration of oxidative stress in differentiated neuronal cells by rutin regulated by a concentration switch. Biomed Pharmacother. 2018;108:15–26.
162. Subash S, Essa MM, Al-Asmi A, et al. Pomegranate from Oman Alleviates the Brain Oxidative Damage in Transgenic Mouse Model of Alzheimer's Disease. J Tradit Complement Med. 2014;4(4):232–238.
163. Cao Z, Wang F, Xiu C, Zhang J, Li Y. *Hypericum perforatum* extract attenuates behavioral, biochemical, and neurochemical abnormalities in aluminum chloride-induced Alzheimer's disease rats. Biomed Pharmacother. 2017;91:931–937.
164. An F, Liu Z, Xuan X, Liu Q, Wei C. Sanweidoukou decoction, a Chinese herbal formula, ameliorates β-amyloid protein-induced neuronal insult via modulating MAPK/NF-κB signaling pathways: Studies *in vivo* and *in vitro*. J Ethnopharmacol. 2021;273:114002.
165. Ragheb SR, El Wakeel LM, Nasr MS, Sabri NA. Impact of Rutin and Vitamin C combination on oxidative stress and glycemic control in patients with type 2 diabetes. Clin Nutr ESPEN. 2020;35:128–135.
166. Manach C, Scalbert A, Morand C, Rémésy C, Jiménez L. Polyphenols: Food sources and bioavailability. Am J Clin Nutr. 2004;79(5):727–747.
167. Parisi OI, Puoci F, Restuccia D, Farina G, Iemma F, Picci N. Polyphenols and Their Formulations. Polyphenols in Human Health and Disease. 2014;1:29–45.
168. Wilson B, Geetha KM. Neurotherapeutic applications of nanomedicine for treating Alzheimer's disease. J Control Release. 2020;325:25–37.
169. Almuhayawi MS, Ramadan WS, Harakeh S, et al. The potential role of pomegranate and its nano-formulations on cerebral neurons in aluminum chloride induced Alzheimer rat model. Saudi J Biol Sci. 2020;27(7):1710–1716.

Index

Note: Locators in *italics* represent figures and **bold** indicate tables in the text.

A

Abelmoschus esculentus (L.) (Lady's finger), 124, *124*
Acacia arabica (babul), 302
Achyranthes aspera (chaff flower), 297
Activator protein 3 (AP-3), 367
Active components, 44
Acupuncture, 24
Acute respiratory infections (ARI), 341
Acute toxicity, 44
Adenosine triphosphate (ATP), 168
Adipocytes' differentiation, 234
Adipose tissue, 177
Adult-onset diabetes, 143
Alanine transaminase (ALT), 152
Alkaloids, 351
All age groups
  calcium and phosphorus recommendations, **58**
  iodine recommendations, **60**
  iron recommendations, **59**
  magnesium recommendation, **58**
  protein requirements, **57**
  sodium and potassium, **59**
  zinc recommendations, **59**
Allium cepa (onion), *122*, 122, 302
Allium sativum (garlic), *122*, 122–123, 302
Aloe vera, 22
Alpha-linolenic acid (ALA), 210
Alternative medicine, 19–20
Alzheimer's disease (AD), 178–179, 187–188
  AChE-inhibitors, 380
  early onset AD (EOAD), 379
  late onset AD (LOAD), 379
  neuro-inflammatory process, 379
  oral mucosal lesions, 304–306
  oxidative stress, 379
  polyphenols, 380–381
  progressive neurodegenerative disease, 378
  tau protein, 379
American Autoimmune Related Diseases Association (AARDA), 84
American Diabetes Association (ADA), 143
The American Journal of Public Health, 13
American Medical Association (AMA), 12
The American thoracic society, 308
Angiotensin converting enzyme (ACE), 140, 201
Anorexigenic neurons, 170
Anthroposophic therapy (ABCW), 27
Antioxidants, 151–152
Apolipoprotein B (apo A), 178
Arctostaphylos uva-ursi (uva-ursi), 327–328
Armoracia rusticana (horseradish), 329
Arthritis, 304
Aspartate transaminase (AST), 152
Asthma, oral mucosal lesions, 308–309
Atkins diet, 20
Autoimmune disorders
  description, 82
  and vitamin D, 82–83
Autophagy-related PI3K/Akt/mTOR pathway, 368
Auxiliary factors, 5
Avocado and soybean unsaponifiables (ASU), 369

B

Baicalein, 394–395
Baicalin, 394–395
Barberry (*Berberries vulgaris*), 329
Basal and resting metabolism (BMR), 172–173
Berberine, yellow-colored plant alkaloid, 305, 307

Beta-sitosterol, 199
Bilberry fruit, 304
Bioactive components, 45–46
Biofeedback technique, 23
Black cohosh, 22
Blueberries, 327
Blue-Ribbon Association, 55
Body mass index (BMI), 231; *see also* Obesity
categories, **235**, 235
Bodywork treatments, 24
Bone health *see* Metabolic bone diseases (MBD)
Boswellia serrata (Indian frankincense, salai guggul, or shallaki), 310
Brain derived neurotropic factor (BDNF), 379
Brain regulation, 170
Brassica oleracea (Broccoli), *128*, 128–129
Brown adipose tissue (BAT), 233

C

Caffeic acid (CA), 305
Calcium, 335
Cancers
cells, 174–175
contribution of, 86
etiology, pathophysiology and health consequences, 84
and vitamin D, 85–86
Capsicum annuum (Chili), 124, *125*
Capsicum frutescens (Chili), 124, *125*
Carbohydrate responsive element-binding protein (ChREBP), 171
Carcinogenicity studies, 45
Cardiovascular disease (CVD), 6, 170–171, 175–176, 184
cocoa, source of flavanols, 201–202
garlic (*Allium sativum*), 199–200
group of disorders, 78
heart and blood vessels, 197
oral mucosal lesions, 306–308
plant sterols, 198–199
risk factors, 198
vitamin D, 78–80
whole grains, 202–203
Carnitine, 153–154
Carotenoids, 181
Carrots, *121*, 121
Catechins, 386–387
Catheter-associated asymptomatic bacteriuria, 321
Catheter-associated UTI (CAUTI), 321
Cauliflower, cruciferous vegetable, *127*, 128
Centella asiatica (gotu kola), 296, 298
Centers for Disease Control (CDC), 12, 148
Cervical lymphadenopathy, 308
Chamomile, 22
Chasteberry, 334–335
Chebulic myrobalan (Terminalia chebula Retz.), 329
Chili (*Capsicum annuum* and *Capsicum frutescens*), 124, *125*
Chiropractic adjustments, 24
Chondroitin sulfate, 364
Chronic hyperglycemia, 301
Chronic liver disease, 173–174, 182, 183
Chronic obstructive pulmonary disease (COPD), 6, 176
Chronic renal failure (CRF), 169, 174
Chylomicrons, 174
Cinnamomum verum (cinnamon), 328
Citrus fixed seed oil (CFSO), 258–260
Colchicine, 311
Complementary and alternative medicine (CAM), 24
Complementary medicine
acupuncture, 24
Atkins diet, 20
bodywork treatments, 24
chiropractic adjustments, 24
description, 19–20
herbal supplements, 22–23
Ketogenic diet, 21
Mediterranean diet, 21
mind-body practices, 23–24
natural treatments, 20
nutritional therapy, 21–22
probiotics, 23
reflexology, 24
therapeutic massage, 24
vegetarians, 21
Zone diet, 20
Conventional medicine, 19
501c organizations, 4
Coronaviruses, 10–11
Cosmetic Ingredient Review Expert Panel, 258–259
Covid-19 infection
etiology, pathophysiology and health consequences, 88–89
and vitamin D, 89–90
Cranberry (*Vaccinium macrocarpon*), 326–327
C-reactive protein (CRP), 324
Cucumber seed oil (CSO), 267–268
Cucurbita maxima (pumpkin), 302
Curcuma longa (turmeric), 298
Curcumin, 216, 367–368, 383–385
Cystic fibrosis (CF), 308

D

Degenerative diseases, 179–180
Depression *see* Major depressive disorder (MDD)
Diabetes mellitus (DM), 143, 176–177, 185–186
  description, 75
  oral mucosal lesions, 301–302
  type 1 and 2, 76–78
  and vitamin D, 76–78
Diastolic blood pressure (DBP), 200
Dietary fiber, 145–146, 148, 150–151, 208–210
Dietary-induced thermogenesis, 173
Dietary intake (DRI), 70
Dietary patterns, 344–345
Dietary phytosterols, 198
Dietary Supplement Health and Education Act of 1994 (DSHEA), 40
Dietary Supplement Ingredient Advisory List, **56**, 56
Dioxins, 210
Disability-adjusted life years (DALYs), 6, 69
Dysbiosis, 232
Dysmenorrhea
  adolescent females, 329
  description, 329
  non-pharmacological treatment, 331–332
  pathophysiology, 329–330
  primary, 329
  signs and symptoms, 330
  treatment, 330–331

E

Early onset AD (EOAD), 379
Echinacea, 22
Edible oils
  chemical composition, 251
  citrus fixed seed oil (CFSO), 258–260
  cucumber seed oil (CSO), 267–268
  description, 250
  evening primrose oil (EPO), 265–266
  Kenaf seed oil (KSO), 268–269
  melon seed oil (MeSO), 267
  Moringa seed oil (MSO), 262–263
  Nigella sativa seed oil (NSO), 260–262
  Pistachio oil (PO), 263–264
  pumpkin seed oil (PSO), 251–253
  watermelon seed oil (WSO), 254–255, 254–257
Effective food matrices, 49
Eggplant, *125*, 125
Eicosapentaenoic acid (EPA), 348
Electron transport chain's (ETC's) Complex IV, 179
Emblica officinalis (amla), 300
Endocrine disorders
  contemporary human afflictions, 138
  dietary intervention, 138
  nutraceuticals (*see* Nutraceuticals)
  prevalence and incidence of, 138
Endogenous biosynthesis pathway, *71*
End-Stage Renal Disease (ESRD), 174
Energy
  defined, 168
Energy metabolism (EM)
  Alzheimer's disease, 178–179
  basal and resting metabolism, 172–173
  brain regulation, 170
  cancer cells, 174–175
  cardiovascular diseases, 175–176
  cardiovascular regulation, 170–171
  chronic liver disease, 173–174
  chronic obstructive pulmonary disease (COPD), 176
  chronic renal failure, 174
  diabetes mellitus (DM), 176–177
  dietary-induced thermogenesis, 173
  energy intake, 173
  gut microbiota regulation, 171
  obesity and metabolic illnesses, 177
  physical activity, 172
  skeletal muscle regulation, 171–172
  thyroid disease (TH), 177–178
Environmental Protection Association (EPA), 10
Epigallocatechin gallate (EGCG), 352
Erythrocyte sedimentation rate (ESR), 324
Erythroplakia, 295
Estimated average requirements (EAR), 56
Eugenia jambolana (jamun), 302
The European Organization for Research and Treatment of Cancer quality-of-life questionnaire core 30 (EORTC QLQ-C30), 27
Evening primrose oil (EPO), 265–266, 335

F

Fats and oils
  carbohydrate, 58
  fiber, 57
  minerals, 58
  WHO recommendations, 57
Fatty liver disease
  antioxidants, 151–152
  excess fat accumulation, 151

herbal medicine, 152
turmeric, 152–153
Fibers, 57
The American population, 51
defined, 51
extraction and processing, 52–53
ß-glucan, 51–52
soluble, 51
types, 51
First-line antibiotics, 325
Fisetin, 396–397
Fish oil, 210–211, 348, 362–364
Flavones, 181
Flavonoids, 144, 146–147, 188, 351
digestive enzymes, 142
green tea supplementation, 141
polyphenolic compounds, 141
Flaxseed, 22
The Food and Agriculture Organization, 42, 118
Food Safety and Standards Act (FSSA), 56
Foods for specified health uses (FOSHU), 40
Forced expiratory volume in one second (FEV1), 350
Forced vital capacity (FVC), 350
Foundation for Innovation in Medicine (FIM), 325
Framingham risk score (FRS), 27
Fruits and vegetables, 203–205, 345–346
Functional Assessment of Cancer Therapy (FACT-G), 26
Functional foods, 343–344; *see also* Cerebrovascular disease (CVD); Nutraceuticals; Plant sterols; Vegetables
defined, 38
dietary components, 198
dietary recommendations, 118
evidence-based research in US Market, **41**
health benefits of, **41**
human health and lowering illness risk, 117
nutraceuticals, 38–7
physiological/metabolic function, 40
safety concerns, 40–41
safety of (*see also* Safety of functional foods)
scientific intelligence, 39

G

Garlic (*Allium sativum*), *122*, 122–123, 199–200
Generally recognized as safe (GRAS), 258
Genetic predispositions, 4
Genistein, 305
Gestational diabetes mellitus (GDM)
defined, 148
dietary fiber, 150–151
flavonoids, 148–149
vitamin C, 149–150
vitamin D, 149
vitamin E, 150
Ghrelin, peptide hormone, 236
Ginger (*Zingiber officinale Roscoe*), *123*, 124, 365–366
Ginko, 22
Global Burden of Disease (GBD); *see also* Autoimmune disorders; Covid-21 infection; Depression
and bone health (*see* Bone health)
cancers (*see* Cancers)
CVD (*see* Cardiovascular disease (CVD))
DALYs, 69
diabetes mellitus (*see* Diabetes mellitus (DM))
obesity (*see* Obesity)
studies, 69
Global health
paradigm shifts, 3
and regional burden of disease, 4–5
stakeholders, 3
Globalization, 9–10
Global Report on Traditional and Complementary Medicine (2019), 25
β-Glucan, 51–52
Glycogen synthase kinase- 3 (GSK), 380
Glycolysis (GLC)
LAC residues, 169
OP, 168
PYR units, 168
Glycosaminoglycans (GAGs), 364
Gonadotropin releasing hormone agonist (GnRH agonist), 334
Governing entities, 4
Grassroots organizations, 4
Graves' disease (GD), 178
Green beans (*Phaseolus vulgaris*), 126–127, *127*
Gross domestic product (GDP) charts, 4
Gut microbiota regulation, 171
Gymnema sylvestre (gurmarbooti, gurmar), 302–303

H

Hangeshashinto, traditional medicine of Japanese, 299
Hashimoto's thyroiditis (HT), 178

Health Belief Model, 204
healthcare systems, 18
Health-related quality of life (HRQoL), 26
Heerfordt Waldenstrom syndrome, 309
Hemoglobin A1C (HbA1c) levels, 141
Hepatic lipogenesis, 140
Herbal medicine, 152
Herbal supplements
  aloe vera, 22
  black cohosh, 22
  chamomile, 22
  description, 22
  echinacea, 22
  flaxseed, 22
  ginko, 22
  peppermint oil, 22
  soy, 22
  St. John's Wort, 23
  tea tree oil, 23
Herbal tea, 352
Hesperetin (aglycone flavonoid), 388–389
Hesperidin (glycone flavonoid), 388–389
Hexane-extracted sweet orange seed fixed oil, 260
Hidden hunger, 69
High-density lipoprotein-C (HDL-C), 144
Homeostatic Model Assessment for Insulin Resistance (HOMA-IR), 141
Humanistic perspective and chronic diseases, 8–9
Hunter-gatherer humans, 70
Hyaluronic acid (HA), 364–365
Hydroxytyrosol, 367
Hypnotherapy, 23
Hypogeusia, 300
Hypothalamic-pituitary-adrenal (HPA) axis dysfunction, 379
Hypothalamus, 234
Hypoxia-inducible factor-1alpha (HIF-1), 175

I

Icariin (*Herba Epimedii*), 306
Immuno-nutrition, 346
Impurities, 44
The Indian Council of Medical Research (ICMR), 56
Insoluble fiber, 208
Integrative medicine
  complementary, 19
  conventional, 19
  defined, 25
  definition, 19
  health-promoting services, 19
  randomized controlled trial, 26
  shortage of investigation, 25
International Diabetes Federation (IDF), 306
International Food Information Council (IFIC), 325–326
International Obesity Task Force (IOTF), 232
Interstitial lung disease (ILD), 6
Inulin
  efficacious dose, 54–55
  fructo-oligosaccharides, 53–54
  prebiotic, 54
Iodine, 154
The Iowa Women's Health Study (IWHS), 205
36-Item Short Form Survey (SF-36), 26

J

Jasminum grandiflorum (Spanish jasmin), 296
Juniperus communis (juniper), 328

K

Kaempferol, flavonoid component, 129
Kalahari oil, 254
Kenaf seed oil (KSO), 268–269
Ketogenic diet, 21, 344–345

L

Lactic acid bacteria (LAB), 215
Lady's finger (*Abelmoschus esculentus* (L.)), *124*, 124
Lagerstroemia speciosa (banaba) leaves, 303
L-ascorbic acid, 368–369
Late-onset AD (LOAD), 178, 379
Leptin, peptide hormone, 236
Leukoplakia, 295
Life satisfaction questionnaire (LSQ), 27
Lignans, 184
Lipopolysaccharides, 171
Lipoprotein lipase (LPL) nutritional deficiencies, 174
Long-chain fatty acids (LCFA), 170
Long-term repeat-dose toxicity, 44
Low-density lipoprotein cholesterol (LDL-C), 48, 139
Low-density lipoproteins (LDL), 178
Luteolin, 390–391
Lycopene, 309, 351

M

Magnesium, 331–332
Major depressive disorder (MDD)
 cognition impairments and emotional dysregulation, 86
 description, 86
 and vitamin D, 87–88
Mangiferin, 392–393
α-Mangostin, polyphenol xanthone molecule, 386
Manufacturing procedure, 44
Marshmallow root (*Althaea officinalis*), 329
Matrix metalloproteinases (MMPs), 367
Maxillary and mandibular jawbones, 309
Medicalization and sick role, 12–13
Medical nutrition therapy, 20
Medical science, 18
Medicare Payment Advisory Commission, 5
Meditation technique, 23–24
Mediterranean diet, 21, 188, 304–305, 307, 344
Melon seed oil (MeSO), 267
Metabolic bone diseases (MBD)
 description, 80
 osteoporosis, 80
 rickets, VDD outcome, 80
 and vitamin D, 81
Metabolic disorders *see* Endocrine disorders
Metabolic syndrome (MetS), 306
 defined, 139
 diabetes mellitus (*see* Diabetes mellitus (DM))
 flavonoids, 141–142
 gestational diabetes mellitus (GDM) (*see* Gestational diabetes mellitus (GDM))
 lifestyle modifications, 139
 omega-3 fatty acids, 139–140
 probiotics, 140–141
 type 1 diabetes (T1DM) (*see* Type 1 diabetes (T1DM))
 type 2 diabetes mellitus (T2DM) (*see* Type 2 diabetes mellitus (T2DM))
 vitamins, 142–143
Micronutrient deficiency (MND), 69, 90
Middle East respiratory syndrome coronavirus (MERS-CoV), 11
Mind-body practices
 biofeedback technique, 23
 hypnotherapy, 23
 meditation technique, 23–24
 Reiki (energy healing) technique, 24
 Yoga and tai chi, 24
Minerals, 58
Momordica charantia (bitter melon), 303
Monitoring and Actualisation of Noetic Trainings (MANTRA), 26
Monoamine oxidase enzyme (MAO), 379
Moringa oleifera, 120, 121
Moringa seed oil (MSO), 262–263
Mortality, 4
Mucositis, 299
Multidimensional Fatigue Inventory (MFI)-mental fatigue subscale, 28
Multimorbidity, 6
Multiple sclerosis (MS), 69, 83
Murraya koenigii, 120, 120
Myoinositol, 153
Myricetin, 392

N

Naringenin, 389–390
Naringin, 389–390
National Health and Nutrition Examination Survey (NHANES), 309
National Institute of Nutrition (NIN), 56
Natural killer cells (NKs), 10
Natural treatments, 20
Nidus vespae, traditional Chinese medicine, 298
Nigella sativa seed oil (NSO), 260–262
N-methyl-D-aspartate receptors (NMDARs), 178
Nonalcoholic fatty liver disease (NAFLD), 142, 173–174
Noncommercial plant-based edible oils, *250*, 250, **270–282**
Noncommunicable diseases (NCDs), 5, 6, 70, 249
Non-insulin-dependent diabetes, 143
No observable adverse effect levels (NOAELs), 44
NPY/AgRP neurons, 234
Nuclear factor kappa B (NF-kB) pathway, 367
Nutraceutical product
 defined, 294
Nutraceuticals, 343–344
 chronic diseases, 179
 classification of, 39
 defined, 38–39, 138
 degenerative diseases, 179–180
 description, 180
 food ingredients and pharmaceuticals, **39**
 food products, 138

health benefits
Alzheimer's disease, 187–188
antioxidant tocopherol, 182
cancer, 183–184
chronic liver disease, 182, 183
diabetes mellitus (DM), 185–186
obesity, 186–187
thyroid disease, 187
metabolic syndrome (*see* Metabolic syndrome)
mode of action, 39
oxidative stress, 180
phytochemicals, 179
Nutritional therapy, 21–22
Nutrition and respiratory tract, 342–343
Nuts, 205–206

O

Obesity, 186–187
anti-obesity potential, 241–243, **242**
BMI categories, **235**, 235
causes, 234
commercial nuts, *240*
complications, 236
data and information, 233
defined, 74
diagnoses of, 235
dietary and medicinal mushrooms, 241
dietary carbohydrates, *238*, 238
dysbiosis, 232
etiology of, **233**, 233–234
excess energy intake, 74
functional foods/nutraceuticals, 233
medical management, 236–237, *237*
and metabolic illnesses, 177
and metabolic syndrome, 232
myths of, 243–244
natural products, *239*, 239–240
noncommunicable diseases, 231–232
pathogenesis, *235*, 235–236
polyphenols, 241
prevalence of, 232
related osteoarthritis (OA), 236
and vitamin D, 74–75
Ocimum sanctum (Tulsi), 303, 310
Odor-free phenolic compound, 200
Olive oil, 211–212, 366–367
Omega-3 fatty acids, 139–140, 210, 332
Onions (*Allium cepa*), *122*, 122
Ootanga oil, 254
Oral cavity, 293
Oral contraceptives (OCs), 334
Oral mucosal lesions
alcohol consumption, 298
Alzheimer's disease (AD), 304–306
antioxidants, 296
asthma, 308–309
black/brown discoloration, 296
caffeinated coffee intake, 296
cardiovascular disease (CVD), 306–308
chemical products and toxins, 297
chemotherapeutic agents, 299
developmental risks, 294–295
diabetes mellitus (DM), 301–302
erythroplakia, 295
food supplements, 297
leukoplakia, 295
mucositis, 299
oral cancer, 295
pneumonia, 309–314
with smoking, 294
with tobacco use, 295
vitamins, 297
Xianhuayin, 296
Osteoarthritis (OA)
ayurvedic formulation, 365
chondroitin sulfate, 364
curcumin, 367–368
degenerative disease, 361
fish oil, 362–364
ginger, 365–366
glycosaminoglycans (GAGs), 364
hyaluronic acid (HA), 364–365
olive oil, 366–367
synovial fluid inflammation, 361
vitamin C, 368–369
Osteoporosis, 80
Osteoradionecrosis (ORN), 301
Overweight, 231
Oxidative phosphorylation (OP), 168

P

Pan American Health Organization, 25
Partial pressure of arterial carbon dioxide (PETCO2), 345
Patient-centered medical homes, 18
Peppermint oil, 22
Periodontitis, 306
Peroxisome proliferator-activated receptor (PPAR), 139
Phaseolus vulgaris (Green beans), 126–127, *127*
Phenolic acids, 350
Phenolic compounds, 349–352
Phosphocreatine (PCr), 168
Phytochemicals, 184
Phytomedicine, 352
Piper longum (Indian long pepper), 310

Pistachio oil (PO), 263–264
Plantago major (common plantain), 299
Plant cereals and breads, 49
Plant sterols, 48, 198–199
biochemical structure, 48
cholesterol levels, 48–49
dose required for efficacy, 49
functional ingredients, 50–51
timing for efficacy, 49–50
Pneumonia, 309–314
Polychlorinated biphenyls (PCBs), 210
Polyherbal formulations, 304
Polyphenols, 185, 350, 380–381, 398–399
Polyunsaturated fatty acids (PUFA), 210–211, 251
Population trends, 5–6
Premenstrual dysphoric disorder (PMDD)
ovulation inhibitors, 332
pathophysiology, 333
physical and mental problems, 332
signs and symptoms, 333
treatment, 333–334
Premenstrual syndrome *see* Premenstrual dysphoric disorder (PMDD)
Prickly acacia, 302
Primary dysmenorrhea, 329–330
Primary oral tuberculosis, 308
Private sector institutive, 4
Probiotics, 23, 140–141, 348–349
benefits of, 215
consumption, 215
defined, 214
lactic acid bacteria (LAB), 215
products, 214
Prostaglandins, 330
Pterocarpus marsupium (vijayasar), 303
Public health endeavors, 4
Pumpkin seed oil (PSO), 251–253
Pycnogenol (French maritime pine bark), 303
Pyruvate dehydrogenase complex (PDHC), 179

Q

Quercetin, natural flavonoid, 299, 311, 385

R

Racial Segregation and the Black-White Test Score Gap., 8
Randomized controlled trials (RCTs), 144
Recommended Dietary Allowance (RDA), 56–57
Recommended dietary intake (RDI), 41
Red yeast rice (RYR), 307
Reference body weight, 56
Reflexology, 24
Reiki (energy healing) technique, 24
Renal dysfunction, 174
Reproductive toxicology, 44–45
Resistant starch, 51
Respiratory diseases (RSDs), 341
Respiratory exchange ratio (RER), 345
Resting metabolic rate (RMR), 171
Resveratrol, 305, 311, 351–352, 381–383
Rethinking American Poverty, 9
Rheumatoid arthritis (RA), 69, 362
Rutin, 397–398

S

Safety of functional foods
evaluation of, 42–43
modern diets, 42
protocols, 42
Salacia reticulata (salacia, kotalahimbatu), 303
Sarcoidosis, 309
SARS-CoV vaccines, 11
Saturated fatty acids (SFA), 251
School retention rates, 5
Scotland, high-income country, 6
Secondary bile acids, 171
Second-line antibiotics, 325
Selective serotonin reuptake inhibitors (SSRIs), 334
Selenium, 347
Self-efficacy (SES), 28
SF-12 physical subdomain scores, 29
Short-chain fatty acids (SCFAs), 141, 171
Signal transducer and activator of transcription (STAT) proteins, 175
Silybum marianum (Milk thistle), 152, 303, 305
Skeletal muscle regulation, 171–172
Social cognitive theory, 204
Socioeconomic status (SES), 4
Soft tissue necrosis, 301
Solanum melongena Linnaeus (brinjal), *125*, 125
Solanum xanthocarpum (Kantkari), 310
Soluble fibers, 51, 208
Soy Isoflavones, 22, 154, 212–214
antioxidant and lipid status, 46–47
components, 46
functional ingredient, 47–48
plant sterols and stanols, 48
soy-derived isoflavones, 46
Spinach, leafy green vegetable, *119*, 119

Spirulina platensis (spirulina), 296
Stages of Change Model, 204
Stanols *see* Plant sterols
Steroidal compound, 200
Sterol regulatory element-binding transcription factor 1 (SREBP1), 171
St. John's Wort, 23
Subclinical hypothyroidism (SH), 178

T

Talcott Parsons' theory, 12–13
Tannic acid, 305
Tea (*Camellia sinesis*), 206–208
Tea tree oil, 23
Thealfavin, natural phenol, 352
Theory of Planned Behavior, 204
Therapeutic massage, 24
Thrombocytopenia, 300t
Thyroid disease (TH), 177–178, 187
Thyroid disorders
  carnitine, 153–154
  hormone synthesis, 153
  iodine, 154
  myoinositol, 153
  soy, 154
Tinospora cordifolia (guduchi), 300, 303
Traditional, complementary, and integrative medicine (TCIM), 25
Trigonella Foenum-graecum (fenugreek), 303
Trimethoprim sulfamethoxazole (TMP-SMX), 324
Trismus (lockjaw), 301
Tuberculosis, 342
Turmeric, traditional spice of India, 152–153
Tylophora indica (Antamoo), 310
Type 1 diabetes (T1DM), 76–78
  definition by WHO, 146
  dietary fiber, 148
  flavonoids, 146–147
  vitamin C, 147
  vitamin D, 147
  vitamin E, 147–148
Type 2 diabetes mellitus (T2DM), 76–78
  dietary fiber, 145–146
  dietary interventions, 143
  flavonoids, 144
  hormone-like compounds, 143
  non-insulin-dependent diabetes, 143
  vitamin C, 145
  vitamin D, 144
  vitamin E, 145

U

Unconventional medicine, 20
Urinary pathway
  antibiotics, 320
  drugs, 319
  female reproductive system, 320
  function of, 318
  genitourinary diseases, 319
  homeostatic mechanisms, 318
  male and female, *319*
  male reproductive system, 320
  system, *318*
  upper and lower tracts, 318
Urinary tract infection (UTI); *see also* Dysmenorrhea
  Arctostaphylos uva-ursi (uva-ursi), 327–328
  Armoracia rusticana (horseradish), 329
  bacterial diseases, 319
  Barberry (*Berberries vulgaris*), 329
  blueberries, 327
  causative agent, 321–322, *322*
  Chebulic myrobalan (Terminalia chebula Retz.), 329
  Cinnamomum verum (cinnamon), 328
  complicated/uncomplicated, 321
  cranberry (*Vaccinium macrocarpon*), 326–327
  defined, 320–321
  Juniperus communis (juniper), 328
  life-threatening sepsis, 321
  Marshmallow root (*Althaea officinalis*), 329
  non-pharmacological treatments, 325–326
  pathophysiology, 322–323, *323*
  signs and symptoms, 323–324
  treatment, 324–325
  urethritis, 321
US Food and Drug Administration (FDA), 40

V

Vaccinium myrtillus (bilberry), 303–304
Vanillic acid (VA), 395–396
Vegetables
  brinjal (*Solanum melongena Linnaeus*), *125*, 125
  Broccoli (*Brassica oleracea*), *128*, 128–129
  carrots, *121*, 121
  cauliflower, cruciferous vegetable, *127*, 128

Chili (*Capsicum annuum* and *Capsicum frutescens*), 124, *125*
garlic (*Allium sativum*), *122*, 122–123
ginger (*Zingiber officinale Roscoe*), *123*, 124
green beans (*Phaseolus vulgaris*), 126–127, *127*
high consumption, 119
human health, benefits of, 118–119
Lady's finger (*Abelmoschus esculentus* (L.)), *124*, 124
Moringa oleifera, *120*, *121*
Murraya koenigii, *120*, 120
onions (*Allium cepa*), *122*, 122
spinach, *119*, 119
Vegetarians, 21
Very-low-density lipoprotein (VLDL), 174
Very low-density lipoprotein cholesterol (VLDL-C) secretion, 140
Visual analogue scale (VAS), 26
Vitamin A, 346
Vitamin C, 145, 147, 346, 368–369
Vitamin D, 86, 144, 147, 149, 346
fortification, 91–92
supplementation, 90–91
Vitamin D and Omega-3 Trial (VITAL), 86
Vitamin D deficiency (VDD); *see also* Global Burden of Disease (GBD)
assessment methods, 73
deficiency, 73
functions, 72
hormone, 70
metabolic bone disorders, 69
metabolism, 70–72, *71*
micronutrient deficiency (MND), 90
requirement, 73–74
Vitamin E, 145, 147–148, 150, 346–347
Vitamin K, 347
Vitamins, 142–143

W

Water-insoluble fibers, 51
Watermelon seed oil (WSO), 254–255, 254–257
Water-soluble fibers, 51
Western diet, 202, 344
White adipose tissue (WAT), 170, 233–234
Whole grains, 202–203
World Bank, 6
World Health Organization (WHO), 5, 6, 10, 17, 69, 118, 197–198, 231, 341
World Health Organization Quality of Life (WHOQOL), 26

X

Xerostomia, 300

Y

Yoga and tai chi, 24

Z

Zinc, 347
Zingiber officinale Roscoe (Ginger), *123*, 124
Zone diet, 20

Printed in the United States
by Baker & Taylor Publisher Services